Mathematics for Computation (M4C)

Mathematics for Computation (M4C)

Editors

Marco Benini
Università degli Studi dell'Insubria, Italy

Olaf Beyersdorff
Friedrich-Schiller-Universität Jena, Germany

Michael Rathjen
University of Leeds, UK

Peter Schuster
Università degli Studi di Verona, Italy

World Scientific

NEW JERSEY • LONDON • SINGAPORE • BEIJING • SHANGHAI • HONG KONG • TAIPEI • CHENNAI • TOKYO

Published by

World Scientific Publishing Co. Pte. Ltd.
5 Toh Tuck Link, Singapore 596224
USA office: 27 Warren Street, Suite 401-402, Hackensack, NJ 07601
UK office: 57 Shelton Street, Covent Garden, London WC2H 9HE

Library of Congress Control Number: 2022050612

British Library Cataloguing-in-Publication Data
A catalogue record for this book is available from the British Library.

MATHEMATICS FOR COMPUTATION (M4C)

ISBN 978-981-124-521-3 (hardcover)
ISBN 978-981-124-522-0 (ebook for institutions)
ISBN 978-981-124-523-7 (ebook for individuals)

For any available supplementary material, please visit
https://www.worldscientific.com/worldscibooks/10.1142/12500#t=suppl

Desk Editors: Soundararajan Raghuraman/Nijia Liu

Typeset by Stallion Press
Email: enquiries@stallionpress.com

https://doi.org/10.1142/9789811245220_fmatter

Preface

The over-all topic of the present volume, *Mathematics for Computation* (M4C), is mathematics taking crucially into account the aspect of computation, investigating the interaction of mathematics with computation, bridging the gap between mathematics and computation wherever desirable and possible, and otherwise explaining why not. M4C has become necessary by the conceptual turn of mathematics around 1900 and the resulting foundational crisis, but has gained more and more interest since the advent of the computer, later also by its use to produce and verify proofs.

Recently, abstract mathematics has proved to have more computational content than ever expected. In fact, the axiomatic method, originally intended to do away with concrete computations, seems to suit surprisingly well the programs-from-proofs paradigm, with abstraction furthering not only clarity but also efficiency. Unlike computational mathematics, which rather focusses on objects of computational nature such as algorithms, the scope of MC4 encompasses all of mathematics, including abstract concepts such as functions. The purpose of M4C actually is a strongly theory-based and therefore more reliable and sustainable approach to actual computation, up to the systematic development of verified software.

While M4C is situated within mathematical logic and the related area of theoretical computer science, in principle it involves all branches of mathematics, especially those which prompt computational considerations. In traditional terms, the topics of M4C include proof theory, constructive mathematics, complexity theory, reverse mathematics, type theory, category theory and domain theory.

The aim of this volume is to provide a point of reference by presenting up-to-date contributions by some of the most active scholars in each field. A variety of approaches and techniques are represented to give as wide a view as possible and promote cross-fertilisation between different styles and traditions.

This volume emerged from the homonymous international workshop Mathematics for Computation (M4C) held from 8th to 13th May 2016 in the Benedictine Abbey St. Mauritius in Niederaltaich, Lower Bavaria, on the occasion of Douglas S. Bridges's 70th birthday in 2015. Accordingly, the volume begins with Bridges's reflections on 50 years of constructive research, followed by Fred Richman's thoughts on computational mathematics.

Como, Jena, Leeds and Verona, January 2023
Marco Benini
Olaf Beyersdorff
Michael Rathjen
Peter Schuster

https://doi.org/10.1142/9789811245220_fmatter

About the Editors

Olaf Beyersdorff is Professor of Theoretical Computer Science at Friedrich Schiller University Jena. Earlier he was Professor of Computational Logic at the University of Leeds. He obtained his PhD from Humboldt University Berlin and his Habilitation from Leibniz University Hanover. He was also visiting professor at Sapienza University Rome. His principal research interests are in algorithms, complexity, computational logic and, in particular, proof complexity.

Marco Benini is Assistant Professor for Mathematical Logic at the University of Insubria. After a doctorate in computer science at the University of Milano, he became Assistant Professor for Computer Science at the University of Insubria and then Marie Curie Fellow in Mathematical Logic at the University of Leeds. Within constructive mathematics in general, his principal research interests are in structural proof theory on types, the relation of computation and proofs in formal systems as well as point-free mathematics.

Michael Rathjen is Professor of Pure Mathematics at the University of Leeds. He has worked in general proof theory and constructivism for around 20 years, particularly on ordinal analysis of strong theories, models and extensions of Martin–Löf type theory, and constructive set theories. From 2002 to 2005, he was Professor of Mathematics at Ohio State University, having previously held a Heisenberg Fellowship and appointments at Leeds, Stanford, Ohio State University and Münster.

Peter Schuster is a Professor for Mathematical Logic at the University of Verona. After both doctorate and habilitation in mathematics, he was Privatdozent at the University of Munich, and Lecturer at the University of Leeds. Apart from constructive mathematics at large, his principal research interests are about Hilbert's programme in abstract mathematics, especially the computational content of classical proofs in algebra and related fields in which transfinite methods such as Zorn's Lemma are invoked.

https://doi.org/10.1142/9789811245220_fmatter

Acknowledgements

The homonymous workshop this volume has emerged from was generously supported by the John Templeton Foundation within the project "Abstract Mathematics for Actual Computation: Hilbert's Program in the 21st Century" (Grant ID 48138).[a] Further support came through the project "Correctness by Construction (CORCON)" funded by the European Commission (FP7-PEOPLE-2013-IRSES), from the Excellence Initiative of the Ludwig-Maximilians-Universität (LMU) in Munich and from the JSPS Core-to-Core Program: (A) Advanced Research Network. All editors are thankful for the great patience exhibited by World Scientific; Beyersdorff, Rathjen and Schuster are indebted to Benini for taking care of the text editing. Last but not least, Hannes Diener coined both the title and the acronym.

[a]The opinions expressed in this book are those of the authors and editors, and do not necessarily reflect the views of the John Templeton Foundation.

https://doi.org/10.1142/9789811245220_fmatter

Contents

Preface v

About the Editors vii

Acknowledgements ix

1. Reflections on 50 Years of Constructive Research 1
Douglas S. Bridges

2. Thoughts on Computational Mathematics 25
Fred Richman

3. Logic for Exact Real Arithmetic: Multiplication 39
Helmut Schwichtenberg

4. Information Systems with Witnesses: The Function Space Construction 71
Dieter Spreen

5. A Constructive Version of Carathéodory's Convexity Theorem 133
Josef Berger and Gregor Svindland

6. Varieties of the Weak Kőnig Lemma and the Disjunctive Dependent Choice 143
Josef Berger, Hajime Ishihara and Takako Nemoto

7. Intermediate Goodstein Principles 165
David Fernández-Duque, Oriola Gjetaj and Andreas Weiermann

8. Infinite Horizon Extensive Form Games, Coalgebraically 195
Matteo Capucci, Neil Ghani, Clemens Kupke, Jérémy Ledent and Fredrik Nordvall Forsberg

9. Concurrent Gaussian Elimination 223
Ulrich Berger, Monika Seisenberger, Dieter Spreen and Hideki Tsuiki

10. A Herbrandised Interpretation of Semi-Intuitionistic Second-Order Arithmetic with Function Variables 251
João Enes and Fernando Ferreira

11. More or Less Uniform Convergence 273
Henry Towsner

12. Constructive Theory of Ordinals 287
Thierry Coquand, Henri Lombardi and Stefan Neuwirth

13. No Speedup for Geometric Theories 319
Michael Rathjen

14. Domain Theory and Realisability over Scott's D_∞ in Constructive Set Theory 335
Eman Dihoum, Michael Rathjen and Avi Silterra

15. Proof Complexity of Quantified Boolean Logic — A Survey 397

Olaf Beyersdorff

16. Subject Reduction in Multi-Universe Type Theories 441

Marco Benini

Index 461

https://doi.org/10.1142/9789811245220_0001

Chapter 1

Reflections on 50 Years of Constructive Research

Douglas S. Bridges
School of Mathematics,
University of Canterbury,
Christchurch, New Zealand
douglas.bridges@canterbury.ac.nz

When one reaches the Biblical allotment of life, there may be an expectation and tolerance of two things that are considered unbecoming in younger mortals: a tendency to grumpiness, and a habit of reminiscence. Putting aside, at least for the moment, the first of these, in this talk I will unashamedly indulge in the second by describing some of my own results that have brought me particular pleasure, in the hope that they will do the same for the reader.

The framework of the talk is, of course, Bishop's constructive mathematics (BISH), which, superficially, we can think of as *mathematics with intuitionistic logic* plus some appropriate foundation such as the constructive set theory (CST) of Aczel and Rathjen [1], the type theory (ML) of Martin-Löf [2], or the constructive Morse set theory (CMST) under development by Alps and Bridges [3]. We shall also count dependent choice as part of our foundation.

1. Beginnings

My very first paper [4] began with a theorem that arose out of need for a constructive version of the Banach–Stone theorem on *-isomorphisms between function spaces (which is discussed in my second paper [5, Theorem 3, Corollary]):

Theorem 1.1. (The Backward Uniform Continuity Theorem). *Let X be a metric space, and h a mapping of X onto a compact metric space Y such that $f \circ h$ is uniformly continuous for each uniformly continuous mapping $f : Y \to \mathbb{R}$. Then f is uniformly continuous.*

A natural analogue is the classically provable **Forward Uniform Continuity Theorem:**

> Let X be a compact metric space, and h a mapping of X into a metric space Y such that $f \circ h$ is uniformly continuous for each uniformly continuous mapping $f : Y \to \mathbb{R}$. Then f is uniformly continuous.

We shall return later to the general case of this theorem. In the mean time, we prove it constructively in the special case where Y is locally compact — that is, every bounded subset of Y is contained in a compact set. Fixing a in X, we see that the mapping[a] $x \rightsquigarrow \rho(h(x), h(a))$ is uniformly continuous on X; so $h(X)$ is a bounded subset of Y and therefore there exists a compact set $K \subset Y$ such that $h(X) \subset K$. For each uniformly continuous $f : K \to \mathbb{R}$, the Tietze Extension Theorem (see [6, p. 107] or [7, pp. 120–121]) yields an extension $f^\sharp$ of f to Y that is uniformly continuous on bounded sets and whose restriction to K is f. Then the function

$$f_1 : y \rightsquigarrow f^\sharp(y) \max(0, 1 - \rho(y, K))$$

is uniformly continuous on Y, so $f_1 \circ h$, and therefore $f \circ h$, is uniformly continuous on X. It now follows from the Backward Uniform Continuity Theorem that h is uniformly continuous on X.

[a] We use ρ to denote the metric on any metric space.

On a different tack, we recall that a topological space is **pseudocompact** if every pointwise continuous, real-valued mapping thereon is bounded. The remainder of [4] deals with a constructive counterpart to Hewitt's classical theorem that a pseudocompact, normal topological space is countably compact [8, Theorem 30].

Theorem 1.2. *Let X be a pseudocompact, separable metric space, and f a pointwise continuous mapping of X into a metric space X'. Then $f(X)$ is totally bounded.*

Proof sketch. Let $(a_n)_{n\geq 1}$ be a dense sequence in X, f a pointwise continuous mapping of X into X', and $\varepsilon > 0$.

For each positive integer n, let

$$g_n(x) = \min_{1\leq k\leq n} \max\left\{0, \rho(f(x), f(a_k)) - \frac{\varepsilon}{3}\right\}.$$

Then g_n is a pointwise continuous mapping of X into $\mathbb{R}$ and $0 \leq g_{n+1} \leq g_n$. For each $x \in X$, there exists $\delta > 0$ such that $\rho(f(x), f(y)) < \varepsilon/3$ whenever $y \in X$ and $\rho(x, y) < \delta$; if $\rho(x, a_n) < \delta$, then $g_m(x) = 0$ for all $m \geq n$. It follows that

$$h \equiv \sum_{n=1}^{\infty} \min\left\{\frac{\varepsilon}{3}, g_n\right\}$$

is a well-defined mapping of X into $\mathbb{R}$. After showing that h is pointwise continuous, we apply pseudocompactness to compute a positive integer N such that $0 \leq f(x) < N\varepsilon/3$ for all $x \in X$. Were $g_N(x) > \varepsilon/3$, we would have the contradiction

$$f(x) \geq \sum_{n=1}^{N} \frac{\varepsilon}{3} = \frac{N\varepsilon}{3} > f(x).$$

Hence, $g_N(x) \leq \varepsilon/3$, so there exists $k \leq N$ such that $\rho(f(x), f(a_k)) - \varepsilon/3 < 2\varepsilon/3$. It follows that $\{f(a_k) : k \leq N\}$ is an ε-approximation to $f(X)$. □

A Brouwerian counterexample shows that were we able to add "and complete, therefore compact" to the conclusion of Theorem 1.2, we could prove the omniscience principle

LPO: for each binary sequence $(a_n)_{n\geq 1}$, either $a_n = 0$ for all n or else there exists n such that $a_n = 1$.

Exciting as it is to see one's first paper — however, as in this case, small — in print, it has to be admitted that unlike Theorem 1.1, which has at least one application and is a weak constructive form of the full uniform continuity theorem with no dependence on Brouwer's principles, the technically trickier Theorem 1.2 appeared to be a dead-end result.

However, 32 years later, it was resurrected as an essential part of our proof, in [9, Theorem 10], of the connection between pseudoboundedness and the uniform continuity theorem:

Theorem 1.3. *The interval* $[0,1]$ *is pseudocompact if and only if every pointwise continuous, real-valued mapping thereon is uniformly continuous.*

Our proof of this theorem is in the informal style of the analyst; a different, formal proof had been given previously by Loeb [10].

There is a moral here, particularly for younger researchers: you may, in later times, have relatively little regard for your early research efforts, but you never know when and where they may prove to be the key to something more substantial as your career unfolds.

2. Sequential Compactness?

It is trivial to prove that the sequential compactness of $\{0,1\}$ is equivalent to LPO relative to BISH. Moreover, if we extend BISH by adding the Church–Markov–Turing thesis, thereby obtaining recursive constructive mathematics (RUSS), we can produce a strong counterexample to the sequential compactness of the interval $[0,1]$.

Recall that a sequence $(x_n)_{n\geq 1}$ in a metric space (X,ρ) is

- **eventually bounded away from the subset** S of X if there exist $N \in \mathbb{N}$ and $r > 0$ such that $\rho(x_n, x) \geq r$ for all $x \in S$ and all $n \geq N$;
- **eventually bounded away from the point** x of X if it is eventually bounded away from the subset $\{x\}$ of S.

In RUSS, we have **Specker's theorem** [11][b]:

> There exists a strictly increasing sequence in $[0,1]$ that is eventually bounded away from each point of that interval.

In light of the foregoing, it might appear that sequential continuity is of no significance whatsoever in constructive mathematics. The following definitions enable us to show that this is not the case.

A compact interval I in $\mathbb{R}$ is said to have the **anti-Specker property**, AS_I, if every sequence in $\mathbb{R}$ that is eventually bounded away from each point of I is eventually bounded away from I. If there exists a compact interval in $\mathbb{R}$ with the anti-Specker property, then every compact interval in $\mathbb{R}$ has that property.

The property $\mathsf{AS}_{[0,1]}$ is classically equivalent to the sequential compactness of $[0,1]$ and, relative to BISH, equivalent to Brouwer's fan theorem for a certain type of fan, the c-fan [14].

Generalising, we say that a metric space X has the **(unrelativised) anti-Specker property** if there exists a one-point extension[c] $X' \equiv X \cup \{\zeta\}$ of X such that every sequence (x_n) in X' that is eventually bounded away from each x in X is eventually bounded away from X — that is, $x_n = \zeta$ for all sufficiently large n. If this property holds for some one-point extension X' of X, then it holds for all one-point extensions of X [15, Proposition 1]. For a compact interval I in $\mathbb{R}$, the unrelativised anti-Specker property is equivalent to the original anti-Specker property AS_I.

For any metric space X, the unrelativised anti-Specker property is classically equivalent to sequential compactness. Given the power of sequential compactness in classical analysis, one might well ask if the anti-Specker property can be used with similar ease and power in constructive analysis. To show that it can, first we define a mapping $f : X \to Y$ between metric spaces to be **uniformly**

---

[b]For Richman's elegant presentation of Specker's theorem, see either [12, Theorem 5.1] or [13, pp. 58–60].

[c]In such an extension, ζ is bounded away from X in the metric space $X \cup \{\zeta\}$.

sequentially continuous if for all sequences $(x_n), (x'_n)$ in X such that $\lim_{n\to\infty} \rho(x_n, x'_n) = 0$, we have $\lim_{n\to\infty} \rho(f(x_n), f(x'_n)) = 0$.

Proposition 2.1. *Let X be a metric space with the anti-Specker property, and f a pointwise continuous mapping of X into a metric space Y. Then f is uniformly sequentially continuous.*

Proof. Construct a one-point extension $Z \equiv X \cup \{\zeta\}$ of X, where $\rho(\zeta, X) > 1$. Let $(x_n)_{n\geq 1}$ and $(x'_n)_{n\geq 1}$ be sequences in X such that $\lim_{n\to\infty} \rho(x_n, x'_n) = 0$. Given $\varepsilon > 0$, define a binary sequence (λ_n) such that

$$\begin{aligned} \lambda_n = 0 &\Rightarrow \rho(f(x_n), f(x'_n)) > \varepsilon/2, \\ \lambda_n = 1 &\Rightarrow \rho(f(x_n), f(x'_n)) < \varepsilon. \end{aligned}$$

If $\lambda_n = 0$, set $\zeta_n = x_n$; if $\lambda_n = 1$, set $\zeta_n = \zeta$. Fixing a in X, choose δ in $(0, 1)$ such that if $x \in X$ and $\rho(x, a) < \delta$, then $\rho(f(x), f(a)) < \varepsilon/4$. There exists N such that $\rho(x_n, x_{n'}) < \delta/2$ for all $n \geq N$. For such n, if $\rho(a, x_n) < \delta/2$, then $\rho(a, x'_n) < \delta$, so

$$\rho(f(x_n), f(x'_n)) \leq \rho(f(a), f(x_n)) + \rho(f(a), f(x'_n)) < \varepsilon/2,$$

whence $\lambda_n = 1, \zeta_n = \zeta$, and therefore $\rho(\zeta_n, a) > 1$. Thus, if $\lambda_n = 0$, then $\rho(x_n, a) > \delta/4$.

We now see that $\rho(\zeta_n, a) > \delta/4$ for all $n \geq N$. Hence, $(\zeta_n)_{n\geq 1}$ is bounded away from each point of X. By the anti-Specker property, there exists $\nu \in \mathbb{N}$ such that $\zeta_n = \zeta$, and therefore $\rho(f(x_n), f(x'_n)) < \varepsilon$, for all $n \geq \nu$. □

We can now shed more light on the Forward Uniform Continuity Theorem.

Theorem 2.2. *Let X be a compact metric space with the anti-Specker property, and h a mapping of X into a metric space Y such that $f \circ h$ is uniformly continuous for each uniformly continuous mapping $f : Y \to \mathbb{R}$. Then f is uniformly continuous.*

This is a special case of [16, Proposition 17].

3. Approximation Theory

Let X be a metric space and V a located subspace of X. Given $a \in X$, by a **best approximation**, or **closest point**, to a in V we mean an element b of V such that

$$\rho(a,b) = \rho(a,V) \equiv \inf\{\rho(a,v) : v \in V\}.$$

We say that V is **proximinal** in X if every element of X has a best approximation in V.

The classical fundamental theorem of approximation theory states that

> Every finite-dimensional subspace of a real normed linear space is proximinal.

This classical theorem is essentially nonconstructive, since it implies the omniscience property

> LLPO: For each binary sequence $(a_n)_{n\geq 1}$ with at most one term equal to 0, either $a_n = 0$ for all even n or else $a_n = 0$ for all odd n [17, page 42].

Can we find a constructively provable statement that is both useful and classically equivalent to the classical fundamental theorem? To that end, we say that

- an element a of our metric space X has **at most one best approximation** in a subspace V if for all distinct points v, v' of V there exists $v'' \in V$ such that

  $$\max\{\rho(a,v), \rho(a,v')\} > \rho(a,v''),$$

- V is **quasiproximinal** in X if each element of X with at most one best approximation in V actually has one (and therefore only one).

It is trivial to prove that quasiproximinality is classically equivalent to proximinality. Thus, the following **constructive fundamental theorem of approximation theory** is classically equivalent to the classical fundamental theorem.

Theorem 3.1. *Finite-dimensional subspaces of a real normed linear space are quasiproximinal* [18, Theorem] or [7, Chapter 7, (2.12)].

This theorem can be applied in analytic practice — for example, to Chebyshev and Haar approximation.[d] Its proof depends on a lemma that covers, and strengthens, the 1-dimensional case and is applied again in the induction step of the proof. For details, see [7, Chapter 7, (2.8) and (2.12)].

A rather different approximation problem arose from research with Wang Yuchuan, my first doctoral student, when we were dealing with constructive solutions of the Dirichlet Problem. It seems visually reasonable that if J is a plane Jordan curve whose curvature is bounded away from 0, then there is a neighbourhood of J within which any point has a unique closest point on the curve. In spite of this reasonableness, we found no trace of such a property in the literature, classical or constructive.

We say that a Jordan curve satisfies the **twin tangent ball condition** if there exists $R > 0$ such that for each $z \in J$ there exist points a, b on opposite sides of J with

$$\overline{B}(a, R) \cap \overline{B}(b, R) = \{z\}.$$

It is straightforward to show that if J has uniformly continuous curvature, then the twin tangent ball condition implies that the radius of curvature of J is at least R.

Theorem 3.2. *Let J be a differentiable Jordan curve that satisfies the twin tangent ball condition. Then there exists $r > 0$ such that any point z of $\mathbb{R}^2$ that lies within r of J has a unique closest point on J. More precisely, if $\rho(z, J) < r$, then there exists $v \in J$ such that $\|z - v\| < \|z - u\|$ for all $u \in J$ with $u \neq v$* [19, Theorem 1].

4. Complex Analysis

In my first, year-long visit to New Mexico State University — a visit that was essentially an apprenticeship with the master, Fred Richman — in order to produce constructive versions of the Picard theorem on poles and essential singularities, we needed to prove that

[d]However, those subjects are best approached directly, rather than by applying the fundamental theorem [20]. For a different (logical) analysis of the constructive content of Chebyshev approximation theory, see [21].

if f is analytic and nonvanishing on an open set $U \subset \mathbb{C}$, then $1/f$ is analytic on U. This is relatively straightforward once we have the following at hand (where $\overline{B}_{\mathbb{C}}(0,1)$ denotes the closed unit ball of $\mathbb{C}$).

> Let $f : \overline{B}_{\mathbb{C}}(0,1) \to \mathbb{C}$ be (uniformly) differentiable, with $|f(0)| > 0$. Then for each $\varepsilon > 0$ there exists r such that $1-\varepsilon < r < 1$ and
>
> $$\inf\{|f(z)| : |z| = r\} > 0. \qquad (*)$$

Although, as Errett Bishop later pointed out to me, we can prove (*) using estimates in Chapter 5 of his book [6], my approach was rather different.

For $0 < \rho < 1/2$, define

$$\alpha(\nu,\rho) \equiv \left(1 + \frac{1}{\rho^{1/\nu}(1-\rho)}\right)^{-1} \quad \text{and} \quad \beta(\nu,\rho) \equiv \frac{2\rho^{\nu+1}}{(1-\rho)^{\nu}}.$$

Theorem 4.1. *Let ν be a positive integer, $0 < \rho < 1/2$, and $r_0, r_1, \ldots, r_\nu$ distinct real numbers with $\alpha(\nu,\rho) \leq r_k \leq \rho$ $(0 \leq k \leq \nu)$. Let f be a differentiable mapping of $\overline{B}_{\mathbb{C}}(0,1)$ into itself such that $|f(0)| > \beta(\nu,\rho)$. Then* $\max_{0\leq k\leq\nu} \inf\{|f(z)| : z = r_k\} > 0$ [22, Theorem 2].

Thus, under the hypotheses of this theorem, if we test *any* $\nu + 1$ radii r_k between $\alpha(\nu,\rho)$ and ρ, and *any* differentiable mapping f of the closed unit ball of $\mathbb{C}$ into itself such that $|f(0)| > \beta(\nu,\rho)$, we are guaranteed to find k such that f is bounded away from 0 on the circle with centre 0 and radius r_k. This leads to

Theorem 4.2. *Let $f : \overline{B}_{\mathbb{C}}(0,1) \to \mathbb{C}$ be differentiable and not identically* 0, *and let* $0 < \rho < 1$. *Then* $\inf\{|f(z)| : z = r\} > 0$ *for all but finitely many r with* $0 < r < 1$ [22, Proposition 1].

The proof of Theorem 4.2 requires two lemmas in addition to Theorem 4.1, and can be found in [22]; a correction to the first of those lemmas is given in [23].

There is a cautionary tale — to be told to university administrators — associated with one of the lemmas preliminary to the proof of Theorem 4.1. When thinking about how to prove (*), I spent some time in the NMSU Library consulting complex analysis books to see if there were any known results connected to the lemma. Eventually

I found just what I needed in a book from the 1930s. This might not have been possible today, in view of the current fashion for getting rarely borrowed books off the shelves and into storage, if not oblivion. Yet old books often contain treasures that are overlooked in their more modern counterparts. The ability to browse library shelves for inspiration should not be hampered in the interests of some bizarre notion of re-using shelf space.

5. Mathematical Economics

My first academic position was at the University College at Buckingham,[e] an institution that was blessed with a small, but particularly strong, group of researchers in economics. Interactions with that group stimulated me to look constructively at aspects of mathematical microeconomics, notably preference [39,40], utility and demand.[f] For the first two, see my recent generalisation, from Euclidean space to locally compact spaces, of the Arrow–Hahn construction of utility functions [24, especially Section 4 and References].

Turning to demand theory, consider a standard configuration for an individual consumer in classical microeconomics. This consists of

- a compact, convex *consumption set* $X \subset \mathbb{R}^N$ of *consumption bundles*,
- a strictly convex, binary *preference relation* $\succ$ on X, and
- an *initial endowment* $w \in \mathbb{R}$ of wealth.

Associated with the preference relation is a preference-indifference relation $\succcurlyeq$ defined by

$$(\mathbf{x} \succcurlyeq \mathbf{y} \equiv \forall_{\mathbf{z} \in X}(\mathbf{y} \succ \mathbf{z} \Rightarrow \mathbf{x} \succ \mathbf{z}).$$

This turns out to be precisely the negation of $\succ$.

A *price vector* is an element of $\mathbf{p} \in \mathbb{R}^N$ such that p_i is the unit cost of the ith commodity; so the price of the whole consumption bundle $\mathbf{x}$ is $\sum_{i=1}^N p_i x_i$, the inner product $\langle \mathbf{p}, \mathbf{x} \rangle$. The consumption

[e]Since 1983 the University of Buckingham.

[f]Not all my economics papers are constructive: see, for example, [25].

bundles available to the consumer lie in his[g] **budget set**

$$\beta(\mathbf{p}, w) \equiv \{\mathbf{x} \in X : \langle \mathbf{p}, \mathbf{x} \rangle \leq w\},$$

where w is his wealth endowment.

Classical compactness arguments show that

(1) If $\beta(\mathbf{p}, w)$ is nonempty, then the corresponding *demand set*, comprising all $\succcurlyeq$-maximal elements of $\beta(\mathbf{p}, w)$, contains a unique element $F(\mathbf{p}, w)$;
(2) Under reasonable conditions, the resulting *demand function* F is pointwise continuous.

What conditions on $X, \succ, \mathbf{p}$ and w ensure that the demand function is constructively defined and continuous? More precisely, let S be a compact set of pairs $(\mathbf{p}, w)$, consisting of a price vector $\mathbf{p} \in \mathbb{R}^N$ and a wealth endowment $w \in \mathbb{R}$ such that $\beta(\mathbf{p}, w)$ is inhabited. Are there conditions which provide:

(i) an algorithm for converting each $(\mathbf{p}, w) \in S$ to a unique corresponding value $F(\mathbf{p}, w)$ of the demand function, and
(ii) an algorithm which, applied to $\varepsilon > 0$, produces $\delta > 0$ such that if $(\mathbf{p}, w) \in S$, $(\mathbf{p}', w') \in S$, and $\|(\mathbf{p}, w) - (\mathbf{p}', w')\| < \delta$, then

$$\left\| F(\mathbf{p}, w) - F(\mathbf{p}', w') \right\| < \varepsilon?$$

To show that there are, we need two definitions.

Let X be a subset of a normed linear space V. We say that X is a **uniformly rotund set** if

> for each $\varepsilon > 0$ there exists $\delta > 0$ such that if $x, y \in X$ and $\|x - y\| \geq \varepsilon$, then $\frac{1}{2}(x + y) + z \in X$ whenever $z \in V$ and $\|z\| < \delta$. (**)

Let $\succ$ be a preference relation on such an X. We say that $\succ$ is a **uniformly rotund preference relation** if we can add to the

[g]In the interests of fairness, consumers are given masculine designation, and producers feminine.

conclusion of (**) the property that for each $z \in V$ with $\|z\| < \delta$, either $\frac{1}{2}(x+y)+z \succ x$ or $\frac{1}{2}(x+y)+z \succ y$.

Theorem 5.1.[h] *Let $\succ$ be a uniformly rotund preference relation on a compact, uniformly rotund subset X of $\mathbb{R}^N$, let P be a compact set of nonzero vectors in $\mathbb{R}^N$, and let S be a subset of $\mathbb{R}^N \times \mathbb{R}$ such that for each $(\mathbf{p}, w) \in S$,*

- $\mathbf{p} \in P$
- *$\beta(\mathbf{p}, w)$ is inhabited, and*
- *there exists $\xi \in X$ such that $\xi \succ \mathbf{x}$ for all $\mathbf{x} \in \beta(\mathbf{p}, w)$.*

Then for each $(\mathbf{p}, w) \in S$ there exists a unique element $F(\mathbf{p}, w)$ of $\beta(\mathbf{p}, w)$ such that $F(\mathbf{p}, w) \not\prec \mathbf{x}$ for all $\mathbf{x} \in \beta(\mathbf{p}, w)$. Moreover, $\langle \mathbf{p}, F(\mathbf{p}, w)\rangle = w$ and the function F is uniformly continuous on S [26, Theorem 1].

For more on constructive mathematical economics, the reader is directed to the relevant chapter in the forthcoming *Handbook of Constructive Mathematics* [28].

6. Measurability and Locatedness

When dealing with representations of preference relations by utility functions, I was sidetracked into measure theory — specifically, connecting the measurability of a convex body in $\mathbb{R}^N$ with its locatedness.

Theorem 6.1. *An inhabited, located, convex subset of $\mathbb{R}^N$ is Lebesgue measurable* [29, Theorem 1].

> The proof of this theorem requires several lemmas on convexity, notably:

Lemma 6.2. *Let S be a compact, convex subset of $\mathbb{R}^N$ containing 0 in its interior, and let $\varepsilon > 0$. Then there exists a Lebesgue integrable convex polyhedron P such that $S \subset P \subset (1+\varepsilon)S$* [29, Lemma 6].

[h] This theorem has been improved by M. Hendtlass [27, Sections 2.3.1 and 2.3.2].

Theorem 6.3. *A bounded, Lebesgue measurable, convex subset of* $\mathbb{R}^N$ *with positive measure is located* [29, Theorem 2].[i]

A Brouwerian example shows that in the case $N = 2$, if "with positive measure" is removed from the hypotheses of Theorem 6.3, then we can prove the omniscience principle LLPO. However, it can be removed in the case $N = 1$ [29, Theorem 3].

7. Operator Spaces

Let $B(H)$ be the space of (bounded linear) operators on a Hilbert space H. The **weak-operator topology** τ_w on $B(H)$ is the topology induced by the seminorms $T \rightsquigarrow |\langle Tx, y\rangle|$, where $x, y \in H$. It is the weakest topology on $B(H)$ with respect to which $T \rightsquigarrow \langle Tx, y\rangle$ is continuous for all $x, y \in H$.

An important result in classical operator-algebra theory states that all weak-operator continuous linear functionals on a subspace $\mathfrak{R}$ of $B(H)$ have the form

$$T \rightsquigarrow \sum_{n=1}^{N} \langle Tx_n, y_n\rangle$$

for some N and some vectors $\mathbf{x}$ and $\mathbf{y}$ in the direct sum $H_N \equiv \bigoplus_{n=1}^{N} H$. The classical proof uses nonconstructive versions of the Hahn–Banach theorem and the Riesz representation theorem.

It proved remarkably tricky to find a constructivisation of the theorem even in the case $\mathfrak{R} = B(H)$. A natural approach was to try to emulate Bishop's proof of [7, Chapter 7, Corollary (6.9)], but it took several attempts, starting with my thesis in 1975, until I found my way there in 2010.

Theorem 7.1 ([30]). Let H be a nontrivial Hilbert space, and u a nonzero weak-operator continuous linear functional on $B(H)$. Let δ be a positive number, $x_1, \ldots, x_N$ linearly independent vectors in H,

[i] Note that "bounded", is omitted in the statement of this theorem in [29]. The omission was pointed out by M. Hendtlass.

and $y_1, \ldots, y_N$ nonzero vectors in H, such that

$$|u(T)| \leqslant \delta \sum_{n=1}^{N} |\langle Tx_n, y_n \rangle| \quad (T \in B(H)).$$

Then there exist $\alpha_1, \ldots, \alpha_N \in \mathbb{C}$ such that

$$u(T) = \sum_{n=1}^{N} \alpha_i \langle Tx_n, y_n \rangle \quad (T \in B(H)).$$

In contrast to the classical theorem, in this one the initially given vectors x_n, y_n appear explicitly in the conclusion.

8. Constructive Topology

Let me begin this part of the paper with two quotes from Errett Bishop, the first coming from page 63 of [6].

> Very little is left of general topology after that vehicle of classical mathematics has been taken apart and reassembled constructively. With some regret, plus a large measure of relief, we see this flamboyant engine collapse to constructive size.

> The problem of finding a suitable constructive framework for general topology is important and elusive.[j]

In 2000, Luminiţa Vîţă and I began to develop a constructive alternative to general topology, based on a primitive notion of *apartness*. This led to some technically tricky mathematics and, eventually, the monograph [31].

Let X be an inhabited set equipped with an inequality $\neq_X$: that is, a binary relation on X such that for all $x, y \in X$,

$$x \neq_X y \Rightarrow (\text{not}(x = y) \quad \text{and} \quad y \neq_X x) .$$

A special case of this is the denial inequality, defined by

$$x \overset{d}{\neq} y \quad \text{if and only if } \neg(x = y).$$

[j]This comment is in a letter, dated 14 April 1975, from Bishop to me.

We define the **complement** and **logical complement** of a subset S of X to be, respectively,

$$\sim S \equiv \{x \in X : \forall s \in S (x \neq_X s)\}$$

and

$$\neg S \equiv \{x \in S : \forall s \in S \, \neg(x = s)\}.$$

A binary relation $\bowtie$ on the subsets of X is called a (set-set) **pre-apartness** if it has the following axiomatic properties:

B1 $X \bowtie \varnothing$
B2 $S \bowtie T \Rightarrow S \subset \sim T$
B3 $R \bowtie (S \cup T) \Leftrightarrow R \bowtie S \wedge R \bowtie T$
B4 $S \bowtie T \Rightarrow T \bowtie S$

The pair $(X, \bowtie)$ — or, loosely, just X itself — is called a **pre-apartness space**. We then have a corresponding **point-set pre-apartness** defined, for $x \in X$ and $S \subset X$, by

$$x \bowtie S \text{ if and only if } \{x\} \bowtie S.$$

The **apartness complement** of S is

$$-S \equiv \{x \in X : x \bowtie S\}.$$

If also the set-set pre-apartness $\bowtie$ satisfies the **axiom of local decomposability**,

B5 $x \in -S \Rightarrow \exists_T (x \in -T \wedge \forall_y (y \in -S \vee y \in T))$,

then we call it a (set-set) **apartness** and X an **apartness space**.

The canonical example of a set-set apartness space is a quasi-uniform space[k] $(X, \mathcal{U})$ with the **quasi-uniform apartness** defined by

$$S \bowtie T \text{ if and only if } \exists_{U \in \mathcal{U}} (S \times T \subset \sim U).$$

A special case of this occurs when the quasi-uniform structure $\mathcal{U}$ arises from a metric ρ on X; the **metric apartness** on X then satisfies

[k]See [31, pages 77ff.] for basic properties of quasi-uniform spaces.

$$S \bowtie T \Leftrightarrow \exists_{\varepsilon>0} \forall_{s\in S} \forall_{t\in T} \left(\rho\left(s,t\right) \geq \varepsilon\right).$$

Which apartness relations are induced by a uniform structure? Classically: an apartness arises from a uniform structure if and only if it satisfies the **Efremovič condition**,

$$S \bowtie T \Rightarrow \exists_{E\subset X} \left(S \bowtie \neg E \wedge E \bowtie T\right).$$

To see that this does not hold constructively, we need an amusing result.

Proposition 8.1. *If there exists a pre-apartness space X such that*

$$A \bowtie B \Rightarrow \forall_{x\in X} \left(x \notin A \vee x \notin B\right),$$

and any two disjoint subsets of a singleton are apart (that is, if $x \in X$, $A \subset \{x\}$, $B \subset \{x\}$, and $A \cap B = \varnothing$, then $A \bowtie B$), then the weak law of excluded middle, $\forall_P(\neg\neg P \vee \neg P)$, holds [31, Proposition 3.2.28].

Note that the eccentric second hypothesis in this proposition holds constructively if the apartness on X is induced by a quasi-uniform structure.

Proposition 8.2. *Let X be an inhabited set with the denial inequality. Then*

$$A \bowtie_0 B \text{ if and only if } \left(A = \varnothing \vee B = \varnothing\right)$$

defines a set-set apartness on X that satisfies the Efremovič condition. If this apartness is induced by a uniform structure, then the weak law of excluded middle holds [31, Corollary 3.3.30].

In view of this proposition, the constructive theory of apartness spaces with the Efremovič condition is more than just the theory of uniform spaces in another guise.

We turn now to morphisms in the category of apartness spaces. A mapping f between apartness spaces X, Y is said to be **strongly continuous** if $A \bowtie B$ whenever $A, B \subset X$ and $f(A) \bowtie f(B)$. This is a kind of contrapositive of the idea that if two subsets of the domain are near each other, then so are their images under f.

For uniform apartness spaces — in particular, metric ones — uniform continuity implies strong continuity. What about the converse? The statement,

> Every strongly continuous mapping between metric spaces is uniformly continuous. (†)

is classically true [32, Corollary (12.20)]. Ishihara and Schuster have shown that a strongly continuous mapping f between metric spaces is uniformly continuous if either X or $f(X)$ is totally bounded [33, Theorem 5 and Proposition 1]; for counterpart results in the context of uniform spaces, see [31, Propositions 3.3.9 and 3.3.18].

Sequences $(x_n)_{n\geq 1}$ and $(y_n)_{n\geq 1}$ in a uniform space $(X,\mathcal{U})$ are said to be **eventually close** if for each $U \in \mathcal{U}$ there exists N such that $(x_n, y_n) \in U$ for all $n \geq N$. A mapping f of X into a uniform space $(X',\mathcal{U}')$ is **uniformly sequentially continuous** if the sequences $(f(x_n))_{n\geq 1}$ and $(f(y_n))_{n\geq 1}$ are eventually close in X' whenever $(x_n)_{n\geq 1}$ and $(y_n)_{n\geq 1}$ are eventually close in X. When X and X' are metric spaces, the latter definition is easily seen to be equivalent to the one we gave in Section 2.

Uniform continuity implies uniform sequential continuity. In the context of metric spaces, sequential uniform continuity implies uniform continuity classically; but this implication is derivable constructively if and only if we allow the use of Ishihara's principle BD-$\mathbb{N}$ [34, Theorem 11], which is known to be independent of BISH [35].

We now have a constructive substitute for, and classical equivalent of, the statement (†).

Theorem 8.3. *A strongly continuous mapping between uniform spaces is uniformly sequentially continuous* [31, Theorem 3.3.11].

Our proof of this result depends on a series of lemmas that amounts to a new proof technique in the setting of uniform spaces; that technique is used also in the context of uniform sequential convergence of sequences of nets of functions between two uniform spaces. See [31, pp. 98–103 and 108–110].

9. Constructive Morse Set Theory

Here is another quote from Errett Bishop:

> The only attempt I know of to develop a completely formal language which is useful for human communication is due to A.P. Morse. Of course, Morse's language is not constructive, so there is no question of using it as a programming language [36].

In 1972, towards the end of my first year as a graduate student at Oxford, I came across Morse's development [37] of classical set theory, and was soon struck by the possibility that a constructive development, using his precisely defined rules of language and notation, might be an ideal means of formalising Bishop's mathematics. My first steps in constructive Morse set theory (CMST) were part of my thesis. In 2013, it occurred to me that it might be worth putting my work on CMST into LaTeX and making it available online. Instead, this became the ongoing project of an expanded, corrected and vastly improved development of CMST [3], a project that was soon joined by Robert Alps, who had worked on classical Morse set theory for his Chicago doctorate in the early 1970s.[1]

Apart from Morse's rules of language and notation, which Alps has radically improved upon, CMST has the following significant features:

- it is, of course, based on intuitionistic logic;
- it draws no distinction between terms and formulae: every term is a formula, and vice versa;
- it has a universe, U, membership of which correlates with our intuition of a set being well constructed;
- it is written in a form of pseudocode and is therefore ripe for computer implementation;
- more importantly, it appears to suffice for the formalisation of BISH.

Space constraints make it impossible to present in detail our rules of language and notation; but we do need some bits of the latter:

[1]Alps and R. Neveln developed a proof-checker for classical Morse set theory. They are currently doing the same for CMST.

notably, schemators such as $\underline{\mathsf{u}}, \underline{\mathsf{u}}', \underline{\mathsf{v}}', \underline{\mathsf{w}}'', \ldots$. We read $\underline{\mathsf{u}}x$ either as "x has the property $\underline{\mathsf{u}}x$" or else as "the set corresponding to x under the assignment $\underline{\mathsf{u}}$". In the latter interpretation, we regard x as running through some index set I and $\underline{\mathsf{u}}x$ as the corresponding element of a family of sets indexed by I. Likewise, we read $\underline{\mathsf{v}}''xyz$ either as "the sets x, y and z (in that order) have the property $\underline{\mathsf{v}}''$ or else as the set corresponding to x, y and z (in that order) under the assignment $\underline{\mathsf{v}}''$.

The logic axioms of CMST are more-or-less standard intuitionistic ones. Those of set theory begin with the axiom of truth[m]:

$$(x \leftrightarrow (0 \in x)),$$

presumably derived from experience by Morse. The axioms include ones that enable us to treat entities like "x belongs to y" and "x implies y" as terms: namely,

$$((t \in \mathsf{U}) \rightarrow ((t \in (x \in y)) \leftrightarrow (x \in y)))$$

and

$$((t \in a) \rightarrow ((t \in (x \rightarrow y)) \leftrightarrow ((t \in x) \rightarrow (t \in y)))) .$$

The remaining axioms in the first set deal with unions, intersections and equality, and are fairly natural. Later we need axioms that ensure that certain sets are well-constructed (belong to U). Among such sets are: the union of a family of well-constructed sets when the index set is well constructed; the union of two well-constructed sets; the singleton of a well-constructed set; the set ω of natural numbers; and the set $\mathsf{map}(A, B)$ of mappings between well-constructed sets A and B.

The classical Zermelo–Frænkel axiom of foundation,

$$\bigwedge A(x \in A \rightarrow \bigvee y \in A(y \cap A = 0)),$$

is known to imply the law of excluded middle.[n] In 1972–1973 my substitute had its origin in Richman's constructive definition of "ordinal"

---

[m]Note that Morse uses "$\wedge$", "$\vee$", "$\rightarrow$", and "$\leftrightarrow$" to denote "for all", "there exists", "implies", and "if and only if", respectively.

[n]This seems to have been noted first by John Myhill [38] in 1971 and later, independently, in 1972–1973 as part of my doctoral research.

[12, Section 2]:

$$(\bigwedge x(((x \in \mathsf{U}) \wedge (x \subset S)) \to (x \in S)) \to (S = \mathsf{U})).$$

It is constructively equivalent to the property of **set induction**:

$$\bigwedge A(S \subset A \wedge \bigwedge x \in A(x \cap A \subset S \to x \in S) \to S = A)),$$

and is classically equivalent to the above ZF axiom. The remaining axiom of CMST is a version of dependent choice, about which there is little to say that is particularly novel.

An appealing feature of Morse's approach is its ability to define the term "$\mathsf{The}\ x\ \underline{\mathsf{u}}x$" to denote the unique x with the property $\underline{\mathsf{u}}x$. (Note that if there is no unique x such that $\underline{\mathsf{u}}x$, then $(\mathsf{The}\ x\ \underline{\mathsf{u}}x = \mathsf{U})$.) Unsurprisingly, this term appears frequently in practice. One place is in Morse's powerful induction theorem, which depends on a notion of

$$\mathsf{Induced}\ Rxy\ \underline{\mathsf{u}}'xy \quad \mathsf{on}\ A,$$

(read as 'R is induced on A by $\underline{\mathsf{u}}'xy$ in x and y') and this definition:

$$(\mathsf{Ndc}\ Axy\underline{\mathsf{u}}'xy \equiv \mathsf{The}\ R(\mathsf{Induced}\ Rxy\ \underline{\mathsf{u}}'xy\ \mathsf{on}\ A)).$$

The **General Induction Theorem** states that if A is an ordinal, then

$$(\mathsf{Induced}\ Rxy\ \underline{\mathsf{u}}'xy\ \mathsf{on}\ A \leftrightarrow R = \mathsf{Ndc}\ Axy\underline{\mathsf{u}}'xy).$$

Using this, we can derive the standard theorems of simple, primitive and other recursions over ω.

The theory of ordinals leads to the rather limited constructive theory of cardinals. In between those we have the usual notions of comparison of size for sets. We write $(A\ \mathsf{eq}\ B)$ to signify that A is equinumerous with B (that is, there exists a one–one function from A onto B) and define **cardinal inclusion** of A in B by

$$((A \precsim B) \equiv \bigvee S((A\ \mathsf{eq}\ S) \wedge (S \subset B))).$$

Denote the set of detachable subsets of ω — those for which membership is decidable over ω — by $\mathsf{dch}\ \omega$. Since this set is in one–one correspondence with the well-constructed set $\mathsf{map}(\omega, 2)$ of binary sequences, it belongs to U. We can now define the **Continuum Hypothesis** by°

$$(\mathsf{CH} \equiv \bigwedge A(((\omega \precapprox A) \wedge (A \precapprox \mathsf{dch}\ \omega) \wedge \neg(A\ \mathsf{eq}\ \omega)) \rightarrow (A\ \mathsf{eq}\ \mathsf{dch}\ \omega))).$$

This version of the Continuum Hypothesis is classically equivalent to the standard classical one. By considering the set

$$(A = \omega \cup \{S : S \in \mathsf{dch}\ \omega \wedge (p \vee \neg p)\}),$$

we can prove that CH implies the law of excluded middle.

Letting

$$(G = \{S : S \in \mathsf{dch}\ \omega \wedge (\mathsf{CH} \vee \neg\mathsf{CH})\}),$$

we have

$$(G = 0 \rightarrow \neg(\mathsf{CH} \vee \neg\mathsf{CH}) \rightarrow 0),$$

so

$$\neg(G = 0).$$

On the other hand,

$$(x \in G \rightarrow \mathsf{CH} \vee \neg\mathsf{CH}).$$

But the *constructive* interpretation of $(\mathsf{CH} \vee \neg\mathsf{CH})$ is that either we have a proof of CH or we have a proof of $\neg\mathsf{CH}$, which runs counter to the classical Gödel–Cohen proof that CH is independent of Zermelo–Frænkel set theory. Thus, provided CH is independent of CMST, G is an explicit example of *a set that is not empty but for which it is impossible to construct a member.*

This, in itself, is not uninteresting; but it also gives a new insight into the constructive status of CH. With $(C = \omega \cup G)$ the proof that CH implies the law of excluded middle yields

$$(\omega \precapprox C \wedge C \precapprox \mathsf{dch}\ \omega \wedge \neg(C\ \mathsf{eq}\ \omega) \wedge \neg\neg(C\ \mathsf{eq}\ \mathsf{dch}\ \omega)), \qquad (**)$$

and

$$(C\ \mathsf{eq}\ \mathsf{dch}\ \omega \rightarrow \mathsf{CH} \vee \neg\mathsf{CH}).$$

We conclude that, again provided CH is independent of CMST, C is an explicit example of a set that has the properties (**) *and that cannot constructively be proved cardinally similar to* $\mathsf{dch}\ \omega$.

°Note that this presentation of CH does not require the power set.

10. Concluding Remarks

The foregoing are some of my favourites among the theorems to whose proofs I have contributed over 50-odd years as a constructive mathematician. There are many other problems that I have looked at with at best partial success. Perhaps they will be solved by some attendees at the M4C conference, or by mathematicians from the next generation. Let us hope so and that "The end of a matter is better than its beginning" (Ecclesiastes 7:8).

Acknowledgements

My thanks go to Marco Benini and Peter Schuster for organising the M4C conference (Niederaltaich, 2015) at which the initial draft of this chapter was presented. Thanks are also due to the sponsors of that conference: The Templeton Foundation; the European Commission; Ludwig-Maximilians-Universität, München; and the JSPS Core-to-Core Program.

References

[1] P. Aczel and M. Rathjen, *Constructive Set Theory*, in preparation.

[2] P. Martin-Löf, *Intuitionistic Type Theory* (Notes by Giovanni Sambin of a series of lectures given in Padua, June 1980), Bibliopolis, Napoli, 1984.

[3] R.A. Alps and D.S. Bridges, *Morse Set Theory as a Foundation for Constructive Mathematics*, in preparation 2022; partial preprint available on email request.

[4] D.S. Bridges, Some notes on continuity in constructive analysis, *Bull. London Math. Soc.* **8**, 179–182 (1976).

[5] D.S. Bridges, On continuous mappings between locally compact metric spaces, *Bull. London Math. Soc.* **10**, 201–208 (1978).

[6] E. Bishop, *Foundations of Constructive Analysis*. McGraw-Hill, New York, 1967.

[7] E. Bishop and D.S. Bridges, *Constructive Analysis*, Grundlehren der Math. Wissenschaften, vol. 279, Springer-Verlag, Heidelberg–Berlin–New York, 1985.

[8] E. Hewitt, Rings of continuous functions — I, *Trans. Amer. Math. Soc.* **64**, 54–99 (1948).

[9] D.S. Bridges, H. Diener, The pseudocompactness of $[0, 1]$ is equivalent to the uniform continuity theorem, *J. Symb. Logic.* **72**(4), 1379–1383 (2007).

[10] I. Loeb, Equivalents of the (weak) fan theorem, *Ann. Pure and Appl. Logic.* **132**, 51–66 (2005).

[11] E. Specker, Nicht konstructiv beweisbare Sätze der Analysis, *J. Symbolic Logic.* **14**, 145–158 (1949).

[12] F. Richman, The constructive theory of countable abelian p-groups, *Pacific J. Math.* **45**(2), 621–637 (1973).

[13] D.S. Bridges, F. Richman, *Varieties of Constructive Mathematics*, London Math. Soc. Lecture Notes, vol. 97, Cambridge Univ. Press, Cambridge, U.K., 1987.

[14] J. Berger and D.S. Bridges: A fan-theoretic equivalent of the antithesis of Specker's theorem, *Proc. Koninklijke Nederlandse Akad. Weten. (Indag. Math., N.S.)* **18**(2), 195–202 (2007).

[15] D.S. Bridges, H. Diener, The anti-Specker property, positivity, and total boundedness, *Math. Logic Quarterly* **56**(4), 434–441 (2010).

[16] D.S. Bridges, Reflections on function spaces, *Ann. Pure Appl. Logic.* **163**, 101–110 (2010).

[17] D.S. Bridges, Recent progress in constructive approximation theory. In *The L.E.J. Brouwer Centenary Symposium*, A.S. Troelstra and D. van Dalen (eds.), North–Holland, Amsterdam, pp. 41–50, 1982.

[18] D.S. Bridges, A constructive proximinality property of finite-dimensional linear spaces, *Rocky Mountain J. Math.* **11**(4), 491–497 (1981).

[19] D.S. Bridges, Wang Yuchuan, Constructing best approximations on a Jordan curve, *J. Approx. Theory* **94**, 222–234 (1998).

[20] D.S. Bridges, A constructive development of Chebyshev approximation theory, *J. Approx. Theory* **30**(2), 99–120 (1981).

[21] U. Kohlenbach, Effective moduli from ineffective uniqueness proofs. An unwinding of de La Vallée Poussin's proof for Chebycheff approximation, *Ann. Pure Appl. Logic.* **64**, 27–94 (1993).

[22] D.S. Bridges, On the isolation of zeroes of an analytic function, *Pacific J. Math.* **96**(1), 13–22 (1981).

[23] D.S. Bridges, Correction to 'On the isolation of zeroes of an analytic function', *Pacific J. Math.* **97**(2), 487–488 (1981).

[24] D.S. Bridges, The Arrow–Hahn construction in a locally compact metric space, in *Mathematical Topics on Representations of Ordered Structures and Utility Theory*, G. Bosi *et al.*, (eds.), Studies in Systems, Decision and Control, vol. 263, Springer Nature Switzerland AG, 2020, pp. 281–299.

[25] D.S. Bridges, A numerical representation of preferences with intransitive indifference, *J. Math. Econ.* **11**, 25–42 (1983).
[26] D.S. Bridges, The construction of a continuous demand function for uniformly rotund preferences, *J. Math. Econ.* **21**, 217–227 (1992).
[27] M. Hendtlass, *Constructing Fixed Points and Economic Equilibria*, Ph.D. thesis, University of Leeds, 2013.
[28] D.S. Bridges, Constructive mathematical economics. In *Handbook of Constructive Mathematics*, Cambridge Univ. Press, (2022).
[29] D.S. Bridges, Locatedness, convexity, and Lebesgue measurability, *Quart. J. Math. Oxford* (2) **39**(4), 411–421 (1988).
[30] D.S. Bridges, Characterising weak-operator continuous linear functionals on $B(H)$ constructively, *Documenta Math.* **16**, 597–617 (2011).
[31] D.S. Bridges, L.S. Vîţă, Apartness and uniformity — A constructive development. In CiE series *Theory and Applications of Computability*, Springer-Verlag, Heidelberg, (2011).
[32] S.A. Naimpally and B.D. Warrack, *Proximity Spaces*, Cambridge Univ. Press, (1970).
[33] H. Ishhara and P.M. Schuster, A constructive uniform continuity theorem, *Quart. J. Math.* **53**, 185–193 (2002).
[34] D.S. Bridges, H. Ishihara, P.M. Schuster, L.S. Vîţă, Strong continuity implies uniform sequential continuity, *Archive for Math. Logic.* **44**(7), 887–895 (2005).
[35] P. Lietz, *From Constructive Mathematics to Computable Analysis via the Realizability Interpretation*, Dr. rer. nat. thesis, Technische Universität, Darmstadt, Germany, 2004.
[36] E. Bishop, *How to Compile Mathematics into Algol.* Unpublished manuscript, (1970).
[37] A.P. Morse, *A Theory of Sets*, Academic Press, New York, 1965.
[38] J. Myhill, Some properties of intuitionistic Zermelo–Frænkel set theory, *Lecture Notes in Mathematics*, vol. 337, Springer-Verlag Berlin–Heidelberg, 1973, pp. 206–231.
[39] D.S. Bridges, The constructive theory of preference orderings on a locally compact space, *Proc. Koninklijke Nederlandse Akad. Wetenschappen (Indag. Math.)* **92**(2), 141–165 (1989).
[40] D.S. Bridges, The constructive theory of preference orderings on a locally compact space II, *Math. Soc. Sci.* **27**, 1–9 (1994).

https://doi.org/10.1142/9789811245220_0002

Chapter 2

Thoughts on Computational Mathematics*

Fred Richman

Department of Mathematical Sciences, Florida Atlantic University, Boca Raton, FL, USA

richman@fau.edu

Idiosyncratic ramblings about mathematics and computation. Topics include computable functions, Bishop's constructive mathematics, the countable axiom of choice, intuitionistic logic, rigorous numerics, the mentality of computer programming and the importance of terminology.

1. Computable Functions

I have modified the title of this workshop a little to conform with my interest which is mathematics rather than computation. Most people agree that mathematics is based on computation; the few exceptions are mathematicians. My focus in this chapter is on what a computational mathematics looks like. I am interested in pure mathematics — general theorems and proofs. Certainly the applied mathematics of our day is heavily involved with computation, due in large part to the

*To my friend and colleague, Douglas Bridges, on his 70th birthday.

fact that our computational tools are now very powerful and readily available.

I was sorry to see the term "computable function" become a synonym for "Turing-computable function". Granted that it is annoying to have to choose among the equivalent notions of "Turing computable", "λ-computable" and "Markov computable", nevertheless, I like the idea of an informal notion of computability so that we can say, for example, that Errett Bishop considered all functions to be computable. Church's thesis says that the informal notion of computability may be identified with the formal notion of Turing computability. Hartley Rogers [1, page 21] used the term "proof by Church's thesis" to indicate a proof where you use informal methods to show that a function is computable rather than formal ones to show that it is Turing computable. That is much the way Bishop operated.

To me, a crucial feature of Bishop's constructive mathematics was his effort to make it accessible to traditional mathematicians. So he prefered the common notation $a \neq b$ to the strange $a \# b$ despite the danger of reading $a \neq b$ as simply the denial of $a = b$. And he used the common term "nonvoid" (I like "nonempty") instead of "inhabited". Now, I think that "inhabited" is a lovely word for this notion, so I have an inclination to use it. But that is wrongheaded. I recently had a student writing a dissertation in constructive mathematics. In a presentation to his committee, he talked about inhabited sets. A traditional mathematician on the committee took umbrage at that term. Well, why wouldn't he? Should he have to learn a new language in order to understand what the student is talking about? "Nonempty" is the traditional word and there is no real reason not to use it. We will not be confused if we agree that a set is *nonempty*, as opposed to *not empty*, if it contains an element. But "inhabited" is a barrier to communicating with ordinary mathematicians, and we need to do that — not to "spread the word" but to keep ourselves in the mainstream.

2. Countable Choice

For some time now I have advocated rejecting countable choice. This rejection might seem natural to the computing community; after all, if you write a programme to compute something, you don't normally allow your computer any discretion in choosing among alternatives. If you eschew countable choice, you will find that you have little reason to use sequences at all. For example, if you define a real number to be a Cauchy sequence of rational numbers, then you will be unable to show that every Cauchy sequence of real numbers converges. So you will be inclined to abandon that definition in favour of one that is not based on sequences, and you will be led to a nonsequential definition of completeness also. A unified, choiceless, construction of the completion of the rational numbers, and of an arbitrary metric space, is given in [2].

Personally, I think that being forced to downplay sequences is an argument in favour of rejecting countable choice, perhaps the most important one. That is, I have come to think that sequences are undesirable, so I don't want to use countable choice, rather than that I don't want to use countable choice, so I must abandon sequences. This also addresses the argument that when we ostensibly appeal to the axiom of countable choice, we are really not invoking that axiom but are instead relying on a particular view of how mathematics is done. That's the best interpretation I can give to Bishop's notorious statement that "A choice function exists in constructive mathematics *because a choice is implied by the very meaning of existence*".

I believe that Brouwer was also bothered by countable choice. His solution was to formulate a more general notion of a sequence that allowed choices. Those sequences that did not allow choices, what we might call *deterministic sequences*, were said to be "lawlike". I find that move quite mysterious. Nevertheless, I appreciate what the motivation might be. Most constructivists reject the idea that every sequence of integers should be assumed Turing computable, so why do we insist that such a sequence be based on a rule? Brouwer's choice sequences are abstractions from the notion of a rule-based sequence. The internals of a sequence are not something we should be concerned about. That's why Brouwer's notion of an infinitely proceeding sequence is so attractive. As long as we know that for each m we get an integer n, the mechanism for getting that integer n is a matter of indifference to us.

Bishop's sequences are definitely lawlike: a sequence is a function on the positive integers, and a function is a well-defined finite procedure that assigns to each element of its domain an element of its codomain. So how can Bishop make moves that appear to be applications of the axiom of countable choice while using only lawlike sequences? I have some ideas, as do other people, but I don't think the question is important. We don't need to know what Bishop's rationale was; we can explain his moves as appeals to countable choice even if he did not believe he was doing that. There is no reason to doubt that he believed that the axiom of countable choice holds. One of the virtues of the axiomatic method is that you can describe the behaviour of mathematicians without inquiring into their thought processes.

I had occasion to examine a book designed for a course in elementary real analysis for undergraduates. The author talks about limits of sequences and proves some theorems about them. He then addresses "functional limits": the limit of $f(x)$ as x goes to 0. He wants to prove the same theorems in that context, so, for efficiency (I suppose), he reduces the latter convergence to the former. Of course, that can't be done constructively, nor can it be done classically in a larger context. Surely one is not supposed to think of such a functional limit in terms of sequences whatever the context. The good idea, that $f(x)$ is close to L if x is small, should not be reduced to the bad idea, that if a sequence s_n converges to 0, then the sequence $f(s_n)$ converges to L.

3. Choice and the Excluded Middle

It's hard not to be disappointed upon first seeing the Goodman–Myhill proof [3] that the full axiom of choice implies the law of excluded middle. They look at the set $S = \{0, 1\}$ with the usual equality, and the set $T = \{0, 1\}$ with equality defined by $0 = 1$ if P, where P is an arbitrary proposition. There is a natural function f from S onto T. The full axiom of choice says that there is a function $g : T \to S$ such that fg is the identity on T. If $g(0) = 1$ or if $g(1) = 0$, then P holds. Otherwise, $\neg P$ holds.

The odd thing here is that if we had constructed such a Brouwerian counterexample to some other mathematical statement, we would have tried to reformulate that statement in a classically equivalent form so that the Brouwerian counterexample didn't apply. Why don't we do this with the axiom of choice? The counterexample has almost nothing to do with the axiom of choice in a classical setting. Indeed, from a classical point of view, both S and T are finite.

Actually, an example of such a statement is the axiom of countable choice itself. We cannot reject countable choice with a Brouwerian example because so many constructivists use it. But the Goodman–Myhill example could be considered to be a refutation of countable choice. The set T is countable, so we must phrase countable choice in terms of the natural numbers, a *discrete* countable set, rather than in terms of an arbitrary countable set. Why don't we do that with the full axiom of choice? I'm not suggesting that this is a potentially fruitful direction to go, I'm only saying that the Goodman–Myhill example has nothing to do with the lack of countability, and everything to do with the lack of discreteness.

4. Constructive Mathematics

Here are some things that I've always liked about constructive mathematics:

- New **insights**. Because constructive mathematics is a *generalisation* of classical mathematics, and generalisation normally results in new insights.
- More **meaningful**. Because of the computational interpretation.
- More **satisfying**. Because of the insights and the increase in meaning. Moreover, some constructive developments are arguably better, even when interpreted purely classically, than their classical counterparts.
- Different notion of **what is interesting**. Mathematics is not about proving theorems, it's about proving *interesting* theorems. Constructive mathematics leads you to formulate (classical) theorems that wouldn't occur to you in the classical context. Of course, certain statements become less interesting — like the continuum hypothesis. That's also good.

I've never had much sympathy with the idea of finding a systematic method to unpack classical proofs and turn them into constructive proofs, even though much of the traditional constructive program is an attempt to do this informally on a case-by-case basis. Suppose you could actually pull off this feat. What would happen to the four aspects of constructive mathematics that I find so appealing? This is a dream of a classical logician, not of a constructive mathematician.

I'm still with the programme of looking at classical results and trying to recast them in a constructive framework. Sometimes this is thought of as relegating constructive mathematics to a sort of scavenger role, feeding off the results of classical mathematics. Yet, we do the same thing within classical mathematics: reading the literature, trying to ascertain what the theorems are really saying, and recasting and proving them in more general and more appropriate contexts.

The kind of results I like in constructive mathematics are those that a classical mathematician can appreciate as classical mathematics. An example is Krull dimension in rings. This is traditionally defined in terms of prime ideals: the Krull dimension of a ring is the supremum of the lengths of chains of prime ideals. Lombardi and Coquand [4,5], in a constructive context, gave a nice arithmetic characterisation of Krull dimension. To see what I mean by "arithmetic", consider the classically well-known arithmetic characterisation of a zero-dimensional ring, that is, a ring in which every prime ideal is maximal: A ring is zero dimensional exactly when for each x there is an n such that x^n is divisible by x^{n+1}. Now anyone can appreciate arithmetic characterisations versus characterisations using prime ideals. When I presented Lombardi's arithmetic characterisation of Krull dimension d to a classical audience in a seminar, one person in the audience was so charmed by this formulation that she came up afterwards to thank me for showing it to her. Similar considerations apply to the standard arithmetic characterisation of the Jacobson radical, the intersection of all maximal ideals. You don't have to be a constructive mathematician to appreciate these.

5. Ascending Tree Condition

As an algebraic example of operating without choice, consider the classical theorem that you can diagonalise a matrix over a principal ideal domain, that is, every principal ideal domain is an **elementary divisor ring**. Of course, the usual classical definition of a principal ideal domain is useless from a constructive point of view because you can't even prove that every ideal in the two-element field is principal. The algorithmic part of the definition is that finitely generated ideals are principal. The ring of integers satisfies this condition as does a polynomial ring in one variable over a field. That's because both have a division algorithm — they are Euclidean rings. An integral domain in which every finitely generated ideal is principal is called a **Bezout domain**.

Suppose we want to diagonalise an arbitrary matrix A over a Bezout domain. Consider the following matrix equation where $sa + tb = d = \gcd(a, b) \neq 0$:

$$\begin{pmatrix} a & b \end{pmatrix} \begin{pmatrix} s & -b/d \\ t & a/d \end{pmatrix} = \begin{pmatrix} d & 0 \end{pmatrix}.$$

The 2-by-2 matrix is invertible because its determinant is 1. This equation shows how to postmultiply an arbitrary matrix A by a unimodular matrix so that the greatest common divisor of the elements of the first row of A appears in the upper left corner with the remaining entries in that row equal to zero. We can premultiply by a unimodular matrix to do the same for the first column at the expense of messing up the first row. But the corner entry is a divisor of the previous corner entry. So postmultiply again to clear the first row, and continue. If, at any time, the corner entry divides all the other entries in the uncleared first row or column, we can clear both the first row and the first column, and we are done by induction on the size of the matrix.

It looks like we can finish the proof by using the **divisor chain condition**: given a sequence of elements $a_1, a_2, \ldots$ in the ring, such that a_{i+1} divides a_i for each i, there exists i such that a_i divides a_{i+1}. We apply this condition to the sequence of upper-left-corner elements that we generate by the process indicated above. That is why a constructive formulation of a principal ideal domain might be that it is a Bezout domain with the divisor chain condition. Any Euclidean ring has those two properties.

However, there is a problem from a choiceless point of view. Choices are made in the generation of the sequence of upper-left-corner elements. That's because in a Bezout domain, the s and t such that $sa + tb = d$ are not singled out — there may be many. Indeed, the d itself is not unique, so there is a choice involved in the construction of the upper-left-corner element to which we want to apply the divisor chain condition.

How can we get around this? One solution is the **ascending tree condition** [6]. Think of the diagonalisation process as a tree, indexed by level, whose root is the matrix A, and also indexed at nodes of level greater than 0, by the unimodular matrix needed to get there from the parent node. Every node has a child — that is the main construction. We then decorate the tree with the corner element of the result of processing the matrix A. Note that the decoration at each node is divisible by the decoration at each of its children. Note also that the branching on the tree can be infinite.

So we are drawn to the ascending tree condition on a partially ordered set P: if we decorate a tree like this with elements of P such that the decorations ascend as you go up the tree, then there exists a node that has the same decoration as one of its children. It's easy to see that, in the presence of countable choice, this condition is equivalent to the ascending chain condition on P.

For Euclidean domains, this condition is met without choice. Also for Dedekind–Hasse domains. Are there more interesting examples, or is this possibly an exercise in superfluous generality?

Applying the ascending tree condition to finitely generated ideals, you can define a Noetherian ring in a way that allows the general theory to develop without choice. Are there interesting examples?

I once thought that this analysis of the proof that a PID is an elementary divisor ring established the superiority of ATC over ACC in a choiceless environment. In fact, it turned out that I was just looking at the wrong proof of that theorem. Helmer [7] also worries about using ACC to prove that a ring is an elementary divisor ring. He defines the notion of an **adequate** Bezout domain that does not refer to a chain condition. Here is a slightly stronger and simpler definition: given a and c, we can write $a = rs$ such that $(r, c) = 1$ and s divides c^n for some n.

Theorem 5.1. *A Bezout domain with ACC is adequate. An adequate Bezout domain is an elementary divisor ring.*

6. Rigorous Numerics

Computer-assisted proof often refers to computer manipulation of symbolic descriptions of mathematics rather than computer participation in the mathematics itself. This is a logician's game rather than a mathematician's game. (A notable exception was the solution of the four-colour problem.) In the field of "rigorous numerics", on the other hand, the computer is involved in proving by doing the kinds of computations referred to in the theorem. The following characterisation is taken from [8]: "In a nutshell, rigorous computations are mathematical theorems formulated in such a way that the assumptions can be rigorously verified on a computer". They go on to say, "This complements the field of scientific computing, where the goal is to achieve highly reliable results for very complicated problems. In rigorous computing, one is after absolutely reliable results for somewhat less complicated (but still hard) problems".

This seems to me like the sort of activity that would appeal to a constructive mathematician. Computations are used to verify theorems, and those computations deal directly with the numbers and functions referred to in the theorems. I am less attracted by interval analysis, which is touted by many in the field, as that seems to be more relevant to applications than to theory.

7. Proof by Contradiction

Classical mathematicians have no coherent notion of a negative proposition. They cannot distinguish between proving $\neg P$ by deriving a contradiction from P, and proving P by deriving a contradiction from $\neg P$. Constructive mathematicians can make that distinction, and some use the term "proof by contradiction" only for the second technique. That strikes me as a bad idea.

Even today, the prototype proof by contradiction is still the proof that $\sqrt{2}$ is irrational, in the sense that $\sqrt{2}$ is not rational. That is, we are trying to prove $\neg P$, where P is the proposition that $2 = m/n$ for integers m and n. I can't see why we should deny that this is a proof by contradiction just because it is a proof that we accept. The idea that this is a proof *of* a contradiction, rather than a proof *by* contradiction, as some suggest, seems silly to me; it puts too much of a load on a preposition, not to mention that a proof of a contradiction is a demonstration that mathematics is inconsistent.

I am equally unimpressed by the suggestion that the technique that we accept be called "proof of a negation". The justification for that seems to be that we have all so completely internalised the technique of proof by contradiction that the very definition of $\neg P$ is that P implies a contradiction.

So what should we call the second technique to indicate that it is unacceptable? Note that it is only unacceptable as a proof of P; we accept it as a proof of $\neg\neg P$. I suggest that we call it an *indirect proof*: instead of proving P directly, it proves $\neg\neg P$ instead, which constitutes, in classical mathematics, a proof of P. Indeed it is an indirect proof even in classical mathematics.

For some insightful comments on proof by contradiction by a classical mathematician, I recommend [9, Section 3.8]. There Gillman defines a *questionable* proof by contradiction, where in order to prove P implies Q, you ostensibly assume P and $\neg Q$, and derive a contradiction, whereas you actually prove the contrapositive $\neg Q$ implies $\neg P$. He suggests that you should say that you are proving the contrapositive. He also defines a *spurious* proof by contradiction as one where you assume $\neg P$, then give a direct proof of P (without using the assumption), and then say that you have proved the contradiction P and $\neg P$. We have all seen that kind of proof.

8. Well-posed Problems

The (classical) notion of a well-posed problem sheds light on constructive mathematics as only a well-posed problem can be shown constructively to have a solution. Of course, we need the right definition of a well-posed problem. Hadamard [10] called a problem "correctly set" if there is exactly one solution to it. He argued that this entailed that the solution depends continuously on the data. He said that, in a correctly set problem, if the data were *known*, then the solution "would be determined without any possible ambiguity. But, in any concrete application, 'known' of course signifies 'known with a certain approximation'". He then went on to say that if the dependence on the data were not continuous, then "Everything takes place, physically speaking, as if the knowledge of Cauchy's data would *not* determine the unknown function". This is much like the rationale for Brouwer's continuity principle. Nowadays, everyone says that a problem is **well posed** if it has a unique solution that depends continuously on the data, and attributes this definition to **Hadamard**.

To be a little more definite, suppose D and S are metric spaces: think of D as the space of data points and S as the space of potential solutions. Our problem is to show that for all $d \in D$ there is $s \in S$ such that $(d, s) \in R$, where R is some given subset of $D \times S$. This problem is well posed if for each $d \in D$ there is exactly one $s \in S$ such that $(d, s) \in R$, and this s depends continuously on d. That is, R is (the graph of) a continuous function from D to S.

For Brouwer's fixed point theorem, D is the set of all uniformly continuous functions from the unit ball S in $\mathbb{R}^n$ to itself, and $R = \{(d, s) \in D \times S : d(s) = s\}$. For a restricted version of the fundamental theorem of algebra, S is the complex numbers, D is the set of all monic complex polynomials of degree $n > 1$ and $R = \{(d, s) \in D \times S : d(s) = 0\}$. Neither of these problems is well posed for the somewhat trivial reason that the solution need not be unique. There is no way to prove Brouwer's fixed point theorem, but countable choice allows the construction a root of a monic polynomial of degree $n > 1$ (see [11], for example). How can we arrange for the latter problem to be well posed? (A theorem is something we want to be true, so we arrange our definitions accordingly [12].)

The simplest way to eliminate the uniqueness requirement is to consider R as a function from D to subsets of S, to require that those subsets be nonempty, and to require that R be continuous for the Hausdorff extended metric on nonempty subsets of S. That is, if A and B are nonempty subsets of S, then

$$\rho(A, B) = \sup_{a \in A} \inf_{b \in B} \rho(a, b) \vee \sup_{b \in B} \inf_{a \in A} \rho(a, b),$$

where we allow ∞ to be a supremum. Note that $\rho(\{a\}, \{b\}) = \rho(a, b)$, and that $\rho(A, B) = 0$ exactly when A and B have the same closure. Is the local version (number 2 in the following) weaker?

Here are two local versions. They define what it means for the problem to be well posed at a point $d_0 \in D$. The first is weaker than Hadamard's condition, and the second is weaker than the first.

(1) For each d_0 in D there is a neighbourhood B of d_0 and a continuous function $f : B \to S$ such that $(d, f(d)) \in R$ for all $d \in B$.
(2) For each $d_0 \in D$, there is s_0 in S such that $(d_0, s_0) \in R$, and for each neighbourhood V of s_0 there is a neighbourhood U of d_0 such that for all $d \in U$ there is $s \in V$ such that $(d, s) \in R$.

For the (classical) step function σ that is 0 on $(-\infty, 0]$ and 1 on $(0, \infty)$, evaluation of σ is not a well-posed problem even in sense 2. Here $D = S = \mathbb{R}$, and $R = \{(d, s) \in D \times S : \sigma(d) = s\}$. This is not a well-posed problem for $d_0 = 0$.

Brouwer's fixed point theorem is not a well-posed problem in sense 2 for $d_0 = \mathrm{id}_S$. To see this, let C be a bounded convex subset of a normed linear space, and fix distinct points a and b in C. For $x \in C$ and $t \in J = [-1, 0] \cup [0, 1]$, let

$$f_t(x) = \begin{cases} ta + (1-t)x & \text{if } t \geq 0 \\ -tb + (1+t)x & \text{if } t \leq 0 \end{cases}.$$

Then $f_0 = \mathrm{id}_C\, x$, and if $t \neq 0$, then f_t has a unique fixed point: a if $t > 0$ and b if $t < 0$. Fixing $a \neq b$ shows that the Brouwer fixed point theorem is not well-posed in sense 2 for $d_0 = \mathrm{id}_C$.

The fundamental theorem of algebra is well posed in sense 2 but not in sense 1. To see that it is well posed in sense 2, let d_0 be a monic complex polynomial of degree $n > 1$. Then (classically) d_0 has a root s_0, and if A is any neighbourhood of s_0, then there is a neighbourhood B of d_0 such that if $d \in B$, then there is $s \in A$ with $d(s) = 0$. Indeed, the *set* of roots of d is a continuous function of d near d_0 in an appropriate metric (see [13]).

However, the fundamental theorem of algebra is not well posed in sense 1. Consider the family of polynomials $z^2 - a$, and let d_0 be the polynomial z^2. For the fundamental theorem of algebra to be well posed in sense 1, we need (at least) a complex-valued continuous function f on a neighbourhood B of 0 in the complex numbers such that $f(a)^2 = a$ for all $a \in B$. But this is impossible even if we restrict a to lie on some fixed circle around 0 contained in B.

References

[1] H. Rogers, *Theory of Recursive Functions and Effective Computability.* MIT Press, 1987, McGraw-Hill, 1967.

[2] F. Richman, Real numbers and other completions, *Math. Logic Quart.* **54**, 98–108, (2008).

[3] N. Goodman and J. Myhill, Choice implies excluded middle, *Z. Math. Logik Grundlag. Math.* **24**, 461, (1978).

[4] T. Coquand and H. Lombardi, Hidden constructions in abstract algebra (3) Krull dimension of distributive lattices and commutative rings. In *Commutative Ring Theory and Applications*, Marco Fontana, Salah-Edine Kabbaj, and Sylvia Wiegand (eds.), *Lecture Notes in Pure and Applied Mathematics*, vol. 231, Marcel Dekker, 2002, pp. 477–499.

[5] H. Lombardi, Dimension de Krull, Nullstellensätze et Évaluation dynamique, *Math. Zeit.* **242**, 23–46 (2002).

[6] F. Richman, The ascending tree condition: Constructive algebra without countable choice, *Commun. Algebra* **31**, 1993–2002 (2003).

[7] O. Helmer, The elementary divisor theorem for certain rings without chain condition, *Bull. Amer. Math. Soc.* **49**, 225–236 (1943).

[8] J.B. van den Berg and J.-P. Lessard, Rigorous numerics in dynamics, *Notices of the AMS* **62**, 1057–1061 (2015).

[9] L. Gillman, *Writing Mathematics Well: A Manual for Authors.* MAA, 1987.

[10] J. Hadamard, *Lectures on Cauchy's Problem in Linear Partial Differential Equations.* Yale University Press, 1923.

[11] D.S. Bridges, F. Richman, and P. Schuster, A weak countable choice principle, *Proc. Amer. Math. Soc.* **128**, 2749–2752 (2000).

[12] S.S. Abhyankar, *Algebraic Geometry for Scientists and Engineers.* AMS, 1990.

[13] F. Richman, The fundamental theorem of algebra: A constructive development without choice, *Pac. J. Math.* **196**, 213–230 (2000).

https://doi.org/10.1142/9789811245220_0003

Chapter 3

Logic for Exact Real Arithmetic: Multiplication

Helmut Schwichtenberg

Mathematisches Institut, Ludwig-Maximilians-Universität München, Theresienstr., München, Germany

schwicht@math.lmu.de

Real numbers in the exact (as opposed to floating-point) sense can be defined as Cauchy sequences (of rationals, with modulus). However, for computational purposes it is better to see them as coded by streams of signed digits $\{1, 0, -1\}$. A variant representation is by binary reflected streams (or Gray code [1,2]) explained in what follows. Our goal is to obtain formally verified algorithms (given by terms in our language) operating on stream represented real numbers. Suppose we have an informal idea of how the algorithm should work. There are two methods how this can be achieved.

(1) Formulate (using corecursion) the algorithm in the term language of a suitable theory, and then formally prove that this term satifies the specification.
(2) Find a formal existence proof M (using coinduction) for the object the algorithm should return. Then apply realisability to extract M's computational content as a term involving corecursion.

A general advantage of (II) over (I) is that one does not need to begin with a detailed formulation of the algorithm, but instead can stay on a more abstract level when proving the (existential) specification.

Method (I) was used in [3]. The present chapter employs method (II) and continues previous work [4–7] by a case study on multiplication. It is shown how a multiplication algorithm for real numbers represented as either signed digit or binary reflected streams can be extracted from an appropriate formal proof. Algorithms arise from these proofs via a realisability interpretation. The main benefit of this proof-theoretic approach to the construction of algorithms is that it yields formal proofs that the algorithms are correct. This is a consequence of the general soundness theorem for the realisability interpretation in the underlying formal theory.

We will work with constructive existence proofs in the style of [8], but in such a way that we can switch on and off the availability of input data for the constructions implicit in the proof, as in [9]. In the present context, this will be applied to real numbers as input data: we do not want to make use of the Cauchy sequence for the constructions to be done, but only the computational content of an appropriate coinductive predicate to which the real number is supposed to belong. We consider multiplication of real numbers as a nontrivial case study; it has been dealt with in [3] using method (I). Based on the algorithmic idea in [3], we employ method (II) to extract signed digit and binary reflected stream-based algorithms from proofs that the reals are closed under multiplication.

Recall that our goal is to extract stream algorithms for multiplication (both for signed digit and binary reflected streams) from constructive proofs. These proofs will be done in a theory TCF (see [10, Section 7.1]), a variant of HA^ω with inductive and coinductive definitions and their fixed point axioms. We use coinductive predicates ${}^{\mathrm{co}}I$ (for signed digit streams) and ${}^{\mathrm{co}}G$, ${}^{\mathrm{co}}H$ (for binary reflected streams). These are computationally relevant predicates, chosen in such a way that (in the signed digit case) a term t realising ${}^{\mathrm{co}}Ix$ defines a signed digit stream representing the real number x. For binary reflected code, we proceed similarly with ${}^{\mathrm{co}}G$ and ${}^{\mathrm{co}}H$. From a given formal proof M of $\forall_{x,y \in {}^{\mathrm{co}}I}(x \cdot y \in {}^{\mathrm{co}}I)$, we extract a TCF-term $\mathrm{et}(M)$ realising this formula, i.e., with the property that for all realisers u, v of ${}^{\mathrm{co}}Ix$, ${}^{\mathrm{co}}Iy$ the applicative term $\mathrm{et}(M)uv$ realises ${}^{\mathrm{co}}I(x \cdot y)$. Since "$u$ realises ${}^{\mathrm{co}}Ix$" is a (noncomputational) predicate in TCF as well, this property can be formulated in TCF. In fact, by TCF's soundness theorem (see [10, p. 340]) it is provable.

The chapter is organised as follows. In Section 1, stream representations of real numbers are introduced, both for signed digit and for binary reflected streams. Sections 2 and 3 derive average and multiplication algorithms for streams of signed digits, and Sections 4 and 5 do the same for binary reflected streams. Section 6 concludes.

1. Stream Representations of Real Numbers

From a computational point of view, a real number x is best seen as a device producing for a given accuracy $\frac{1}{2^n}$ a rational number a_n being $\frac{1}{2^n}$-close to x, i.e., $|a_n - x| \leq \frac{1}{2^n}$. For simplicity, we only consider real numbers in the interval $[-1, 1]$ (rather than $[-2^n, 2^n]$), and dyadic rationals $\sum_{n<m} \frac{d_n}{2^{n+1}}$ ($d_n \in \{1, \bar{1}\}$; we write $\bar{1}$ for -1); see Figure 1.

Note that adjacent dyadics can differ in many digits:

$$\frac{7}{16} \sim 1\bar{1}11, \quad \frac{9}{16} \sim 11\bar{1}\bar{1}.$$

A possible cure is to *flip* after an occurrence of 1; see Figure 2. The result is called binary reflected (or Gray-) code. Then two dyadics of the same length are adjacent if and only if they differ in exactly one digit. For instance, we now have

$$\frac{7}{16} \sim \text{RRRL}, \quad \frac{9}{16} \sim \text{RLRL}.$$

This is a desirable property of stream-coded real numbers.

However, a problem with "productivity" remains: we cannot determine what the first digit of $\bar{1}111\ldots + 1\bar{1}\bar{1}\bar{1}\ldots$ (or LRLL... +

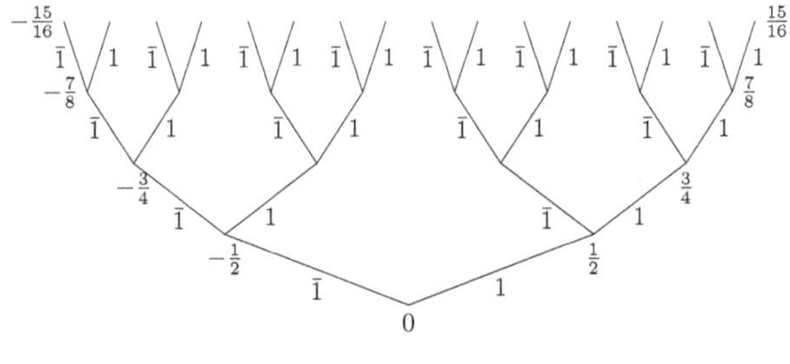

Fig. 1. Dyadic rationals.

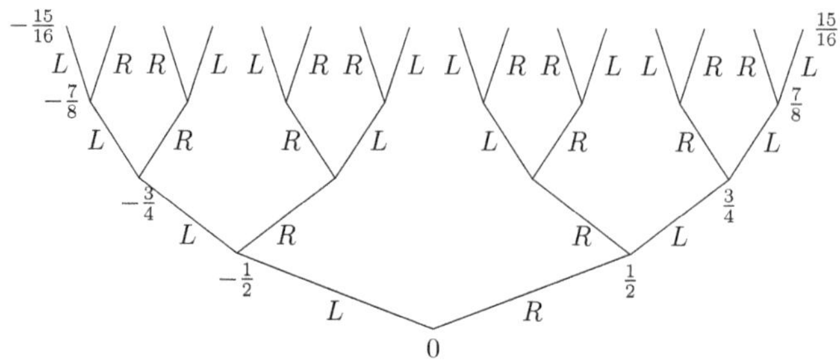

Fig. 2. Gray code.

RRRL...) should be. The cure is to add *delay* digits. For dyadic rationals, we add the digit 0 and obtain *signed digit code* widely used in numerical computation:

$$\sum_{i<k} \frac{d_i}{2^{i+1}} \quad \text{with } d_i \in \{-1, 0, 1\}.$$

We have a lot of redundancy here: for instance, $\bar{1}1$ and $0\bar{1}$ both denote $-\frac{1}{4}$.

For binary reflected code, we add digits U (for undefined) and D (for delay). This will be explained in Section 4.

2. Average for Signed Digit Streams

We now tackle our goal to extract stream algorithms from proofs, and as an example we consider a proof that the average of two real numbers in $[-1, 1]$ is in $[-1, 1]$ again. We first deal with the representation of reals as signed digit streams; in Section 4, we solve the corresponding problem for binary reflected code. To start, we need to accommodate streams in our logical framework.

We model infinite sequences of signed digits (or streams for short) as objects[a] in the (free) algebra $\mathbb{S}$ given by just one constructor $\mathrm{C}\colon \mathbb{D} \to \mathbb{S} \to \mathbb{S}$, where $\mathbb{D} := \{\mathrm{SdR}, \mathrm{SdM}, \mathrm{SdL}\}$ is a formal representation of signed digits. Each such object can be decomposed into

[a]More precisely, cototal ideals; cf. [7,10]

its head (an object in $\mathbb{D}$) and tail (another stream). Intuitively, the stream $d_0, d_1, d_2, \ldots$ represents the real number

$$\sum_{i=0}^{\infty} \frac{d_i}{2^{i+1}} \quad \text{with } d_i \in \{1, 0, -1\}.$$

We inductively define a predicate I by the single clause

$$\forall_{d,x,x'} \Big(d \in \mathrm{Sd} \to x' \in I \to x = \frac{x' + d}{2} \to x \in I \Big). \tag{1}$$

Here (and later) x ranges over real numbers and d over integers. Sd is a (formally inductive) predicate expressing that the integer d is a signed digit, i.e., $|d| \leq 1$.

We have chosen (1) rather than the simpler

$$\forall_{d,x} \Big(d \in \mathrm{Sd} \to x \in I \to \frac{x + d}{2} \in I \Big), \tag{2}$$

since we want I to be compatible with the defined equality $=$ on real numbers

$$\forall_{x,y} (x = y \to x \in I \to y \in I), \tag{3}$$

which follows from (1) (with reflexivity, symmetry and transitivity of $=$). Using (3), we then obtain (2) from (1) as a lemma, called IClosure.

At this point the present chapter deviates from [5]. In the latter work real numbers including their equality were viewed as given axiomatically. Desired well-known properties of reals were just taken as axioms, provided they have no computational content and hence do not influence the terms extracted from proofs. However, one also needs compatibilities like (3), which do have computational content (in fact, identities). In the formal development the present chapter reports on such axioms having been replaced by proofs. This required a full development of the number systems (unary and binary natural numbers, integers, rationals and reals). A real number x is taken as a pair $((a_n)_{n \in \mathbb{N}}, M)$ with $a_n \in \mathbb{Q}$ and $M \colon \mathbb{Z}^+ \to \mathbb{N}$ such that $(a_n)_n$ is a Cauchy sequence with modulus M, that is

$$|a_n - a_m| \leq \frac{1}{2^p} \quad \text{for } n, m \geq M(p).$$

Two reals $x := ((a_n)_n, M)$, $y := ((b_n)_n, N)$ are equal (written $x = y$) if

$$|a_{M(p+1)} - b_{N(p+1)}| \leq \frac{1}{2^p} \quad \text{for all } p \in \mathbb{Z}^+.$$

Dually to I we define a predicate ${}^{\text{co}}I$ coinductively by the closure axiom

$$\forall_x \Big(x \in {}^{\text{co}}I \to \exists_{d \in \text{Sd}} \exists_{x' \in {}^{\text{co}}I} \Big(x = \frac{x'+d}{2} \Big) \Big). \tag{4}$$

Therefore, the coinduction (or greatest-fixed-point) axiom is

$$\forall_{x \in P} \exists_{d \in \text{Sd}} \exists_{x' \in {}^{\text{co}}I \cup P} \Big(x = \frac{x'+d}{2} \Big) \to P \subseteq {}^{\text{co}}I. \tag{5}$$

The quantifiers $\forall_x$, $\exists_x$ deserve a special comment. As in [9], they are considered to be "noncomputational". This means that the type of the computational content of a formula involving them is independent of x and hence in particular of the concrete representations of integers and real numbers in the underlying theory. Computational content *only* arises from inductive and coinductive predicates, here Sd, I and ${}^{\text{co}}I$. Therefore, the type of I's single clause (1) is $\mathbb{D} \to \mathbb{S} \to \mathbb{S}$, i.e., the type of $\mathbb{S}$'s, constructor C.

Consider the problem to compute the average of two real numbers coded by streams. To this end we will prove

$$\forall_{x,y} \Big(x \in {}^{\text{co}}I \to y \in {}^{\text{co}}I \to \frac{x+y}{2} \in {}^{\text{co}}I \Big).$$

The computational content of this proof will be the desired algorithm.

We give an informal proof, following [11]. Consider two sets of averages, the second one with a "carry" $i \in \mathbb{Z}$

$$P := \Big\{ \frac{x+y}{2} \mid x, y \in {}^{\text{co}}I \Big\}, \quad Q := \Big\{ \frac{x+y+i}{4} \mid x, y \in {}^{\text{co}}I, i \in \text{Sd}_2 \Big\},$$

where Sd_2 is a (formally inductive) predicate expressing that the integer i is an extended signed digit, i.e., $|i| \leq 2$. It suffices to show that Q satisfies the closure axiom for ${}^{\text{co}}I$, i.e.,

$$\forall_x \Big(x \in Q \to \exists_{d \in \text{Sd}} \exists_{x' \in Q} \Big(x = \frac{x'+d}{2} \Big) \Big),$$

for then by the greatest-fixed-point axiom for ${}^{\text{co}}I$ we have $Q \subseteq {}^{\text{co}}I$. Since we also have $P \subseteq Q$, we obtain $P \subseteq {}^{\text{co}}I$, which is our claim.

Throughout this chapter, we will use functions $J, K\colon \mathbb{Z} \to \mathbb{Z}$ such that

$$\forall_i(i = J(i) + 4K(i)) \quad (\text{with } \forall_i(|J(i)| \leq 2),\ \forall_i(|i| \leq 6 \to |K(i)| \leq 1)). \tag{6}$$

We use the following notation for streams u of type $\mathbb{S}$:

$$\begin{aligned} (u)_n &:= n\text{th entry in } u, \\ Tu &:= \text{rest of } u \text{ after removal of its first entry.} \end{aligned}$$

Clearly, $(T^n u)_m = (u)_{n+m}$, where T^n denotes the nth iteration of T.

Lemma 2.1 (CoIAvToAvc).

$$\forall_{x,y}\Big(x \in {}^{\mathrm{co}}I \to y \in {}^{\mathrm{co}}I \to \exists_{i\in \mathrm{Sd}_2}\exists_{x',y'\in {}^{\mathrm{co}}I}\Big(\frac{x+y}{2} = \frac{x'+y'+i}{4}\Big)\Big).$$

Proof. By the closure axiom for ${}^{\mathrm{co}}I$, we can write $x = \frac{x'+d}{2}$ and $y = \frac{y'+e}{2}$ for some $d, e \in \mathrm{Sd}$ and $x', y' \in {}^{\mathrm{co}}I$. Then

$$\frac{x+y}{2} = \frac{x'+y'+d+e}{4}.$$

□

The computational content[b] of this proof is a function

$$\begin{aligned} &\mathrm{Av}_{\mathrm{init}}\colon \mathbb{S} \to \mathbb{S} \to \mathbb{D}_2 \times \mathbb{S} \times \mathbb{S} \\ &\mathrm{Av}_{\mathrm{init}}(u, v) := \langle (u)_0 + (v)_0, Tu, Tv \rangle. \end{aligned}$$

Lemma 2.2 (CoIAvcSatCoICl).

$$\forall_{i,x,y}\left(i \in \mathrm{Sd}_2 \wedge x, y \in {}^{\mathrm{co}}I \to \exists_{d\in \mathrm{Sd}}\exists_{j\in \mathrm{Sd}_2}\exists_{x',y'\in {}^{\mathrm{co}}I} \times \left(\frac{x+y+i}{4} = \frac{\frac{x'+y'+j}{4} + d}{2}\right)\right).$$

---

[b]We use $\mathbb{D}_2 := \{\mathrm{RT}, \mathrm{RR}, \mathrm{MT}, \mathrm{LT}, \mathrm{LL}\}$ as a formal representation of the set Sd_2 of extended signed digits. – Formally $d :: u$ is not correct: we have $d \in \mathbb{Z}$, but the constructor :: has type $\mathbb{D} \to \mathbb{S} \to \mathbb{S}$. First d has to be converted (by IntToSd) into an element of $\mathbb{D} := \{\mathrm{SdR}, \mathrm{SdM}, \mathrm{SdL}\}$. For readability such conversions are suppressed. However, they will show up in the extracted terms.

Proof. By the closure axiom for ${}^{\mathrm{co}}I$, we can write $x = \frac{x'+d}{2}$ and $y = \frac{y'+e}{2}$ for some $d, e \in \mathrm{Sd}$ and $x', y' \in {}^{\mathrm{co}}I$. Then

$$\frac{x+y+i}{4} = \frac{x'+y'+d+e+2i}{8}.$$

Since $|d+e+2i| \leq 6$, we can write $d+e+2i = j+4k$ with $|j| \leq 2$ and $|k| \leq 1$ by the JK-property (6). Therefore,

$$\frac{x+y+i}{4} = \frac{x'+y'+j+4k}{8} = \frac{\frac{x'+y'+j}{4}+k}{2}.$$

□

The computational content of this proof is a function

$$\begin{aligned}
&\mathrm{Av}_{\mathrm{step}}\colon \mathbb{D}_2 \times \mathbb{S} \times \mathbb{S} \to \mathbb{D} \times \mathbb{D}_2 \times \mathbb{S} \times \mathbb{S}\\
&\mathrm{Av}_{\mathrm{step}}\langle i, u, v\rangle := \langle K((u)_0 + (v)_0 + 2i), J((u)_0 + (v)_0 + 2i), Tu, Tv\rangle.
\end{aligned}$$

Lemma 2.3 (CoIAvcToCoI).

$$\forall_z \Big(\exists_{i \in \mathrm{Sd}_2} \exists_{x,y \in {}^{\mathrm{co}}I} \Big(z = \frac{x+y+i}{4}\Big) \to z \in {}^{\mathrm{co}}I\Big).$$

Proof. By coinduction from Lemma 2.2. □

Proposition 2.4 (CoIAverage).

$$\forall_{x,y} \Big(x \in {}^{\mathrm{co}}I \to y \in {}^{\mathrm{co}}I \to \frac{x+y}{2} \in {}^{\mathrm{co}}I\Big).$$

Proof. Immediate from Lemmas 2.1 and 2.3. □

The extracted term is short (19 lines in `minlog/examples/analysis/`, file `sdavaux.scm`), and involves the corecursion operator. It represents the function

$$\mathrm{cCoIAverage}\colon \mathbb{S} \to \mathbb{S} \to \mathbb{S},$$

defined using corecursion, as follows.

(1) From $u, v \in \mathbb{S}$ form an initial triple $\mathrm{Av}_{\mathrm{init}}(u, v) \in \mathbb{D}_2 \times \mathbb{S} \times \mathbb{S}$.
(2) Iterate $\mathrm{Av}_{\mathrm{step}}\colon \mathbb{D}_2 \times \mathbb{S} \times \mathbb{S} \to \mathbb{D} \times \mathbb{D}_2 \times \mathbb{S} \times \mathbb{S}$ starting with $\mathrm{Av}_{\mathrm{init}}(u, v)$.
(3) Return the stream of the generated $d \in \mathbb{D}$.

3. Multiplication for Signed Digit Streams

Our goal in this section is to prove that $[-1, 1]$ is closed under multiplication, w.r.t. the representation of reals as signed digit streams. Here, we follow Ciaffaglione and Di Gianantonio [3], who found a nice way to reduce this problem to the one for the average function.

Our goal is

$$\forall_{x,y}(x \in {}^{\mathrm{co}}I \to y \in {}^{\mathrm{co}}I \to x \cdot y \in {}^{\mathrm{co}}I). \tag{7}$$

The computational content of this proof will be the desired algorithm. We begin with an informal proof. Consider the two sets

$$P := \{\, x \cdot y \mid x, y \in {}^{\mathrm{co}}I \,\}, \quad Q := \left\{ \frac{x \cdot y + z + i}{4} \middle| x, y, z \in {}^{\mathrm{co}}I, i \in \mathrm{Sd}_2 \right\}.$$

It suffices to show that Q satisfies the closure axiom for ${}^{\mathrm{co}}I$, i.e.,

$$\forall_x\Big(x \in Q \to \exists_{d \in \mathrm{Sd}} \exists_{x' \in Q}\Big(x = \frac{x' + d}{2}\Big)\Big),$$

for then by the greatest-fixed-point axiom for ${}^{\mathrm{co}}I$, we have $Q \subseteq {}^{\mathrm{co}}I$. Since again we have $P \subseteq Q$, we obtain $P \subseteq {}^{\mathrm{co}}I$, which is our claim.

For $P \subseteq Q$ — which is the following Lemma 3.4 — we need auxiliary lemmata.

Lemma 3.1 (CoIUMinus). $\forall_x(-x \in {}^{\mathrm{co}}I \to x \in {}^{\mathrm{co}}I)$.

Proof. By coinduction, using properties of the unary minus function. □

Lemma 3.2 (CoIZero). $0 \in {}^{\mathrm{co}}I$.

Proof. Let $P := \{\, x \mid x = 0 \,\}$. It suffices to show that CoIClosure (4) holds for P

$$\forall_x\Big(x \in P \to \exists_{d,x'}\Big(d \in \mathrm{Sd} \wedge x' \in P \wedge x = \frac{x' + d}{2}\Big)\Big),$$

for then by the greatest-fixed-point axiom for ${}^{\mathrm{co}}I$, we have $P \subseteq {}^{\mathrm{co}}I$ and, hence, $0 \in {}^{\mathrm{co}}I$. Assume $x \in P$. Choose $d = 0$ and $x' = 0$. Then $d \in \mathrm{Sd}$, $x' \in P$ and also $x = \frac{x'+d}{2}$, since 0 is the only element of P. □

Lemma 3.3 (CoISdTimes). $\forall_{d,x}(d \in \mathrm{Sd} \to x \in {}^{\mathrm{co}}I \to dx \in {}^{\mathrm{co}}I)$.

Proof. By the definition of Sd, using Lemma 3.1 (CoIUMinus) and in the zero case Lemma 3.2 (CoIZero). □

The computational content of this proof is a function $f^*\colon \mathbb{D} \to \mathbb{S} \to \mathbb{S}$:

$$\begin{aligned} f^*(1,u) &= u, & -(\mathrm{SdR} :: u) &:= \mathrm{SdL} :: (-u),\\ f^*(0,u) &= \mathrm{SdM} :: \mathrm{SdM} :: \dots, \quad \text{where } & -(\mathrm{SdM} :: u) &:= \mathrm{SdM} :: (-u),\\ f^*(-1,u) &= -u & -(\mathrm{SdL} :: u) &:= \mathrm{SdR} :: (-u). \end{aligned}$$

Lemma 3.4 (CoIMultToMultc).

$$\forall_{x,y}\Bigl(x \in {}^{\mathrm{co}}I \to y \in {}^{\mathrm{co}}I \to \exists_{y' \in {}^{\mathrm{co}}I}\exists_{i \in \mathrm{Sd}_2}\exists_{x',z \in {}^{\mathrm{co}}I}\Bigl(xy = \frac{x'y'+z+i}{4}\Bigr)\Bigr).$$

Proof. By the closure axiom for ${}^{\mathrm{co}}I$, we can write

$$x = \frac{x'+d}{2} \quad y = \frac{y'+e}{2} \quad \text{with } x', y' \in {}^{\mathrm{co}}I \text{ and } d, e \in \mathrm{Sd}.$$

Then

$$\frac{ex'+dy'}{2} \in {}^{\mathrm{co}}I,$$

using Proposition 2.4 (CoIAverage) and Lemma 3.3 (CoISdTimes). Now again from the closure axiom for ${}^{\mathrm{co}}I$, we obtain z, d_0 such that

$$\frac{ex'+dy'}{2} = \frac{z+d_0}{2}.$$

Therefore,

$$\frac{(x'+d)(y'+e)}{4} = \frac{x'y'+(ex'+dy')+de}{4} = \frac{x'y'+z+(d_0+de)}{4},$$

which is of the required form. □

The computational content of this proof is a function

$$\begin{aligned} &\mathrm{Mul_{init}}\colon \mathbb{S} \to \mathbb{S} \to \mathbb{S} \times \mathbb{D}_2 \times \mathbb{S} \times \mathbb{S}\\ &\mathrm{Mul_{init}}(u,v) := \langle v, d_0 + (u)_0 \cdot (v)_0, Tu, w\rangle, \end{aligned}$$

where d_0 and w are head and tail of cCoIAverage($f^*((v)_0, Tu)$, $f^*((u)_0, Tv)$).

Lemma 3.5 (CoIMultcSatCoICl).

$$\forall_y \left(y \in {}^{\mathrm{co}}I \to \forall_{i,x,z} \left(i \in \mathrm{Sd}_2 \wedge x \in {}^{\mathrm{co}}I \wedge z \in {}^{\mathrm{co}}I \right.\right.$$

$$\left.\left. \to \exists_{d \in \mathrm{Sd}} \exists_{j \in \mathrm{Sd}_2} \exists_{x',z' \in {}^{\mathrm{co}}I} \left(\frac{xy+z+i}{4} = \frac{\frac{x'y+z'+j}{4}+d}{2} \right) \right) \right).$$

Proof. By the closure axiom for ${}^{\mathrm{co}}I$, we can write

$$x = \frac{x_1+d_1}{2} \qquad z = \frac{z_0+d_0}{2} \qquad \text{with } x_1, z_0 \in {}^{\mathrm{co}}I \text{ and } d_1, d_0 \in \mathrm{Sd}.$$

Then

$$\begin{aligned} \frac{xy+z+i}{4} &= \frac{(x_1+d_1)y+(z_0+d_0)+2i}{8} \\ &= \frac{x_1y+(z_0+d_1y+i)+d_0+i}{8}. \end{aligned}$$

We have $d_1y \in {}^{\mathrm{co}}I$ by Lemma 3.3 (CoISdTimes) and $\frac{z_0+d_1y+i}{4} := v \in {}^{\mathrm{co}}I$ by Lemma 2.3 (CoIAvcToCoI). Using v, we can continue the chain of equations by

$$= \frac{x_1y+4v+d_0+i}{8}.$$

Because of $v \in {}^{\mathrm{co}}I$, we can write

$$v = \frac{z_1+e_0}{2} = \frac{\frac{z_2+e}{2}+e_0}{2} \qquad \text{with } z_1, z_2 \in {}^{\mathrm{co}}I \text{ and } e_0, e \in \mathrm{Sd}.$$

Therefore,

$$= \frac{x_1y+(z_2+e+2e_0)+d_0+i}{8}.$$

Using again the functions J, K with the JK-property (6), we can write $e+2e_0+d_0+i$ as $j+4d$ with $j := J(e+2e_0+d_0+i)$ and $d := K(e+2e_0+d_0+i)$. Hence,

$$= \frac{x_1y+z_2+j+4d}{8} = \frac{\frac{x_1y+z_2+j}{4}+d}{2}.$$

□

The computational content of this proof is a function

$$\mathrm{Mul}_{\mathrm{step}}\colon \mathbb{S} \to \mathbb{D}_2 \times \mathbb{S} \times \mathbb{S} \to \mathbb{D} \times \mathbb{D}_2 \times \mathbb{S} \times \mathbb{S},$$

defined by

$$\begin{aligned}\mathrm{Mul}_{\mathrm{step}}(v, \langle i, u, w\rangle) := \langle & K((u'')_1 + 2(u'')_0 + (w)_0 + i), \\ & J((u'')_1 + 2(u'')_0 + (w)_0 + i), Tu, T^2u''\rangle,\end{aligned}$$

with $u'' := \mathrm{cCoIAvcToCoI}(i, f^*((u)_0, v), Tw)$ (here cCoIAvcToCoI is the computational content of Lemma 2.3).

Lemma 3.6 (CoIMultcToCoI).

$$\forall_z \Big(\exists_{y\in{}^{\mathrm{co}}I}\exists_{i\in\mathrm{Sd}_2}\exists_{x,z_0\in{}^{\mathrm{co}}I}\Big(z = \frac{xy + z_0 + i}{4}\Big) \to z \in {}^{\mathrm{co}}I\Big).$$

Proof. By coinduction from Lemma 3.5. □

Proposition 3.7 (CoIMult).

$$\forall_{x,y}(x \in {}^{\mathrm{co}}I \to y \in {}^{\mathrm{co}}I \to xy \in {}^{\mathrm{co}}I).$$

Proof. Immediate from Lemmas 3.4 and 3.6. □

The term extracted from this proof is short (6 lines in `sdmult.scm`), and involves the corecursion operator. It represents the function

$$\mathrm{cCoIMult}\colon \mathbb{S} \to \mathbb{S} \to \mathbb{S},$$

defined using corecursion, as follows.

(1) $\mathrm{Mul}_{\mathrm{init}}$ applied to the two input streams returns a quadruple $\langle v, t, u, w\rangle$.
(2) Fix v and iterate $\mathrm{Mul}_{\mathrm{step}}(v, \cdot)\colon \mathbb{D}_2 \times \mathbb{S} \times \mathbb{S} \to \mathbb{D} \times \mathbb{D}_2 \times \mathbb{S} \times \mathbb{S}$ starting with $\langle t, u, w\rangle$.
(3) Return the stream of the generated $d \in \mathbb{D}$.

The terms extracted from proofs can be translated into Scheme or Haskell programmes. However, since they may involve corecursion operators the lazy evaluation of Haskell is more appropriate. As an example, we have run the Haskell programme obtained from the proof of Proposition 3.7 on approximations of 1/2 and 1/3. It returns the first 20 (signed) digits in 0.04 seconds.

This example can also be used to show that cCoIMult is not commutative, w.r.t. the (nonunique) representation of reals as strems of signed digits. This is to be expected since in the proof of Lemma 3.5 the arguments x and y are treated differently: y is viewed as a parameter and kept fixed. Indeed, after swapping the arguments in our example, the resulting stream differs from the previous one in the 14th digit.

4. Average for Binary Reflected Code

We now consider the problem to compute the average of two real numbers given in binary reflected code. Again, when doing arithmetical operations on binary reflected streams, the problem of productivity arises, which is dealt with in a somewhat different way. The idea is to introduce two "modes" when generating the stream, and flip from one mode to the other whenever we encounter the digit R (cf. Figure 2). Instead of the predicate I, we now use two predicates G, H (G for Gray, H next character) and flip from one to the other after reading 1. They are defined by simultaneous induction, with clauses (the ones containing both G and H indicate flipping)

$$\forall_{d\in\mathrm{Psd}}\forall_{x\in G}\Big(-d\frac{x-1}{2}\in G\Big),\quad \forall_{x\in H}\Big(\frac{x}{2}\in G\Big),$$
$$\forall_{d\in\mathrm{Psd}}\forall_{x\in G}\Big(d\frac{x+1}{2}\in H\Big),\quad \forall_{x\in H}\Big(\frac{x}{2}\in H\Big).$$

Here Psd (for proper signed digit) is the (inductive) predicate consisting of the integers $1, -1$. We require that Psd has computational content, in the booleans $\mathbb{B}=\{\mathtt{tt},\mathtt{ff}\}$. For brevity, from now on we write $\forall_{x\in G}A$ for $\forall_x(x\in G\to A)$, and similarly for other predicates. Note that there are no nullary clauses (as for I), since they are not needed. We are only interested in the duals ${}^{\mathrm{co}}G, {}^{\mathrm{co}}H$, coinductively defined by the simultaneous closure axioms

$$\begin{aligned}&\forall_{x\in{}^{\mathrm{co}}G}\Big(\exists_{d\in\mathrm{Psd}}\exists_{x'\in{}^{\mathrm{co}}G}\Big(x=-d\frac{x'-1}{2}\Big)\vee\exists_{x'\in{}^{\mathrm{co}}H}\Big(x=\frac{x'}{2}\Big)\Big),\\&\forall_{x\in{}^{\mathrm{co}}H}\Big(\exists_{d\in\mathrm{Psd}}\exists_{x'\in{}^{\mathrm{co}}G}\Big(x=d\frac{x'+1}{2}\Big)\vee\exists_{x'\in{}^{\mathrm{co}}H}\Big(x=\frac{x'}{2}\Big)\Big),\end{aligned}\tag{8}$$

and the simultaneous coinduction axioms

$$\begin{aligned}
&\forall_{x\in P}\Big(\exists_{d\in \mathrm{Psd}}\exists_{x'\in {}^{\mathrm{co}}G\cup P}\Big(x=-d\frac{x'-1}{2}\Big)\vee\exists_{x'\in {}^{\mathrm{co}}H\cup Q}\Big(x=\frac{x'}{2}\Big)\Big)\\
&\quad\to\forall_{x\in Q}\Big(\exists_{d\in \mathrm{Psd}}\exists_{x'\in {}^{\mathrm{co}}G\cup P}\Big(x=d\frac{x'+1}{2}\Big)\vee\exists_{x'\in {}^{\mathrm{co}}H\cup Q}\Big(x=\frac{x'}{2}\Big)\Big)\\
&\quad\to P\subseteq {}^{\mathrm{co}}G\quad\text{and the same with conclusion } Q\subseteq {}^{\mathrm{co}}H.
\end{aligned}\tag{9}$$

We require that the predicates G, H and therefore ${}^{\mathrm{co}}G, {}^{\mathrm{co}}H$ as well have computational content, in simultaneously defined algebras $\mathbb{G}$ and $\mathbb{H}$ given by the constructors

$$\begin{aligned}
&\mathrm{Lr}\colon \mathbb{B}\to\mathbb{G}\to\mathbb{G}, &&\mathrm{U}\colon \mathbb{H}\to\mathbb{G} &&\text{for } \mathbb{G},\\
&\mathrm{Fin}\colon \mathbb{B}\to\mathbb{G}\to\mathbb{H}, &&\mathrm{D}\colon \mathbb{H}\to\mathbb{H} &&\text{for } \mathbb{H}.
\end{aligned}$$

We write $\mathrm{Lr}_{\mathtt{tt}}(u)$ for $\mathrm{Lr}(\mathtt{tt}, u)$ and $\mathrm{Lr}_{\mathtt{ff}}(u)$ for $\mathrm{Lr}(\mathtt{ff}, u)$, and similarly for Fin. Then $\mathrm{Lr}_{\mathtt{tt}/\mathtt{ff}}$ means left/right, $\mathrm{Fin}_{\mathtt{tt}/\mathtt{ff}}$ means finally left/right. The delay constructors are U ("undefined") for $\mathbb{G}$ and D ("delay") for $\mathbb{H}$. In Figure 3, bold lines and circles indicate G, and dashed lines and $\circ$ indicate H.

The computational content of the closure axioms (8) are the destructors $\mathcal{D}_{\mathbb{G}}\colon \mathbb{G}\to(\mathbb{B}\times\mathbb{G})+\mathbb{H}$ and $\mathcal{D}_{\mathbb{H}}\colon \mathbb{H}\to(\mathbb{B}\times\mathbb{G})+\mathbb{H}$ defined by

$$\begin{aligned}
&\mathcal{D}_{\mathbb{G}}(\mathrm{Lr}_b(u))=\mathrm{InL}\langle b,u\rangle, &&\mathcal{D}_{\mathbb{G}}(\mathrm{U}v)=\mathrm{InR}(v),\\
&\mathcal{D}_{\mathbb{H}}(\mathrm{Fin}_b(u))=\mathrm{InL}\langle b,u\rangle, &&\mathcal{D}_{\mathbb{H}}(\mathrm{D}v)=\mathrm{InR}(v).
\end{aligned}$$

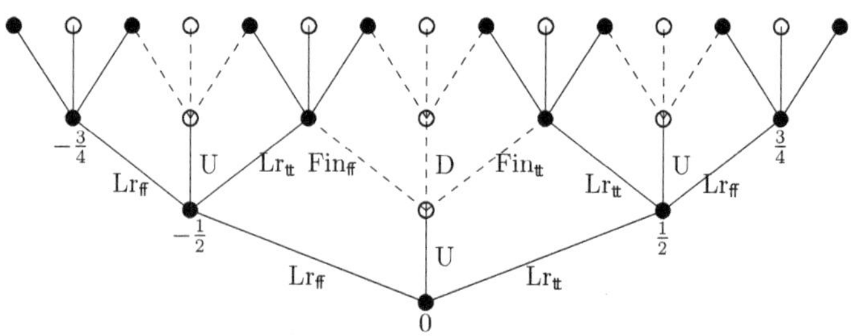

Fig. 3. Binary reflected code with delay.

The computational content of the coinduction axioms (9) are instances of the simultaneous corecursion operators

$$\begin{aligned}
{}^{\mathrm{co}}\mathcal{R}_{\mathbb{G}}^{(\mathbb{G},\mathbb{H}),(\sigma,\tau)} &: \sigma \to \delta_{\mathbb{G}} \to \delta_{\mathbb{H}} \to \mathbb{G},\\
{}^{\mathrm{co}}\mathcal{R}_{\mathbb{H}}^{(\mathbb{G},\mathbb{H}),(\sigma,\tau)} &: \tau \to \delta_{\mathbb{G}} \to \delta_{\mathbb{H}} \to \mathbb{H},
\end{aligned} \tag{10}$$

with step types

$$\begin{aligned}
\delta_{\mathbb{G}} &:= \sigma \to \mathbb{B} \times (\mathbb{G} + \sigma) + (\mathbb{H} + \tau),\\
\delta_{\mathbb{H}} &:= \tau \to \mathbb{B} \times (\mathbb{G} + \sigma) + (\mathbb{H} + \tau).
\end{aligned}$$

The type $\mathbb{B}\times(\mathbb{G}+\sigma)+(\mathbb{H}+\tau)$ appears since $\mathbb{G}$ has the two constructors $\mathrm{Lr}\colon \mathbb{B} \to \mathbb{G} \to \mathbb{G}$ and $\mathrm{U}\colon \mathbb{H} \to \mathbb{G}$, and $\mathbb{H}$ has the two constructors $\mathrm{Fin}\colon \mathbb{B} \to \mathbb{G} \to \mathbb{H}$ and $\mathrm{D}\colon \mathbb{H} \to \mathbb{H}$. Omitting the upper indices of ${}^{\mathrm{co}}\mathcal{R}$, the computation rules for the terms ${}^{\mathrm{co}}\mathcal{R}_{\mathbb{G}}sgh$ and ${}^{\mathrm{co}}\mathcal{R}_{\mathbb{H}}tgh$ are

$$\begin{aligned}
{}^{\mathrm{co}}\mathcal{R}_{\mathbb{G}}sgh &= \begin{cases}
\mathrm{Lr}(b,u) & \text{if } g(s) \equiv \mathrm{InL}\langle b, \mathrm{InL}_{\mathbb{G}+\sigma}u\rangle\\
\mathrm{Lr}(b, {}^{\mathrm{co}}\mathcal{R}_{\mathbb{G}}s'gh) & \text{if } g(s) \equiv \mathrm{InL}\langle b, \mathrm{InR}_{\mathbb{G}+\sigma}s'\rangle\\
\mathrm{U}(v) & \text{if } g(s) \equiv \mathrm{InR}(\mathrm{InL}_{\mathbb{H}+\tau}v)\\
\mathrm{U}({}^{\mathrm{co}}\mathcal{R}_{\mathbb{H}}tgh) & \text{if } g(s) \equiv \mathrm{InR}(\mathrm{InR}_{\mathbb{H}+\tau}t)
\end{cases}\\
{}^{\mathrm{co}}\mathcal{R}_{\mathbb{H}}tgh &= \begin{cases}
\mathrm{Fin}(b,u) & \text{if } h(t) \equiv \mathrm{InL}\langle b, \mathrm{InL}_{\mathbb{G}+\sigma}u\rangle\\
\mathrm{Fin}(b, {}^{\mathrm{co}}\mathcal{R}_{\mathbb{G}}sgh) & \text{if } h(t) \equiv \mathrm{InL}\langle b, \mathrm{InR}_{\mathbb{G}+\sigma}s\rangle\\
\mathrm{D}(v) & \text{if } h(t) \equiv \mathrm{InR}(\mathrm{InL}_{\mathbb{H}+\tau}v)\\
\mathrm{D}({}^{\mathrm{co}}\mathcal{R}_{\mathbb{H}}t'gh) & \text{if } h(t) \equiv \mathrm{InR}(\mathrm{InR}_{\mathbb{H}+\tau}t'),
\end{cases}
\end{aligned} \tag{11}$$

with s of type σ and t of type τ.

Our goal is to prove that $[-1,1]$ is closed under average, w.r.t. the representation of reals in binary reflected code. We proceed essentially as for the signed digit case. The main difference is that the simultaneous definition of ${}^{\mathrm{co}}G$ and ${}^{\mathrm{co}}H$ makes it necessary to use simultaneous coinduction.

Lemma 4.1 (CoGCoHMinus, CoHCoGMinus). *${}^{\mathrm{co}}G$ and ${}^{\mathrm{co}}H$ are closed under minus:*

$$\forall_{x\in {}^{\mathrm{co}}G}(-x \in {}^{\mathrm{co}}G), \quad \forall_{x\in {}^{\mathrm{co}}H}(-x \in {}^{\mathrm{co}}H).$$

Proof. We prove both claims simultaneously, by coinduction:

$$\forall_{x \in P}\Big(\exists_{d \in \mathrm{Psd}}\exists_{x' \in {}^{\mathrm{co}}G \cup P}\Big(x = -d\frac{x'-1}{2}\Big) \vee \exists_{x' \in {}^{\mathrm{co}}H \cup Q}\Big(x = \frac{x'}{2}\Big)\Big)$$
$$\to \forall_{x \in Q}\Big(\exists_{d \in \mathrm{Psd}}\exists_{x' \in {}^{\mathrm{co}}G \cup P}\Big(x = d\frac{x'+1}{2}\Big) \vee \exists_{x' \in {}^{\mathrm{co}}H \cup Q}\Big(x = \frac{x'}{2}\Big)\Big)$$
$$\to \forall_x(x \in P \to x \in {}^{\mathrm{co}}G) \wedge \forall_x(x \in Q \to x \in {}^{\mathrm{co}}H).$$

We choose $P := \{\, x \mid -x \in {}^{\mathrm{co}}G \,\}$ and $Q := \{\, x \mid -x \in {}^{\mathrm{co}}H \,\}$. Since ${}^{\mathrm{co}}G$ and ${}^{\mathrm{co}}H$ are invariant under real equality (also proven by coinduction), we have to prove the two premises. We only consider the first one; the second is proved similarly. Let $x \in P$ be given. The goal is

$$\exists_{d \in \mathrm{Psd}}\exists_{x' \in {}^{\mathrm{co}}G \cup P}\Big(x = -d\frac{x'-1}{2}\Big) \vee \exists_{x' \in {}^{\mathrm{co}}H \cup Q}\Big(x = \frac{x'}{2}\Big).$$

From $x \in P$, we get $-x \in {}^{\mathrm{co}}G$ and, therefore,

$$\exists_{d \in \mathrm{Psd}}\exists_{x' \in {}^{\mathrm{co}}G}\Big(-x = -d\frac{x'-1}{2}\Big) \vee \exists_{x' \in {}^{\mathrm{co}}H}\Big(-x = \frac{x'}{2}\Big),$$

by the closure axiom. If the right-hand side of this disjunction holds, we are done by definition of Q. If the left-hand side holds, we have $x = d\frac{x'-1}{2}$ for some $d \in \mathrm{Psd}$ and $x' \in {}^{\mathrm{co}}G$. To get the left-hand side of the disjunction in the goal formula, we take $d := -d$ and $x' := x'$. □

The functions we get from the proof take a binary reflected stream and flip all $\mathtt{tt}$ to $\mathtt{ff}$ and the other way around. We denote both functions by $-$. Then the computation rules are

$$\begin{aligned}
-(\mathrm{Lr}_{\mathtt{tt}}(u)) &= \mathrm{Lr}_{\mathtt{ff}}(-u), & -(\mathrm{Fin}_{\mathtt{tt}}(u)) &= \mathrm{Fin}_{\mathtt{ff}}(-u),\\
-(\mathrm{Lr}_{\mathtt{ff}}(u)) &= \mathrm{Lr}_{\mathtt{tt}}(-u), & -(\mathrm{Fin}_{\mathtt{ff}}(u)) &= \mathrm{Fin}_{\mathtt{tt}}(-u)),\\
-(\mathrm{U}(v)) &= \mathrm{U}(-v), & -(\mathrm{D}(v)) &= \mathrm{D}(-v).
\end{aligned}$$

Lemma 4.2 (CoGToCoH, CoHToCoG). *${}^{\mathrm{co}}G$ and ${}^{\mathrm{co}}H$ are equivalent:*

$$\forall_{x \in {}^{\mathrm{co}}G}(x \in {}^{\mathrm{co}}H), \quad \forall_{x \in {}^{\mathrm{co}}H}(x \in {}^{\mathrm{co}}G).$$

Proof. In the coinduction axiom, we choose $P := {}^{\mathrm{co}}H$ and $Q := {}^{\mathrm{co}}G$, and the conclusion in the coinduction axiom is our goal. Therefore, we have to prove the premises of the coinduction axiom.

We only consider the first one; the second one is proved similarly. Let $x \in {}^{\mathrm{co}}H$ be given. By the closure axiom of ${}^{\mathrm{co}}H$ we have

$$\exists_{d \in \mathrm{Psd}} \exists_{x' \in {}^{\mathrm{co}}G} \Big(x = d\frac{x'+1}{2} \Big) \vee \exists_{x' \in {}^{\mathrm{co}}H} \Big(x = \frac{x'}{2} \Big).$$

If the second part of this disjunction holds, we get directly the second part of the first premise. If the first part of the disjunction holds, we have $x = d\frac{x'+1}{2}$ for some $d \in \mathrm{Psd}$ and $x' \in {}^{\mathrm{co}}G$. Then $x = -d\frac{(-x')-1}{2}$ and using Lemma 4.1, we get the first part of disjunction in the premise. □

As computational content, we have two functions cCoGToCoH: $\mathbb{G} \to \mathbb{H}$ and cCoHToCoG: $\mathbb{H} \to \mathbb{G}$, with computation rules

$$\mathrm{cCGoGToCoH}(\mathrm{Lr}_b(u)) = \mathrm{Fin}_b(-u), \quad \mathrm{cCoHToCoG}(\mathrm{Fin}_b(u)) = \mathrm{Lr}_b(-u),$$
$$\mathrm{cCGoGToCoH}(\mathrm{U}(v)) = \mathrm{D}(v), \quad \mathrm{cCoHToCoG}(\mathrm{D}(v)) = \mathrm{U}(v).$$

Note that no corecursive call is involved.

The proof of the existence of the average w.r.t. reals in binary reflected code is similar to the proof in Section 2 of the existence of the average w.r.t. signed digit stream coded reals. We will need

Lemma 4.3 (CoGPsdTimes). $\forall_{a \in \mathrm{Psd}} \forall_{x \in {}^{\mathrm{co}}G} (ax \in {}^{\mathrm{co}}G)$.

Proof. By the definition of Psd, using Lemma 4.1. □

Lemma 4.4 (CoGAvToAvc).

$$\forall_{x,y \in {}^{\mathrm{co}}G} \exists_{i \in \mathrm{Sd}_2} \exists_{x',y' \in {}^{\mathrm{co}}G} \Big(\frac{x+y}{2} = \frac{x'+y'+i}{4} \Big).$$

Proof. Let $x, y \in {}^{\mathrm{co}}G$. With the closure axioms (8) the argument is easy. There are two cases (Lr and U) for each of x, y. □

The computational content of this proof is a function f defined in the following. We use f^* for cCoGPsdTimes (the computational content of Lemma 4.3) and s for cCoHToCoG (part of the computational content of Lemma 4.2)

$$\begin{aligned}
f(\mathrm{Lr}_a(u), \mathrm{Lr}_{a'}(u')) &= (a + a', f^*(-a, u), f^*(-a', u')), \\
f(\mathrm{Lr}_a(u), \mathrm{U}(v)) &= (a, f^*(-a, u), s(v)), \\
f(\mathrm{U}(v), \mathrm{Lr}_a(u)) &= (a, s(v), f^*(-a, u)), \\
f(\mathrm{U}(v), \mathrm{U}(v')) &= (0, s(v), s(v')).
\end{aligned}$$

Lemma 4.5 (CoGAvcSatCoICl).

$$\forall_{i\in\mathrm{Sd}_2}\forall_{x,y\in{}^{\mathrm{co}}G}\exists_{j\in\mathrm{Sd}_2}\exists_{d\in\mathrm{Sd}}\exists_{x',y'\in{}^{\mathrm{co}}G}\left(\frac{x+y+i}{4}=\frac{\frac{x'+y'+j}{4}+d}{2}\right).$$

Proof. Let $x,y\in{}^{\mathrm{co}}G$. Again by (8), there are two cases (Lr and U) for each of x, y. Since the proof is very similar to the proof of Lemma 2.2, we omit it. However, we need the functions J,K defined there. □

The computational content of Lemma 4.5 is a function f defined as follows:

$$\begin{aligned}
f(i,\mathrm{Lr}_a(u),\mathrm{Lr}_{a'}(u')) &= (J(a{+}a'{+}2i),K(a{+}a'{+}2i),\\
&\qquad f^*(-a,u),f^*(-a',u')),\\
f(i,\mathrm{Lr}_a(u),\mathrm{U}(v)) &= (J(a+2i),K(a+2i),f^*(-a,u),s(v)),\\
f(i,\mathrm{U}(v),\mathrm{Lr}_a(u)) &= (J(a+2i),K(a+2i),s(v),f^*(-a,u)),\\
f(i,\mathrm{U}(v),\mathrm{U}(v')) &= (J(2i),K(2i),s(v),s(v')).
\end{aligned}$$

By coinduction from Lemma 4.5, we obtain

Lemma 4.6 (CoGAvcToCoG).

$$\forall_z\left(\exists_{i\in\mathrm{Sd}_2}\exists_{x,y\in{}^{\mathrm{co}}G}\left(z=\frac{x+y+i}{4}\right)\to z\in{}^{\mathrm{co}}G\right).$$

Proof. Let $Q\subseteq{}^{\mathrm{co}}G$ abbreviate its claim. We prove it simultaneously with $Q\subseteq{}^{\mathrm{co}}H$ (defined similarly), by coinduction. Let $z_1\in Q$. We first show

$$\exists_{z\in{}^{\mathrm{co}}G\cup Q}\exists_{a\in\mathrm{Psd}}\left(z_1=-a\frac{z-1}{2}\right)\vee\exists_{z\in{}^{\mathrm{co}}H\cup Q}\left(z_1=\frac{z}{2}\right).\qquad(12)$$

Lemma 4.5 applied to $z_1\in Q$ gives us $x_1,y_1\in{}^{\mathrm{co}}G$ and i_1,d_1 such that

$$z_1=\frac{\frac{x_1+y_1+i_1}{4}+d_1}{2}.$$

Case $d_1=0$. Go for the right-hand side of (12) with $z:=(x_1+y_1+i_1)/4\in Q$. *Case* $d_1=\pm1$. Go for the left-hand side of (12) with $a:=d_1$

and $z := (-ax_1 - ay_1 - ai_1)/4 \in Q$. Then

$$-a\frac{z-1}{2} = -a\frac{4z-4}{8} = \frac{x_1 + y_1 + i_1 + 4a}{8} = z_1.$$

This finishes the proof of (12). The rest of the proof is similar, and we omit it. □

The computational content of this proof consists of two functions g and h defined in the following. In the proof, we used the following lemma:

$$\text{SdDisj}\colon \forall_{d \in \text{Sd}}(d = 0 \vee^{\text{r}} \exists_{a \in \text{Psd}}(d = a)).$$

Here $\vee^{\text{r}}$ is an (inductively defined) variant of $\vee$ where only the content of the right-hand side is kept.

$$\begin{aligned}
g(i, u, u') = \text{let } &(i_1, d, u_1, u_1') = \text{cCoGAvcSatCoICl}(i, u, u') \text{ in} \\
&\text{case cSdDisj}(d) \text{ of} \\
&\quad 0 \to \text{U}(h(i, u_1, u_1')) \\
&\quad a \to \text{Lr}_a(g(-ai, f^*(-a, u_1), f^*(-a, u_1'))),
\end{aligned}$$

$$\begin{aligned}
h(i, u, u') = \text{let } &(i_1, d, u_1, u_1') = \text{cCoGAvcSatCoICl}(i, u, u') \text{ in} \\
&\text{case cSdDisj}(d) \text{ of} \\
&\quad 0 \to \text{D}(h(i, u_1, u_1')) \\
&\quad a \to \text{Fin}_a(g(-ai, f^*(-a, u_1), f^*(-a, u_1'))).
\end{aligned}$$

Proposition 4.7 (CoGAverage).

$$\forall_{x,y \in {}^{\text{co}}G}\left(\frac{x+y}{2} \in {}^{\text{co}}G\right).$$

Proof. Compose the proofs of Lemma 4.4 (CoGAvToAvc) and Lemma 4.6 (CoGAvcToCoG). □

The computational context of each of the two lemmata has been spelled out above, after their proofs.

5. Multiplication for Binary Reflected Code

Finally, we consider a proof that $[-1, 1]$ is closed under multiplication, w.r.t. the representation of reals in binary reflected code. We will proceed as for the signed digit case, and make use of material from Section 4. Our goal is to find a proof of

$$\forall_{x,x'}(x, x' \in {}^{\mathrm{co}}G \to x \cdot x' \in {}^{\mathrm{co}}G), \tag{13}$$

which will give us the desired algorithm. Consider the two sets

$$P := \{x \cdot y \mid x, y \in {}^{\mathrm{co}}G\},$$
$$Q := \left\{\frac{x \cdot y + z + i}{4} \mid x, y, z \in {}^{\mathrm{co}}G, i \in \mathrm{Sd}_2\right\}.$$

It again suffices to show that Q satisfies the closure axiom (8) for ${}^{\mathrm{co}}G$, for then by the greatest-fixed-point axiom (9) for ${}^{\mathrm{co}}G$, we have $Q \subseteq {}^{\mathrm{co}}G$. Since we also have $P \subseteq Q$, we obtain $P \subseteq {}^{\mathrm{co}}G$, which is our claim.

We need an auxiliary lemma

Lemma 5.1 (CoGZero). $0 \in {}^{\mathrm{co}}G$.

Proof. By coinduction. The proof is similar to the one for Lemma 3.2 (CoIZero), but we must prove the claim $0 \in {}^{\mathrm{co}}G$ simultaneously with $0 \in {}^{\mathrm{co}}H$. However, since in both cases we can use the second alternative in the closure axiom (8) for ${}^{\mathrm{co}}G$, the proof is even simpler. □

Lemma 5.2 (CoGMultToMultc).

$$\forall_{x,y \in {}^{\mathrm{co}}G} \exists_{i \in \mathrm{Sd}_2} \exists_{x',y',z \in {}^{\mathrm{co}}G}\left(xy = \frac{x'y' + z + i}{4}\right).$$

Proof. We distinguish cases on $x \in {}^{\mathrm{co}}G$ and $y \in {}^{\mathrm{co}}G$ according to the disjunction in the closure axiom (8) for ${}^{\mathrm{co}}G$.

Case Lrx, Lry. Then we can write

$$x = -a\frac{x' - 1}{2} \quad y = -b\frac{y' - 1}{2} \quad \text{with } x', y' \in {}^{\mathrm{co}}G \text{ and } a, b \in \mathrm{Psd}.$$

By Proposition 4.7 (CoGAverage) and Lemma 4.3 (CoGPsdTimes), we have

$$\frac{-abx' - aby'}{2} =: z \in {}^{\mathrm{co}}G.$$

We distinguish cases according to the disjunction in (8). *Subcase* Lrz. Then we can write

$$z = -c\frac{z'-1}{2} \quad \text{with } z' \in {}^{\text{co}}G \text{ and } c \in \text{Psd}.$$

Therefore,

$$\begin{aligned} xy &= \frac{(ax'-a)(by'-b)}{4} = \frac{ax'by' - ab(x'+y') + ab}{4} \\ &= \frac{ax'by' - cz' + (c+ab)}{4}, \end{aligned}$$

which is of the required form. *Subcase* Uz. Then we can write

$$z = \frac{z'}{2} \quad \text{with } z' \in {}^{\text{co}}H.$$

Therefore,

$$\begin{aligned} xy &= \frac{(ax'-a)(by'-b)}{4} = \frac{ax'by' - ab(x'+y') + ab}{4} \\ &= \frac{ax'by' + z' + ab}{4}, \end{aligned}$$

which is of the required form because of Lemma 4.2.

Case Lrx, Uy. Then we can write

$$x = -a\frac{x'-1}{2} \quad y = \frac{y'}{2} \quad \text{with } x' \in {}^{\text{co}}G,\ y' \in {}^{\text{co}}H \text{ and } a \in \text{Psd}.$$

Therefore,

$$xy = \frac{(-ax'+a)y'}{4} = \frac{-ax'y' + ay'}{4},$$

which is of the required form because of Lemma 4.2.

Case Ux, Lry. Then we can write

$$x = \frac{x'}{2} \quad y = -b\frac{y'-1}{2} \quad \text{with } x' \in {}^{\text{co}}H,\ y' \in {}^{\text{co}}G \text{ and } b \in \text{Psd}.$$

Therefore,

$$xy = \frac{x'(-by'+b)}{4} = \frac{-bx'y' + bx'}{4},$$

which is of the required form because of Lemma 4.2.

Case Ux, Uy. Then we can write

$$x = \frac{x'}{2} \quad y = \frac{y'}{2} \quad \text{with } x', y' \in {}^{\mathrm{co}}H.$$

Therefore,

$$xy = \frac{x'y'}{4},$$

which is of the required form because of Lemma 5.1 (CoGZero). □

The computational content of Lemma 5.2 is spelled out in what follows. We use s for cCoHToCoG. For brevity, from now on we omit f^* (for cCoGPsdTimes) and simply write au for $f^*(a, u)$.

$$\begin{aligned} g(\mathrm{Lr}_a(u), \mathrm{Lr}_b(v)) &= \text{case cCoGAverage}(-abu, -abv) \text{ of} \\ &\qquad \mathrm{Lr}_c(w) \to (c + ab, au, bv, -cw) \\ &\qquad \mathrm{U}(w) \to (ab, au, bv, s(w)) \\ g(\mathrm{Lr}_a(u), \mathrm{U}(v))) &= (0, -au, s(v), as(v)) \\ g(\mathrm{U}(u), \mathrm{Lr}_b(v)) &= (0, s(u), -bv, bs(u)) \\ g(\mathrm{U}(u), \mathrm{U}(v)) &= (0, s(u), s(v), \text{cCoGZero}). \end{aligned}$$

Lemma 5.3 (CoGMultcSatCoICl).

$$\forall_{y \in {}^{\mathrm{co}}G} \forall_{i \in \mathrm{Sd}_2} \forall_{x,z \in {}^{\mathrm{co}}G} \exists_{d \in \mathrm{Sd}} \exists_{j \in \mathrm{Sd}_2} \exists_{x',z' \in {}^{\mathrm{co}}G} \left(\frac{xy + z + i}{4} = \frac{\frac{x'y + z' + j}{4} + d}{2} \right).$$

Proof. We distinguish cases on $x \in {}^{\mathrm{co}}G$ and $z \in {}^{\mathrm{co}}G$ according to the disjunction in the closure axiom in (8).

Case Lrx, Lrz. We can write

$$x = -a_1 \frac{x_1 - 1}{2} \quad z = -a_0 \frac{z_1 - 1}{2}$$
$$\text{with } x_1, z_1 \in {}^{\mathrm{co}}G \text{ and } a_1, a_0 \in \mathrm{Psd}.$$

Then

$$\begin{aligned} \frac{xy + z + i}{4} &= \frac{(-a_1x_1 + a_1)y - a_0z_1 + a_0 + 2i}{8} \\ &= \frac{-a_1x_1y + (a_1y - a_0z_1 + i) + a_0 + i}{8}. \end{aligned}$$

We have $a_1y, -a_0z_1 \in {}^{\mathrm{co}}G$ by Lemma 4.3 (CoGPsdTimes) and $\frac{a_1y - a_0z_1 + i}{4} =: v \in {}^{\mathrm{co}}G$ by Lemma 4.6 (CoGAvcToCoG). Hence, we can continue

$$= \frac{-a_1x_1y + 4v + a_0 + i}{8} = \frac{\frac{-a_1x_1y}{4} + v + \frac{a_0+i}{4}}{2} = \frac{\frac{-a_1x_1y+z+j}{4} + d}{2},$$

with $z \in {}^{\mathrm{co}}G$, $j \in \mathrm{Sd}_2$ and $d \in \mathrm{Sd}$, by the following Lemma 5.4 (JKLr).

Case Lrx, Uz. We can write

$$x = -a_1\frac{x_1 - 1}{2} \quad z = \frac{z_1}{2} \quad \text{with } x_1 \in {}^{\mathrm{co}}G,\ z_1 \in {}^{\mathrm{co}}H \text{ and } a_1 \in \mathrm{Psd}.$$

Then

$$\begin{aligned} \frac{xy + z + i}{4} &= \frac{(-a_1x_1 + a_1)y + z_1 + 2i}{8} \\ &= \frac{-a_1x_1y + (a_1y + z_1 + i) + i}{8}. \end{aligned}$$

We have $a_1y \in {}^{\mathrm{co}}G$ by Lemma 4.3 (CoGPsdTimes), $z_1 \in {}^{\mathrm{co}}G$ by (8) and $\frac{a_1y+z_1+i}{4} =: v \in {}^{\mathrm{co}}G$ by Lemma 4.6 (CoGAvcToCoG). Hence, we can continue

$$= \frac{-a_1x_1y + 4v + i}{8} = \frac{\frac{-a_1x_1y}{4} + v + \frac{i}{4}}{2} = \frac{\frac{-a_1x_1y+z+j}{4} + d}{2},$$

with $z \in {}^{\mathrm{co}}G$, $j \in \mathrm{Sd}_2$ and $d \in \mathrm{Sd}$, by the following Lemma 5.5 (JKU).

Case Ux, Lrz. We can write

$$x = \frac{x_1}{2} \quad z = -a_0\frac{z_1 - 1}{2} \quad \text{with } x_1 \in {}^{\mathrm{co}}H,\ z_1 \in {}^{\mathrm{co}}G \text{ and } a_0 \in \mathrm{Psd}.$$

Then

$$\frac{xy + z + i}{4} = \frac{x_1y - a_0z_1 + a_0 + 2i}{8} = \frac{x_1y + (0 - a_0z_1 + i) + a_0 + i}{8}.$$

We have $0 \in {}^{\mathrm{co}}G$ by Lemma 5.1 (CoGZero), $-a_0z_1 \in {}^{\mathrm{co}}G$ by Lemma 4.3 (CoGPsdTimes) and $\frac{0-a_0z_1+i}{4} =: v \in {}^{\mathrm{co}}G$ by Lemma 4.6 (CoGAvcToCoG). Hence, we can continue

$$= \frac{x_1y + 4v + a_0 + i}{8} = \frac{\frac{x_1y}{4} + v + \frac{a_0+i}{4}}{2} = \frac{\frac{x_1y+z+j}{4} + d}{2},$$

with $z \in {}^{\mathrm{co}}G$, $j \in \mathrm{Sd}_2$ and $d \in \mathrm{Sd}$, by Lemma 5.4 (JKLr).

Case Ux, Uz. We can write

$$x = \frac{x_1}{2} \quad z = \frac{z_1}{2} \quad \text{with } x_1, z_1 \in {}^{\text{co}}H.$$

Then

$$\frac{xy + z + i}{4} = \frac{x_1 y + z_1 + 2i}{8} = \frac{x_1 y + (0 + z_1 + i) + i}{8}.$$

We have $0 \in {}^{\text{co}}G$ by Lemma 5.1 (CoGZero), $z_1 \in {}^{\text{co}}G$ by (8) and $\frac{0+z_1+i}{4} =: v \in {}^{\text{co}}G$ by Lemma 4.6 (CoGAvcToCoG). Hence, we can continue

$$= \frac{x_1 y + 4v + i}{8} = \frac{\frac{x_1 y}{4} + v + \frac{i}{4}}{2} = \frac{\frac{x_1 y + z + j}{4} + d}{2},$$

with $z \in {}^{\text{co}}G$, $j \in \text{Sd}_2$ and $d \in \text{Sd}$, by Lemma 5.5 (JKU). □

The computational content of Lemma 5.3 is the following function g. We use z for cCoGZero and s for cCoHToCoG.

$$\begin{aligned}
&g(v, i, \text{Lr}_a(u), \text{Lr}_c(w_1)) \\
&\quad = \text{let } (j, d, w) = \text{cJKLr}(i, c, \text{cCoGAvcToCoG}(i, av, -cw_1)) \\
&\qquad \text{in } (d, j, -au, w) \\
&g(v, i, \text{Lr}_a(u), \text{U}(w_1)) \\
&\quad = \text{let } (j, d, w) = \text{cJKU}(i, \text{cCoGAvcToCoG}(i, av, s(w_1))) \\
&\qquad \text{in } (d, j, -au, w) \\
&g(v, i, \text{U}(u), \text{Lr}_c(w_1)) \\
&\quad = \text{let } (j, d, w) = \text{cJKLr}(i, c, \text{cCoGAvcToCoG}(i, z, -cw_1)) \\
&\qquad \text{in } (d, j, s(u), w) \\
&g(v, i, \text{U}(u), \text{U}(w_1)) \\
&\quad = \text{let } (j, d, w) = \text{cJKU}(i, \text{cCoGAvcToCoG}(i, z, s(w_1))) \\
&\qquad \text{in } (d, j, s(u), w)
\end{aligned}$$

Lemma 5.4 (JKLr).

$$\forall_{i \in \text{Sd}_2} \forall_{a \in \text{Psd}} \forall_{v \in {}^{\text{co}}G} \exists_{j \in \text{Sd}_2} \exists_{d \in \text{Sd}} \exists_{z \in {}^{\text{co}}G} \Big(v + \frac{a + i}{4} = \frac{z + j}{4} + d \Big).$$

Proof. We distinguish cases on $v \in {}^{co}G$ according to the disjunction in (8).

Case Lrv. Then $v = -b_0\frac{w-1}{2}$ with $w \in {}^{co}G$ and $b_0 \in \mathrm{Psd}$.

Subcase Lrw. Then we can write

$$v = -b_0\frac{w-1}{2} = -b_0\frac{-b\frac{z-1}{2} - 1}{2} \quad \text{with } w, z \in {}^{co}G \text{ and } b_0, b \in \mathrm{Psd}.$$

Hence,

$$v + \frac{a+i}{4} = \frac{b_0bz - b_0b + 2b_0 + a + i}{4}.$$

Using the functions J, K with the JK-property (6), we can write $-b_0b + 2b_0 + a + i$ as $j + 4d$ with $j := J(-b_0b + 2b_0 + a + i)$ and $d := K(-b_0b + 2b_0 + a + i)$. Hence,

$$= \frac{b_0bz + j + 4d}{4} = \frac{b_0bz + j}{4} + d.$$

Subcase Lrv, Uw. Then, we can write

$$v = -b_0\frac{w-1}{2} = -b_0\frac{\frac{z}{2} - 1}{2} \quad \text{with } w \in {}^{co}G,\ z \in {}^{co}H \text{ and } b_0 \in \mathrm{Psd}.$$

Hence,

$$v + \frac{a+i}{4} = \frac{-b_0z + 2b_0 + a + i}{4}.$$

Using the functions J, K with the JK-property (6) we can write $2b_0 + a + i$ as $j + 4d$ with $j := J(2b_0 + a + i)$ and $d := K(2b_0 + a + i)$. Hence,

$$= \frac{-b_0z + j + 4d}{4} = \frac{-b_0z + j}{4} + d.$$

Because of (8), we obtain the required form.

Subcase Uv, Finw. Then we can write

$$v = \frac{w}{2} = \frac{b\frac{z+1}{2}}{2} \quad \text{with } w \in {}^{co}H,\ z \in {}^{co}G \text{ and } b \in \mathrm{Psd}.$$

Hence,

$$v + \frac{a+i}{4} = \frac{bz+b+a+i}{4}.$$

Using the functions J, K with the JK-property (6), we can write $b+a+i$ as $j+4d$ with $j := J(b+a+i)$ and $d := K(b+a+i)$. Hence,

$$= \frac{bz+j+4d}{4} = \frac{bz+j}{4} + d.$$

Subcase Uv, Dw. Then we can write

$$v = \frac{w}{2} = \frac{\frac{z}{2}}{2} \quad \text{with } w, z \in {}^{\mathrm{co}}H.$$

Hence,

$$v + \frac{a+i}{4} = \frac{z+a+i}{4}.$$

Using the functions J, K with the JK-property (6), we can write $a+i$ as $j+4d$ with $j := J(a+i)$ and $d := K(a+i)$. Hence,

$$= \frac{z+j+4d}{4} = \frac{z+j}{4} + d.$$

Because of (8), we obtain the required form. □

The computational content of Lemma 5.4 is the following function g. We use s for cCoHToCoG.

$$\begin{aligned}
g(i, a, \mathrm{Lr}_{b_0}(\mathrm{Lr}_b(w))) &= (J(-b_0b + 2b_0 + a + i), \\
&\qquad K(-b_0b + 2b_0 + a + i), b_0bw) \\
g(i, a, \mathrm{Lr}_{b_0}(\mathrm{U}(w))) &= (J(2b_0 + a + i), K(2b_0 + a + i), -b_0s(w)) \\
g(i, a, \mathrm{U}(\mathrm{Lr}_b(w))) &= (J(b + a + i), K(b + a + i), bw) \\
g(i, a, \mathrm{U}(\mathrm{U}(w))) &= (J(a + i), K(a + i), s(w)).
\end{aligned}$$

Lemma 5.5 (JKU).

$$\forall_{i\in\mathrm{Sd}_2}\forall_{v\in{}^{\mathrm{co}}G}\exists_{j\in\mathrm{Sd}_2}\exists_{d\in\mathrm{Sd}}\exists_{z\in{}^{\mathrm{co}}G}\Big(v + \frac{i}{4} = \frac{z+j}{4} + d\Big).$$

Proof. As in the previous lemma, we need to consider four subcases, this time with a missing. In detail: we distinguish cases on $v \in {}^{\mathrm{co}}G$ according to the disjunction in (8).

Case Lrv. Then $v = -b_0 \frac{w-1}{2}$ with $w \in {}^{\mathrm{co}}G$ and $b_0 \in \mathrm{Psd}$.

Subcase Lrw. Then we can write

$$v = -b_0 \frac{w-1}{2} = -b_0 \frac{-b\frac{z-1}{2} - 1}{2} \quad \text{with } w, z \in {}^{\mathrm{co}}G \text{ and } b_0, b \in \mathrm{Psd}.$$

Hence,

$$v + \frac{i}{4} = \frac{b_0 bz - b_0 b + 2b_0 + i}{4}.$$

Using the functions J, K with the JK-property (6), we can write $-b_0 b + 2b_0 + i$ as $j + 4d$ with $j := J(-b_0 b + 2b_0 + i)$ and $d := K(-b_0 b + 2b_0 + i)$. Hence,

$$= \frac{b_0 bz + j + 4d}{4} = \frac{b_0 bz + j}{4} + d.$$

Subcase Lrv, Uw. Then we can write

$$v = -b_0 \frac{w-1}{2} = -b_0 \frac{\frac{z}{2} - 1}{2} \quad \text{with } w \in {}^{\mathrm{co}}G,\ z \in {}^{\mathrm{co}}H \text{ and } b_0 \in \mathrm{Psd}.$$

Hence,

$$v + \frac{i}{4} = \frac{-b_0 z + 2b_0 + i}{4}.$$

Using the functions J, K with the JK-property (6), we can write $2b_0 + i$ as $j + 4d$ with $j := J(2b_0 + i)$ and $d := K(2b_0 + i)$. Hence,

$$= \frac{-b_0 z + j + 4d}{4} = \frac{-b_0 z + j}{4} + d.$$

Because of (8), we obtain the required form.

Subcase Uv, Finw. Then we can write

$$v = \frac{w}{2} = \frac{b\frac{z+1}{2}}{2} \quad \text{with } w \in {}^{\mathrm{co}}H,\ z \in {}^{\mathrm{co}}G \text{ and } b \in \mathrm{Psd}.$$

Hence,

$$v + \frac{i}{4} = \frac{bz + b + i}{4}.$$

Using the functions J, K with the JK-property (6), we can write $b+i$ as $j + 4d$ with $j := J(b+i)$ and $d := K(b+i)$. Hence,

$$= \frac{bz + j + 4d}{4} = \frac{bz + j}{4} + d.$$

Subcase Uv, Dw. Then we can write

$$v = \frac{w}{2} = \frac{\frac{z}{2}}{2} \quad \text{with } w, z \in {}^{\text{co}}H.$$

Hence,

$$v + \frac{i}{4} = \frac{z + i}{4}.$$

Using the functions J, K with the JK-property (6), we can write i as $j + 4d$ with $j := J(i)$ and $d := K(i)$. Hence,

$$= \frac{z + j + 4d}{4} = \frac{z + j}{4} + d.$$

Because of (8), we obtain the required form. □

The computational content of Lemma 5.5 is the following function g. We use s for cCoHToCoG.

$$\begin{aligned}
g(i, \mathrm{Lr}_{b_0}(\mathrm{Lr}_b(w))) &= (J(-b_0 b + 2b_0 + i), K(-b_0 b + 2b_0 + i), b_0 b w) \\
g(i, \mathrm{Lr}_{b_0}(\mathrm{U}(w))) &= (J(2b_0 + i), K(2b_0 + i), -b_0 s(w)) \\
g(i, \mathrm{U}(\mathrm{Lr}_b(w))) &= (J(b + i), K(b + i), bw) \\
g(i, \mathrm{U}(\mathrm{U}(w))) &= (J(i), K(i), s(w)).
\end{aligned}$$

By coinduction from Lemma 5.3 (CoGMultcSatCoICl), we obtain

Lemma 5.6 (CoGMultcToCoG).

$$\forall_{z_0} \Big(\exists_{i \in \mathrm{Sd}_2} \exists_{x,y,z \in {}^{\text{co}}G} \Big(z_0 = \frac{xy + z + i}{4} \Big) \to z_0 \in {}^{\text{co}}G \Big).$$

Proof. Let $Q \subseteq {}^{\mathrm{co}}G$ abbreviate its claim. We prove it simultaneously with $Q \subseteq {}^{\mathrm{co}}H$ (defined similarly), by coinduction. Let $z_0 \in Q$. We first show

$$\exists_{z \in {}^{\mathrm{co}}G \cup Q} \exists_{a \in \mathrm{Psd}} \Big(z_0 = -a\frac{z-1}{2} \Big) \vee \exists_{z \in {}^{\mathrm{co}}H \cup Q} \Big(z_0 = \frac{z}{2} \Big). \tag{14}$$

Lemma 5.3 (CoGMultcSatCoICl) applied to $z_0 \in Q$ gives us $x_1, y_1, z_1 \in {}^{\mathrm{co}}G$ and i_1, d_1 such that

$$z_0 = \frac{\frac{x_1 y_1 + z_1 + i_1}{4} + d_1}{2}.$$

Case $d_1 = 0$. Go for the right-hand side of (14) with $z := (x_1 y_1 + z_1 + i_1)/4 \in Q$. *Case* $d_1 = \pm 1$. Go for the left-hand side of (14) with $a := d_1$ and $z := (-a x_1 y_1 - a z_1 - a i_1)/4 \in Q$. Then

$$-a\frac{z-1}{2} = -a\frac{4z-4}{8} = \frac{x_1 y_1 + z_1 + i_1 + 4a}{8} = z_0.$$

This finishes the proof of (14). The rest of the proof is similar, and we omit it. □

The computational content of Lemma 5.6 is the following function g.

$$\begin{aligned} g(i,u,u',u'') = {} & \mathrm{let}\ (d,j,u_1,u_1') = \mathrm{cCoGMultcSatCoICl}(u',i,u,u'')\ \mathrm{in} \\ & \mathrm{case}\ \mathrm{cSdDisj}(d)\ \mathrm{of} \\ & \quad 0 \to \mathrm{U}(h(j,u_1,u',u_1')) \\ & \quad a \to \mathrm{Lr}_a(g(-aj,u_1,-au',-au_1')), \end{aligned}$$

$$\begin{aligned} h(i,u,u',u'') = {} & \mathrm{let}\ (d,j,u_1,u_1') = \mathrm{cCoGMultcSatCoICl}(u',i,u,u'')\ \mathrm{in} \\ & \mathrm{case}\ \mathrm{cSdDisj}(d)\ \mathrm{of} \\ & \quad 0 \to \mathrm{D}(h(j,u_1,u',u_1')) \\ & \quad a \to \mathrm{Fin}_a(g(aj,u_1,au',au_1')). \end{aligned}$$

Proposition 5.7 (CoGMult).

$$\forall_{x,y}(x \in {}^{\mathrm{co}}G \to y \in {}^{\mathrm{co}}G \to xy \in {}^{\mathrm{co}}G).$$

Proof. Compose Lemma 5.2 (CoGMultToMultc) and Lemma 5.6 (CoGMultcToCoG). □

The computational context of each of the two lemmata has been spelled out above, after their proofs.

6. Conclusion

All these proofs have been formalised in the proof assistant Minlog.[c] The formalisation of Propositions 3.7 (CoIMult) and 5.7 (CoGMult) closely followed the informal proofs in Sections 2–5. In particular, the structure of the proofs by splitting them up into suitable lemmata is kept. The extracted term of each lemma represents its computational content; it involves recursion and corecursion operators where the original proof used induction or coinduction axioms. This computational content has been explained informally in each case above. In addition, for each lemma a soundness proof is generated. This possibility to automatically obtain formal proofs of the correctness of programs is an important aspect of the present approach (cf. (II) in the introduction).

An alternative approach (cf. (I) in the introduction) would be to directly write a programme term and then formally prove that it satisfies the specification. An advantage of (II) over (I) is that it can be difficult to find such a programme term, particularly when — as in the present case — it involves corecursion operators. When generating a proof using a proof assistant (here Minlog), one works on an intuitively clearer level. For instance, in an application of coinduction the proof assistant can generate the next proof obligations, which often helps.

Acknowledgement

This work was supported by the International Research Staff Exchange Scheme (IRSES) Nr. 612638 CORCON.

References

[1] P.D. Gianantonio, An abstract data type for real numbers, *Theor. Comp. Sci.*, **221**(1–2), 295–326 (1999).

[c]See http://www.minlog-system.de/, which gives instructions on how to download (or clone) the system. The formalisations can be found in files `sdmult.scm` and `graymult.scm` of the directory `minlog/examples/analysis/`.

[2] H. Tsuiki, Real number computation through Gray code embedding, *Theor. Comp. Sci.*, **284**, 467–485 (2002).

[3] A. Ciaffaglione and P.D. Gianantonio, A certified, corecursive implementation of exact real numbers, *Theor. Comp. Sci.*, **351**, 39–51 (2006).

[4] U. Berger, From coinductive proofs to exact real arithmetic. In E. Grädel and R. Kahle, (eds.), *Computer Science Logic*, vol. 5771 of *LNCS*. Springer Verlag, Berlin, Heidelberg, New York, 2009, pp. 132–146.

[5] U. Berger, K. Miyamoto, H. Schwichtenberg, and H. Tsuiki, Logic for Gray-code computation. In D. Probst and P. Schuster, (eds.), *Concepts of Proof in Mathematics, Philosophy, and Computer Science.* De Gruyter, 2016, pp. 69–110.

[6] K. Miyamoto and H. Schwichtenberg, Program extraction in exact real arithmetic, *Math. Struct. Comp. Sci.*, **25**, 1692–1704 (2015).

[7] H. Schwichtenberg and F. Wiesnet, Logic for exact real arithmetic, *Logical Meth. Comp. Sci.*, **17**(2) (2021). arxiv.org/abs/1904.12763.

[8] E. Bishop, *Foundations of Constructive Analysis.* McGraw-Hill, New York, 1967.

[9] U. Berger, Program extraction from normalization proofs. In M. Bezem and J. Groote, (eds.), *Typed Lambda Calculi and Applications, vol. 664 of LNCS.* Springer Verlag, Berlin, Heidelberg, New York, 1993, pp. 91–106.

[10] H. Schwichtenberg and S.S. Wainer, *Proofs and Computations.* Perspectives in Logic. Association for Symbolic Logic and Cambridge University Press, 2012.

[11] U. Berger and M. Seisenberger, Proofs, programs, processes. In F. Ferreira *et al.*, (eds.), *Proceedings CiE 2010, vol. 6158 of LNCS.* Springer Verlag, Berlin, Heidelberg, New York, 2010, pp. 39–48.

https://doi.org/10.1142/9789811245220_0004

Chapter 4

Information Systems with Witnesses: The Function Space Construction

Dieter Spreen

Department of Mathematics, University of Siegen
Siegen, Germany
Department of Decision Sciences, University of South Africa,
Pretoria, South Africa
spreen@math.uni-siegen.de

Information systems with witnesses have been introduced in Ref. [1] as a logic-style representation of L-domains: The category of such information systems with approximable mappings as morphisms is equivalent to the category of L-domains with Scott continuous functions, which is known to be Cartesian closed. In the present chapter, a direct proof of the Cartesian closure of the category of information systems with witnesses and approximable mappings is given. As is shown, the collection of approximable mappings between two information systems with witnesses comes with a natural information system structure.

1. Introduction

In a recent paper (Ref. [1]), the author introduced information systems with witnesses as a logic-style representation of L-domains. L-domains have independently been introduced by Coquand (cf. Ref. [2]) and Jung (cf. Refs. [3,4]) and are known to form one of the two maximal Cartesian closed full subcategories of the continuous domains. They generalise the bounded-complete ones: whereas in a bounded-complete domain every bounded subset has a global least upper bound, in an L-domain such sets may have different local least upper bounds depending on the upper bounds they have.

The idea to represent classes of domains via logical calculi goes back to Dana Scott's seminal 1982 paper (cf. Ref. [5]), in which he introduced information systems to capture the bounded-complete algebraic domains. An information system consists of a set of tokens to be thought of as atomic statements about a computational process, a consistency predicate telling us which finite sets of such statements contain consistent information, and an entailment relation saying what atomic statements are entailed by which consistent sets of these. Theories of such a logic, also called states, i.e., finitely consistent and entailment-closed sets of atomic statements, form a domain with respect to set inclusion. A state represents consistent information. So, any finite collection of substates must contain consistent information as well, and this fact is witnessed by any of its upper bounds.

Whereas in Scott's approach the consistency witnesses are hidden, in the new approach they are made explicit. This allows to consider the more general situation in which a finite set of tokens may have different consistency witnesses, and the result of entailment may depend on them. As was shown in Ref. [1], the theories, or states, of such a more general information system form an L-domain, and, up to isomorphism, each L-domain can be obtained in this way. Moreover, there is an equivalence between the categories of information systems with witnesses and L-domains.

The category of information systems has approximable mappings as morphisms. These are relations between the consistent subsets of one information system and the tokens of another. Entailment is a particular approximable mapping.

As mentioned earlier, the category of L-domains and Scott continuous functions is Cartesian closed. Because of the equivalence of this category with the category of information systems with witnesses, we know that the latter one is Cartesian closed as well. However, this means that in concrete situations we have to pass back and forth between information systems and domains in order to construct the exponent of two information systems with witnesses. In this chapter, we present a direct proof of the Cartesian closure of the category of information systems with witnesses. In particular, we present a construction of the exponent. This will be an information system with witnesses the states of which are exactly the approximable mappings between the information systems under consideration.

Whereas for Scott's information systems capturing the bounded-complete algebraic domains, the function space construction is straightforward and well understood, the situation is rather intricate in the present case. This is due to the fact that consistency is only locally defined and we have to deal with consistency witnesses explicitly. Moreover, as is known from Hoofman's work (cf. Ref. [6]), in the continuous case entailment is required to allow for interpolation.

The chapter is organised as follows: Section 2 contains basic definitions and results from domain theory. In Section 3, relevant definitions and facts about information systems with witnesses are recalled from Ref. [1]. Approximable mappings between such information systems are considered in Section 4. Section 5 is concerned with the function space construction: an information system with witnesses is presented the states of which are exactly the approximable mappings between two given systems. As is shown in Section 6, this information system is indeed the exponent of the two given ones. The chapter closes with the discussion of an application of the results obtained in proof assistants and programme extraction.

2. Domains: Basic Definitions and Results

For any set A, we write $X \subseteq_{\text{fin}} A$ to mean that X is a finite subset of A. The collection of all subsets of A will be denoted by $\mathcal{P}(A)$ and that of all finite subsets by $\mathcal{P}_f(A)$. Moreover, for two sets A_1 and A_2, we let pr_1 and pr_2, respectively, be the canonical projections of $A_1 \times A_2$ onto the first and second component. For $\nu, \mu \in \{1, 2\}$, set $\text{pr}_{\nu,\mu} = \text{pr}_\nu \circ \text{pr}_\mu$.

Let $(D, \sqsubseteq)$ be a poset. D is *pointed* if it contains a least element $\bot$. For an element $x \in D$, $\downarrow x$ denotes the principal ideal generated by x, i.e., $\downarrow x = \{ y \in D \mid y \sqsubseteq x \}$. A subset S of D is called *consistent* if it has an upper bound. S is *directed* if it is nonempty and every pair of elements in S has an upper bound in S. D is a *directed-complete partial order* (*dcpo*) if every directed subset S of D has a least upper bound $\bigsqcup S$ in D, and D is *bounded-complete* if every consistent subset has a least upper bound.

Assume that x, y are elements of a poset D. Then x is said to *approximate* y, written $x \ll y$, if for any directed subset S of D, the least upper bound of which exists in D, the relation $y \sqsubseteq \bigsqcup S$ always implies the existence of some $u \in S$ with $x \sqsubseteq u$. Moreover, x is *compact* if $x \ll x$. A subset B of D is a *basis* of D if for each $x \in D$ the set $\Downarrow_B x = \{ u \in B \mid u \ll x \}$ contains a directed subset with least upper bound x. Note that the set of all compact elements of D is included in every basis of D. A directed-complete partial order D is said to be *continuous* (or a *domain*) if it has a basis and it is called *algebraic* (or an *algebraic domain*) if its compact elements form a basis. A pointed bounded-complete domain is called *bc-domain.* Standard references for domain theory and its applications are [6–11].

Lemma 2.1. *In a poset D, the following statements hold for all $x, y, z \in D$:*

(1) *The approximation relation $\ll$ is transitive.*
(2) $x \ll y \Longrightarrow x \sqsubseteq y$.
(3) $x \ll y \sqsubseteq z \Longrightarrow x \ll z$.
(4) *If D has a least element $\bot$, then $\bot \ll x$.*
(5) *If $F \subseteq \downarrow x \cap \downarrow y$ such that the least upper bounds $\bigsqcup^x F$ and $\bigsqcup^y F$, respectively, exist relative to $\downarrow x$ and $\downarrow y$, then*

$$x, y \sqsubseteq z \Longrightarrow \bigsqcup^x F = \bigsqcup^y F.$$

(6) *If D is a continuous domain with basis B, and $M \subseteq_{\text{fin}} D$, then*

$$M \ll x \Longrightarrow (\exists v \in B) M \ll v \ll x,$$

where $M \ll x$ means that $m \ll x$, for any $m \in M$.

Property 6 is known as the *interpolation law.*

Definition 2.2. Let D and D' be posets. A function $f\colon D \to D'$ is *Scott continuous* if it is monotone and for any directed subset S of D with existing least upper bound,

$$\bigsqcup f(S) = f(\bigsqcup S).$$

With respect to the pointwise order, the set $[D \to D']$ of all Scott continuous functions between two dcpos D and D' is a dcpo again. Observe that it need not be continuous even if D and D' are. This is the case, however, if D' is an L-domain (cf. Ref. [6]).

Definition 2.3. A pointed[a] domain D is an *L-domain* if each pair $x, y \in D$ bounded above by $z \in D$ has a least upper bound $x \sqcup^z y$ in $\downarrow z$.

Obviously, every bc-domain is an L-domain. As has been shown by Jung (cf. Refs. [3,4]), the category **L** of L-domains is one of the two maximal Cartesian closed full subcategories of the category $\mathbf{CONT}_\perp$ of pointed domains and Scott continuous maps. The same holds for the category **aL** of algebraic L-domains with respect to the category $\mathbf{ALG}_\perp$ of pointed algebraic domains. The one-point domain is the terminal object in these categories and the categorical product $D \times E$ of two domains D and E is the Cartesian product of the underlying sets ordered coordinatewise.

For domains D and D' and basic elements $d \in D$ and $d' \in D'$, the *single-step function* $(d \searrow d')\colon D \to D'$ is defined by

$$(d \searrow d')(x) = \begin{cases} d' & \text{if } d \ll x, \\ \perp' & \text{otherwise.} \end{cases}$$

As is well known, every Scott continuous function $f\colon D \to D'$ is the least upper-bound of all single-step functions approximating it:

$$f = \bigsqcup \{\, (d \searrow d') \mid d' \ll f(d) \,\}.$$

In general, however, the set of these single-step functions is not directed. A way to get out of this problem is to require the existence of joins of bounded finite collections of single-step functions. Such joins are called *step functions.*

If D' is bounded-complete, the pointwise least upper bound $\bigsqcup_{\nu=1}^{n}(d_\nu \searrow d'_\nu)$ exists if the set $\{\, d'_\nu \mid 1 \le \nu \le n \wedge d_\nu \ll x \,\}$ is bounded for all $x \in D$. The cost of generalising this to the case of L-domains is at least the burden of bookkeeping where least upper

[a]Note that in Ref. [8] pointedness is not required.

bounds are taken. In particular, if $(d_\nu \searrow d'_\nu)$, $\nu = 1, \ldots, n$, are single-step functions below f, then their least upper bound in $\downarrow f$, written $\bigsqcup^f_{1\le\nu\le n}(d_\nu \searrow d'_\nu)$, is given by

$$\bigsqcup^f_{1\le\nu\le n}(d_\nu \searrow d'_\nu)(x) = \bigsqcup^{f(x)}_{\nu:\, d_\nu \ll x} d'_\nu.$$

3. Information Systems with Witnesses

In this section, the ideas outlined in the introduction are made precise: An information system with witnesses consists of a set A of tokens, a consistency predicate CON classifying which finite sets of tokens are consistent with which token as witness, and an entailment relation between pairs of consistent sets and associated witnesses on the one side, and arbitrary tokens on the other. The conditions that have to be satisfied are grouped. There are requirements which consistency predicate and entailment relation have to meet in which the consistency witness is kept fixed. They are well known from Scott's information systems and Hoofman's extension of this notion to the continuous case. In addition, we find conditions that specify the interplay between consistency witnesses.

Note that we sometimes write $X \in \mathrm{CON}(i)$ instead of $(i, X) \in \mathrm{CON}$. Proofs of the results can be found in Ref. [1].

Definition 3.1. Let A be a set, $\Delta \in A$, $\mathrm{CON} \subseteq A \times \mathcal{P}_f(A)$, and $\vdash \subseteq \mathrm{CON} \times A$. Then $(A, \mathrm{CON}, \vdash, \Delta)$ is an *information system with witnesses* if the following conditions hold, for all $i, j, a \in A$ and all finite subsets X, Y of A:

(1) $\{i\} \in \mathrm{CON}(i)$
(2) $Y \subseteq X \wedge X \in \mathrm{CON}(i) \Rightarrow Y \in \mathrm{CON}(i)$
(3) $(i, \emptyset) \vdash \Delta$
(4) $X \in \mathrm{CON}(i) \wedge (i, X) \vdash Y \Rightarrow Y \in \mathrm{CON}(i)$
(5) $X, Y \in \mathrm{CON}(i) \wedge X \subseteq Y \wedge (i, X) \vdash a \Rightarrow (i, Y) \vdash a$
(6) $X \in \mathrm{CON}(i) \wedge (i, X) \vdash Y \wedge (i, Y) \vdash a \Rightarrow (i, X) \vdash a$
(7) $(i, X) \vdash a \Rightarrow (\exists Z \in \mathrm{CON}(i))(i, X) \vdash Z \wedge (i, Z) \vdash a$
(8) $(i, X) \vdash Y \Rightarrow (\exists e \in A)(i, X) \vdash e \wedge Y \in \mathrm{CON}(e)$
(9) $\{i\} \in \mathrm{CON}(j) \Rightarrow \mathrm{CON}(i) \subseteq \mathrm{CON}(j)$

(10) $\{i\} \in \mathrm{CON}(j) \wedge X \in \mathrm{CON}(i) \wedge (i, X) \vdash a \Rightarrow (j, X) \vdash a$
(11) $\{i\} \in \mathrm{CON}(j) \wedge X \in \mathrm{CON}(i) \wedge (j, X) \vdash a \Rightarrow (i, X) \vdash a.$

All requirements are very natural: Each token witnesses its own consistency (1). If the consistency of some set is witnessed by i, the same holds for all of its subsets (2). Δ is entailed by any set of information and each of its witnesses, i.e., it represents global truth (3). By Condition (4), entailment preserves consistency. If a set X with consistency witness i entails a, so does any bigger set with the same witness (5). For fixed witness, entailment is transitive (6). Consistency and entailment are preserved when moving from a witness i to another one j with respect to which i is consistent (9, 10). Moreover, entailment is conservative in that case: what is entailed with respect to witness j from a set consistent with respect to i is already entailed with respect to i (11). Conditions (7) and (8) are both interpolation properties. They can be combined into one, called *Global Interpolation Property.*

Lemma 3.2. *Let A be a set, $\Delta \in A$, $\mathrm{CON} \subseteq A \times \mathcal{P}_f(A)$ and $\vdash \subseteq \mathrm{CON} \times A$ such that Axioms* 3.1(4, 5, 9–11) *are satisfied. Then Axioms* 3.1(7, 8) *hold if, and only if, for all $i \in A$, $X \in \mathrm{Con}_i$ and $F \subseteq_{\mathrm{fin}} A$,*

$$(i, X) \vdash F \Rightarrow (\exists (j, Y) \in \mathrm{CON})(i, X) \vdash (j, Y) \wedge (j, Y) \vdash F, \qquad \text{(GIP)}$$

where $(i, X) \vdash (j, Y)$ means that $(i, X) \vdash j$ and $(i, X) \vdash Y$.

The next result extends Condition 3.1(7).

Lemma 3.3. *Let A be a set, $\Delta \in A$, $\mathrm{CON} \subseteq A \times \mathcal{P}_f(A)$ and $\vdash \subseteq \mathrm{CON} \times A$ such that Axioms* 3.1(4, 5, 7) *are satisfied. Then the following rule holds, for all $a \in A$, $F \subseteq_{\mathrm{fin}} A$ and $(i, X) \in \mathrm{CON}$,*

$$(i, X) \vdash a \Rightarrow (\exists Z \in \mathrm{CON}(i))(i, X) \vdash Z \wedge (i, Z) \vdash F.$$

Proof. Let $b \in F$. By Axiom 3.1(7), there is some $Z_b \in \mathrm{CON}(i)$ with $(i, X) \vdash Z_b$ and $(i, Z_b) \vdash b$. Set $Z = \bigcup \{ Z_b \mid b \in F \}$. Then $(i, X) \vdash Z$. Hence, $Z \in \mathrm{CON}(i)$, by Condition 3.1(4). Because of Axiom 3.1(5), we therefore have that $(i, Z) \vdash F$. $\square$

Sometimes a stronger requirement than 3.1(6) is needed which reverses Condition 3.1(7).

Lemma 3.4. *Let* $(A, \mathrm{CON}, \vdash, \Delta)$ *be an information system with witnesses. Then the following rule holds, for all* $a \in A$ *and* $(i, X), (j, Y) \in \mathrm{CON}$,

$$(i, X) \vdash (j, Y) \wedge (j, Y) \vdash a \Rightarrow (i, X) \vdash a.$$

To relate information systems to domains, the notion of state is required.

Definition 3.5. Let $(A, \mathrm{CON}, \vdash, \Delta)$ be an information system with witnesses. A subset x of A is a *state* of $(A, \mathrm{CON}, \vdash, \Delta)$ if the following three conditions hold:

(1) $(\forall F \subseteq_{\mathrm{fin}} x)(\exists i \in x) F \in \mathrm{CON}(i)$.
(2) $(\forall i \in x)(\forall X \subseteq_{\mathrm{fin}} x)(\forall a \in A)[X \in \mathrm{CON}(i) \wedge (i, X) \vdash a \Rightarrow a \in x]$.
(3) $(\forall a \in x)(\exists i \in x)(\exists X \subseteq_{\mathrm{fin}} x) X \in \mathrm{CON}(i) \wedge (i, X) \vdash a$.

As follows from the definition, states are subsets of tokens that are *finitely consistent* (1) and *closed under entailment* (2). Furthermore, each token in a state is *derivable* (3), i.e., for each token the state contains a consistent set and its witness entailing the token.

By Condition 3.5(1), states are never empty: Choose F to be the empty set. Then the state contains some i with $\emptyset \in \mathrm{CON}(i)$.

Note that Conditions (1, 3) in Definition 3.5 can be replaced by a single requirement.

Proposition 3.6. *Let* $(A, \mathrm{CON}, \vdash, \Delta)$ *be an information system with witnesses and* x *be a subset of* A*. Then Conditions* 3.5(1) *and* (3) *together are equivalent to the following statement:*

$$(\forall F \subseteq_{\mathrm{fin}} x)(\exists i \in x)(\exists X \subseteq_{\mathrm{fin}} x) X \in \mathrm{CON}(i) \wedge (i, X) \vdash F. \qquad \text{(ST)}$$

With respect to set inclusion, the states of A form a directed-complete partially ordered set, denoted by $|A|$. Moreover, the consistent subsets of A generate a canonical basis of $|A|$. For $(i, X) \in \mathrm{CON}$, let

$$[X]_i = \{ a \in A \mid (i, X) \vdash a \}.$$

Then $[X]_i$ is a state and for every $z \in |A|$, the set of all $[X]_i$ with $\{i\} \cup X \subseteq z$ is directed and z is its union.

This result allows characterising the approximation relation on A in terms of the entailment relation. The characterisation nicely reflects the intuition that $x \ll y$ if x is covered by a "finite part" of y.

Proposition 3.7. *For* $x, y \in |A|$,

$$x \ll y \iff (\exists (i, V) \in \text{CON})\{i\} \cup V \subseteq y \wedge (i, V) \vdash x.$$

Because of Axioms 3.1(1, 2), we have that $\emptyset \in \text{CON}(i)$, for all $i \in A$. Moreover, with Axioms 3.1(3, 4), it follows that $\{\Delta\} \in \text{CON}(j)$, for all $j \in A$. As is easily verified, $[\emptyset]_i = [\{\Delta\}]_j$, for all $i, j \in A$, and $[\emptyset]_\Delta \subseteq x$, for all $x \in |A|$.

Local least upper bounds can be computed in a similar way as directed least upper bounds. Let $x, y, z \in |A|$ so that $x, y \subseteq z$. Then

$$\bigcup \{ [Z]_k \mid (k, Z) \in \text{CON} \wedge k \in z \wedge Z \subseteq_{\text{fin}} x \cup y \}$$

is the least upper bound of x and y in $\downarrow z$.

Theorem 3.8. *Let* $(A, \text{CON}, \vdash, \Delta)$ *be an information system with witnesses. Then* $\mathcal{L}(A) = (|A|, \subseteq, [\emptyset]_\Delta)$ *is an L-domain with basis* $\widehat{\text{CON}} = \{ [X]_i \mid (i, X) \in \text{CON} \}$.

Let us see next when $\mathcal{L}(A)$ is algebraic.

Definition 3.9. Let $(A, \text{CON}, \vdash, \Delta)$ be an information system with witnesses. An element $(j, V) \in \text{CON}$ is called *reflexive* if $(j, V) \vdash (j, V)$.

Obviously, $[V]_j$ is compact, for every reflexive $(j, V) \in \text{CON}$. We denote the subset of reflexive elements of CON by CON_{refl}.

Theorem 3.10. *Let* $(A, \text{CON}, \vdash, \Delta)$ *be an information system with witnesses. Then* $\mathcal{L}(A)$ *is algebraic if, and only if, the information system* A *satisfies Condition* (ALG) *saying that for all* $(i, X) \in \text{CON}$ *and* $F \subseteq_{\text{fin}} A$,

$$(i, X) \vdash F \Rightarrow (\exists (j, V) \in \text{CON}_{\text{refl}})(i, X) \vdash (j, V) \wedge (j, V) \vdash F. \tag{ALG}$$

Condition (ALG) is a global interpolation requirement. Similarly to Lemma 3.2, it is equivalent to a local condition.

Lemma 3.11. *Condition (ALG) holds if, and only if, the following Condition (SALG) is satisfied for all* $(i, X) \in \mathrm{CON}$ *and* $a \in A$,

$$(i, X) \vdash a \Rightarrow (\exists Z \in \mathrm{CON}(i))(i, X) \vdash Z \wedge (i, Z) \vdash Z \wedge (i, Z) \vdash a. \tag{SALG}$$

In Scott's information systems, a finite set of tokens is consistent if it has a consistency witness, independently of which token this might be. This provides us with a condition forcing an information system with witnesses to generate a bounded-complete domain.

Theorem 3.12. *Let* $(A, \mathrm{CON}, \vdash, \Delta)$ *be an information system with witnesses. Then* $\mathcal{L}(A)$ *is bounded-complete, and hence a bc-domain, if the information system* A *satisfies Condition (BC) saying that for all* $X \subseteq_{\mathrm{fin}} A$ *and* $i, j \in A$,

$$(i, X), (j, X) \in \mathrm{CON} \Rightarrow (\forall a \in A)[(i, X) \vdash a \Leftrightarrow (j, X) \vdash a]. \tag{BC}$$

It is unknown whether the requirement on A is also necessary.

So far, we have seen that information systems with witnesses generate L-domains. But the converse holds as well: Every L-domain defines a canonical information system with witnesses such that the L-domain generated by it is isomorphic to the given domain.

Let $(D, \sqsubseteq)$ be an L-domain with basis B and least element $\bot$. Set $\mathcal{I}(D) = (B, \mathrm{CON}, \vdash, \bot)$ with

$$\mathrm{CON} = \{\, (i, X) \mid i \in B \wedge X \subseteq_{\mathrm{fin}} {\downarrow} i \cap B \,\} \text{ and } (i, X) \vdash a \Longleftrightarrow a \ll \bigsqcup^{i} X.$$

Theorem 3.13. *Let* D *be an L-domain. Then* $\mathcal{I}(D)$ *is an information system with witnesses such that* D *and* $\mathcal{L}(\mathcal{I}(D))$ *are isomorphic. In addition,*

(1) *D is algebraic if, and only if, the information system $\mathcal{I}(D)$ satisfies Condition (ALG).*
(2) *D is bounded-complete if, and only if, Condition (BC) holds in $\mathcal{I}(D)$.*

4. Approximable Mappings

In the next step, we want to consider the appropriate morphisms between information systems with witnesses. They will be relations between the consistent sets and their consistency witnesses of one information system with witnesses and the tokens of another, just as the entailment relations.

Definition 4.1. An *approximable mapping* H between information systems with witnesses $(A, \mathrm{CON}, \vdash, \Delta)$ and $(A', \mathrm{CON}', \vdash', \Delta')$, written $H\colon A \Vdash A'$, is a relation between CON and A' satisfying the following nine conditions, for all $i, j \in A$, $X, X' \subseteq_{\mathrm{fin}} A$, $b, k \in A'$ and $Y, F \subseteq_{\mathrm{fin}} A'$ with $X \in \mathrm{CON}(i)$ and $Y \in \mathrm{CON}'(k)$:

(1) $(\Delta, \emptyset) H \Delta'$
(2) $X' \in \mathrm{CON}(i) \wedge X \subseteq X' \wedge (i, X)Hb \Rightarrow (i, X')Hb$
(3) $(i, X) \vdash X' \wedge (i, X')Hb \Rightarrow (i, X)Hb$
(4) $(i, X)Hb \Rightarrow (\exists U \in \mathrm{CON}(i))(i, X) \vdash U \wedge (i, U)Hb$
(5) $(i, X)H(k, Y) \wedge (k, Y) \vdash' b \Rightarrow (i, X)Hb$
(6) $(i, X)Hb \Rightarrow (\exists (d, V) \in \mathrm{CON}')(i, X)H(d, V) \wedge (d, V) \vdash' b$
(7) $(i, X)HF \Rightarrow (\exists e \in A')(i, X)He \wedge F \in \mathrm{CON}'(e)$.
(8) $\{i\} \in \mathrm{CON}(j) \wedge (i, X)Hb \Rightarrow (j, X)Hb$
(9) $\{i\} \in \mathrm{CON}(j) \wedge (j, X)Hb \Rightarrow (i, X)Hb$.

Here, $(i, X)HY$ means that $(i, X)Hc$, for all $c \in Y$, and $(i, X)H(k, Y)$ that $(i, X)Hk$ as well as $(i, X)HY$.

Note that because of Condition 3.1(5), Condition 4.1(2) follows from Conditions 4.1(3, 4). The Left Interpolation Rule 4.1(4) together with the Conservativity Requirement 4.1(9) can be replaced by one rule.

Lemma 4.2. *Let* $(A, \mathrm{CON}, \vdash, \Delta)$ *and* $(A', \mathrm{CON}', \vdash', \Delta')$ *be information systems with witnesses. Then, for any* $H \subseteq \mathrm{CON} \times A'$ *satisfying Conditions* 4.1(2, 3, 8), *any* $(i, X) \in \mathrm{CON}$, *and all* $F \subseteq_{\mathrm{fin}} A'$, *Conditions* 4.1(4, 9) *together are equivalent to the following requirement:*

$$(i, X)HF \Rightarrow (\exists (c, U) \in \mathrm{CON})(i, X) \vdash (c, U) \wedge (c, U)HF. \tag{1}$$

Proof. Assume that Requirement (1) holds. Then Condition 4.1(4) follows as a special case because of Condition 4.1(8). For Condition 4.1(9), suppose that $\{i\} \in \mathrm{CON}(j)$ and $(j, X)Hb$. With Requirement (1), we obtain that there is some $(c, U) \in \mathrm{CON}$ so that $(j, X) \vdash (c, U)$ and $(c, U)Hb$. Since $X \in \mathrm{CON}(i)$, it follows with Axiom 3.1(11) that $(i, X) \vdash (c, U)$. Thus, $\{c\} \in \mathrm{CON}(i)$ and, therefore, $(i, U)Hb$, by Condition 4.1(8). With Condition 4.1(3), we finally obtain that $(i, X)Hb$.

Now, conversely, suppose that Conditions 4.1(4) and 4.1(9) hold. Moreover, assume that $(i, X)HF$. Then $(i, X)Hb$, for all $b \in F$. Thus, there are $U_b \in \mathrm{CON}(i)$ with $(i, X) \vdash U_b$ and $(i, U_b)Hb$. Set $U = \{\, u_b \mid b \in F \,\}$. It follows that $(i, X) \vdash U$. Hence, $U \in \mathrm{CON}(i)$. With Condition 4.1(2), we therefore have that $(i, U)HF$. By Axiom 3.1(8), we moreover obtain from $(i, X) \vdash U$ that there is some $c \in A$ with $(i, X) \vdash c$ and $U \in \mathrm{CON}(c)$. Thus, $\{c\} \in \mathrm{CON}(i)$. Consequently, $(c, U)HF$, because of Condition 4.1(9). □

Similarly, the Right Interpolation Rule 4.1(6) and the Witness Generation Rule 4.1(7) can be combined into one rule.

Lemma 4.3. *Let* $(A, \mathrm{CON}, \vdash, \Delta)$ *and* $(A', \mathrm{CON}', \vdash', \Delta')$ *be information systems with witnesses. Then, for any* $H \subseteq \mathrm{CON} \times A'$, $(i, X) \in \mathrm{CON}$ *and* $F \subseteq_{\mathrm{fin}} A'$, *Conditions* 4.1(6, 7) *together are equivalent to the following requirement:*

$$(i, X)HF \Rightarrow (\exists (e, V) \in \mathrm{CON}')(i, X)H(e, V) \wedge (e, V) \vdash' F. \qquad (2)$$

Proof. Assume that Requirement (2) holds. Then Condition 4.1(6) follows as a special case. For Condition 4.1(7), suppose that $(i, X)HF$. Then there is some $(e, V) \in \mathrm{CON}')$ so that $(i, X)H(e, V)$ and $(e, V) \vdash' F$. It follows that $(i, X)He$ and $F \in \mathrm{CON}'(e)$.

Next, conversely, suppose that Conditions 4.1(6) and 4.1(7) hold. Moreover, assume that $(i, X)HF$. Then, for all $b \in F$, there exist $(d_b, V_b) \in \mathrm{CON}'$ such that $(i, X)H(d_b, V_b)$ and $(d_b, V_b) \vdash' b$. Let $V = \{\, V_b \cup \{d_b\} \mid b \in F \,\}$. Since we have that $(i, X)HV$, it follows with Requirement 4.1(7) that there is some $e \in A'$ with $(i, X)He$ and $V \in \mathrm{CON}'(e)$. Hence, $\{d_b\} \in \mathrm{CON}'(e)$ and thus $(e, V_b) \vdash' b$, from which we obtain that $(e, V) \vdash' F$. In addition, we have that $(i, X)H(e, V)$. □

Finally, the extended Left and Right Interpolation Rules (1) and (2) together can be exchanged for one rule.

Lemma 4.4. *Let* $(A, \mathrm{CON}, \vdash, \Delta)$ *and* $(A', \mathrm{CON}', \vdash', \Delta')$ *be information systems with witnesses. Then, for any* $H \subseteq \mathrm{CON} \times A'$, $(i, X) \in \mathrm{CON}$ *and* $F \subseteq_{\mathrm{fin}} A'$, *Conditions* (1) *and* (2) *together are equivalent to the following rule:*

$$(i, X)HF \Rightarrow (\exists(c, U) \in \mathrm{CON})(\exists(e, V) \in \mathrm{CON}') \\ (i, X) \vdash (c, U) \wedge (c, U)H(e, V) \wedge (e, V) \vdash' F. \quad (3)$$

For a proof see Ref. [1].

Similarly to Lemma 3.4, a strengthening of Axiom 4.1(3) can be derived. It reverses the implication in Condition 4.1(4).

Lemma 4.5. *Let* H *be an approximable mapping between information systems* A *and* A' *with witnesses. Then for all* $(i, X), (j, Y) \in \mathrm{CON}$ *and* $b \in A'$,

$$(i, X) \vdash (j, Y) \wedge (j, Y)Hb \Rightarrow (i, X)Hb.$$

As has already been mentioned, entailment relations are special approximable mappings. For $(i, X) \in \mathrm{CON}$ and $a \in A$, set $(i, X)\,\mathrm{Id}_A\, a$ if $(i, X) \vdash a$. Then $\mathrm{Id}\colon A \Vdash A$ such that for all $H\colon A \Vdash A'$, $H \circ \mathrm{Id}_{A'} = H = \mathrm{Id}_A \circ H$, where for approximable mappings $H\colon A \Vdash A'$ and $G\colon A' \Vdash A''$ their composition $H \circ G\colon A \Vdash A''$ is defined by

$$(i, X)(H \circ G)c \iff (\exists(j, Y) \in \mathrm{CON}')(i, X)H(j, Y) \wedge (j, Y)Gc.$$

Let **ISW** be the category of information systems with witnesses and approximable mappings and **aISW**, **bcISW** and **abcISW**, respectively, be the full categories of information systems with witnesses that satisfy Condition (ALG), Condition (BC) or both of them.

Theorem 4.6. *The category* **ISW** *of information systems with witnesses and approximable mappings is equivalent to the category* **L** *of L-domains and Scott continuous functions.*

Corollary 4.7. *The categories* **aISW**, **bcISW** *and* **abcISW**, *respectively, of information systems with witnesses satisfying Conditions (ALG), (BC) or both of them, and approximable mappings* *are equivalent to the categories* **aL**, **BC** *and* **aBC** *of algebraic L-domain, bc-domains and algebraic bc-domains with Scott continuous functions.*

5. The Function Space Construction

As mentioned earlier, the categories **L** and **aL** are Cartesian closed. The same is true for **BC** and **aBC**. Because of the equivalence of theses categories with **ISW**, **aISW**, **bcISW** and **abcISW**, respectively, we know that the latter categories are Cartesian closed as well. In this and the next section, we present a direct proof of the Cartesian closure of **ISW** and its just mentioned full subcategories. To this end we first show that the collection of approximable mappings between two information systems with witnesses comes itself with a natural information system structure. We start with some preliminary definitions and then discuss what will be the tokens of this information system.

Definition 5.1. Let $(A, \mathrm{CON}, \vdash, \Delta)$ be an information system with witnesses and $X \subseteq_{\mathrm{fin}} A$. Two witnesses $i, j \in A$ are called *X-equivalent*, written $i \sim j\,[X]$, if there are $n \in \omega$ and $k_1, \ldots, k_n, a_0, \ldots, a_n \in A$ with $a_0 = i$, $a_n = j$, $X \in \mathrm{CON}(a_{\nu-1}) \cap \mathrm{CON}(a_\nu)$ and $\{a_{\nu-1}\}, \{a_\nu\} \in \mathrm{CON}(k_\nu)$, for $\nu = 1, \ldots, n$.

If the information system is generated from an L-domain as in Theorem 3.13, the witnesses i and j are basic elements and X is a finite subset of such, then $i \sim j\,[X]$ implies that the suprema of X with respect to i and j, respectively, coincide.

Lemma 5.2. *For any* $X \subseteq_{\mathrm{fin}} A$, $\sim [X]$ *is a partial equivalence relation. Moreover, the following five statements hold, for all* $i, j, k \in A$ *and* $U \subseteq_{\mathrm{fin}} A$*:*

(1) $i \sim \Delta\,[\emptyset]$
(2) $i \sim j\,[X] \Longrightarrow (\forall a \in A)[(i, X) \vdash a \Leftrightarrow (j, X) \vdash a]$
(3) $i \sim j\,[X] \wedge U \subseteq X \Longrightarrow i \sim j\,[U]$
(4) $i \sim j\,[X] \wedge (i, X) \vdash U \Longrightarrow i \sim j\,[U]$
(5) $i \sim j\,[X] \wedge (j, X) \vdash (k, U) \Longrightarrow i \sim k\,[U]$.

Proof. (1) is obvious.

For the remaining statements let $i \sim j\,[X]$. Then there are $n \in \omega$ and $k_1, \ldots, k_n, a_0, \ldots, a_n \in A$ such that $a_0 = i$, $a_n = j$, $X \in \mathrm{CON}(a_{\nu-1}) \cap \mathrm{CON}(a_\nu)$ and $\{a_{\nu-1}\}, \{a_\nu\} \in \mathrm{CON}(k_\nu)$, for $\nu = 1, \ldots, n$.

(2) With Conditions 3.1(10, 11), it follows that

$$(a_{\nu-1}, X) \vdash a \Longleftrightarrow (k_\nu, X) \vdash a \Longleftrightarrow (a_\nu, X) \vdash a,$$

and consequently that $(i, X) \vdash a$, exactly if $(j, X) \vdash a$, for all $a \in A$.

(4) Let $1 \le \nu \le n$. Then we have that $a_\nu \sim i\,[X]$ and hence, by Statement (2), that $(a_\nu, X) \vdash U$. Thus, $U \in \mathrm{CON}(a_\nu)$. This shows that for all $\nu = 1, \ldots, n$, $U \in \mathrm{CON}(a_{\nu-1}) \cap \mathrm{CON}(a_\nu)$ Since also $\{a_{\nu-1}\}, \{a_\nu\} \in \mathrm{CON}(k_\nu)$, we obtain that $i \sim j\,[U]$.

(3) follows similarly.

(5) We have that $(j, X) \vdash U$ and hence that $i \sim j\,[U]$, by Rule (3). Thus, there are $m \in \omega$ and $r_1, \ldots, r_m, b_0, \ldots, b_m$ with $b_0 = i$, $b_m = j$, $U \in \mathrm{CON}(b_{\mu-1}) \cap \mathrm{CON}(b_\mu)$ and $\{b_{\mu-1}\}, \{b_\mu\} \in \mathrm{CON}(r_\mu)$, with $\mu = 1, \ldots, m$. By assumption, $(j, X) \vdash k$. Hence, $\{k\} \in \mathrm{CON}(j)$. As $\{j\} \in \mathrm{CON}(r_m)$, we obtain that $\{k\} \in \mathrm{CON}(r_m)$. Now, set $b_m = k$. Then it follows that $i \sim k\,[U]$. □

Lemma 5.3. *Let $(A', \mathrm{CON}', \vdash', \Delta')$ be a further information system with witnesses and H be an approximable mapping between A and A'. Moreover, let $X \subseteq_{\mathrm{fin}} A$, $i, j \in A$, and $b \in A'$. Then*

$$i \sim j\,[X] \wedge (i, X)Hb \Longrightarrow (j, X)Hb.$$

Proof. Let $i \sim j\,[X]$ and $(i, X)Hb$. By Lemma 4.4, there is some $(c, U) \in \mathrm{CON}$ such that $(i, X) \vdash (c, U)$ and $(c, U)Hb$. Because of Lemma 5.2(2), it follows that also $(j, X) \vdash (c, U)$. With Lemma 4.5, we therefore obtain that $(j, X)Hb$. □

For $\mathfrak{X} \subseteq_{\mathrm{fin}} \mathrm{CON}$ and $(a, S) \in \mathrm{CON}$, define

$$\mathrm{CL}((a, S), \mathfrak{X}) = \{\, (i, X) \in \mathfrak{X} \mid (\exists j \in A) i \sim j\,[X] \wedge (a, S) \vdash (j, X) \,\}.$$

Then $\mathrm{CL}((a, S), \mathfrak{X})$ is finite. Let in addition

$$\overline{\mathrm{CL}}((a, S), \mathfrak{X}) = \bigcup \{\, \{j\} \cup X \mid (\exists i \in A)(i, X) \in \mathfrak{X} \wedge j \sim i\,[X] \wedge (a, S) \vdash (j, X) \,\}.$$

Lemma 5.4. *Let $(a, S), (c, T) \in \mathrm{CON}$ such that $(a, S) \vdash T$ and $a \sim c\,[T]$. Then*

$$\mathrm{CL}((c, T), \mathfrak{X}) \subseteq \mathrm{CL}((a, S), \mathfrak{X}).$$

Proof. Let $(i, X) \in \mathfrak{X}$ and $i \sim j\,[X]$ so that $(c, T) \vdash (j, X)$. Then also $(a, T) \vdash (j, X)$ and, hence, $(a, S) \vdash (j, X)$. □

Next, set

$$W(\mathfrak{X}) = \{\, a \in A \mid \bigcup \mathrm{pr}_2(\mathfrak{X}) \in \mathrm{CON}(a) \wedge (\forall (i, X) \in \mathfrak{X}) a \sim i\,[X] \,\},$$

and let $(A', \mathrm{CON}', \vdash', \Delta')$ be a further information system with witnesses.

Definition 5.5. For $\mathfrak{V} \subseteq_{\mathrm{fin}} \mathrm{CON} \times A'$, $\mathfrak{U} \subseteq \mathfrak{V}$ and $a \in W(\mathrm{pr}_1(\mathfrak{V}))$, $(a, \mathfrak{U})$ is $\mathfrak{V}$-*maximal* if the following condition holds, for all $((i, X), j) \in \mathfrak{V}$:

$$X \subseteq \bigcup \mathrm{pr}_{2,1}(\mathfrak{U}) \wedge a \sim i\,[X] \Longrightarrow ((i, X), j) \in \mathfrak{U}.$$

Lemma 5.6. *Let $(a, \mathfrak{U})$ be $\mathfrak{V}$-maximal and $c \in W(\mathrm{pr}_1(\mathfrak{U}))$ with $c \sim a\,[\bigcup \mathrm{pr}_{2,1}(\mathfrak{U})]$. Then $(c, \mathfrak{U})$ is $\mathfrak{V}$-maximal as well.*

Proof. Let $((i, X), j) \in \mathfrak{V}$ with $X \subseteq \bigcup \mathrm{pr}_{2,1}(\mathfrak{U})$ with $c \sim i\,[X]$. Then $a \sim c\,[X]$, by Lemma 5.2(3), and, hence, $a \sim i\,[X]$. Therefore, $((i, X), j) \in \mathfrak{U}$, as $(a, \mathfrak{U})$ is $\mathfrak{V}$-maximal. □

As in the case of Scott's information systems, the tokens of the function space $A \to A'$ will be finite subsets of $\mathrm{CON} \times A'$. Let $\mathfrak{V} = \{((i_1, X_1), c_1), \ldots, ((i_n, X_n), c_n)\}$ be such a set and $(a, S) \in \mathrm{CON}$. Assume that $J \subseteq \{1, \ldots, n\}$ with $(a, S) \vdash (i_\nu, X_\nu)$ exactly if $\nu \in J$. Then we need a witness for the set $\{\, c_\nu \mid \nu \in J \,\}$. Moreover, if $J_\mu \subseteq \{1, \ldots, n\}$, for $\mu = 1, \ldots, m$, with $(a, S) \vdash (i_\nu, X_\nu)$, for $\nu \in J_\mu$ and $1 \le \mu \le m$, and t_μ is a witness for $\{\, c_\nu \mid \nu \in J_\mu \,\}$, then we also need a witness for $\{\, t_\mu \mid 1 \le \mu \le m \,\}$.

Definition 5.7. $\mathfrak{W} \subseteq \mathrm{CON} \times A'$ is an *associate* of $\mathfrak{V}$ if for each $J \subseteq \{1, \ldots, n\}$, $R(\{\, (i_\nu, X_\nu) \mid \nu \in J \,\})$ is a system of representatives of $W(\{\, (i_\nu, X_\nu) \mid \nu \in J \,\})$ with respect to $\sim [\bigcup_{\nu \in J} X_\nu]$ so that $R(\emptyset) = \{\Delta\}$ and $R(\{(i_\nu, X_\nu\}) = \{i_\nu\}$, and there exists an increasing chain $(\mathfrak{W}^{(\kappa)})_{\kappa \in \omega}$ of subsets of $\mathrm{CON} \times A'$ such that

(1) $\mathfrak{W} = \bigcup_{\kappa \in \omega} \mathfrak{W}^{(\kappa)}$,

(2) $\mathfrak{W}^{(0)} = \begin{cases} \{((\Delta, \emptyset), \Delta')\} & \text{if } (\Delta, \emptyset) \text{ is } \mathfrak{V}\text{-maximal}, \\ \emptyset & \text{otherwise}, \end{cases}$

(3) and for all $\kappa \geq 1$,

$$\mathfrak{W}^{(\kappa)} = \mathfrak{W}^{(\kappa-1)} \cup \widehat{\mathfrak{W}}^{(\kappa)},$$

where $\widehat{\mathfrak{W}}^{(\kappa)}$ satisfies the following conditions:

(a) For all $J \subseteq \{1, \ldots, n\}$ with $\|J\| = \kappa$ and all $e \in R(\{\, (i_\nu, X_\nu) \mid \nu \in J \,\})$ for which $(e, \{\, ((i_\nu, X_\nu), c_\nu) \mid \nu \in J \,\})$ is $\mathfrak{V}$-maximal, there is exactly one $j \in A'$ so that

i. $\{t \in A' \mid (\exists a \in A)(\exists K \subseteq \{1, \ldots, n\}) \bigcup_{\rho \in K} X_\rho \subseteq \bigcup_{\nu \in J} X_\nu \wedge a \sim e[\bigcup_{\rho \in K} X_\rho] \wedge [(\|K\| = 1 \wedge ((a, \bigcup_{\rho \in K} X_\rho), t) \in \mathfrak{V}) \vee (a \in R(\{\, (i_\rho, X_\rho) \mid \rho \in K \,\}) \wedge ((a, \bigcup_{\rho \in K} X_\rho), t) \in \mathfrak{W}^{(\kappa-1)})] \,\} \in \mathrm{CON}'(j)$,

ii. $((a, \bigcup_{\nu \in J} X_\nu), j) \in \widehat{\mathfrak{W}}^{(\kappa)}$.

(b) For all $((a, T), b) \in \widehat{\mathfrak{W}}^{(\kappa)}$ there are $J \subseteq \{1, \ldots n\}$ and $a \in R(\{\, (i_\nu, X_\nu) \mid \nu \in J \,\})$ so that $\|J\| = \kappa$, $(a, \{\, ((i_\nu, X_\nu), c_\nu) \mid \nu \in J \,\})$ is $\mathfrak{V}$-maximal, $T = \bigcup_{\nu \in J} X_\nu$, and b satisfies Condition (3(a)i).

Let $J \subseteq \{1, \ldots, n\}$ and $a \in W(\{\, (i_\nu, X_\nu) \mid \nu \in J \,\})$ so that $(a, \{\, ((i_\nu, X_\nu), c_\nu) \mid \nu \in J \,\})$ is $\mathfrak{V}$-maximal. If $J = \emptyset$, then both sets $\{\, (i_\nu, X_\nu) \mid \nu \in J \,\}$ and $\{\, c_\nu \mid \nu \in J \,\}$ are empty as well. Moreover, $\Delta \in W(\emptyset)$ and $\emptyset \in \mathrm{CON}'(\Delta')$. This explains Condition 5.7(2). Note that $(\Delta, \emptyset)$ is not $\mathfrak{V}$-maximal if for some $1 \leq \kappa \leq n$, $X_\kappa = \emptyset$.

For the larger cardinalities, it follows from Condition 5.7(3(a)i) that the second components of pairs in $\mathfrak{W}$ are the consistency witnesses for the sets of second components in $\mathfrak{V}$ we were looking for above.

Note in Condition 5.7(3(a)i) that if $J \subseteq \{1, \ldots, n\}$ with $\|J\| \geq 1$ and

$$((a_J, \bigcup_{\nu \in J} X_\nu), t_J) \in \mathfrak{W},$$

for some $a_j \in A$ and $t_J \in A'$, then $(a_J, \{\, ((i_\nu, X_\nu), c_\nu) \mid \nu \in J \,\})$ must be $\mathfrak{V}$-maximal, by Condition 5.7(3b). Thus, if for some $K \subseteq \{1, \ldots, n\}$ one has that $\bigcup_{\kappa \in K} X_\kappa \subseteq \bigcup_{\nu \in J} X_\nu$, then $K \subseteq J$.

Let $\mathrm{AC}(\mathfrak{V})$ be the set of all associates of $\mathfrak{V}$. We write $\langle \mathfrak{W} | \mathfrak{V} \rangle$ with $\mathfrak{V} \subseteq_{\mathrm{fin}} \mathrm{CON} \times A'$ to mean that $\mathfrak{W} \in \mathrm{AC}(\mathfrak{V})$. Similarly, for sets $\mathcal{V} \subseteq_{\mathrm{fin}} \mathcal{P}_f(\mathrm{CON} \times A')$ and $\mathcal{W} \subseteq_{\mathrm{fin}} \mathcal{P}(\mathrm{CON} \times A')$ of equal cardinality, say $\mathcal{V} =$

$\{\mathfrak{V}_1, \ldots, \mathfrak{V}_m\}$ and $\mathcal{W} = \{\mathfrak{W}_1, \ldots, \mathfrak{W}_m\}$, such that $\mathfrak{W}_\nu \in \mathrm{AC}(\mathfrak{V}_\nu)$, for $\nu = 1, \ldots, m$, we write $\langle \mathcal{W} | \mathcal{V} \rangle$ for the set $\{ \langle \mathfrak{W}_\nu | \mathfrak{V}_\nu \rangle \mid 1 \leq \nu \leq m \}$. In this sense we also say that $\langle \mathcal{W} | \mathcal{V} \rangle \subseteq_{\mathrm{fin}} \mathcal{P}((\mathrm{CON} \times A')^2)$.

Lemma 5.8. *For $(c, U) \in \mathrm{CON}$ and $Z \subseteq_{\mathrm{fin}} A'$, let $\mathfrak{D} = \{ ((c, U), d) \mid d \in Z \}$. Moreover, let $b \in A'$ with $Z \cup \{\Delta'\} \in \mathrm{CON}'(b)$ if U is not empty, and $Z \in \mathrm{CON}'(b)$ otherwise, and define*

$$\mathfrak{E} = \begin{cases} \{((\Delta, \emptyset), \Delta'), ((c, U), b)\} & \text{if } U \neq \emptyset, \\ \{((c, U), b)\} & \text{otherwise.} \end{cases}$$

Then $\mathfrak{E} \in \mathrm{AC}(\mathfrak{D})$.

The proof is a straightforward exercise. Note that for any nonempty subset $V \subseteq Z$, $(c, \{ ((c, U), d) \mid d \in V \})$ is $\mathfrak{D}$-maximal, exactly if $V = Z$. If U is empty, this is even true for any subset V.

For $(a, S) \in \mathrm{CON}$,

$$\mathrm{SP}((a, S), \mathfrak{V}) = \{ ((i, X), j) \in \mathfrak{V} \mid a \sim i\,[X] \wedge X \subseteq S \},$$

is the *$\mathfrak{V}$-spectrum* of (a, S). Set $\mathrm{DS}((a, S), \mathfrak{V}) = \mathrm{pr}_1(\mathrm{SP}((a, S), \mathfrak{V})$ and $\mathrm{RS}((a, S), \mathfrak{V}) = \mathrm{pr}_2(\mathrm{SP}((a, S), \mathfrak{V})$.

Assume again that $\mathfrak{V} = \{((i_1, X_1), c_1), \ldots, ((i_n, X_n), c_n)\}$ and let $\mathfrak{W} \in \mathrm{AC}(\mathfrak{V})$. If $((c, T), d) \in \mathfrak{W}$, if follows from Condition 5.7(3b) that there is exactly one subset $J \subseteq \{1, \ldots, n\}$ such that $(c, \{ ((i_\nu, X_\nu), c_\nu) \mid \nu \in J \})$ is $\mathfrak{V}$-maximal, $c \in W(\{ (i_\nu, X_\nu) \mid \nu \in J \})$, and $T = \bigcup_{\nu \in J} X_\nu$. Then $\mathrm{DS}((c, T), \mathfrak{V}) = \{ (i_\nu, X_\nu) \mid \nu \in J \}$ and $\mathrm{RS}((c, T), \mathfrak{V}) = \{ c_\nu \mid \nu \in J \}$.

Define

$$\begin{aligned} \mathrm{AP}((a, S), \mathfrak{V}) = \{ c \in A' \mid (\exists (i, X) \in \mathrm{CL}((a, S), \mathrm{pr}_1(\mathfrak{V}))) \\ \times ((i, X), c) \in \mathfrak{V} \}. \end{aligned}$$

As a consequence of Lemma 5.4, we have for $(c, T) \in \mathrm{CON}$ with $(a, S) \vdash T$ and $a \sim c\,[T]$ that $\mathrm{AP}((c, T), \mathfrak{V}) \subseteq \mathrm{AP}((a, S), \mathfrak{V})$.

Let $M \subseteq \{1, \ldots, n\}$, such that

$$\mathrm{CL}((a, S), \mathrm{pr}_1(\mathfrak{V})) = \{ (i_\kappa, X_\kappa) \mid \kappa \in M \}.$$

Then $(a, \{ ((i_\kappa, X_\kappa), c_\kappa) \mid \kappa \in M \})$ is $\mathfrak{V}$-maximal. It follows that

$$\bigcup \mathrm{pr}_2(\mathrm{CL}((a,S), \mathrm{pr}_1(\mathfrak{V}))) = \bigcup\nolimits_{\kappa \in M} X_\kappa \text{ and } \mathrm{AP}((a,S), \mathfrak{V}) = \{ c_\kappa \mid \kappa \in M \}.$$

Moreover, there exists exactly one pair (e, j) with $e \in W(\{ (i_\kappa, X_\kappa) \mid \kappa \in M \})$ and $j \in A'$ so that

$$a \sim e \left[\bigcup_{\kappa \in M} X_\kappa \right], \ \{c_\kappa \mid \kappa \in M\} \in \mathrm{CON}'(j), \text{ and } \left(\left(e, \bigcup_{\kappa \in M} X_\kappa \right), j \right) \in \mathfrak{W}.$$

Set

$$\mathfrak{W}(a,S) = \left(\left(e, \bigcup_{\kappa \in M} X_\kappa \right), j \right).$$

Then we also write

$$\mathfrak{W}_1(a,S) = \left(e, \bigcup_{\kappa \in M} X_\kappa \right), \ \mathfrak{W}_{1,1}(a,S) = e, \ \mathfrak{W}_{1,2}(a,S) = \bigcup_{\kappa \in M} X_\kappa,$$

and

$$\mathfrak{W}_2(a,S) = j.$$

The next lemma is a consequence of Lemma 5.4 and Condition 5.7(3(a)i)

Lemma 5.9. *Let $(a,S), (c,T) \in \mathrm{CON}$. Then the following statements hold:*

(1) *If $(a,S) \vdash T$ and $a \sim c\,[T]$. Then $\{\mathfrak{W}_2(c,T)\} \in \mathrm{CON}'(\mathfrak{W}_2(a,S))$.*
(2) *If $T \subseteq S$ and $a \sim c\,[T]$. Then $\{\mathfrak{W}_2(c,T)\} \in \mathrm{CON}'(\mathfrak{W}_2(a,S))$.*
(3) *If $(a,S) \vdash (c,T)$ and $(c,T) \vdash \overline{\mathrm{CL}}((a,S), \mathrm{pr}_1(\mathfrak{V}))$. Then*

$$\mathrm{CL}((a,S), \mathrm{pr}_1(\mathfrak{V})) = \mathrm{CL}((c,T), \mathrm{pr}_1(\mathfrak{V})) \textit{ and } \mathfrak{W}_2(c,T) = \mathfrak{W}_2(a,S).$$

Definition 5.10. Let $(A, \mathrm{CON}, \vdash, \Delta)$ and $(A', \mathrm{CON}', \vdash', \Delta')$ be information systems with witnesses. Define

(1) $A \to A' = \{\, \langle \mathfrak{W} | \mathfrak{V} \rangle \in \mathcal{P}(\mathrm{CON} \times A') \times \mathcal{P}_f(\mathrm{CON} \times A') \mid \mathfrak{W} \in \mathrm{AC}(\mathfrak{V}) \,\}$
(2) For $\langle \mathfrak{W} | \mathfrak{V} \rangle \in A \to A'$ and $\langle \mathcal{W} | \mathcal{V} \rangle \subseteq_{\mathrm{fin}} A \to A'$, $\langle \mathcal{W} | \mathcal{V} \rangle \in \mathrm{CON}_{\to}(\langle \mathfrak{W} | \mathfrak{V} \rangle)$ if for all $(a, S) \in \mathrm{CON}$,
   (a) $(\forall \langle \mathfrak{G} | \mathfrak{F} \rangle \in \langle \mathcal{W} | \mathcal{V} \rangle)(\exists k \in A')\, \mathfrak{G}_2(a, S) \sim k\, [\mathrm{AP}((a, S), \mathfrak{F})] \wedge \{k\} \in \mathrm{CON}'(\mathfrak{W}_2(a, S))$
   (b) $(\forall r \in A')(r \sim \mathfrak{W}_2(a, S)\, [\mathrm{AP}((a, S), \mathfrak{V})] \Rightarrow r \sim \mathfrak{W}_2(a, S)\, [\mathrm{AP}((a, S), \bigcup \mathcal{V})])$.
(3) For $\langle \mathfrak{W} | \mathfrak{V} \rangle, \langle \mathfrak{B} | \mathfrak{A} \rangle \in A \to A'$ and $\langle \mathcal{W} | \mathcal{V} \rangle \in \mathrm{CON}_{\to}(\langle \mathfrak{W} | \mathfrak{V} \rangle)$, $(\langle \mathfrak{W} | \mathfrak{V} \rangle, \langle \mathcal{W} | \mathcal{V} \rangle) \vdash_{\to} \langle \mathfrak{B} | \mathfrak{A} \rangle$ if
   (a) $(\forall ((e, Y), f) \in \mathfrak{A})(\mathfrak{W}_2(e, Y), \mathrm{AP}((e, Y), \bigcup \mathcal{V})) \vdash' f$
   (b) $(\forall ((a, Z), b) \in \mathfrak{B})(\exists k \in A')\, b \sim k\, [\mathrm{RS}((a, Z), \mathfrak{A})] \wedge (\mathfrak{W}_2(a, Z), \bigcup \{\, \mathrm{AP}((e, Y), \bigcup \mathcal{V}) \mid (e, Y) \in \mathrm{DS}((a, Z), \mathfrak{A}) \,\}) \vdash' (k, \mathrm{RS}((a, Z), \mathfrak{A}))$.
(4) $\Delta_{\to} = \langle \{((\Delta, \emptyset), \Delta')\} | \emptyset \rangle$.

In Condition 5.10(3a), one needs that

$$\mathrm{AP}\left((e, Y), \bigcup \mathcal{V}\right) \in \mathrm{CON}'(\mathfrak{W}_2(e, Y)).$$

This is an easy consequence of Condition 5.10(2b): choose $r = \mathfrak{W}_2(e, Y)$ and note that by definition of $\mathfrak{W}_2(e, Y)$, $\mathrm{AP}((e, Y), \mathfrak{V}) \in \mathrm{CON}'(\mathfrak{W}_2(e, Y))$.

For Condition 5.10(3b), observe that

$$\bigcup \left\{\mathrm{AP}\left((e, Y), \bigcup \mathcal{V}\right) \mid (e, Y) \in \mathrm{DS}((a, Z), \mathfrak{A})\right\} \subseteq \mathrm{AP}\left((a, Z), \bigcup \mathcal{V}\right).$$

Because of Condition 3.1(2), we therefore have that

$$\bigcup \left\{\mathrm{AP}\left((e, Y), \bigcup \mathcal{V}\right) \mid (e, Y) \in \mathrm{DS}((a, Z), \mathfrak{A})\right\} \in \mathrm{CON}'(\mathfrak{W}_2(a, Z)).$$

Lemma 5.11. *Let* $\langle \mathcal{W} | \mathcal{V} \rangle \in \mathrm{CON}_{\to}(\langle \mathfrak{W} | \mathfrak{V} \rangle)$ *and* $\langle \mathcal{G} | \mathcal{F} \rangle \subseteq_{\mathrm{fin}} A \to A'$ *such that Condition* 5.10(3a) *holds for all* $\langle \mathfrak{G} | \mathfrak{F} \rangle \in \langle \mathcal{G} | \mathcal{F} \rangle$. *Then for all* $(a, S) \in \mathrm{CON}$,

$$\left(\mathfrak{W}_2(a, S), \bigcup \left\{\mathrm{AP}((c, T), \bigcup \mathcal{V}) \mid (c, T) \in \mathrm{CL}\left((a, S), \mathrm{pr}_1\left(\bigcup \mathcal{F}\right)\right)\right\}\right)$$
$$\vdash' \mathrm{AP}\left((a, S), \bigcup \mathcal{F}\right).$$

Proof. Let $b \in \mathrm{AP}((a,S), \bigcup \mathcal{F})$. Then there are $(j,U) \in \mathrm{CON}$ and $j' \in A$ such that $((j,U),b) \in \bigcup \mathcal{F}$, $j \sim j'\,[U]$, and $(a,S) \vdash (j',U)$. As a consequence of our assumption, we have that $(\mathfrak{W}_2(j,U), \mathrm{AP}((j,U), \bigcup \mathcal{V})) \vdash' b$. Then $\mathfrak{W}_2(j,U) = \mathfrak{W}_2(j',U)$ and $\{\mathfrak{W}_2(j,U)\} \in \mathrm{CON}'(\mathfrak{W}_2(a,S))$. It follows that also $(\mathfrak{W}_2(a,S), \mathrm{AP}((j,U), \bigcup \mathcal{V})) \vdash' b$. Since moreover $\mathrm{AP}((j,U), \bigcup \mathcal{V}) \subseteq \bigcup \{\, \mathrm{AP}((c,T), \bigcup \mathcal{V}) \mid (c,T) \in \mathrm{CL}((a,S), \mathrm{pr}_1(\bigcup \mathcal{F}))\,\}$, we finally obtain that $(\mathfrak{W}_2(a,S), \bigcup\{\, \mathrm{AP}((c,T), \bigcup \mathcal{V}) \mid (c,T) \in \mathrm{CL}((a,S), \mathrm{pr}_1(\bigcup \mathcal{F}))\,\}) \vdash' b$. □

Lemma 5.12. *Let* $\langle \mathcal{W} | \mathcal{V} \rangle \in \mathrm{CON}_{\rightarrow}(\langle \mathfrak{W} | \mathfrak{V} \rangle)$ *and* $\langle \mathfrak{B} | \mathfrak{A} \rangle \in A \to A'$ *so that Condition* 5.10(3b) *holds. Then for all* $(a,S) \in \mathrm{CON}$, $\mathfrak{B}_2(a,S) \sim \mathfrak{W}_2(a,S)\,[\mathrm{AP}((a,S), \mathfrak{A})]$.

Proof. By definition of $\mathfrak{B}(a,S)$,

$$\mathrm{AP}((a,S), \mathfrak{A}) = \mathrm{RS}(\mathfrak{B}_1(a,S), \mathfrak{A}) \text{ and } (\mathfrak{B}_1(a,S), \mathfrak{B}_2(a,S)) \in \mathfrak{B}.$$

So, Condition 5.10(3b) implies that $\mathfrak{B}_2(a,S) \sim \mathfrak{W}_2(a,S)\,[\mathrm{AP}((a,S), \mathfrak{A})]$. □

Proposition 5.13. *Let* $(A, \mathrm{CON}, \vdash, \Delta)$ *and* $(A', \mathrm{CON}', \vdash', \Delta')$ *be information systems with witnesses. Then* $(A \to A', \mathrm{CON}_{\rightarrow}, \vdash_{\rightarrow}, \Delta_{\rightarrow})$ *is an information system with witnesses as well.*

In the subsequent lemmas, we verify the conditions in Definition 3.1.

Lemma 5.14. $(A \to A', \mathrm{CON}_{\rightarrow}, \vdash_{\rightarrow}, \Delta_{\rightarrow})$ *satisfies Condition* 3.1(1).

Proof. We have to show that $\{\langle \mathfrak{W} | \mathfrak{V} \rangle\} \in \mathrm{CON}_{\rightarrow}(\langle \mathfrak{W} | \mathfrak{V} \rangle)$. Let $(a,S) \in \mathrm{CON}$. Condition 5.10(2b) is vacuously true and for Requirement 5.10(2a), choose $k = \mathfrak{W}_2(a,S)$. □

Lemma 5.15. $(A \to A', \mathrm{CON}_{\rightarrow}, \vdash_{\rightarrow}, \Delta_{\rightarrow})$ *satisfies Condition* 3.1(2).

Proof. Let $\langle \mathcal{W} | \mathcal{V} \rangle \in \mathrm{CON}_{\rightarrow}(\langle \mathfrak{W} | \mathfrak{V} \rangle)$ and $\langle \mathcal{G} | \mathcal{F} \rangle \subseteq \langle \mathcal{W} | \mathcal{V} \rangle$. We need to verify that $\langle \mathcal{G} | \mathcal{F} \rangle \in \mathrm{CON}_{\rightarrow}(\langle \mathfrak{W} | \mathfrak{V} \rangle)$: Condition 5.10(2a) holds trivially. For Requirement 5.10(2b), apply Lemma 5.2(3). □

Lemma 5.16. $(A \to A', \mathrm{CON}_{\rightarrow}, \vdash_{\rightarrow}, \Delta_{\rightarrow})$ *satisfies Condition* 3.1(3).

Proof. We have to prove that $(\langle \mathfrak{W} | \mathfrak{V} \rangle, \emptyset) \vdash_{\rightarrow} \Delta_{\rightarrow}$. Condition 5.10(3a) is vacuously satisfied and for Condition 5.10(3b), choose $k = \Delta'$. □

Lemma 5.17. *$(A \to A', \mathrm{CON}_{\to}, \vdash_{\to}, \Delta_{\to})$ satisfies Condition* 3.1(4).

Proof. Let $\langle \mathcal{W}|\mathcal{V}\rangle \in \mathrm{CON}_{\to}(\langle \mathfrak{W}|\mathfrak{V}\rangle)$ and $\langle \mathcal{G}|\mathcal{F}\rangle \subseteq_{\mathrm{fin}} A \to A'$ with $(\langle \mathfrak{W}|\mathfrak{V}\rangle, \langle \mathcal{W}|\mathcal{V}\rangle) \vdash_{\to} \langle \mathcal{G}|\mathcal{F}\rangle$. We must verify that $\langle \mathcal{G}|\mathcal{F}\rangle \in \mathrm{CON}_{\to}(\langle \mathfrak{W}|\mathfrak{V}\rangle)$.

Let to this end $(a, S) \in \mathrm{CON}$ and $\langle \mathfrak{B}|\mathfrak{A}\rangle \in \langle \mathcal{G}|\mathcal{F}\rangle$. As $(\langle \mathfrak{W}|\mathfrak{V}\rangle, \langle \mathcal{W}|\mathcal{V}\rangle) \vdash_{\to} \langle \mathcal{G}|\mathcal{F}\rangle$, there is some $k \in A'$ so that $\mathfrak{B}_2(a, S) \sim k\,[\mathrm{AP}((a, S), \mathfrak{A})]$ and

$$(\mathfrak{W}_2(a, S), \bigcup \left\{\mathrm{AP}((c, T), \bigcup \mathcal{V}) | (c, T) \in \mathrm{CL}((a, S), \mathrm{pr}_1(\mathfrak{A}))\right\}$$
$$\vdash' (k, \mathrm{AP}((a, S), \mathfrak{A})).$$

Thus, $\{k\} \in \mathrm{CON}'(\mathfrak{W}_2(a, S))$, which proves Condition 5.10(2a).

For Condition 5.10(2b), let $r \in A'$ with $r \sim \mathfrak{W}_2(a, S)$ $[\mathrm{AP}((a, S), \mathfrak{V})]$. Then

$$r \sim \mathfrak{W}_2(a, S) \left[\mathrm{AP}\left((a, S), \bigcup \mathcal{V}\right)\right], \tag{4}$$

as $\langle \mathcal{W}|\mathcal{V}\rangle \in \mathrm{CON}_{\to}(\langle \mathfrak{W}|\mathfrak{V}\rangle)$. With Lemma 5.11, we moreover have that

$$\left(\mathfrak{W}_2(a, S), \mathrm{AP}\left((a, S), \bigcup \mathcal{V}\right)\right) \vdash' \mathrm{AP}\left((a, S), \bigcup \mathcal{F}\right).$$

By Condition 3.1(7), there is thus some $k \in A'$ with

$$\left(\mathfrak{W}_2(a, S), \mathrm{AP}\left((a, S), \bigcup \mathcal{V}\right)\right) \vdash' \left(k, \mathrm{AP}((a, S), \bigcup \mathcal{F})\right).$$

Because of Statement (4), we obtain with Lemma 5.2(5) that

$$k \sim \mathfrak{W}_2(a, S) \left[\mathrm{AP}((a, S), \bigcup \mathcal{F})\right],$$

and similarly that $k \sim r\,[\mathrm{AP}((a, S), \bigcup \mathcal{F})]$. Therefore,

$$r \sim \mathfrak{W}_2(a, S) \left[\mathrm{AP}((a, S), \bigcup \mathcal{F})\right].$$

□

Requirement 3.1(5) is obvious, because for $\langle \mathcal{W}|\mathcal{V}\rangle \subseteq \langle \mathcal{G}|\mathcal{F}\rangle$, we have that $\mathrm{AP}((a, s), \bigcup \mathcal{V}) \subseteq \mathrm{AP}((a, s), \bigcup \mathcal{F})$.

Lemma 5.18. $(A \to A', \mathrm{CON}_{\to}, \vdash_{\to}, \Delta_{\to})$ *satisfies Condition* 3.1(6).

Proof. Let $\langle \mathcal{W}|\mathcal{V}\rangle \in \mathrm{CON}_{\to}(\langle \mathfrak{W}|\mathfrak{V}\rangle)$, $\langle \mathcal{G}|\mathcal{F}\rangle \subseteq_{\text{fin}} A \to A'$ and $\langle \mathfrak{B}|\mathfrak{A}\rangle \in A \to A'$ with $(\langle \mathfrak{W}|\mathfrak{V}\rangle, \langle \mathcal{W}|\mathcal{V}\rangle) \vdash_{\to} \langle \mathcal{G}|\mathcal{F}\rangle$ and $(\langle \mathfrak{W}|\mathfrak{V}\rangle, \langle \mathcal{G}|\mathcal{F}\rangle) \vdash_{\to} \langle \mathfrak{B}|\mathfrak{A}\rangle$. We need to show that

$$(\langle \mathfrak{W}|\mathfrak{V}\rangle, \langle \mathcal{W}|\mathcal{V}\rangle) \vdash_{\to} \langle \mathfrak{B}|\mathfrak{A}\rangle.$$

The first condition to be verified is a consequence of Lemma 5.11. For the second condition, let $((a, Z), b) \in \mathfrak{B}$. By our assumption, there is some $k \in A'$ so that

$$k \sim b\,[\mathrm{RS}((a, Z), \mathfrak{A})],$$

and

$$\left(\mathfrak{W}_2(a, Z), \bigcup \left\{\mathrm{AP}((e, Y), \bigcup \mathcal{F}) \mid (e, Y) \in \mathrm{DS}((a, Z), \mathfrak{A})\right\}\right) \vdash' (k, \mathrm{RS}((a, Z), \mathfrak{A})).$$

Moreover, with Lemma 5.11, we obtain that

$$\Big(\mathfrak{W}_2(a, Z), \bigcup \Big\{\mathrm{AP}\left((c, T), \bigcup \mathcal{V}\right) \mid \left(\exists (e, Y) \in \mathrm{DS}((a, Z), \mathfrak{A})\right)(c, T) \in \mathrm{CL}\left((e, Y), \mathrm{pr}_1\left(\bigcup \mathcal{F}\right)\right)\Big\}\Big) \vdash' \bigcup \left\{\mathrm{AP}((e, Y), \bigcup \mathcal{F}) \mid (e, Y) \in \mathrm{DS}((a, Z), \mathfrak{A})\right\}.$$

Since

$$\bigcup \Big\{\mathrm{AP}((c, T), \bigcup \mathcal{V}) \mid \left(\exists (e, Y) \in \mathrm{DS}((a, Z), \mathfrak{A})\right)(c, T) \in \mathrm{CL}\left((e, Y), \mathrm{pr}_1\left(\bigcup \mathcal{F}\right)\right)\Big\} \subseteq \bigcup \left\{\mathrm{AP}((e, Y), \bigcup \mathcal{V}) \mid (e, Y) \in \mathrm{DS}((a, Z), \mathfrak{A})\right\},$$

it follows that

$$\left(\mathfrak{W}_2(a, Z), \bigcup \left\{\mathrm{AP}((e, Y), \bigcup \mathcal{V}) \mid (e, Y) \in \mathrm{DS}((a, Z), \mathfrak{A})\right\}\right) \vdash' (k, \mathrm{RS}((a, Z), \mathfrak{A})),$$

as was to be demonstrated. □

Lemma 5.19. $(A \to A', \mathrm{CON}_{\to}, \vdash_{\to}, \Delta_{\to})$ *satisfies Condition* 3.1(7).

Proof. Let $\langle \mathfrak{W}|\mathfrak{V}\rangle, \langle \mathfrak{G}|\mathfrak{F}\rangle \in A \to A'$ and $\langle \mathcal{W}|\mathcal{V}\rangle \in \mathrm{CON}_{\to}(\langle \mathfrak{W}|\mathfrak{V}\rangle)$ with

$$(\langle \mathfrak{W}|\mathfrak{V}\rangle, \langle \mathcal{W}|\mathcal{V}\rangle) \vdash_{\to} \langle \mathfrak{G}|\mathfrak{F}\rangle.$$

We have to show that there exists $\langle \mathcal{E}|\mathcal{D}\rangle \subseteq_{\mathrm{fin}} A \to A'$ so that

$$(\langle \mathfrak{W}|\mathfrak{V}\rangle, \langle \mathcal{W}|\mathcal{V}\rangle) \vdash_{\to} \langle \mathcal{E}|\mathcal{D}\rangle \text{ and } (\langle \mathfrak{W}|\mathfrak{V}\rangle, \langle \mathcal{E}|\mathcal{D}\rangle\}) \vdash_{\to} \langle \mathfrak{G}|\mathfrak{F}\rangle. \tag{5}$$

If $\mathfrak{F}$ is empty, set $\langle \mathcal{E}|\mathcal{D}\rangle = \{\Delta_{\to}\}$. Otherwise, let

$$\mathfrak{F} = \{((a_1, T_1), d_1), \ldots, ((a_m, T_m), d_m)\}.$$

By assumption, we know that for $1 \le \nu \le m$,

$$(\mathfrak{W}_2(a_\nu, T_\nu), \mathrm{AP}((a_\nu, T_\nu), \bigcup \mathcal{V})) \vdash' d_\nu. \tag{6}$$

Because of the Global Interpolation Property (GIP), there is therefore some $(k_\nu, Z_\nu) \in \mathrm{CON}'$ with

$$\left(\mathfrak{W}_2(a_\nu, T_\nu), \mathrm{AP}\left((a_\nu, T_\nu), \bigcup \mathcal{V}\right)\right) \vdash' (k_\nu, Z_\nu), \tag{7}$$

$$(k_\nu, Z_\nu) \vdash' d_\nu. \tag{8}$$

Since

$$(a_\nu, T_\nu) \vdash \overline{\mathrm{CL}}\left((a_\nu, T_\nu), \mathrm{pr}_1\left(\bigcup \mathcal{V}\right)\right), \tag{9}$$

we similarly obtain some $(e_\nu, U_\nu) \in \mathrm{CON}$ such that

$$(a_\nu, T_\nu) \vdash (e_\nu, U_\nu), \tag{10}$$

$$(e_\nu, U_\nu) \vdash \overline{\mathrm{CL}}\left((a_\nu, T_\nu), \mathrm{pr}_1\left(\bigcup \mathcal{V}\right)\right). \tag{11}$$

With Lemma 5.9(3) it follows that

$$\mathrm{AP}\left((a_\nu, T_\nu), \bigcup \mathcal{V}\right) = \mathrm{AP}\left((e_\nu, U_\nu), \bigcup \mathcal{V}\right) \text{ and } \mathfrak{W}_2(a_\nu, T_\nu) = \mathfrak{W}_2(e_\nu, U_\nu). \tag{12}$$

Let $\mathfrak{D}_\nu = \{\, ((e_\nu, U_\nu), j) \mid j \in Z_\nu \,\}$ and set

$$\mathfrak{E}_\nu = \begin{cases} \{((\Delta, \emptyset), \Delta'), ((e_\nu, U_\nu), k_\nu)\} & \text{if } U_\nu \neq \emptyset, \\ \{((e_\nu, U_\nu), k_\nu)\} & \text{otherwise.} \end{cases}$$

Then $\mathfrak{E}_\nu \in \mathrm{AC}(\mathfrak{D}_\nu)$ by Lemma 5.8. It remains to show that $\langle \mathcal{E} | \mathcal{D} \rangle$ with $\mathcal{E} = \{\mathfrak{E}_1, \ldots, \mathfrak{E}_m\}$ and $\mathcal{D} = \{\mathfrak{D}_1 \ldots, \mathfrak{D}_m\}$ meets Requirements (5).

As a consequence of Statements (7) and (12), we gain that

$$\left(\mathfrak{W}_2(e_\nu, U_\nu), \mathrm{AP}\left((e_\nu, U_\nu), \bigcup \mathcal{V}\right)\right) \vdash' (k_\nu, Z_\nu). \tag{13}$$

For the verification of Condition 5.10(3b), let $((a, Z), b) \in \mathfrak{E}_\nu$ and set $k = b$. If

$$((a, Z), b) = ((\Delta, \emptyset), \Delta'),$$

it follows that U_ν is not empty and, therefore, $\mathrm{DS}((\Delta, \emptyset), \mathfrak{D}_\nu))$ and $\mathrm{RS}((\Delta, \emptyset), \mathfrak{D}_\nu)$ are both empty. In case that $((a, Z), b) = ((e_\nu, U_\nu), k_\nu)$, we have that $\mathrm{DS}((e_\nu, U_\nu), \mathfrak{D}_\nu) = \{(e_\nu, U_\nu)\}$ and $\mathrm{RS}((e_\nu, U_\nu), \mathfrak{D}_\nu) = Z_\nu$. Thus, in both cases,

$$\left(\mathfrak{W}_2(a, Z), \bigcup \left\{\mathrm{AP}((c, T), \bigcup \mathcal{V}) \mid (c, T) \in \mathrm{DS}((a, Z), \mathfrak{D}_\nu)\right\}\right)$$
$$\vdash' (k, \mathrm{RS}((a, Z), \mathfrak{D}_\nu)),$$

in the first case because of Axiom 3.1(3), and in the other one as a consequence of Statements (7, 11) as well as Axiom 3.1(11). This shows that $(\langle \mathfrak{W} | \mathfrak{V} \rangle, \langle \mathcal{W} | \mathcal{V} \rangle) \vdash_{\rightarrow} \langle \mathcal{E} | \mathcal{D} \rangle$.

It remains to show that $(\langle \mathfrak{W} | \mathfrak{V} \rangle, \langle \mathcal{E} | \mathcal{D} \rangle) \vdash_{\rightarrow} \langle \mathfrak{G} | \mathfrak{F} \rangle$. Note that by Statement (7) and Axiom 3.1(4), $\{k_\nu\} \in \mathrm{CON}(\mathfrak{W}_2(a_\nu, T_\nu))$. Moreover,

$$Z_\nu \subseteq \mathrm{AP}((a_\nu, T_\nu), \mathfrak{D}_\nu) \subseteq \mathrm{AP}\left((a_\nu, T_\nu), \bigcup \mathcal{D}\right).$$

With Lemma 5.17, it follows from what we have just seen that $\langle \mathcal{E} | \mathcal{D} \rangle \in \mathrm{CON}_{\rightarrow}(\langle \mathfrak{W} | \mathfrak{V} \rangle)$. Thus, $\mathrm{AP}((a_\nu, T_\nu), \bigcup \mathcal{D}) \in \mathrm{CON}'(\mathfrak{W}_2(a_\nu, T_\nu))$. As a consequence of Statement (8), we now obtain that

$$\left(\mathfrak{W}_2(a_\nu, T_\nu), \mathrm{AP}\left((a_\nu, T_\nu), \bigcup \mathcal{D}\right)\right) \vdash' d_\nu, \tag{14}$$

for all $1 \leq \nu \leq m$.

For the second condition let $((a, Z), b) \in \mathfrak{G}$. Since $(\langle \mathfrak{W} | \mathfrak{V} \rangle, \langle \mathcal{W} | \mathcal{V} \rangle) \vdash_{\rightarrow} \langle \mathfrak{G} | \mathfrak{F} \rangle$, there is some $k \in A'$ with $k \sim b[\mathrm{RS}((a, Z), \mathfrak{F})]$

and $\{k\} \in \mathrm{CON}'(\mathfrak{W}_2(a,Z))$. With Statement (14), it follows that

$$\left(\mathfrak{W}_2(a,Z), \bigcup\left\{\mathrm{AP}\left((c,T), \bigcup\mathcal{D}\right) \mid (c,T) \in \mathrm{DS}((a,Z),\mathfrak{F})\right\}\right)$$
$$\vdash' \mathrm{RS}((a,Z),\mathfrak{F}).$$

Because of Axiom 3.1(8), there is thus some $r \in A'$ such that

$$\left(\mathfrak{W}_2(a,Z), \bigcup\left\{\mathrm{AP}\left((c,T), \bigcup\mathcal{D}\right) \mid (c,T) \in \mathrm{DS}((a,Z),\mathfrak{F})\right\}\right) \vdash' r,$$

and $\mathrm{RS}((a,Z),\mathfrak{F}) \in \mathrm{CON}'(r)$. Hence, $\{r\} \in \mathrm{CON}'(\mathfrak{W}_2(a,Z))$, from which it ensues that $r \sim k\,[\mathrm{RS}((a,Z),\mathfrak{F})]$. So, $r \sim b\,[\mathrm{RS}((a,Z),\mathfrak{F})]$ and

$$\left(\mathfrak{W}_2(a,Z), \bigcup\left\{\mathrm{AP}\left((c,T), \bigcup\mathcal{D}\right) \mid (c,T) \in \mathrm{DS}((a,Z),\mathfrak{F})\right\}\right)$$
$$\vdash' (r, \mathrm{RS}((a,Z),\mathfrak{F})).$$

□

Lemma 5.20. *$(A \to A', \mathrm{CON}_{\to}, \vdash_{\to}, \Delta_{\to})$ satisfies Condition* 3.1(8).

Proof. Let $\langle \mathfrak{W}|\mathfrak{V}\rangle \in A \to A'$, $\langle \mathcal{G}|\mathcal{F}\rangle \subseteq_{\mathrm{fin}} A \to A'$ and $\langle \mathcal{W}|\mathcal{V}\rangle \in \mathrm{CON}_{\to}(\langle \mathfrak{W}|\mathfrak{V}\rangle)$ with

$$(\langle \mathfrak{W}|\mathfrak{V}\rangle, \langle \mathcal{W}|\mathcal{V}\rangle) \vdash_{\to} \langle \mathcal{G}|\mathcal{F}\rangle.$$

We will construct some $\langle \mathfrak{B}|\mathfrak{A}\rangle \in A \to A'$ so that

$$(\langle \mathfrak{W}|\mathfrak{V}\rangle, \langle \mathcal{W}|\mathcal{V}\rangle) \vdash_{\to} \langle \mathfrak{B}|\mathfrak{A}\rangle \text{ and } \langle \mathcal{G}|\mathcal{F}\rangle \in \mathrm{CON}_{\to}(\langle \mathfrak{B}|\mathfrak{A}\rangle). \tag{15}$$

If $\bigcup\mathcal{F}$ is empty, set $\langle \mathfrak{B}|\mathfrak{A}\rangle = \Delta_{\to}$. Let us now assume that $\bigcup\mathcal{F}$ is nonempty. In particular, let $\bigcup\mathcal{F} = \{((a_1,T_1),d_1),\ldots, ((a_m,T_m),d_m)\}$ and $1 \le \nu \le m$. As in the proof of Lemma 5.19, there exists $Z_\nu \in \mathrm{CON}'(\mathfrak{W}_2(a_\nu, T_\nu))$ with

$$\left(\mathfrak{W}_2(a_\nu, T_\nu), \mathrm{AP}\left((a_\nu, T_\nu), \bigcup \mathcal{V}\right)\right) \vdash' Z_\nu, \tag{16}$$

$$(\mathfrak{W}_2(a_\nu, T_\nu), Z_\nu) \vdash' d_\nu, \tag{17}$$

and $(e_\nu, U_\nu) \in \mathrm{CON}$ such that

$$(a_\nu, T_\nu) \vdash (e_\nu, U_\nu), \tag{18}$$

$$(e_\nu, U_\nu) \vdash \overline{\mathrm{CL}}\left((a_\nu, T_\nu), \mathrm{pr}_1\left(\bigcup \mathcal{V}\right)\right). \tag{19}$$

Set

$$\mathfrak{A} = \{\, ((e_\nu, U_\nu), j) \mid j \in Z_\nu \wedge 1 \leq \nu \leq m \,\}.$$

Then it follows as in the derivation of Lemma 5.19 that

$$(\mathfrak{W}_2(e_\nu, U_\nu), \mathrm{AP}((e_\nu, U_\nu), \bigcup \mathcal{V})) \vdash' Z_\nu. \tag{20}$$

We will now construct some $\mathfrak{B} \in \mathrm{AC}(\mathfrak{A})$ satisfying the Requirements (15). For each $J \subseteq \{1, \ldots, m\}$, let $R(\{\, (e_\nu, U_\nu) \mid \nu \in J \,\})$ be a system of representatives of $W(\{\, (e_\nu, U_\nu) \mid \nu \in J \,\})$ with respect to $\sim [\bigcup_{\nu \in J} U_\nu]$ so that $R(\emptyset) = \{\Delta\}$ and $R(\{(e_\nu, U_\nu)\}) = \{e_\nu\}$, for $1 \leq \nu \leq m$. By Axiom 5.10(3b), there is some $k_{a,Z,\langle \mathfrak{G}|\mathfrak{F}\rangle} \in A'$, for each $\langle \mathfrak{G}|\mathfrak{F}\rangle \in \langle \mathcal{G}|\mathcal{F}\rangle$ and $((a, Z), b) \in \mathfrak{G}$, with $k_{a,Z,\langle \mathfrak{G}|\mathfrak{F}\rangle} \sim b\,[\mathrm{RS}((a, Z), \mathfrak{F})]$ and

$$\left(\mathfrak{W}_2(a, Z), \bigcup \left\{\mathrm{AP}\left((e, Y), \bigcup \mathcal{V}\right) \mid (e, Y) \in \mathrm{DS}((a, Z), \mathfrak{F})\right\}\right)$$
$$\vdash' (k_{a,Z,\langle \mathfrak{G}|\mathfrak{F}\rangle}, \mathrm{RS}((a, Z), \mathfrak{F})). \tag{21}$$

In case $((a, Z), b) = ((\Delta, \emptyset), \Delta')$, we choose $k_{\Delta,\emptyset,\langle \mathfrak{G}|\mathfrak{F}\rangle} = \Delta'$.

Claim 5.21. *For every $\mathfrak{J} \subseteq \mathfrak{A}$ and each $a \in R(\mathrm{pr}_1(\mathfrak{J}))$ such that $(a, \mathfrak{J})$ is $\mathfrak{A}$-maximal, there is some $t_{a,\mathfrak{J}} \in A'$ with the next two properties:*

$$\left(\mathfrak{W}_2(a, \bigcup \mathrm{pr}_{2,1}(\mathfrak{J})), \bigcup \left\{\mathrm{AP}((e, U), \bigcup \mathcal{V}) \mid (e, U) \in \mathrm{pr}_1(\mathfrak{J})\right\}\right)$$
$$\vdash' \{t_{a,\mathfrak{J}}, \Delta'\} \cup \mathrm{pr}_2(\mathfrak{J}) \cup \{\, t_{c,\mathfrak{K}} \mid \mathfrak{K} \subsetneq \mathfrak{J} \wedge c \sim a\,[\bigcup \mathrm{pr}_{2,1}(\mathfrak{K})] \wedge (c, \mathfrak{K})$$
$$\mathfrak{A}\text{-maximal} \,\} \cup \{\, k_{c,Z,\langle \mathfrak{G}|\mathfrak{F}\rangle} \mid (\exists L \subseteq \{1, \ldots, m\})(\exists b \in A')\{\, (e_\mu, U_\mu) \mid$$
$$\mu \in L \,\} \subseteq \mathrm{pr}_1(\mathfrak{J}) \wedge Z = \bigcup_{\mu \in L} T_\mu \wedge c \sim a\,[Z] \wedge \langle \mathfrak{G}|\mathfrak{F}\rangle \in \langle \mathcal{G}|\mathcal{F}\rangle$$
$$\wedge ((c, Z), b) \in \mathfrak{G} \wedge b \sim k_{c,Z,\langle \mathfrak{G}|\mathfrak{F}\rangle}\,[\mathrm{RS}((c, Z), \mathfrak{F})] \,\}, \tag{22}$$

$$\begin{aligned}
&\mathrm{CON}'(t_{a,\mathfrak{J}}) \ni \{\Delta'\} \cup \mathrm{pr}_2(\mathfrak{J}) \cup \{\, t_{c,\mathfrak{K}} \mid \mathfrak{K} \subsetneqq \mathfrak{J} \wedge c \sim a \left[\bigcup \mathrm{pr}_{2,1}(\mathfrak{K})\right] \\
&\quad \wedge (c,\mathfrak{K}) \ \mathfrak{A}\text{-maximal} \,\} \cup \{\, k_{c,Z,\langle \mathfrak{G}|\mathfrak{F}\rangle} \mid (\exists L \subseteq \{1,\ldots,m\})(\exists b \in A') \\
&\quad \{\, (e_\mu, U_\mu) \mid \mu \in L \,\} \subseteq \mathrm{pr}_1(\mathfrak{J}) \wedge Z = \bigcup\nolimits_{\mu \in L} T_\mu \wedge c \sim a\,[Z] \\
&\quad \wedge \langle \mathfrak{G}|\mathfrak{F}\rangle \in \langle \mathcal{G}|\mathcal{F}\rangle \wedge ((c,Z),b) \in \mathfrak{G} \wedge b \sim k_{c,Z,\langle \mathfrak{G}|\mathfrak{F}\rangle}\,[\mathrm{RS}((c,Z),\mathfrak{F})] \,\}.
\end{aligned} \tag{23}$$

With respect to the last set on the left-hand side, note that if $((c,Z),b) \in \mathfrak{G}$, then there is some $\mathfrak{L} \subseteq \mathfrak{F}$ so that $(c,\mathfrak{L})$ is $\mathfrak{F}$-maximal and $Z = \bigcup \mathrm{pr}_{2,1}(\mathfrak{L})$. It follows that $\|\mathfrak{L}\| \leq \|\mathfrak{J}\|$.

The claim is shown by induction on the cardinality κ of the subset $\mathfrak{J}$. The case $\kappa = 0$ is obvious. Set $t_{a,\emptyset} = \Delta'$, if $(a,\emptyset)$ is $\mathfrak{A}$-maximal. All other sets on the right hand side in Statement (22) are empty in this case, except the last one, which is the singleton $\{\Delta'\}$, if for some $\langle \mathfrak{G}|\mathfrak{F}\rangle \in \langle \mathcal{G}|\mathcal{F}\rangle$, $((\Delta,\emptyset),\Delta') \in \mathfrak{G}$.

Assume next that the claim holds for all subsets $\mathfrak{K} \subseteq \mathfrak{A}$ of cardinality κ and let $\mathfrak{J} \subseteq \mathfrak{A}$ with $\|\mathfrak{J}\| = \kappa + 1$. Then

$$\begin{aligned}
&\left(\mathfrak{W}_2(a, \bigcup \mathrm{pr}_{2,1}(\mathfrak{J})), \bigcup \left\{\mathrm{AP}((e,U), \bigcup \mathcal{V}) \mid (e,U) \in \mathrm{pr}_1(\mathfrak{J})\right\}\right) \\
&\quad \vdash' \{\Delta'\} \cup \mathrm{pr}_2(\mathfrak{J}) \cup \{\, t_{c,\mathfrak{K}} \mid \mathfrak{K} \subsetneqq \mathfrak{J} \wedge c \sim a \left[\bigcup \mathrm{pr}_{2,1}(\mathfrak{K})\right] \\
&\quad \wedge (c,\mathfrak{K}) \ \mathfrak{A}\text{-maximal} \,\} \\
&\quad \cup \{\, k_{c,Z,\langle \mathfrak{G}|\mathfrak{F}\rangle} \mid (\exists b \in A')(\exists L \subseteq \{1,\ldots,m\})\{\, (e_\mu,U_\mu) \mid \mu \in L \,\} \\
&\quad \subseteq \mathrm{pr}_1(\mathfrak{J}) \\
&\quad \wedge Z = \bigcup\nolimits_{\mu \in L} T_\mu \wedge c \sim a\,[Z] \wedge \langle \mathfrak{G}|\mathfrak{F}\rangle \in \langle \mathcal{G}|\mathcal{F}\rangle \wedge ((c,Z),b) \in \mathfrak{G} \\
&\quad \wedge b \sim k_{c,Z,\langle \mathfrak{G}|\mathfrak{F}\rangle}\,[\mathrm{RS}((c,Z),\mathfrak{F})] \,\},
\end{aligned}$$

because of Statement (20), Axiom 5.10(3b), and the induction hypothesis. By Condition 3.1(8), there is hence some $t_{a,\mathfrak{J}} \in A'$, so that Properties (22) and (23) hold.

With help of the claim, we can now define $\mathfrak{B}$. Let $\mathfrak{B}^{(0)}$ be as in Definition 5.7 and for $\kappa \geq 1$, set

$$\widehat{\mathfrak{B}}^{(\kappa)} = \Big\{((a, \bigcup \mathrm{pr}_{2,1}(\mathfrak{J})), t_{a,\mathfrak{J}}) \mid \mathfrak{J} \subseteq \mathfrak{A} \wedge \|\mathfrak{J}\| = \kappa \wedge \\ a \in R(\mathrm{pr}_1(\mathfrak{J})) \wedge (a,\mathfrak{J}) \ \mathfrak{A}\text{-maximal}\Big\}.$$

Obviously, $\mathfrak{B} \in \mathrm{AC}(\mathfrak{A})$. Moreover, $(\langle \mathfrak{W}|\mathfrak{V}\rangle, \langle \mathcal{W}|\mathcal{V}\rangle) \vdash_{\rightarrow} \langle \mathfrak{B}|\mathfrak{A}\rangle$: Condition 5.10(3a) follows with Statement (20), and Condition 5.10(3b) is a consequence of Statement (22). It remains to show that

$$\langle \mathcal{G}|\mathcal{F}\rangle \in \mathrm{CON}_{\rightarrow}(\langle \mathfrak{B}|\mathfrak{A}\rangle).$$

For Condition 5.10(2a), let $\langle \mathfrak{G}|\mathfrak{F}\rangle \in \langle \mathcal{G}|\mathcal{F}\rangle$ and $(a, S) \in \mathrm{CON}$. Then there is some $M \subseteq \{1, \ldots, m\}$ such that

$$\mathrm{CL}((a, S), \mathfrak{F}) = \{(a_\mu, T_\mu) \mid \mu \in M\}.$$

In addition, there are $e \in A$ and $b \in A'$ with $a \sim e\,[\bigcup_{\mu\in M} T_\mu)]$ so that $((e, \bigcup_{\mu\in M} T_\mu), b) = \mathfrak{G}(a, S)$. Moreover,

$$b \sim k_{e,\bigcup_{\mu\in M} T_\mu,\langle \mathfrak{G}|\mathfrak{F}\rangle} \left[\mathrm{RS}\left(\left(e, \bigcup\nolimits_{\mu\in M} T_\mu\right), \mathfrak{F}\right)\right],$$

that is

$$\mathfrak{G}_2(a, S) \sim k_{e,\bigcup_{\mu\in M} T_\mu,\langle \mathfrak{G}|\mathfrak{F}\rangle} [\mathrm{AP}((a, S), \mathfrak{F})].$$

Let $\mathfrak{J} \subseteq \mathfrak{A}$, such that

$$\mathrm{CL}((a, S), \mathrm{pr}_1(\mathfrak{A})) = \mathrm{pr}_1(\mathfrak{J}).$$

Then $(a, \mathfrak{J})$ is $\mathfrak{A}$-maximal. Hence, there is some $c \in A$ such that $a \sim c\,[\bigcup \mathrm{pr}_{2,1}(\mathfrak{J})]$ and $((c, \bigcup \mathrm{pr}_{2,1}(\mathfrak{J})), t_{c,\mathfrak{J}}) = \mathfrak{B}(a, S)$. Because of Statement (18), we moreover have that $\{\,(e_\mu, U_\mu) \mid \mu \in M\,\} \subseteq \mathrm{pr}_1(\mathfrak{J})$. Thus, $\{k_{e,\bigcup_{\mu\in M} T_\mu,\langle \mathfrak{G}|\mathfrak{F}\rangle}\} \in \mathrm{CON}'(\mathfrak{B}_2(a, S))$, by Statement (23).

For Condition 5.10(2b), finally, let $1 \leq \nu \leq m$. Note that by Statement (18),

$$Z_\nu \subseteq \mathrm{AP}((a_\nu, T_\nu), \mathfrak{A}).$$

With Lemma 5.17, it follows from what we have just seen that $\{\langle \mathfrak{B}|\mathfrak{A}\rangle\} \in \mathrm{CON}_{\rightarrow}(\langle \mathfrak{W}|\mathfrak{V}\rangle)$. Thus, $\mathrm{AP}((a_\nu, T_\nu), \mathfrak{A}) \in \mathrm{CON}'(\mathfrak{W}_2(a_\nu, T_\nu))$. As a consequence of Statement (17), we obtain that

$$(\mathfrak{W}_2(a_\nu, T_\nu), \mathrm{AP}((a_\nu, T_\nu), \mathfrak{A})) \vdash' d_\nu,$$

from which we gain with Lemma 5.11 that for all $(i, X) \in \mathrm{CON}$,

$$(\mathfrak{W}_2(i, X), \mathrm{AP}((i, X), \mathfrak{A})) \vdash' \mathrm{AP}\left((i, X), \bigcup \mathcal{F}\right).$$

Hence,

$$(\mathfrak{B}_2(i, X), \mathrm{AP}\left((i, X), \mathfrak{A})\right) \vdash' \mathrm{AP}((i, X), \bigcup \mathcal{F}\Big),$$

because of Lemmas 5.12 and 5.2(2). Condition 5.10(2b) now follows with Lemma 5.2(4). □

Lemma 5.22. $(A \to A', \mathrm{CON}_{\to}, \vdash_{\to}, \Delta_{\to})$ *satisfies Condition* 3.1(9).

Proof. Let $\{\langle \mathfrak{B}|\mathfrak{A}\rangle\} \in \mathrm{CON}_{\to}(\langle \mathfrak{W}|\mathfrak{V}\rangle)$ and $\langle \mathcal{G}|\mathcal{F}\rangle \in \mathrm{CON}_{\to}(\langle \mathfrak{B}|\mathfrak{A}\rangle)$. We have to show that $\langle \mathcal{G}|\mathcal{F}\rangle \in \mathrm{CON}_{\to}(\langle \mathfrak{W}|\mathfrak{V}\rangle)$.

Let $(a, S) \in \mathrm{CON}$. As $\{\langle \mathfrak{B}|\mathfrak{A}\rangle\} \in \mathrm{CON}_{\to}(\langle \mathfrak{W}|\mathfrak{V}\rangle)$, there is some $k_{\mathfrak{A}} \in A'$ so that $\{k_{\mathfrak{A}}\} \in \mathrm{CON}'(\mathfrak{W}_2(a, S))$ and $k_{\mathfrak{A}} \sim \mathfrak{B}_2(a, S)\,[\mathrm{AP}((a, S), \mathfrak{A})]$. Now, using that $\langle \mathcal{G}|\mathcal{F}\rangle \in \mathrm{CON}_{\to}(\langle \mathfrak{B}|\mathfrak{A}\rangle)$, it follows that $k_{\mathfrak{A}} \sim \mathfrak{B}_2(a, S)\,[\mathrm{AP}((a, S), \bigcup \mathcal{F})]$, from which we obtain with Lemma 5.2(3) that for every $\langle \mathfrak{G}|\mathfrak{F}\rangle \in \langle \mathcal{G}|\mathcal{F}\rangle$, $k_{\mathfrak{A}} \sim \mathfrak{B}_2(a, S)\,[\mathrm{AP}((a, S), \mathfrak{F})]$.

Moreover, there is some $k_{\mathfrak{F}} \in A'$ so that $k_{\mathfrak{F}} \sim \mathfrak{G}_2(a, S)\,[\mathrm{AP}((a, S), \mathfrak{F})]$ and $\{k_{\mathfrak{F}}\} \in \mathrm{CON}'(\mathfrak{B}_2(a, S))$. It follows that

$$\mathfrak{G}_2(a, S) \sim \mathfrak{B}_2(a, S)\,[\mathrm{AP}((a, S), \mathfrak{F})].$$

This shows that

$$\mathfrak{G}_2(a, S) \sim k_{\mathfrak{A}}\,[\mathrm{AP}((a, S), \mathfrak{A})].$$

As said, $\{k_{\mathfrak{A}}\} \in \mathrm{CON}'(\mathfrak{W}_2(a, S))$.

Thus, the first of the two conditions in Def. 5.10(2) holds. For the second one, let $r \in A'$ with $r \sim \mathfrak{W}_2(a, S)\,[\mathrm{AP}((a, S), \mathfrak{V})]$. Then

$$r \sim \mathfrak{W}_2(a, S)\,[\mathrm{AP}((a, S), \mathfrak{A})],$$

as $\{\langle \mathfrak{B}|\mathfrak{A}\rangle\} \in \mathrm{CON}_{\to}(\langle \mathfrak{W}|\mathfrak{V}\rangle)$. Since $\{k_{\mathfrak{A}}\} \in \mathrm{CON}'(\mathfrak{W}_2(a, S))$, it follows that $r \sim k_{\mathfrak{A}}\,[\mathrm{AP}((a, S), \mathfrak{A})]$, and because $k_{\mathfrak{A}} \sim \mathfrak{B}_2(a, S)\,[\mathrm{AP}((a, S), \mathfrak{A})]$, we obtain that $r \sim \mathfrak{B}_2(a, S)\,[\mathrm{AP}((a, S), \mathfrak{A})]$. Moreover,

$$\mathfrak{B}_2(a, S) \sim \mathfrak{W}_2(a, S)\,[\mathrm{AP}((a, S), \mathfrak{A})].$$

Now, note that for any $s \in A'$ with $s \sim \mathfrak{B}_2(a, S)\,[\mathrm{AP}((a, S), \mathfrak{A})]$, we have that

$$s \sim \mathfrak{B}_2(a, S)\left[\mathrm{AP}\left((a, S), \bigcup \mathcal{F}\right)\right].$$

As a consequence, we gain that $r \sim \mathfrak{W}_2(a, S)\,[\mathrm{AP}((a, S), \bigcup \mathcal{F})]$. □

Lemma 5.23. $(A \to A', \mathrm{CON}_{\to}, \vdash_{\to}, \Delta_{\to})$ *satisfies Condition* 3.1(10).

Proof. Let $\{\langle \mathfrak{B}|\mathfrak{A}\rangle\} \in \mathrm{CON}_{\rightarrow}(\langle \mathfrak{W}|\mathfrak{V}\rangle)$, $\langle \mathcal{G}|\mathcal{F}\rangle \in \mathrm{CON}_{\rightarrow}(\langle \mathfrak{B}|\mathfrak{A}\rangle)$ and $\langle \mathfrak{Z}|\mathfrak{Y}\rangle \in A \rightarrow A'$ so that $(\langle \mathfrak{B}|\mathfrak{A}\rangle, \langle \mathcal{G}|\mathcal{F}\rangle) \vdash_{\rightarrow} \langle \mathfrak{Z}|\mathfrak{Y}\rangle$. We must prove that $(\langle \mathfrak{W}|\mathfrak{V}\rangle, \langle \mathcal{G}|\mathcal{F}\rangle) \vdash_{\rightarrow} \langle \mathfrak{Z}|\mathfrak{Y}\rangle$.

Let $((c,T),b) \in \mathfrak{Y}$ and choose $k_{\mathfrak{A}}$ according to Condition 5.10(2a). Then $\mathfrak{B}_2(c,T) \sim k_{\mathfrak{A}}\,[\mathrm{AP}((c,T),\mathfrak{A})]$ and $\{k_{\mathfrak{A}}\} \in \mathrm{CON}'(\mathfrak{W}_2(c,T))$. Moreover, we have that

$$\left(\mathfrak{B}_2(c,T), \mathrm{AP}\left((c,T), \bigcup \mathcal{F}\right)\right) \vdash' b.$$

With Lemma 5.2(2), it follows that also $(k_{\mathfrak{A}}, \mathrm{AP}((c,T),\mathfrak{A})) \vdash' b$. Hence,

$$(\mathfrak{W}_2(c,T), \mathrm{AP}((c,T),\mathfrak{A})) \vdash' b,$$

by Axiom 3.1(10). The other condition follows similarly. □

The verification of Condition 3.1(11) proceeds in an analogous way.

Proposition 5.24. *Let $(A, \mathrm{CON}, \vdash, \Delta)$ and $(A', \mathrm{CON}', \vdash', \Delta')$ be information systems with witnesses satisfying Condition (ALG). Then $(A \rightarrow A', \mathrm{CON}_{\rightarrow}, \vdash_{\rightarrow}, \Delta_{\rightarrow})$ satisfies (ALG) as well.*

Proof. As follows from a slight modification of the construction in Lemma 5.19, Condition (SALG) is satisfied: Because of Condition (ALG), $(e_\nu, U_\nu) \in \mathrm{CON}$ and $(k_\nu, Z_\nu) \in \mathrm{CON}'$ can be chosen as reflexive, for each $1 \leq \nu \leq m$. Moreover, as a consequence of Statement (13), $\{k_\nu\} \in \mathrm{CON}'(\mathfrak{W}_2(e_\nu, U_\nu))$. Therefore, we have that $(\mathfrak{W}_2(e_\nu, U_\nu), Z_\nu) \vdash' (k_\nu, Z_\nu)$. By definition of $\langle \mathcal{E}|\mathcal{D}\rangle$, $Z_\nu \subseteq \mathrm{AP}((e_\nu, U_\nu), \bigcup \mathcal{D})$. Hence,

$$\left(\mathfrak{W}_2(e_\nu, U_\nu), \mathrm{AP}\left((e_\nu, U_\nu), \bigcup \mathcal{D}\right)\right) \vdash' \{k_\nu\} \cup Z_\nu,$$

which implies that $(\langle \mathfrak{W}|\mathfrak{V}\rangle, \langle \mathcal{E}|\mathcal{D}\rangle) \vdash_{\rightarrow} \langle \mathcal{E}|\mathcal{D}\rangle$. □

Condition (BC) as well is inherited to the function space.

Proposition 5.25. *Let $(A, \mathrm{CON}, \vdash, \Delta)$ and $(A', \mathrm{CON}', \vdash', \Delta')$ be information systems with witnesses. If $(A', \mathrm{CON}', \vdash', \Delta')$ satisfies Condition (BC), so does $(A \rightarrow A', \mathrm{CON}_{\rightarrow}, \vdash_{\rightarrow}, \Delta_{\rightarrow})$.*

Proof. Let $\langle \mathfrak{B}|\mathfrak{A}\rangle, \langle \mathfrak{W}|\mathfrak{V}\rangle \in A \to A'$ and $\langle \mathcal{G}|\mathcal{F}\rangle \in \mathrm{CON}_{\to}(\langle \mathfrak{B}|\mathfrak{A}\rangle) \cap \mathrm{CON}_{\to}(\langle \mathfrak{W}|\mathfrak{V}\rangle)$. We need to show that for any $\langle \mathfrak{I}|\mathfrak{H}\rangle \in A \to A'$,

$$(\langle \mathfrak{B}|\mathfrak{A}\rangle, \langle \mathcal{G}|\mathcal{F}\rangle) \vdash_{\to} \langle \mathfrak{I}|\mathfrak{H}\rangle \iff (\langle \mathfrak{W}|\mathfrak{V}\rangle, \langle \mathcal{G}|\mathcal{F}\rangle) \vdash_{\to} \langle \mathfrak{I}|\mathfrak{H}\rangle. \tag{24}$$

As a consequence of Axiom 5.10(2b), we have that for any $(a, S) \in \mathrm{CON}$,

$$\mathrm{AP}\left((a, S), \bigcup \mathcal{F}\right) \in \mathrm{CON}'(\mathfrak{B}_2(a, S)) \cap \mathrm{CON}'(\mathfrak{W}_2(a, S)).$$

Then also

$$\bigcup \left\{ \mathrm{AP}\left((e, Y), \bigcup \mathcal{F}\right) \mid (e, Y) \in \mathrm{DS}((a, S), \mathfrak{H}) \right\} \\ \in \mathrm{CON}'(\mathfrak{B}_2(a, S)) \cap \mathrm{CON}'(\mathfrak{W}_2(a, S)).$$

Because of Condition (BC), it follows for all $((e, Y), f) \in \mathfrak{H}$ that

$$\left(\mathfrak{B}_2(e, Y), \mathrm{AP}\left((e, Y), \bigcup \mathcal{F}\right)\right) \vdash' f \\ \iff \left(\mathfrak{W}_2(e, Y), \mathrm{AP}\left((e, Y), \bigcup \mathcal{F}\right)\right) \vdash' f,$$

and for all $((c, Z), b) \in \mathfrak{I})$ and $k \in A'$ that

$$\left(\mathfrak{B}_2(c, Z), \bigcup \left\{\mathrm{AP}((e, Y), \bigcup \mathcal{F}) \mid (e, Y) \in \mathrm{DS}((c, Z), \mathfrak{H})\right\}\right) \\ \vdash' (k, \mathrm{RS}((c, Z), \mathfrak{H})) \\ \Updownarrow \\ \left(\mathfrak{W}_2(c, Z), \bigcup \left\{\mathrm{AP}((e, Y), \bigcup \mathcal{F}) \mid (e, Y) \in \mathrm{DS}((c, Z), \mathfrak{H})\right\}\right) \\ \vdash' (k, \mathrm{RS}((c, Z), \mathfrak{H})),$$

which implies the Statement (24). □

Let $f \in |A \to A'|$. Then $f \subseteq A \to A'$, i.e., f is a set of pairs in $\mathcal{P}(\mathrm{CON} \times A') \times \mathcal{P}_f(\mathrm{CON} \times A')$ such that the first component is an associate of the second. We will now show that the states of $A \to A'$ correspond to the approximable mappings between A and A' in a one-to-one way.

Lemma 5.26. *For* $f \in |A \to A'|$, *let* $\mathrm{AM}(f) = \bigcup \mathrm{pr}_2(f)$. *Then* $\mathrm{AM}(f) : A \Vdash A'$.

Proof. In order to show that $\mathrm{AM}(f)$ is an approximable mapping, we need to verify Conditions (4.1)(1–9).

(1) Is a consequence of the fact that because of Axiom 3.1(3) and Condition 3.5(2), $\Delta_{\to}$ is contained in every state.

(2) Let $i \in A$, $X, X' \in \mathrm{CON}(i)$ and $b \in A'$ so that $X \subseteq X'$ and $(i, X)\,\mathrm{AM}(f)b$. We need to show that $(i, X')\,\mathrm{AM}(f)b$.

Since $(i, X)\,\mathrm{AM}(f)b$, there is some $\langle\mathfrak{B}|\mathfrak{A}\rangle \in f$ with $((i, X), b) \in \mathfrak{A}$. Because of Condition 3.5(3), it follows that $(\langle\mathfrak{W}|\mathfrak{V}\rangle, \langle\mathcal{W}|\mathcal{V}\rangle) \vdash_{\to} \langle\mathfrak{B}|\mathfrak{A}\rangle$, for some $\langle\mathfrak{W}|\mathfrak{V}\rangle \in f$ and $\langle\mathcal{W}|\mathcal{V}\rangle \subseteq_{\mathrm{fin}} f$ with $\langle\mathcal{W}|\mathcal{V}\rangle \in \mathrm{CON}_{\to}(\langle\mathfrak{W}|\mathfrak{V}\rangle)$. Thus, $(\mathfrak{W}_2(i, X), \mathrm{AP}((i, X), \bigcup\mathcal{V})) \vdash' b$. By Lemma 5.9(2), we have that $\{\mathfrak{W}_2(i, X)\} \in \mathrm{CON}'(\mathfrak{W}_2(i, X'))$. Moreover, $\mathrm{AP}((i, X), \bigcup\mathcal{V}) \subseteq \mathrm{AP}((i, X'), \bigcup\mathcal{V})$. Therefore,

$$\left(\mathfrak{W}_2(i, X'), \mathrm{AP}\left((i, X'), \bigcup\mathcal{V}\right)\right) \vdash' b. \tag{25}$$

Because of Condition 3.1(3), we also obtain that

$$\left(\mathfrak{W}_2(i, X'), \mathrm{AP}\left((i, X'), \bigcup\mathcal{V}\right)\right) \vdash' \Delta'.$$

Consequently, by Axiom 3.1(8), there is some $e \in A'$ with

$$\left(\mathfrak{W}_2(i, X'), \mathrm{AP}\left((i, X'), \bigcup\mathcal{V}\right)\right) \vdash' e,$$

and $\{b, \Delta'\} \in \mathrm{CON}'(e)$.

Now, set $\mathfrak{D} = \{((i, X'), b)\}$ and let $\mathfrak{E} \in \mathrm{AC}(\mathfrak{D}$ be as in Lemma 5.8. Then Statement (25) means that $(\langle\mathfrak{W}|\mathfrak{V}\rangle, \langle\mathcal{W}|\mathcal{V}\rangle) \vdash_{\to} \langle\mathfrak{E}|\mathfrak{D}\rangle$: for Condition 5.10(3b) choose $k = e$ or $k = \Delta'$, respectively. By Condition 3.5(2), it follows that $\langle\mathfrak{E}|\mathfrak{D}\rangle \in f$. Hence, $(i, X')\,\mathrm{AM}(f)b$.

(3) and (8) follow similarly.

(5) Let $(i, X) \in \mathrm{CON}$, $(k, Y) \in \mathrm{CON}'$ and $b \in A'$ so that $(i, X)\,\mathrm{AM}(f)(k, Y)$ and $(k, Y) \vdash' b$. We must show that $(i, X)\,\mathrm{AM}(f)b$.

Since $(i, X)\,\mathrm{AM}(f)(k, Y)$, we have again that for all $c \in \{k\} \cup Y$ there are tokens $\langle\mathfrak{B}_c|\mathfrak{A}_c\rangle \in f$ so that $((i, X), c) \in \mathfrak{A}_c$. Let $\langle\mathcal{G}|\mathcal{F}\rangle = \{\,\langle\mathfrak{B}_c|\mathfrak{A})_c\rangle \mid c \in \{k\} \cup Y\,\}$. Because of Condition (ST), there exist $\langle\mathfrak{W}|\mathfrak{V}\rangle \in f$ and $\langle\mathcal{W}|\mathcal{V}\rangle \subseteq_{\mathrm{fin}} f$ so that $(\langle\mathfrak{W}|\mathfrak{V}\rangle, \langle\mathcal{W}|\mathcal{V}\rangle) \vdash_{\to} \langle\mathcal{G}|\mathcal{F}\rangle$. It follows that

$$\left(\mathfrak{W}_2(i, X), \mathrm{AP}\left((i, X), \bigcup\mathcal{V}\right)\right) \vdash' (k, Y).$$

Since $(k, Y) \vdash' b$, we have that $(\mathfrak{W}_2(i, X), \mathrm{AP}((i, X), \bigcup\mathcal{V})) \vdash' b$, from which it follows as above that $(i, X)\,\mathrm{AM}(f)b$.

It remains to verify Conditions 4.1(4, 6, 7, 9). Because of Lemmas 4.2, 4.3 and 4.4, it suffices to verify Requirement (3) instead.

Let $(i, X) \in \mathrm{CON}$ and $F \subseteq_{\mathrm{fin}} A'$ with $(i, X)\,\mathrm{AM}(f)F$. We have to show that there are $(c, U) \in \mathrm{CON}$ and $(e, V) \in \mathrm{CON}'$ so that $(i, X) \vdash (c, U)$, $(c, U)\,\mathrm{AM}(f)(e, V)$ and $(e, V) \vdash' F$.

As we have already seen, there are $\langle \mathfrak{W}|\mathfrak{V}\rangle \in f$ and $\langle \mathcal{W}|\mathcal{V}\rangle \subseteq_{\mathrm{fin}} f$ with

$$\left(\mathfrak{W}_2(i, X), \mathrm{AP}\left((i, X), \bigcup \mathcal{V}\right)\right) \vdash' F.$$

Moreover, by the Global Interpolation Property (GIP), there exists $(c, U) \in \mathrm{CON}$ such that $(i, X) \vdash (c, U)$ and $(c, U) \vdash \overline{\mathrm{CL}}((i, X), \mathrm{pr}_1(\bigcup \mathcal{V}))$. It follows with Lemma 5.9(3) that also $(\mathfrak{W}_2(c, U), \mathrm{AP}((c, U), \bigcup \mathcal{V})) \vdash' F$. Apply the Global Interpolation Property again to obtain some $(e, V) \in \mathrm{CON}'$ so that $(\mathfrak{W}_2(c, U), \mathrm{AP}((c, U), \bigcup \mathcal{V})) \vdash' (e, V)$ and $(e, V) \vdash' F$. Let $V \cup \{e\} = \{b_1, \ldots, b_n\}$, and for $1 \leq \nu \leq n$, set $\mathfrak{D}_\nu = \{((c, U), b_\nu)\}$. Moreover, let $\mathfrak{E}_\nu \in \mathrm{AC}(\mathfrak{D}_\nu)$ be as in Lemma 5.8. Then $(\langle \mathfrak{W}|\mathfrak{V}\rangle, \langle \mathcal{W}|\mathcal{V}\rangle) \vdash_\rightarrow \langle \mathfrak{E}_\nu|\mathfrak{D}_\nu\rangle$. It follows that $\langle \mathfrak{E}_\nu|\mathfrak{D}_\nu\rangle \in f$ and hence that $(c, U)\,\mathrm{AM}(f)(e, V)$. □

Lemma 5.27. *For $H\colon A \Vdash A'$, let $\mathrm{ST}(H)$ be the set of all $\langle \mathfrak{B}|\mathfrak{A}\rangle \in A \rightarrow A'$ such that the following two conditions hold:*

(1) $\mathfrak{A} \subseteq H$
(2) $(\forall ((a, Z), b) \in \mathfrak{B})(\exists j \in A')\,(a, Z)Hj \wedge j \sim b\,[\mathrm{RS}((a, Z), \mathfrak{A})]$.

Then $\mathrm{ST}(H) \in |A \rightarrow A'|$.

Proof. In order to verify that $\mathrm{ST}(H)$ is a state of $|A \rightarrow A'|$, we check Conditions 3.5(2) and (ST).

3.5(2) Let $\langle \mathfrak{W}|\mathfrak{V}\rangle, \langle \mathfrak{B}|\mathfrak{A}\rangle \in \mathrm{CON}_\rightarrow$ and $\langle \mathcal{W}|\mathcal{V}\rangle \subseteq_{\mathrm{fin}} \mathrm{CON}_\rightarrow(\langle \mathfrak{W}|\mathfrak{V}\rangle)$ such that $\{\langle \mathfrak{W}|\mathfrak{V}\rangle\} \cup \langle \mathcal{W}|\mathcal{V}\rangle \subseteq \mathrm{ST}(H)$ and $(\langle \mathfrak{W}|\mathfrak{V}\rangle, \langle \mathcal{W}|\mathcal{V}\rangle) \vdash_\rightarrow \langle \mathfrak{B}|\mathfrak{A}\rangle$. We must show that $\langle \mathfrak{B}|\mathfrak{A}\rangle \in \mathrm{ST}(H)$.

For the first requirement, let $((c, U), d) \in \mathfrak{A}$. Then, we have that

$$\left(\mathfrak{W}_2(c, U), \mathrm{AP}\left((c, U), \bigcup \mathcal{V}\right)\right) \vdash' d.$$

By assumption, $\langle \mathfrak{W}|\mathfrak{V}\rangle \in \mathrm{ST}(H)$. Thus, there is some $j \in A'$ such that $(c, U)Hj$ and $j \sim \mathfrak{W}_2(c, U)\,[\mathrm{AP}((c, U), \mathfrak{V})]$. Since $\langle \mathcal{W}|\mathcal{V}\rangle \in$

$\mathrm{CON}_{\rightarrow}(\langle\mathfrak{W}|\mathfrak{V}\rangle)$, we conclude that $j \sim \mathfrak{W}_2(c,U)\,[\mathrm{AP}((c,U),\bigcup\mathcal{V})]$. With Lemma 5.2(2), we thus obtain that also

$$\left(j, \mathrm{AP}\left((c,U),\bigcup\mathcal{V}\right)\right) \vdash' d.$$

Now, let $b \in \mathrm{AP}((c,U),\bigcup\mathcal{V})$. Then there is $(e,Y) \in \mathrm{CON}$ so that $((e,Y),b) \in \bigcup\mathcal{V}$ and for some $e' \in A$ with $e \sim e'\,[Y]$, $(c,U) \vdash (e',Y)$. As $\bigcup\mathcal{V} \subseteq H$, it follows with Lemma 5.3 that $(e',Y)Hb$ and, hence, with Lemma 4.5 that $(c,U)Hb$. This shows that

$$\left(c,U\right)H(j,\mathrm{AP}\left((c,U),\bigcup\mathcal{V}\right)).$$

Consequently, by Condition 4.1(5), $(c,U)Hd$.

For the second requirement let $((a,Z),b) \in \mathfrak{B}$. By assumption, $(\langle\mathfrak{W}|\mathfrak{V}\rangle, \langle\mathcal{W}|\mathcal{V}\rangle) \vdash_{\rightarrow} \langle\mathfrak{B}|\mathfrak{A}\rangle$. Hence, there is some $k \in A'$ such that $b \sim k\,[\mathrm{RS}((a,Z),\mathfrak{A})]$ and

$$\left(\mathfrak{W}_2(a,Z), \mathrm{AP}\left((a,Z),\bigcup\mathcal{V}\right)\right) \vdash' (k, \mathrm{RS}((a,Z),\mathfrak{A})),$$

from which it follows as above that $(a,Z)Hk$.

(ST) Let $\langle\mathcal{G}|\mathcal{F}\rangle \subseteq_{\mathrm{fin}} \mathrm{ST}(H)$. We need to show that there are $\langle\mathfrak{W}|\mathfrak{V}\rangle \in \mathrm{ST}(H)$ and $\langle\mathcal{W}|\mathcal{V}\rangle \subseteq_{\mathrm{fin}} \mathrm{ST}(H)$ with $\langle\mathcal{W}|\mathcal{V}\rangle \in \mathrm{CON}_{\rightarrow}(\langle\mathfrak{W}|\mathfrak{V}\rangle)$ so that $(\langle\mathfrak{W}|\mathfrak{V}\rangle, \langle\mathcal{W}|\mathcal{V}\rangle) \vdash_{\rightarrow} \langle\mathcal{G}|\mathcal{F}\rangle$. The construction proceeds as in Lemma 5.20. Instead of Condition 5.10(3b), however, we now make use of Requirement 5.27(2).

Let $\bigcup\mathcal{F} = \{\,((c_\nu,T_\nu),e_\nu) \mid 1 \le \nu \le n\,\}$. Then $(c_\nu,T_\nu)He_\nu$. By Rule (3), there are $(i_\nu,Y_\nu) \in \mathrm{CON}$ and $(j_\nu,Z_\nu) \in \mathrm{CON}'$ with

$$(c_\nu,T_\nu) \vdash (i_\nu,Y_\nu), \tag{26}$$

$$(i_\nu,Y_\nu)H(j_\nu,Z_\nu), \tag{27}$$

$$(j_\nu,Z_\nu) \vdash' e_\nu. \tag{28}$$

Set

$$\mathfrak{V} = \{((i_\nu,Y_\nu),d) \mid d \in Z_\nu \wedge 1 \le \nu \le n\}.$$

Then $\mathfrak{V} \subseteq H$. We will next construct $\mathfrak{W} \in \mathrm{AC}(\mathfrak{V})$ such that $\langle\mathfrak{W}|\mathfrak{V}\rangle \in \mathrm{ST}(H)$ and $(\langle\mathfrak{W}\ \mathfrak{V}\rangle, \{\langle\mathfrak{W}\ \mathfrak{V}\rangle\}) \vdash_{\rightarrow} \langle\mathcal{G}|\mathcal{F}\rangle$. If $\bigcup\mathcal{F}$ is empty, set $\langle\mathfrak{W}|\mathfrak{V}\rangle = \Delta_{\rightarrow}$.

Let us now assume that $\bigcup\mathcal{F}$ is nonempty and that, for each $J \subseteq \{1,\ldots,n\}$, $R(\{\,(i_\nu,Y_\nu) \mid \nu \in J\,\})$ is a system of representatives of

$W(\{ (i_\nu, Y_\nu) \mid \nu \in J \})$ with respect to $\sim [\bigcup_{\nu \in J} Y_\nu]$ so that $R(\emptyset) = \{\Delta\}$ and $R(\{(i_\nu, Y_\nu)\}) = \{i_\nu\}$, for $1 \le \nu \le n$. By Condition 5.27(2), there is some $j_{a,Z,\langle\mathfrak{G}|\mathfrak{F}\rangle} \in A'$, for each $\langle\mathfrak{G}|\mathfrak{F}\rangle \in \langle\mathcal{G}|\mathcal{F}\rangle$ and $((a, Z), b) \in \mathfrak{G}$, with $j_{a,Z,\langle\mathfrak{G}|\mathfrak{F}\rangle} \sim b\,[\mathrm{RS}((a, Z), \mathfrak{F})]$ and $(a, Z) H j_{a,Z,\langle\mathfrak{G}|\mathfrak{F}\rangle}$. In case $((c, Z), b) = ((\Delta, \emptyset), \Delta')$, choose $j_{\Delta,\emptyset,\langle\mathfrak{G}|\mathfrak{F}\rangle} = \Delta'$.

Claim 5.28. *For every $\mathfrak{J} \subseteq \mathfrak{V}$ and each $a \in R(\mathrm{pr}_1(\mathfrak{J}))$ so that $(a, \mathfrak{J})$ is $\mathfrak{V}$-maximal, there is some $t_{a,\mathfrak{J}} \in A'$ with the following two properties:*

$$\begin{aligned}
&\{\Delta'\} \cup \bigcup \{\{k_\nu\} \cup Z_\nu \mid 1 \le \nu \le n \wedge (i_\nu, Y_\nu) \in \mathrm{pr}_1(\mathfrak{J})\}\\
&\quad \cup \left\{t_{c,\mathfrak{K}} \mid \mathfrak{K} \subsetneqq \mathfrak{J} \wedge c \sim a \left[\bigcup \mathrm{pr}_{2,1}(\mathfrak{K})\right] \wedge (c, \mathfrak{K})\ \mathfrak{V}\text{-maximal}\right\}\\
&\quad \cup \Big\{ j_{c,Z,\langle\mathfrak{G}|\mathfrak{F}\rangle} \mid\\
&\quad (\exists L \subseteq \{1, \ldots, n\})(\exists b \in A')\, \{(i_\mu, Y_\mu) \mid \mu \in L\} \subseteq \mathrm{pr}_1(\mathfrak{J}) \wedge Z\\
&\quad = \bigcup\nolimits_{\mu \in L} T_\mu \wedge c \sim a\,[Z] \wedge \langle\mathfrak{G}|\mathfrak{F}\rangle \in \langle\mathcal{G}|\mathcal{F}\rangle \wedge ((c, Z), b) \in \mathfrak{G}\\
&\quad \wedge b \sim j_{c,Z,\langle\mathfrak{G}|\mathfrak{F}\rangle}\,[\mathrm{RS}((c, Z), \mathfrak{F})]\Big\} \in \mathrm{CON}'(t_{a,\mathfrak{J}}),
\end{aligned} \tag{29}$$

$$\begin{aligned}
&\left(a, \bigcup \mathrm{pr}_{2,1}(\mathfrak{J})\right) H (t_{a,\mathfrak{J}}, \{\Delta'\} \cup \bigcup \Big\{\{k_\nu\} \cup Z_\nu \mid 1 \le \nu \le n \wedge (i_\nu, Y_\nu)\\
&\quad \in \mathrm{pr}_1(\mathfrak{J})\Big\} \cup \left\{t_{c,\mathfrak{K}} \mid \mathfrak{K} \subsetneqq \mathfrak{J} \wedge c \sim a \left[\bigcup \mathrm{pr}_{2,1}(\mathfrak{K})\right] \wedge (c, \mathfrak{K})\ \mathfrak{V}\text{-maximal}\right\}\\
&\quad \cup \Big\{j_{c,Z,\langle\mathfrak{G}|\mathfrak{F}\rangle} \mid (\exists L \subseteq \{1, \ldots, n\})(\exists b \in A')\{ (i_\mu, Y_\mu) \mid \mu \in L \}\\
&\quad \subseteq \mathrm{pr}_1(\mathfrak{J}) \wedge Z\\
&\quad = \bigcup\nolimits_{\mu \in L} T_\mu \wedge c \sim a\,[Z] \wedge \langle\mathfrak{G}|\mathfrak{F}\rangle \in \langle\mathcal{G}|\mathcal{F}\rangle \wedge ((c, Z), b) \in \mathfrak{G}\\
&\quad \wedge b \sim j_{c,Z,\langle\mathfrak{G}|\mathfrak{F}\rangle}\,[\mathrm{RS}((c, Z), \mathfrak{F})]\Big\}\Big).
\end{aligned} \tag{30}$$

With respect to the last set on the right-hand side, note that if $((c, Z), b) \in \mathfrak{G}$, then there is some $\mathfrak{L} \subseteq \mathfrak{F}$ so that $(c, \mathfrak{L})$ is $\mathfrak{F}$-maximal and $Z = \bigcup \mathrm{pr}_{2,1}(\mathfrak{L})$. It follows that $\|\mathfrak{L}\| \le \|\mathfrak{J}\|$.

The claim is shown by induction on the cardinality κ of the subset $\mathfrak{J}$. The case $\kappa = 0$ is obvious. Set $t_{a,\emptyset} = \Delta'$, if $(a, \emptyset)$ is $\mathfrak{V}$-maximal. All other sets on the left-hand side in Statement (29) are empty in this case, except the last one, which is the singleton $\{\Delta'\}$, if for some $\langle\mathfrak{G}|\mathfrak{F}\rangle \in \langle\mathcal{G}|\mathcal{F}\rangle$, $((\Delta, \emptyset), \Delta') \in \mathfrak{G}$.

Assume next that the claim holds for all subsets $\mathfrak{K} \subseteq \mathfrak{V}$ of cardinality κ and let $\mathfrak{J} \subseteq \mathfrak{V}$ with $\|\mathfrak{J}\| = \kappa + 1$. Then

$$\begin{aligned}
&\Big(a, \bigcup \mathrm{pr}_{2,1}(\mathfrak{J})\Big) H\{\Delta'\} \cup \bigcup \{\{k_\nu\} \cup Z_\nu \mid 1 \leq \nu \leq n \wedge (i_\nu, Y_\nu) \in \mathrm{pr}_1(\mathfrak{J})\} \\
&\quad \cup \{\, t_{c,\mathfrak{K}} \mid \mathfrak{K} \subsetneq \mathfrak{J} \wedge c \sim a \Big[\bigcup \mathrm{pr}_{2,1}(\mathfrak{K})\Big] \wedge (c, \mathfrak{K}) \text{ } \mathfrak{V}\text{-maximal} \,\} \\
&\quad \cup \{\, j_{c,Z,\langle \mathfrak{G} | \mathfrak{F} \rangle} \mid \\
&\quad (\exists L \subseteq \{1, \ldots, n\})(\exists b \in A') \{(i_\mu, Y_\mu) \mid \mu \in L\} \subseteq \mathrm{pr}_1(\mathfrak{J}) \wedge Z \\
&\quad = \bigcup\nolimits_{\mu \in L} T_\mu \wedge c \sim a\,[Z] \wedge \langle \mathfrak{G} | \mathfrak{F} \rangle \in \langle \mathcal{G} | \mathcal{F} \rangle \wedge ((c, Z), b) \in \mathfrak{G} \\
&\quad\quad \wedge b \sim j_{c,Z,\langle \mathfrak{G} | \mathfrak{F} \rangle}\,[\mathrm{RS}((c, Z), \mathfrak{F})] \,\},
\end{aligned}$$

because of Statement (27), Condition 5.27(2) and the induction hypothesis. By Condition 4.1(7), there is hence some $t_{a,\mathfrak{J}} \in A'$ so that Properties (29) and (30) hold.

With the help of Claim 5.28, we can now define $\mathfrak{W}$. Let $\mathfrak{W}^{(0)}$ be as in Definition 5.7, and for $\kappa \geq 1$, set

$$\begin{aligned}
\widehat{\mathfrak{W}}^{(\kappa)} = \Big\{ \Big(\Big(a, \bigcup \mathrm{pr}_{2,1}(\mathfrak{J}), t_{a,\mathfrak{J}}\Big) \mid \mathfrak{J} \subseteq \mathfrak{V} \\
\wedge \|\mathfrak{J}\| = \kappa \wedge a \in R(\mathrm{pr}_1(\mathfrak{J})) \wedge (a, \mathfrak{J}) \text{ } \mathfrak{V}\text{-maximal} \Big\}.
\end{aligned}$$

Obviously, $\mathfrak{W} \in \mathrm{AC}(\mathfrak{V})$.

Claim 5.29. $\langle \mathfrak{W} | \mathfrak{V} \rangle \in St(H)$.

By definition, $\mathfrak{V} \subseteq H$. For Requirement 5.27(2), let $(a, Z), b) \in \mathfrak{W}$. Then there is some subset $\mathfrak{J} \subseteq \mathfrak{V}$ so that $(a, \mathfrak{J})$ is $\mathfrak{V}$-maximal, $a \in R(\mathrm{pr}_1(\mathfrak{J}))$, $Z = \bigcup \mathrm{pr}_{2,1}(\mathfrak{J})$ and $b = t_{a,\mathfrak{J}}$. Hence, $\mathrm{pr}_2(\mathfrak{J}) \in \mathrm{CON}'(b)$ and $(a, Z)Hb$. Note that $\mathrm{pr}_2(\mathfrak{J}) = \mathrm{RS}((a, Z), \mathfrak{V})]$. Thus, $b \sim b\,[\mathrm{RS}((a, Z), \mathfrak{J})]$.

Claim 5.30. $(\langle \mathfrak{W} | \mathfrak{V} \rangle, \{\langle \mathfrak{W} | \mathfrak{V} \rangle\}) \vdash_{\rightarrow} \langle \mathcal{G} | \mathcal{F} \rangle$.

Let $1 \leq \nu \leq n$, Then $Z_\nu \subseteq \mathrm{AP}((c_\nu, T_\nu), \mathfrak{V})$ and $\{k_\nu\} \in \mathrm{CON}'(\mathfrak{W}_2(c_\nu, T_\nu))$. As a consequence of Statement (28), we therefore have that $(\mathfrak{W}_2(c_\nu, T_\nu), \mathrm{AP}((c_\nu, T_\nu), \mathfrak{V})) \vdash' e_\nu$, which means that Condition 5.10(3a) holds.

For Condition 5.10(3b), let $\langle \mathfrak{G} | \mathfrak{F} \rangle \in \langle \mathcal{G} | \mathcal{F} \rangle$ and $((a, Z), b) \in \mathfrak{G}$. As a consequence of Condition 5.10(3a), we have that $(\mathfrak{W}_2(a, Z),$

$\mathrm{AP}((a,Z),\mathfrak{V})) \vdash' \mathrm{RS}((a,Z),\mathfrak{F})$. Because of Condition 3.1(8), there is thus some $k \in A'$ so that $(\mathfrak{W}_2(a,Z), \mathrm{AP}((a,Z),\mathfrak{V})) \vdash' k$ and

$$\mathrm{RS}((a,Z),\mathfrak{F}) \in \mathrm{CON}'(k). \tag{31}$$

It follows that

$$\{k\} \in \mathrm{CON}'(\mathfrak{W}_2(a,Z)). \tag{32}$$

Let $\mathfrak{J} = \{ ((i,Y),d) \in \mathfrak{V} \mid (i,Y) \in \mathrm{CL}((a,Z),\mathrm{pr}_1(\mathfrak{V})) \}$ and $\mathfrak{W}(a,Z) = ((e,U),t)$. Then $(e,\mathfrak{J})$ is $\mathfrak{V}$-maximal, $e \sim a\,[U]$, $U = \bigcup \mathrm{pr}_{2,1}(\mathfrak{J})$, and $\mathfrak{W}_2(a,Z) = t_{e,\mathfrak{J}}$. Since $((a,Z),b) \in \mathfrak{G}$, there is some $\mathfrak{K} \subseteq \mathfrak{F}$ such that $Z = \bigcup \mathrm{pr}_{2,1}(\mathfrak{K})$ and $\{ (i_\mu, Y_\mu) \mid 1 \le \mu \le n \wedge (c_\mu, T_\mu) \in \mathrm{pr}_1(\mathfrak{K}) \} \subseteq \mathrm{pr}_1(\mathfrak{J})$. It follows that

$$\{j_{a,Z,\langle \mathfrak{G}|\mathfrak{F}\rangle}\} \in \mathrm{CON}'(\mathfrak{W}_2(a,Z)). \tag{33}$$

Moreover,

$$b \sim j_{a,Z,\langle \mathfrak{G}|\mathfrak{F}\rangle}\,[\mathrm{RS}((a,Z),\mathfrak{F})], \tag{34}$$

since $\langle \mathfrak{G}|\mathfrak{F}\rangle \in \mathrm{ST}(H)$, by assumption. As a consequence of Statements (31–34), we obtain that $b \sim k\,[\mathrm{RS}((a,Z),\mathfrak{F})]$, as was to be shown. □

In the next two lemmas, we verify that the operators AM and ST between $|A \to A'|$ and the set of all approximable mappings between A and A' are inverse to each other.

Lemma 5.31. *For $f \in |A \to A'|$, $\mathrm{ST}(\mathrm{AM}(f)) = f$.*

Proof. Let $\langle \mathfrak{B}|\mathfrak{A}\rangle \in f$. Then $\mathfrak{A} \subseteq \bigcup \mathrm{pr}_2(f)$. Thus, $\mathfrak{A} \subseteq \mathrm{AM}(f)$, i.e., Condition 5.27(1) holds.

For Condition 5.27(2), let $((a,Z),b) \in \mathfrak{B}$. Since $\langle \mathfrak{B}|\mathfrak{A}\rangle \in f$, there are $\langle \mathfrak{W}|\mathfrak{V}\rangle \in f$ and $\langle \mathcal{W}|\mathcal{V}\rangle \subseteq_{\mathrm{fin}} f$ with $\langle \mathcal{W}|\mathcal{V}\rangle \in \mathrm{CON}_{\rightarrow}(\langle \mathfrak{W}|\mathfrak{V}\rangle)$ and $(\langle \mathfrak{W}|\mathfrak{V}\rangle, \langle \mathcal{W}|\mathcal{V}\rangle) \vdash_{\rightarrow} \langle \mathfrak{B}|\mathfrak{A}\rangle$. It follows by Condition 5.10(3b) that there is some $k \in A'$ such that

$$k \sim b\,[\mathrm{RS}((a,Z),\mathfrak{A})] \text{ and } \left(\mathfrak{W}_2(a,Z), \mathrm{AP}\left((a,Z),\bigcup \mathcal{V}\right)\right) \vdash' (k, \mathrm{RS}((a,Z),\mathfrak{A})). \tag{35}$$

Let $((i,X),c) \in \bigcup \mathcal{V}$ such that $(i,X) \in \mathrm{CL}((a,Z),\mathrm{pr}_1(\bigcup \mathcal{V}))$. Then $(i,X)\,\mathrm{AM}(f)c$, as $\mathcal{V} \subseteq \mathrm{pr}_2(f)$ and, hence, $\bigcup \mathcal{V} \subseteq \bigcup \mathrm{pr}_2(\mathcal{V}) = \mathrm{AM}(f)$.

So, we have that

$$(a, Z)\,\mathrm{AM}(f)\,\mathrm{AP}\left((a, Z), \bigcup \mathcal{V}\right). \tag{36}$$

By Condition 4.1(7), there is now some $j \in A'$ such that

$$(a, Z)\,\mathrm{AM}(f)j, \tag{37}$$

and $\mathrm{AP}((a, Z), \bigcup \mathcal{V}) \in \mathrm{CON}'(j)$. Furthermore, with Lemma 4.2, we obtain some $(c, U) \in \mathrm{CON}$ such that $(a, Z) \vdash (c, U)$ and $(c, U)\,\mathrm{AM}(f)j$. It follows that there is some $\langle \mathfrak{E} | \mathfrak{D} \rangle \in f$ with $((c, U), j) \in \mathfrak{D}$. Moreover, since f is a state, there exists $\langle \mathfrak{G} | \mathfrak{F} \rangle \in f$ with $\{\langle \mathfrak{W} | \mathfrak{V} \rangle, \langle \mathfrak{E} | \mathfrak{D} \rangle\}\,\mathrm{CON}_{\rightarrow}(\langle \mathfrak{G} | \mathfrak{F} \rangle)$. Because of Condition 5.10(2a), there are thus $e_{\langle \mathfrak{W} | \mathfrak{V} \rangle}, e_{\langle \mathfrak{E} | \mathfrak{D} \rangle} \in A'$ so that

$$\mathfrak{W}_2(a, Z) \sim e_{\langle \mathfrak{W} | \mathfrak{V} \rangle}\,[\mathrm{AP}((a, Z), \mathfrak{V})] \text{ and } \{e_{\langle \mathfrak{W} | \mathfrak{V} \rangle}\} \in \mathrm{CON}'(\mathfrak{G}_2(a, Z)), \tag{38}$$

$$\mathfrak{E}_2(a, Z) \sim e_{\langle \mathfrak{E} | \mathfrak{D} \rangle}\,[\mathrm{AP}((a, Z), \mathfrak{D}] \text{ and } \{e_{\langle \mathfrak{E} | \mathfrak{D} \rangle}\} \in \mathrm{CON}'(\mathfrak{G}_2(a, Z)). \tag{39}$$

Note that by construction, $j \in \mathrm{AP}((a, Z), \mathfrak{D})$. With Lemma 5.2(3), we hence obtain from the first of Statements (39) that $\mathfrak{E}_2(a, Z) \sim e_{\langle \mathfrak{E} | \mathfrak{D} \rangle}\,[\{j\}]$. Since $\mathrm{AP}((a, Z), \bigcup \mathcal{V}) \in \mathrm{CON}'(j)$, it follows that $\mathfrak{E}_2(a, Z) \sim e_{\langle \mathfrak{E} | \mathfrak{D} \rangle}\,[\mathrm{AP}((a, Z), \bigcup \mathcal{V})]$.

With Axiom 5.10(2b), we obtain from the first of Statements (38) that also

$$\mathfrak{W}_2(a, Z) \sim e_{\langle \mathfrak{W} | \mathfrak{V} \rangle}\left[\mathrm{AP}\left((a, Z), \bigcup \mathcal{V}\right)\right].$$

So, since $\{e_{\langle \mathfrak{E} | \mathfrak{D} \rangle}\}, \{e_{\langle \mathfrak{W} | \mathfrak{V} \rangle}\} \in \mathrm{CON}'(\mathfrak{G}_2(a, Z))$, we have that

$$\mathfrak{E}_2(a, Z) \sim \mathfrak{W}_2(a, Z)\left[\mathrm{AP}\left((a, Z), \bigcup \mathcal{V}\right)\right].$$

Finally, as $\{j\} \in \mathrm{CON}'(\mathfrak{E}_2(a, Z))$ by Axiom 5.7(3a), we find that $j \sim \mathfrak{W}_2(a, Z)\ \mathrm{AP}((a, Z), [\bigcup \mathcal{V})]$.

Because of Statement (35), it follows that also

$$\left(j, \mathrm{AP}\left((a, Z), \bigcup \mathcal{V}\right)\right) \vdash' (k, \mathrm{RS}((a, Z), \mathfrak{A})). \tag{40}$$

Statements (36, 37, 40) now imply that $(a, Z)\,\mathrm{AM}(f)k$. Moreover, if follows with Statement (40) that $\{k\} \in \mathrm{CON}'(j)$. With Statement (35), we thus have that $j \sim b\,[\mathrm{RS}((a, Z), \mathfrak{A})]$. This shows that $\langle \mathfrak{B} | \mathfrak{A} \rangle \in \mathrm{ST}(\mathrm{AM}(f))$.

For the converse inclusion, let $\langle\mathfrak{B}|\mathfrak{A}\rangle \in \mathrm{ST}(\mathrm{AM}(f))$. Then $\mathfrak{A} \subseteq \mathrm{AM}(f)$. Let $\mathfrak{A} = \{\, ((c_\nu, T_\nu), e_\nu) \mid 1 \le \nu \le n \,\}$. As $((c_\nu, T_\nu), e_\nu) \in \mathrm{AM}(f)$, there is some $\langle\mathfrak{E}^{(\nu)}|\mathfrak{D}^{(\nu)}\rangle \in f$ with $((c_\nu, T_\nu), e_\nu) \in \mathfrak{D}^{(\nu)}$, for each $1 \le \nu \le n$. Because of Condition (ST), there are furthermore some $\langle\mathfrak{W}|\mathfrak{V}\rangle \in f$ and a subset $\langle\mathcal{W}|\mathcal{V}\rangle \subseteq_{\mathrm{fin}} f$ with $(\langle\mathfrak{W}|\mathfrak{V}\rangle, \langle\mathcal{W}|\mathcal{V}\rangle) \vdash_{\rightarrow} \{\, \langle\mathfrak{E}^{(\nu)}|\mathfrak{D}^{(\nu)}\rangle \mid 1 \le \nu \le n \,\}$. Hence, we have for every $1 \le \nu \le n$ that

$$\left(\mathfrak{W}_2(c_\nu, T_\nu), \mathrm{AP}\left((c_\nu, T_\nu), \bigcup \mathcal{V}\right)\right) \vdash' e_\nu. \tag{41}$$

Let $((a, Z), b) \in \mathfrak{B}$. Then it follows that

$$\left(\mathfrak{W}_2(a, Z), \bigcup \left\{\mathrm{AP}((c, T), \bigcup \mathcal{V}) \mid (c, T) \in \mathrm{DS}((a, Z), \mathfrak{A})\right\}\right)$$
$$\vdash' \mathrm{RS}((a, Z), \mathfrak{A}).$$

Because of Condition 3.1(8), there is thus some $k \in A'$ with $\mathrm{RS}((a, Z), \mathfrak{A}) \in \mathrm{CON}'(k)$ and

$$\left(\mathfrak{W}_2(a, Z), \bigcup \left\{\mathrm{AP}((c, T), \bigcup \mathcal{V}) \mid (c, T) \in \mathrm{DS}((a, Z), \mathfrak{A})\right\}\right)$$
$$\vdash' (k, \mathrm{RS}((a, Z), \mathfrak{A})).$$

Hence, $\{k\} \in \mathrm{CON}'(\mathfrak{W}_2(a, Z))$. By Requirement 5.27(2), we furthermore obtain some $j \in A'$ with $j \sim b\,[\mathrm{RS}((a, Z), \mathfrak{A})]$ and $(a, Z)\,\mathrm{AM}(f) j$. As above, it follows that

$$\left(\mathfrak{W}_2(a, Z), \mathrm{AP}\left((a, Z), \bigcup \mathcal{V}\right)\right) \vdash' j.$$

Thus, $\{j\} \in \mathrm{CON}'(\mathfrak{W}_2(a, Z))$ as well. Consequently, $b \sim k\,[\mathrm{RS}((a, Z), \mathfrak{A})]$. With Statement (41), we now obtain that

$$(\langle\mathfrak{W}|\mathfrak{V}\rangle, \langle\mathcal{W}|\mathcal{V}\rangle) \vdash_{\rightarrow} \langle\mathfrak{B}|\mathfrak{A}\rangle.$$

Therefore, $\langle\mathfrak{B}|\mathfrak{A}\rangle \in f$. □

Lemma 5.32. *For $H\colon A \Vdash A'$, $\mathrm{AM}(\mathrm{ST}(H)) = H$.*

Proof. Let $((a, S), b) \in \mathrm{AM}(\mathrm{ST}(H))$. Then $((a, S), b) \in \bigcup \mathrm{pr}_2(\mathrm{ST}(H))$. Thus, there is some $\langle\mathfrak{B}|\mathfrak{A}\rangle \in \mathrm{ST}(H)$ with $((a, S), b) \in \mathfrak{A}$. Since $\mathfrak{A} \subseteq H$, it follows that $((a, S), b) \in H$.

For the converse implication, let $((c,T),e) \in H$. As $(\Delta,\emptyset)H\Delta'$, we have that

$$(c,T)H\{\Delta',e\}.$$

Hence, there is some $d \in A'$ with $(c,T)Hd$ and $\{\Delta',e\} \in \mathrm{CON}'(d)$. Set $\mathfrak{D} = \{((c,T),e)\}$ and let $\mathfrak{E} \in \mathrm{AC}(\mathfrak{D})$ be as in Lemma 5.8. Then $\mathfrak{D} \subseteq H$. Next, let $((a,Z),b) \in \mathfrak{E}$. Then either $((a,Z),b) = ((c,T),d)$ or $((a,Z),b) = ((\Delta,\emptyset),\Delta')$. In the latter case, T is nonempty. Therefore, $\mathrm{RS}((a,Z),\mathfrak{D}) = \emptyset$. Choose $j = \Delta'$ in this case. In the other case, $\mathrm{RS}((a,Z),\mathfrak{D}) = \{d\}$. Now, choose $j = d$. Then, we have in both cases that $j \sim b\,[\mathrm{RS}((a,Z),\mathfrak{D})]$ and $(a,Z)Hj$. This shows that $\langle\mathfrak{E}|\mathfrak{D}\rangle \in \mathrm{ST}(H)$. Thus, $\mathfrak{D} \subseteq \bigcup \mathrm{pr}_2(\mathrm{ST}(H))$, which means that $((c,T),e) \in \mathrm{AM}(\mathrm{ST}(H))$. □

Proposition 5.33. *Let $(A,\mathrm{CON},\vdash,\Delta)$ and $(A',\mathrm{CON}',\vdash',\Delta')$ be information systems with witnesses. Then the states of $A \to A'$ are in a one-to-one correspondence with the approximable mappings between A and A':*

(1) *$\{\bigcup \mathrm{pr}_2(f) \mid f \in |A \to A'|\}$ is the set of all approximable mappings between A and A'.*
(2) *$|A \to A'|$ is the collection of all sets $\{\langle\mathfrak{B}|\mathfrak{A}\rangle \in A \to A' \mid \mathfrak{A} \subseteq H \wedge (\forall((a,Z),b) \in \mathfrak{B})(\exists j \in A')(a,Z)Hj \wedge j \sim b\,[\mathrm{RS}((a,Z),\mathfrak{A})]\}$, where H is an approximable mapping between A and A'.*

In Ref. [1], it was shown how approximable mappings between two information systems with witnesses A and A' as well as Scott continuous functions from $|A|$ to $|A'|$ correspond with each other. As we will show next, this correspondence establishes an isomorphism between the domains $|A \to A'|$ and $[|A| \to |A'|]$.

For $G\colon A \Vdash A'$ and $x \in |A|$, let

$$\mathcal{L}(G)(x) = \{\, b \in A' \mid (\exists(i,X) \in \mathrm{CON})\{i\} \cup X \subseteq x \wedge (i,X)Gb \,\}.$$

Then $\mathcal{L}(G) \in [|A| \to |A'|]$. Since for $f \in |A \to A'|$, $\mathrm{AM}(f)\colon A \Vdash A'$, it follows that $\mathcal{L}(\mathrm{AM}(f)) \in [|A| \to |A'|]$. Set

$$\mathrm{fct}(f) = \mathcal{L}(\mathrm{AM}(f)).$$

Then we have for $x \in |A|$ that

$$\begin{aligned}
&\mathrm{fct}(f)(x) \\
&\quad = \{\, b \in A' \mid (\exists (i,X) \in \mathrm{CON})\{i\} \cup X \subseteq x \wedge (i,X)\,\mathrm{AM}(f) b \,\} \\
&\quad = \{\, b \in A' \mid (\exists (i,X) \in \mathrm{CON})(\exists \langle \mathfrak{B} | \mathfrak{A} \rangle \in f)\{i\} \cup X \subseteq x \\
&\qquad \wedge ((i,X), b) \in \mathfrak{A} \,\} \\
&\quad = \{\, b \in A' \mid (\exists (i,X) \in \mathrm{CON})(\exists \langle \mathfrak{B} | \mathfrak{A} \rangle, \langle \mathfrak{W} | \mathfrak{V} \rangle \in A \to A') \\
&\qquad\quad (\exists \langle \mathcal{W} | \mathcal{V} \rangle \in \mathrm{CON}_{\to}(\langle \mathfrak{W} | \mathfrak{V} \rangle))\{\langle \mathfrak{W} | \mathfrak{V} \rangle\} \cup \langle \mathcal{W} | \mathcal{V} \rangle \subseteq f \\
&\qquad \wedge \{i\} \cup X \subseteq x \wedge \\
&\qquad\quad (\langle \mathfrak{W} | \mathfrak{V} \rangle, \langle \mathcal{W} | \mathcal{V} \rangle) \vdash_{\to} \langle \mathfrak{B} | \mathfrak{A} \rangle \wedge ((i,X), b) \in \mathfrak{A} \,\} \\
&\quad = \{\, b \in A' \mid \{\,\}(\exists (i,X) \in \mathrm{CON})(\exists \langle \mathfrak{B} | \mathfrak{A} \rangle, \langle \mathfrak{W} | \mathfrak{V} \rangle \in A \to A') \\
&\qquad\quad (\exists \langle \mathcal{W} | \mathcal{V} \rangle \in \mathrm{CON}_{\to}(\langle \mathfrak{W} | \mathfrak{V} \rangle))\{\langle \mathfrak{W} | \mathfrak{V} \rangle\} \cup \langle \mathcal{W} | \mathcal{V} \rangle \subseteq f \\
&\qquad \wedge \{i\} \cup X \subseteq x \wedge \\
&\qquad\quad \left(\mathfrak{B}_2(i,X), \mathrm{AP}\left((i,X), \bigcup \mathcal{V}\right)\right) \vdash' b \Big\},
\end{aligned}$$

where the last equality follows as in the second part of the proof of Lemma 5.32. As is now easily seen, $\mathrm{fct} \in [|A \to A'| \to [|A| \to |A'|]]$.

Conversely, let $g \in [|A| \to |A'|]$, and for $(i,X) \in \mathrm{CON}$ and $b \in A'$, define

$$(i,X) H^g b \Longleftrightarrow b \in g([X]_i).$$

Then $H^g \colon A \Vdash A'$. Set

$$\mathrm{st}(g) = \mathrm{ST}(H^g).$$

By Lemma 5.27, $\mathrm{st}(g) \in |A \to A'|$.

Lemma 5.34. *The function* $\mathrm{st} \colon [|A| \to |A'|] \to |A \to A'|$ *is Scott continuous.*

Proof. Obviously, st is monotone. Let $\mathbb{G} \subseteq [|A| \to |A'|]$ be directed and $\langle \mathfrak{B} | \mathfrak{A} \rangle \in \mathrm{st}(\bigsqcup \mathbb{G})$. Then $\mathfrak{A} \subseteq H^{\bigsqcup \mathbb{G}}$. Since $\mathfrak{A}$ is finite and $\mathbb{G}$ directed, we gain some $g \in \mathbb{G}$ so that $\mathfrak{A} \subseteq H^g$.

Now, let $((a,Z), b) \in \mathfrak{B}$. Then $(a,Z) H^g \, \mathrm{RS}((a,Z), \mathfrak{A})$. Because of Condition 4.1(9), there is hence some $j \in A'$ with $(a,Z) H^g j$ and $\mathrm{RS}((a,Z), \mathfrak{A}) \in \mathrm{CON}'(j)$. It follows that $(a,Z) H^{\bigsqcup \mathbb{G}} j$ as well. Since $\langle \mathfrak{B} | \mathfrak{A} \rangle \in \mathrm{st}(\bigsqcup \mathbb{G})$, there is also some $k \in A'$ with $(a,Z) H^{\bigsqcup \mathbb{G}} k$

and $k \sim b\,[\mathrm{RS}((a,Z),\mathfrak{A})]$. Again by Condition 4.1(9), there is some $r \in A'$ with $(a,Z)H^{\bigsqcup \mathbb{G}}r$ and $\{j,k\} \in \mathrm{CON}'(r)$. Consequently, $j \sim b\,[\mathrm{RS}((a,Z),\mathfrak{A})]$, which shows that $\langle \mathfrak{B}|\mathfrak{A}\rangle \in \mathrm{st}(g)$. The converse inclusion follows by monotonicity. □

As shown in Ref. [1], $\mathcal{L}(H^g) = g$ and $H^{\mathcal{L}(G)} = G$. With Lemmas 5.31 and 5.32, we therefore obtain that the two functions fct and st are inverse to each other.

Proposition 5.35. *Let* $(A, \mathrm{CON}, \vdash, \Delta)$ *and* $(A', \mathrm{CON}', \vdash', \Delta')$ *be information systems with witnesses. Then the domains* $|A \to A'|$ *and* $[|A| \to |A'|]$ *are isomorphic.*

6. Cartesian Closure

As was shown in Ref. [1], the category **ISW**, as well as its full subcategories **aISW**, **bcISW** and **abcISW**, possess a terminal object and are closed under taking finite products.

The one-point information system with witnesses $T = (\{\Delta\}, \mathrm{CON}_T, \vdash_T, \Delta)$ with $\mathrm{CON}_T = \{(\Delta,\emptyset),(\Delta,\{\Delta\})\}$ and $\vdash_T = \mathrm{CON}_T \times \{\Delta\}$ is a terminal object.

For $\nu = 1, 2$, let $(A_\nu, \mathrm{CON}_\nu, \vdash_\nu, \Delta_\nu)$ be an information system with witnesses. Set $A_\times = A_1 \times A_2$, $\Delta_\times = (\Delta_1, \Delta_2)$,

$$\mathrm{CON}_\times = \{\, ((i,j),X) \in A_\times \times \mathcal{P}_f(A_\times) \mid \\ \mathrm{pr}_1(X) \in \mathrm{CON}_1(i) \wedge \mathrm{pr}_2(X) \in \mathrm{CON}_2(j)\,\},$$

and for $((i,j),X) \in \mathrm{CON}_\times$ and $(a_1,a_2) \in A_\times$ define

$$((i,j),X) \vdash_\times (a_1,a_2) \Longleftrightarrow (i,\mathrm{pr}_1(X)) \vdash_1 a_1 \wedge (j,\mathrm{pr}_2(X)) \vdash_2 a_2.$$

Moreover, let the relations $\mathrm{Pr}_\nu \subseteq \mathrm{CON}_\times \times A_\nu$, with $\nu = 1, 2$, be given by

$$((i_1,i_2),X)\,\mathrm{Pr}_\nu\, a_\nu \Longleftrightarrow (i_\nu, \mathrm{pr}_\nu(X)) \vdash_\nu a_\nu.$$

Then $(A_\times, \mathrm{Pr}_1, \mathrm{Pr}_2)$ is the categorical product of A_1 and A_2.

The aim of this section is to show for information systems with witnesses A and A' that $|A \to A'|$ is the exponent of A and

A' in the category **ISW**. For $a \in A$, $\langle \mathfrak{B}|\mathfrak{A}\rangle \in A \to A'$, $\mathbb{Z} \in \mathrm{CON}_{(A\to A')\times A}(\langle \mathfrak{B}|\mathfrak{A}\rangle, a)$, and $b \in A'$, let

$$(((\langle \mathfrak{B}|\mathfrak{A}\rangle, a), \mathbb{Z}) \,\mathrm{EV}\, b \iff \Big(\mathfrak{B}_2(a, \mathrm{pr}_2(\mathbb{Z})), \mathrm{AP}\Big((a, \mathrm{pr}_2(\mathbb{Z})), \bigcup \mathrm{pr}_{2,1}(\mathbb{Z})\Big)\Big) \vdash' b.$$

Lemma 6.1. $\mathrm{EV}\colon (A \to A') \times A \Vdash A'$.

Proof. We have to verify the conditions in Def. 4.1. In case of Condition 4.1(1), we have to verify that

$$((\Delta_{\to}, \Delta), \emptyset) \,\mathrm{EV}\, \Delta',$$

i.e., we have to check whether $(\Delta', \mathrm{AP}((\Delta, \emptyset), \emptyset)) \vdash' \Delta'$, which holds by Axiom 3.1(3). Note that $\mathrm{AP}((\Delta, \emptyset), \emptyset)$ is empty.

For Condition 4.1(2), let $\mathbb{Z}, \mathbb{Z}' \in \mathrm{CON}_{(A\to A')\times A}(\langle \mathfrak{B}|\mathfrak{A}\rangle, a))$ with $\mathbb{Z} \subseteq \mathbb{Z}'$. Then $\mathrm{CL}((a, \mathrm{pr}_2(\mathbb{Z}), \mathfrak{A})) \subseteq \mathrm{CL}((a, \mathrm{pr}_2(\mathbb{Z}')), \mathfrak{A})$. Thus,

$$\{\mathfrak{B}_2(a, \mathrm{pr}_2(\mathbb{Z}))\} \in \mathrm{CON}'(\mathfrak{B}_2(a, \mathrm{pr}_2(\mathbb{Z}'))).$$

Moreover,

$$\mathrm{AP}\Big((a, \mathrm{pr}_2(\mathbb{Z})), \bigcup \mathrm{pr}_{2,1}(\mathbb{Z})) \subseteq \mathrm{AP}((a, \mathrm{pr}_2(\mathbb{Z})), \bigcup \mathrm{pr}_{2,1}(\mathbb{Z})\Big).$$

Therefore, the condition follows with Axioms 3.1(5, 9).

For Condition 4.1(3), assume that

$$(((\langle \mathfrak{B}|\mathfrak{A}\rangle, a), \mathbb{Z}) \vdash_{(A\to A')\times A} \mathbb{T},$$

and $(((\langle \mathfrak{B}|\mathfrak{A}\rangle, a), \mathbb{T}) \,\mathrm{EV}\, b$. Then we have that

$$(\langle \mathfrak{B}|\mathfrak{A}\rangle, \mathrm{pr}_1(\mathbb{Z})) \vdash_{\to} \mathrm{pr}_1(\mathbb{T}), \tag{42}$$

$$(a, \mathrm{pr}_2(\mathbb{Z})) \vdash \mathrm{pr}_2(\mathbb{T}), \tag{43}$$

$$\Big(\mathfrak{B}_2(a, \mathrm{pr}_2(\mathbb{T})), \mathrm{AP}((a, \mathrm{pr}_2(\mathbb{T})), \bigcup \mathrm{pr}_{2,1}(\mathbb{T}))\Big) \vdash' b. \tag{44}$$

As a consequence of Statement (43),

$$\{\mathfrak{B}_2(a, \mathrm{pr}_2(\mathbb{T}))\} \in \mathrm{CON}(\mathfrak{B}_2(a, \mathrm{pr}_2(\mathbb{Z}))),$$

and hence,

$$\Big(\mathfrak{B}_2(a, \mathrm{pr}_2(\mathbb{Z})), \mathrm{AP}((a, \mathrm{pr}_2(\mathbb{T})), \bigcup \mathrm{pr}_{2,1}(\mathbb{T}))\Big) \vdash' b.$$

Because of Statements (42) and (43), it moreover follows that

$$\mathrm{AP}\left((a, \mathrm{pr}_2(\mathbb{T})), \bigcup \mathrm{pr}_{2,1}(\mathbb{T})\right) \subseteq \mathrm{AP}\left((a, \mathrm{pr}_2(\mathbb{Z})), \bigcup \mathrm{pr}_{2,1}(\mathbb{Z})\right).$$

Therefore, $\left(\mathfrak{B}_2(a, \mathrm{pr}_2(\mathbb{Z})), \mathrm{AP}((a, \mathrm{pr}_2(\mathbb{Z})), \bigcup \mathrm{pr}_{2,1}(\mathbb{Z}))\right) \vdash' b$, i.e.,

$$(((\langle\mathfrak{B}|\mathfrak{A}\rangle, a), \mathbb{Z})\, \mathrm{EV}\, b.$$

For Conditions 4.1(4, 6, 7, 9), we make use of Lemmas 4.2, 4.3 and 4.4, and verify Requirement (3) instead. Suppose to this end that $(((\langle\mathfrak{B}|\mathfrak{A}\rangle, a), \mathbb{Z})\, \mathrm{EV}\, F$. Then

$$\left(\mathfrak{B}_2(a, \mathrm{pr}_2(\mathbb{Z})), \mathrm{AP}((a, \mathrm{pr}_2(\mathbb{Z})), \bigcup \mathrm{pr}_{2,1}(\mathbb{Z}))\right) \vdash' F.$$

Because of the Global Interpolation Property there is some $(j, Y) \in \mathrm{CON}'$ so that

$$\left(\mathfrak{B}_2(a, \mathrm{pr}_2(\mathbb{Z})), \mathrm{AP}((a, \mathrm{pr}_2(\mathbb{Z})), \bigcup \mathrm{pr}_{2,1}(\mathbb{Z}))\right) \vdash' (j, Y), \tag{45}$$

$$(j, Y) \vdash F. \tag{46}$$

By definition, we moreover have that

$$(a, \mathrm{pr}_2(\mathbb{Z})) \vdash \overline{\mathrm{CL}}\left((a, \mathrm{pr}_2(\mathbb{Z})), \bigcup \mathrm{pr}_{2,1}(\mathbb{Z})\right).$$

Hence, there exists $(i, X) \in \mathrm{CON}$ with

$$(a, \mathrm{pr}_2(\mathbb{Z})) \vdash (i, X), \tag{47}$$

$$(i, X) \vdash \overline{\mathrm{CL}}\left((a, \mathrm{pr}_2(\mathbb{Z})), \bigcup \mathrm{pr}_{2,1}(\mathbb{Z})\right). \tag{48}$$

It follows that $\mathrm{CL}((i, X), \mathrm{pr}_1(\mathfrak{A})) \subseteq \mathrm{CL}((a, \mathrm{pr}_2(\mathbb{Z})), \mathrm{pr}_1(\mathfrak{A}))$ and thus that $\{\mathfrak{B}_2(i, X)\} \in \mathrm{CON}'(\mathfrak{B}_2(a, \mathrm{pr}_2(\mathbb{Z})))$. Moreover, $\mathrm{AP}((i, X), \bigcup \mathrm{pr}_{2,1}(\mathbb{Z})) = \mathrm{AP}((a, \mathrm{pr}_2(\mathbb{Z})), \bigcup \mathrm{pr}_{2,1}(\mathbb{Z}))$. So, we gain that

$$\left(\mathfrak{B}_2(i, X), \mathrm{AP}((i, X), \bigcup \mathrm{pr}_{2,1}(\mathbb{Z}))\right) \vdash' Y \cup \{j\}. \tag{49}$$

By Axiom 3.1(8), there is thus some $\bar{\jmath} \in A'$ such that

$$\left(\mathfrak{B}_2(i, X), \mathrm{AP}((i, X), \bigcup \mathrm{pr}_{2,1}(\mathbb{Z}))\right) \vdash' \bar{\jmath},$$

and $Y \cup \{j, \Delta'\} \in \mathrm{CON}'(\bar{\jmath})$.

Let $\mathfrak{D} = \{\, ((i,X),e) \mid e \in Y \cup \{j\} \,\}$ and $\mathfrak{E} \in \mathrm{AC}(\mathfrak{D})$ be as in Lemma 5.8. Then we have for $((d,V),b) \in \mathfrak{E}$ that either $((d,V),b) = ((i,X),\bar{\jmath})$, $\mathrm{DS}((d,V),\mathfrak{D}) = \{(i,X)\}$ and $\mathrm{RS}((d,V),\mathfrak{D}) = Y \cup \{j\}$, or $((d,V),b) = ((\Delta,\emptyset),\Delta')$, X is not empty, and $\mathrm{DS}((d,V),\mathfrak{D})$ and $\mathrm{RS}((d,V),\mathfrak{D})$ are both empty. It follows that in either case

$$\Big(\mathfrak{B}_2(d,V), \bigcup\Big\{\, \mathrm{AP}((c,U), \bigcup \mathrm{pr}_{2,1}(\mathbb{Z})) \mid (c,U) \in \mathrm{DS}((d,V),\mathfrak{D}) \,\Big\}\Big) \\ \vdash' (b, \mathrm{RS}((c,U),\mathfrak{D})).$$

With Statement (49), we hence obtain that $(\langle\mathfrak{B}|\mathfrak{A}\rangle, \mathrm{pr}_1(\mathbb{Z})) \vdash_{\rightarrow} \langle\mathfrak{E}|\mathfrak{D}\rangle$. By the Global Interpolation Property there are now $\langle\mathfrak{W}|\mathfrak{V}\rangle \in A \to A'$ and $\langle\mathcal{W}|\mathcal{V}\rangle \in \mathrm{CON}_{\rightarrow}(\langle\mathfrak{W}|\mathfrak{V}\rangle)$, such that

$$(\langle\mathfrak{B}|\mathfrak{A}\rangle, \mathrm{pr}_1(\mathbb{Z})) \vdash_{\rightarrow} (\langle\mathfrak{W}|\mathfrak{V}\rangle, \langle\mathcal{W}|\mathcal{V}\rangle) \text{ and } (\langle\mathfrak{W}|\mathfrak{V}\rangle, \langle\mathcal{W}|\mathcal{V}\rangle) \vdash_{\rightarrow} \langle\mathfrak{E}|\mathfrak{D}\rangle.$$

In particular, we have that

$$\Big(\mathfrak{W}_2(i,X), \mathrm{AP}\Big((i,X), \bigcup \mathcal{V}\Big)\Big) \vdash' (j,Y).$$

Set $\mathbb{T} = \langle\mathcal{W}|\mathcal{V}\rangle \times X$. Then

$$\mathbb{T} \in \mathrm{CON}_{(A \to A') \times A}(\langle\mathfrak{W}|\mathfrak{V}\rangle, i), \\ ((\langle\mathfrak{B}|\mathfrak{A}\rangle, a), \mathbb{Z}) \vdash_{(A \to A') \times A} ((\langle\mathfrak{W}|\mathfrak{V}\rangle, i), \mathbb{T}), \\ ((\langle\mathfrak{W}|\mathfrak{V}\rangle, i), \mathbb{T})\, \mathrm{EV}(j,Y),$$

and

$$(j,Y) \vdash' F.$$

Condition 4.1(5) is a consequence of Lemma 3.4.

For Condition 4.1(8), let

$$\{(\langle\mathfrak{B}|\mathfrak{A}\rangle, j)\} \in \mathrm{CON}_{(A \to A') \times A}(\langle\mathfrak{W}|\mathfrak{V}\rangle, i),$$

$\mathbb{Z} \in \mathrm{CON}_{(A \to A') \times A}(\langle\mathfrak{B}|\mathfrak{A}\rangle, j)$, and $((\langle\mathfrak{B}|\mathfrak{A}\rangle, j), \mathbb{Z})\, \mathrm{EV}\, b$. Then we have that

$$\Big(\mathfrak{B}_2(j, \mathrm{pr}_2(\mathbb{Z})), \mathrm{AP}\Big((j, \mathrm{pr}_2(\mathbb{Z})), \bigcup \mathrm{pr}_{2,1}(\mathbb{Z})\Big)\Big) \vdash' b,$$

$\mathrm{pr}_1(\mathbb{Z}) \in \mathrm{CON}_{\rightarrow}(\langle\mathfrak{B}|\mathfrak{A}\rangle)$, $\mathrm{pr}_2(\mathbb{Z}) \in \mathrm{CON}(j)$, $\{\langle\mathfrak{B}|\mathfrak{A}\rangle\} \in \mathrm{CON}_{\rightarrow}(\langle\mathfrak{W}|\mathfrak{V}\rangle)$, and $\{j\} \in \mathrm{CON}(i)$, from which it follows that $\mathrm{pr}_1(\mathbb{Z}) \in$

$\mathrm{CON}_{\to}(\langle\mathfrak{W}|\mathfrak{V}\rangle)$ and $\mathrm{pr}_2(\mathbb{Z}) \in \mathrm{CON}(i)$. Thus, $\mathbb{Z} \in \mathrm{CON}_{(A\to A')\times A}(\langle\mathfrak{W}|\mathfrak{V}\rangle, i)$.

We have to show that

$$\left(\mathfrak{W}_2(i, \mathrm{pr}_2(\mathbb{Z})), \mathrm{AP}((i, \mathrm{pr}_2(\mathbb{Z})), \bigcup \mathrm{pr}_{2,1}(\mathbb{Z}))\right) \vdash' b. \tag{50}$$

Note that because of Axioms 3.1(10, 11),

$$\mathrm{AP}\left((j, \mathrm{pr}_2(\mathbb{Z})), \bigcup \mathrm{pr}_{2,1}(\mathbb{Z})\right) = \mathrm{AP}\left((i, \mathrm{pr}_2(\mathbb{Z})), \bigcup \mathrm{pr}_{2,1}(\mathbb{Z})\right).$$

Similarly, $\mathrm{CL}((j, \mathrm{pr}_2(\mathbb{Z})), \mathrm{pr}_1(\mathfrak{A})) = \mathrm{CL}((i, \mathrm{pr}_2(\mathbb{Z})), \mathrm{pr}_1(\mathfrak{A}))$ which implies that

$$\mathfrak{B}_2(j, \mathrm{pr}_2(\mathbb{Z})) = \mathfrak{B}_2(i, \mathrm{pr}_2(\mathbb{Z})).$$

Thus,

$$\left(\mathfrak{B}_2(i, \mathrm{pr}_2(\mathbb{Z})), \mathrm{AP}((i, \mathrm{pr}_2(\mathbb{Z})), \bigcup \mathrm{pr}_{2,1}(\mathbb{Z}))\right) \vdash' b.$$

Since $\langle\mathfrak{B}|\mathfrak{A}\rangle \in \mathrm{CON}_{\to}(\langle\mathfrak{W}|\mathfrak{V}\rangle)$, there is some $r \in A'$ with

$$r \sim \mathfrak{B}_2(i, \mathrm{pr}_2(\mathbb{Z}))\, [\mathrm{AP}((i, \mathrm{pr}_2(\mathbb{Z})), \mathfrak{A})] \text{ and } \{r\} \in \mathrm{CON}'(\mathfrak{W}_2(i, \mathrm{pr}_2(\mathbb{Z}))). \tag{51}$$

By assumption $\mathrm{pr}_1(\mathbb{Z}) \in \mathrm{CON}_{\to}(\langle\mathfrak{B}|\mathfrak{A}\rangle)$. With Axiom 5.10(2b) it follows that

$$r \sim \mathfrak{B}_2(i, \mathrm{pr}_2(\mathbb{Z}))\left[\mathrm{AP}((i, \mathrm{pr}_2(\mathbb{Z})), \bigcup \mathrm{pr}_{2,1}(\mathbb{Z}))\right]$$

as well. Therefore, also

$$\left(r, \mathrm{AP}((i, \mathrm{pr}_2(\mathbb{Z})), \bigcup \mathrm{pr}_{2,1}(\mathbb{Z}))\right) \vdash' b.$$

With Statements (51), Statement (50) is now a consequence. □

Let $(A'', \mathrm{CON}'', \vdash'', \Delta'')$ be a further information system with witnesses.

Definition 6.2. For $H\colon A \times A' \Vdash A''$, $(a, S) \in \mathrm{CON}$ and $\langle\mathfrak{B}|\mathfrak{A}\rangle \in A' \to A''$ define

$$(a, S)\Lambda(H)\langle\mathfrak{B}|\mathfrak{A}\rangle$$

if

(1) $(\forall((i,X),e) \in \mathfrak{A})((a,i), S \times X)He$
(2) $(\forall((c,T),b) \in \mathfrak{B})(\exists j \in A'')((a,c), S \times T)Hj \wedge j \sim b$[RS $((c,T),\mathfrak{A})$].

Lemma 6.3. $\Lambda(H)\colon A \Vdash A' \to A''$.

Proof. Again, we have to verify the conditions in Def. 4.1.

For Condition 4.1(1), we need to show that $(\Delta,\emptyset)\Lambda(H)\Delta_{\to}$. Condition 6.2(1) holds vacuously. In case of Condition 6.2(2), it only need be shown that $((\Delta,\Delta'),\emptyset)H\Delta''$, which is valid by Axiom 4.1(1). The remaining condition follows with Lemma 5.2(1).

Condition 4.1(2) is obviously satisfied, as it holds for H.

For Condition 4.1(3), assume that $(a,S) \vdash S'$. Moreover, let $(i,X) \in \mathrm{CON}'$ and $c \in A''$ such that $((a,i), S' \times X)Hc$. By Lemma 4.2, there is some $((\bar{a},\bar{\imath}),U) \in \mathrm{CON}_{A\times A'}$ such that

$$((a,i), S' \times X) \vdash_{A\times A'} ((\bar{a},\bar{\imath}),U),$$

and $((\bar{a},\bar{\imath}),U)Hc$. It follows that $(a,S') \vdash (\bar{a},\mathrm{pr}_1(U))$. Hence, $(a,S) \vdash (\bar{a},\mathrm{pr}_1(U))$ and thus, $((a,i), S \times X) \vdash_{A\times A'} ((\bar{a},\bar{\imath}),U)$. Therefore, $((a,i), S \times X)Hc$. It is now an easy consequence that from $(a,S')\Lambda(H)\langle\mathfrak{E}|\mathfrak{D}\rangle$ one obtains $(a,S)\Lambda(H)\langle\mathfrak{E}|\mathfrak{D}\rangle$.

As in the preceding proof, we verify Requirement (3) instead of Conditions 4.1(4, 6, 7, 9). Suppose that

$$(a,S)\Lambda(H)\langle\mathcal{G}|\mathcal{F}\rangle.$$

We have to show that there exists $(\bar{\imath},\overline{U}) \in \mathrm{CON}$ and $(\langle\mathfrak{B}|\mathfrak{A}\rangle,\langle\mathcal{W}|\mathcal{V}\rangle) \in \mathrm{CON}_{A'\to A''}$ so that

$$(a,S) \vdash (\bar{\imath},\overline{U}),$$
$$(\bar{\imath},\overline{U})\Lambda(H)(\langle\mathfrak{B}|\mathfrak{A}\rangle,\langle\mathcal{W}|\mathcal{V}\rangle),$$
$$(\langle\mathfrak{B}|\mathfrak{A}\rangle,\langle\mathcal{W}|\mathcal{V}\rangle) \vdash_{A'\to A''} \langle\mathcal{G}|\mathcal{F}\rangle.$$

If $\bigcup\mathcal{F}$ is empty, let $(\bar{\imath},\overline{U}) = (\Delta,\emptyset)$, $\langle\mathfrak{B}|\mathfrak{A}\rangle = \Delta_{A'\to A''}$ and $\langle\mathcal{W}|\mathcal{V}\rangle = \{\Delta_{A'\to A''}\}$. Otherwise, assume that $\bigcup\mathcal{F} = \{\,((i_\nu,X_\nu),e_\nu) \mid 1 \le \nu \le n\,\}$. Then we have for all $1 \le \nu \le n$ that $((a,i_\nu), S \times X_\nu)He_\nu$.

Thus, for each such ν there are $((\bar{a}_\nu, \bar{\imath}_\nu), M_\nu) \in \mathrm{CON}_{A\times A'}$ and $(\hat{\jmath}_\nu, \widehat{N}_\nu) \in \mathrm{CON}''$ with

$$((a, i_\nu), S \times X_\nu) \vdash_{A\times A'} ((\bar{a}_\nu, \bar{\imath}_\nu), M_\nu), \tag{52}$$

$$((\bar{a}_\nu, \bar{\imath}_\nu), M_\nu) H (\hat{\jmath}_\nu, \widehat{N}_\nu), \tag{53}$$

$$(\hat{\jmath}_\nu, \widehat{N}_\nu) \vdash'' e_\nu. \tag{54}$$

It follows that

$$(a, S) \vdash \bigcup_{\nu=1}^{n} \{\bar{a}_\nu\} \cup \mathrm{pr}_1(M_\nu).$$

Thus, there is $(\bar{a}, \overline{M}) \in \mathrm{CON}$ with $(a, S) \vdash (\bar{a}, \overline{M})$ and $(\bar{a}, \overline{M}) \vdash \bigcup_{\nu=1}^{n} \{\bar{a}_\nu\} \cup \mathrm{pr}_1(M_\nu)$. Moreover, $((a, i_\nu), S \times X_\nu) \vdash_{A\times A'} (\bar{a}, \bar{\imath}_\nu)$.

By definition of product information systems, we have that

$$\mathrm{pr}_1(M_\nu) \times \mathrm{pr}_2(M_\nu) \in \mathrm{CON}_{A\times A'}(\bar{a}, \bar{\imath}_\nu).$$

Since $M_\nu \subseteq \mathrm{pr}_1(M_\nu) \times \mathrm{pr}_2(M_\nu)$, it follows with Statement (53) that

$$((\bar{a}, \bar{\imath}_\nu), \mathrm{pr}_1(M_\nu) \times \mathrm{pr}_2(M_\nu)) H (\hat{\jmath}_\nu, \widehat{N}_\nu). \tag{55}$$

By Statement (52), we have that $(i_\nu, X_\nu) \vdash' (\bar{\imath}_\nu, \mathrm{pr}_2(M_\nu))$. Thus, there is some $(\hat{\imath}_\nu, \widehat{M}_\nu) \in \mathrm{CON}'$ with $(i_\nu, X_\nu) \vdash' (\hat{\imath}_\nu, \widehat{M}_\nu)$ and $(\hat{\imath}_\nu, \widehat{M}_\nu) \vdash' (\bar{\imath}_\nu, \mathrm{pr}_2(M_\nu))$. Here is what we have obtained so far:

$$((a, i_\nu), S \times X_\nu) \vdash_{A\times A'} ((\bar{a}, \hat{\imath}_\nu), \overline{M} \times \widehat{M}_\nu), \tag{56}$$

$$((\bar{a}, \hat{\imath}_\nu), \overline{M} \times \widehat{M}_\nu) \vdash_{A\times A'} ((\bar{a}, \hat{\imath}_\nu), \mathrm{pr}_1(M_\nu) \times \mathrm{pr}_2(M_\nu)). \tag{57}$$

Since $\{(\bar{a}, \bar{\imath}_\nu)\} \in \mathrm{CON}_{A\times A'}(\bar{a}, \hat{\imath}_\nu)$, it follows with Statement (55) that

$$((\bar{a}, \hat{\imath}_\nu), \mathrm{pr}_1(M_\nu) \times \mathrm{pr}_2(M_\nu)) H (\hat{\jmath}_\nu, \widehat{N}_\nu),$$

and hence, because of Statement (57), that

$$((\bar{a}, \hat{\imath}_\nu), \overline{M} \times \widehat{M}_\nu) H (\hat{\jmath}_\nu, \widehat{N}_\nu). \tag{58}$$

Set

$$\mathfrak{A} = \{\, ((\hat{\imath}_\nu, \widehat{M}_\nu), \hat{d}) \mid \hat{d} \in \widehat{N}_\nu \wedge 1 \leq \nu \leq n \,\}.$$

Next, we will construct $\mathfrak{B} \in \mathrm{AC}(\mathfrak{A})$ so that $(\bar{a}, \overline{M}) \Lambda(H)(\langle \mathfrak{B} | \mathfrak{A} \rangle, \{\langle \mathfrak{B} | \mathfrak{A} \rangle\})$ and

$$(\langle \mathfrak{B} | \mathfrak{A} \rangle, \{\langle \mathfrak{B} | \mathfrak{A} \rangle\}) \vdash_{A' \to A''} \langle \mathcal{G} | \mathcal{F} \rangle.$$

For each $J \subseteq \{1, \ldots, n\}$, let $R(\{\, (\hat{\imath}_\nu, \widehat{M}_\nu) \mid \nu \in J \,\})$ be a system of representatives of $W(\{\, (\hat{\imath}_\nu, \widehat{M}_\nu) \mid \nu \in J \,\})$ with respect to

$\sim [\bigcup_{\nu \in J} \widehat{M_\nu}]$ so that $R(\emptyset) = \{\Delta\}$ and $R(\{\hat{\imath}_\nu, \widehat{M_\nu})\}) = \{\hat{\imath}_\nu\}$, for $1 \leq \nu \leq n$. By 6.2(2), there is some $j_{(e,Z),(c,T),\langle \mathfrak{G}|\mathfrak{F}\rangle} \in A''$, for each $(e, Z) \in \text{CON}$, $\langle \mathfrak{G}|\mathfrak{F}\rangle \in \langle \mathcal{G}|\mathcal{F}\rangle$ and $((c, T), b) \in \mathfrak{G}$, with

$$j_{(e,Z),(c,T),\langle \mathfrak{G}|\mathfrak{F}\rangle} \sim b\,[\text{RS}((c, T), \mathfrak{F})],$$

and $((e, c), Z \times T) H j_{(e,Z),(c,T),\langle \mathfrak{G}|\mathfrak{F}\rangle}$. In case that $((c, T), b) = ((\Delta', \emptyset), \Delta'')$, choose $j_{(e,Z),(\Delta',\emptyset),\langle \mathfrak{G}|\mathfrak{F}\rangle} = \Delta''$.

Claim 6.4. *For every* $\mathfrak{J} \subseteq \mathfrak{A}$ *and each* $e \in R(\text{pr}_1(\mathfrak{J}))$ *so that* $(e, \mathfrak{J})$ *is* $\mathfrak{A}$-maximal, *there is some* $t_{e,\mathfrak{J}} \in A''$ *with the following two properties:*

$$\begin{aligned}
&\{\Delta''\} \cup \bigcup \Big\{ \{\hat{\jmath}_\nu\} \cup \widehat{N}_\nu \,\Big|\, 1 \leq \nu \leq n \wedge (\hat{\imath}_\nu, \widehat{M_\nu}) \in \text{pr}_1(\mathfrak{J}) \Big\} \cup \{\, t_{c,\mathfrak{K}} \mid \\
&\qquad \mathfrak{K} \subsetneqq \mathfrak{J} \wedge c \sim e \Big[\bigcup \text{pr}_{2,1}(\mathfrak{K}) \Big] \wedge (c, \mathfrak{K})\ \mathfrak{A}\text{-maximal} \,\} \\
&\qquad\quad \cup \{\, j_{(\bar{a},\overline{M}),(c,T),\langle \mathfrak{G}|\mathfrak{F}\rangle} \,\Big| \\
&\qquad (\exists L \subseteq \{1, \ldots, n\})(\exists b \in A'') \Big\{ (\hat{\imath}_\mu, \widehat{M_\mu}) \mid \mu \in L \Big\} \subseteq \text{pr}_1(\mathfrak{J}) \\
&\qquad \wedge T = \bigcup\nolimits_{\mu \in L} X_\mu \wedge c \sim e\,[T] \wedge \langle \mathfrak{G}|\mathfrak{F}\rangle \in \langle \mathcal{G}|\mathcal{F}\rangle \wedge ((c, T), b) \in \mathfrak{G} \\
&\qquad \wedge b \sim j_{(\bar{a},\overline{M}),(c,T),\langle \mathfrak{G}|\mathfrak{F}\rangle}\, [\text{RS}((c, T), \mathfrak{F})] \,\} \in \text{CON}'(t_{e,\mathfrak{J}}),
\end{aligned} \tag{59}$$

$$\begin{aligned}
&\Big((\bar{a}, e), \overline{M} \times \bigcup \text{pr}_{2,1}(\mathfrak{J}) \Big)\, H(t_{e,\mathfrak{J}}, \{\Delta''\} \cup \bigcup \{\, \{\hat{\jmath}_\nu\} \cup \widehat{N}_\nu \,\Big|\, 1 \leq \nu \leq n \\
&\quad \wedge (\hat{\imath}_\nu, \widehat{M_\nu}) \in \text{pr}_1(\mathfrak{J}) \,\} \cup \{\, t_{c,\mathfrak{K}} \mid \mathfrak{K} \subsetneqq \mathfrak{J} \wedge c \sim e \Big[\bigcup \text{pr}_{2,1}(\mathfrak{K}) \Big] \\
&\quad \wedge (c, \mathfrak{K})\ \mathfrak{A}\text{-maximal} \,\} \cup \{\, j_{(\bar{a},\overline{M}),(c,T),\langle \mathfrak{G}|\mathfrak{F}\rangle} \,\Big| \\
&\quad (\exists L \subseteq \{1, \ldots, n\})(\exists b \in A'') \\
&\quad T = \bigcup\nolimits_{\mu \in L} X_\mu \wedge c \sim e\,[T] \wedge \Big\{ (\hat{\imath}_\mu, \widehat{M_\mu}) \mid \mu \in L \Big\} \subseteq \text{pr}_1(\mathfrak{J}) \wedge \langle \mathfrak{G}|\mathfrak{F}\rangle \\
&\quad \in \langle \mathcal{G}|\mathcal{F}\rangle \wedge ((c, T), b) \in \mathfrak{G} \wedge b \sim j_{(\bar{a},\overline{M}),(c,T),\langle \mathfrak{G}|\mathfrak{F}\rangle}\, [\text{RS}((c, T), \mathfrak{F})] \,\}).
\end{aligned} \tag{60}$$

With respect to the last set on the right-hand side note that if $((c, T), b) \in \mathfrak{G}$, then there is some $\mathfrak{L} \subseteq \mathfrak{F}$ so that $(c, \mathfrak{L})$ is $\mathfrak{F}$-maximal and $T = \bigcup \text{pr}_{2,1}(\mathfrak{L})$. It follows that $\|\mathfrak{L}\| \leq \|\mathfrak{J}\|$.

The claim is demonstrated by induction on the cardinality κ of $\mathfrak{J}$. The case $\kappa = 0$ is obvious. Set $t_{e,\emptyset} = \Delta''$ if $(e, \emptyset)$ is $\mathfrak{A}$-maximal. All other sets on the left-hand side in Statement (59) are empty in this case, except the last one, which is the singleton $\{\Delta''\}$, if for some $\langle\mathfrak{G}|\mathfrak{F}\rangle \in \langle\mathcal{G}|\mathcal{F}\rangle$, $((\Delta', \emptyset), \Delta'') \in \mathfrak{G}$.

Assume next that the claim holds for all $\mathfrak{K} \subseteq \mathfrak{A}$ of cardinality κ and let $\mathfrak{J} \subseteq \mathfrak{A}$ of cardinality $\kappa + 1$. Then

$$\begin{aligned}
&\Big((\bar{a}, e), \overline{M} \times \bigcup \mathrm{pr}_{2,1}(\mathfrak{J})\Big)\, H\{\Delta''\} \cup \bigcup \{\, \{\hat{\jmath}_\nu\} \cup \widehat{N}_\nu \,\Big|\, 1 \leq \nu \leq n \wedge (\hat{\imath}_\nu, \widehat{M}_\nu) \\
&\in \mathrm{pr}_1(\mathfrak{J})\,\} \cup \{\, t_{c,\mathfrak{K}} \mid \mathfrak{K} \subsetneqq \mathfrak{J} \wedge c \sim e \left[\bigcup \mathrm{pr}_{2,1}(\mathfrak{K})\right] \\
&\wedge (c, \mathfrak{K})\ \mathfrak{A}\text{-maximal}\,\} \\
&\cup \{\, j_{(\bar{a},\overline{M}),(c,T),\langle\mathfrak{G}|\mathfrak{F}\rangle} \,\Big|\, (\exists L \subseteq \{1, \ldots, n\})(\exists b \in A'')\, T = \bigcup\nolimits_{\mu \in L} X_\mu \\
&\wedge c \sim e\,[T] \wedge \Big\{(\hat{\imath}_\mu, \widehat{M}_\mu) \mid \mu \in L\Big\} \subseteq \mathrm{pr}_1(\mathfrak{J}) \wedge \langle\mathfrak{G}|\mathfrak{F}\rangle \in \langle\mathcal{G}|\mathcal{F}\rangle \\
&\wedge ((c, T), b) \in \mathfrak{G} \wedge b \sim j_{(\bar{a},\overline{M}),(c,T),\langle\mathfrak{G}|\mathfrak{F}\rangle}\, [\mathrm{RS}((c, T), \mathfrak{F})]\,\},
\end{aligned}$$

because of Statement (58), Axiom 6.2(2) and the induction hypothesis. By Condition 4.1(9), there is hence some $t_{e,\mathfrak{J}} \in A''$ so that Properties (59) and (60) hold.

With help of Claim 6.4 we can now define $\mathfrak{B}$. Let $\mathfrak{B}^{(0)}$ be as in Definition 5.7 and for $\kappa \geq 1$ set

$$\begin{aligned}
\mathfrak{B}^{(\kappa)} &= \{\, ((e, \mathrm{pr}_{2,1}(\mathfrak{J})), t_{e,\mathfrak{J}}) \mid \mathfrak{J} \subseteq \mathfrak{A} \wedge \|\mathfrak{J}\| \\
&= \kappa \wedge e \in R(\mathrm{pr}_1(\mathfrak{J})) \wedge (e, \mathfrak{J})\ \mathfrak{A}\text{-maximal}\,\}.
\end{aligned}$$

Obviously, $\mathfrak{B} \in \mathrm{AC}(\mathfrak{A})$.

Claim 6.5. $(\bar{a}, \overline{M})\Lambda(H)\langle\mathfrak{B}|\mathfrak{A}\rangle$.

Because of Statement (58), Condition 6.2(1) is satisfied. For Condition 6.2(2), let $((e, T), b) \in \mathfrak{B}$. Then there is some $\mathfrak{J} \subseteq \mathfrak{A}$ such that $(e, \mathfrak{J})$ is $\mathfrak{A}$-maximal, $T = \bigcup \mathrm{pr}_{2,1}(\mathfrak{J})$ and $b = t_{e,\mathfrak{J}}$. As follows from Properties (59) and (60), $\mathrm{pr}_2(\mathfrak{J}) \in \mathrm{CON}''(t_{e,\mathfrak{J}})$ and $((\bar{a}, e), \overline{M} \times T)Ht_{e,\mathfrak{J}}$. Note that $\mathrm{pr}_2(\mathfrak{J}) = \mathrm{RS}((e, T), \mathfrak{A})$ and choose $j = t_{e,\mathfrak{J}}$.

Claim 6.6. $(\langle\mathfrak{B}|\mathfrak{A}\rangle, \{\langle\mathfrak{B}|\mathfrak{A}\rangle\}) \vdash_{A' \to A''} \langle\mathcal{G}|\mathcal{F}\rangle$.

Let $\langle\mathfrak{G}|\mathfrak{F}\rangle \in \langle\mathcal{G}|\mathcal{F}\rangle$ and $((i_\nu, X_\nu), e_\nu) \in \mathfrak{F}$. Then $(\hat{\jmath}_\nu, \widehat{N}_\nu) \vdash'' e_\nu$. Moreover, $\widehat{N}_\nu \subseteq \mathrm{AP}((i_\nu, X_\nu), \mathfrak{A})$, and by construction of $\mathfrak{B}$, $\{\hat{\jmath}_\nu\} \in$

$\mathrm{CON}''(\mathfrak{B}_2(i_\nu, X_\nu))$. Thus,

$$(\mathfrak{B}_2(i_\nu, X_\nu), \mathrm{AP}((i_\nu, X_\nu), \mathfrak{A})) \vdash'' e_\nu, \tag{61}$$

which means that Condition 5.10(3a) holds. For Condition 5.10(3b), let $(c,T),b) \in \mathfrak{G}$ and $L \subseteq \{1,\dots,n\}$ such that $\mathrm{DS}((c,T),\mathfrak{F}) = \{\,(i_\nu, X_\nu) \mid \nu \in L\,\}$. Then it follows with Statement (61) that

$$\Big(\mathfrak{B}_2(c,T), \bigcup\{\,\mathrm{AP}((i_\nu, X_\nu), \mathfrak{A}) \mid \nu \in L\,\}\Big) \vdash'' \{\, e_\nu \mid \nu \in L\,\}.$$

Note that $\{\, e_\nu \mid \nu \in L\,\} = \mathrm{RS}((c,T),\mathfrak{F})$. By Axiom 3.1(8), there is thus some $k \in A''$ with $\{\, e_\nu \mid \nu \in L\,\} \in \mathrm{CON}''(k)$ and

$$\Big(\mathfrak{B}_2(c,T), \bigcup\{\,\mathrm{AP}((i_\nu, X_\nu), \mathfrak{A}) \mid \nu \in L\,\}\Big) \vdash'' k.$$

Therefore, $\{k\} \in \mathrm{CON}''(\mathfrak{B}_2(c,T))$.

By the definition of $\mathfrak{B}(c,T)$ there is some $\mathfrak{J} \subseteq \mathfrak{A}$ such that $(\mathfrak{B}_{1,1}(c,T),\mathfrak{J})$ is $\mathfrak{A}$-maximal and $\mathfrak{B}_{1,2}(c,T) = \bigcup \mathrm{pr}_{2,1}(\mathfrak{J})$. Moreover, $\{\,((\hat{\imath}_\nu, \widehat{M_\nu}), d) \mid d \in \widehat{N}_\nu \wedge \nu \in L\,\} \subseteq \mathfrak{J}$ and $\mathfrak{B}_2(c,T) = t_{\mathfrak{B}_{1,1}(c,T),\mathfrak{J}}$. It follows that

$$j_{(\bar{a},\overline{M}),(c,T),\langle\mathfrak{G}|\mathfrak{F}\rangle} \in \mathrm{CON}''(\mathfrak{B}_2(c,T)).$$

Since in addition, $j_{(\bar{a},\overline{M}),(c,T),\langle\mathfrak{G}|\mathfrak{F}\rangle} \sim b\,[\mathrm{RS}((c,T),\mathfrak{F})]$, we obtain that also $k \sim b\,[\mathrm{RS}((c,T),\mathfrak{F})]$.

For Condition 4.1(5), assume that

$$(a,S)\Lambda(H)(\langle\mathfrak{B}|\mathfrak{A}\rangle, \langle\mathcal{W}|\mathcal{V}\rangle), \tag{62}$$
$$(\langle\mathfrak{B}|\mathfrak{A}\rangle, \langle\mathcal{W}|\mathcal{V}\rangle) \vdash_{\rightarrow} \langle\mathfrak{E}|\mathfrak{D}\rangle. \tag{63}$$

We need to show that $(a,S)\Lambda(H)\langle\mathfrak{E}|\mathfrak{D}\rangle$.

Let to this end $((j,Y),d) \in \mathfrak{D}$. As a consequence of Statement (63), we obtain that

$$\Big(\mathfrak{B}_2(j,Y), \mathrm{AP}((j,Y), \bigcup\mathcal{V})\Big) \vdash'' d. \tag{64}$$

Moreover, because of Statement (62), we have for $((i,X),e) \in \mathfrak{A} \cup \bigcup\mathcal{V}$ that

$$((a,i), S \times X)He. \tag{65}$$

Now, suppose that $(i,X) \in \mathrm{CL}\,((j,Y), \mathrm{pr}_1\,(\mathfrak{A} \cup \bigcup\mathcal{V}))$. Then $i \sim j\,[X]$ and hence $(a,i) \sim (a,j)\,[S \times X]$. With Lemma 5.3 it therefore follows with Statement (65) that $((a,j), S\times X)He$. Note that $X \subseteq \overline{\mathrm{CL}}((j,Y), \mathrm{pr}_1(\mathfrak{A}\cup\bigcup\mathcal{V}))$ and $\overline{\mathrm{CL}}((j,Y), \mathrm{pr}_1(\mathfrak{A}\cup\bigcup\mathcal{V})) \in \mathrm{CON}'(j)$. Thus,

$$\left((a,j), S\times\overline{\mathrm{CL}}\left((j,Y),\mathrm{pr}_1\left(\mathfrak{A}\cup\bigcup\mathcal{V}\right)\right)\right) He,$$

by Axiom 4.1(2). Since $(j,Y)\vdash' \overline{\mathrm{CL}}((j,Y), \mathrm{pr}_1(\mathfrak{A}\cup\bigcup\mathcal{V}))$, it ensues that $((a,j), S\times Y)He$, which shows that

$$((a,j), S\times Y)H\,\mathrm{AP}\left((j,Y),\mathfrak{A}\cup\bigcup\mathcal{V}\right).$$

Because of Condition 4.1(9), there is hence some $r\in A''$ with

$$((a,j), S\times Y)H\left(r,\mathrm{AP}\left((j,Y),\mathfrak{A}\cup\bigcup\mathcal{V}\right)\right). \tag{66}$$

Furthermore, as a consequence of Axiom 6.2(2) and Statement (62), there is some $\bar{r}\in A''$ with $((a,j), S\times Y)H(\bar{r},\mathrm{AP}((j,Y),\mathfrak{A}))$ and

$$\bar{r}\sim\mathfrak{B}_2(j,Y)\,[\mathrm{AP}((j,Y),\mathfrak{A})].$$

It follows that for some $\hat{r}\in A''$, $((a,j), S\times Y)H\hat{r}$ and $\{r,\bar{r}\}\in \mathrm{CON}''(\hat{r})$. Thus, $r\sim\mathfrak{B}_2(j,Y)\,[\mathrm{AP}((j,Y),\mathfrak{A})]$. As $\langle\mathcal{W}|\mathcal{V}\rangle\in \mathrm{CON}_{\rightarrow}(\langle\mathfrak{B}|\mathfrak{A}\rangle)$, we obtain with Axiom 5.10(2b) that $r\sim\mathfrak{B}_2(j,Y)$ $[\mathrm{AP}((j,Y),\bigcup\mathcal{V})]$. With Statement (64) we therefore have that

$$\left(r,\mathrm{AP}\left((j,Y),\bigcup\mathcal{V}\right)\right)\vdash'' d,$$

from which we gain with Statement (66) that

$$((a,j), S\times Y)Hd.$$

Next, let $((c,T),e)\in\mathfrak{E}$. Then it follows that

$$((a,c), S\times T)H\,\mathrm{RS}((c,T),\mathfrak{D}).$$

Because of Statement (63) and Axiom 5.10(3b), there is some $k\in A''$ so that

$$e\sim k\,[\mathrm{RS}((c,T),\mathfrak{D})],$$

and

$$\left(\mathfrak{B}_2(c,T),\bigcup\left\{\mathrm{AP}\left((j,Y),\bigcup\mathcal{V}\right)\mid (j,Y)\in\mathrm{DS}((c,T),\mathfrak{D})\right\}\right)$$
$$\vdash'' (k,\mathrm{RS}((c,T),\mathfrak{D})).$$

As in the first part of the proof, this implies that $((a,c), S\times T)Hk$. Thus, $(a,S)\Lambda(H)\langle\mathfrak{E}|\mathfrak{D}\rangle$.

For Condition 4.1(8), let $\{a\} \in \mathrm{CON}(b)$, $S \in \mathrm{CON}(a)$, $(i, X) \in \mathrm{CON}'$ and $c \in A''$ so that $((a, i), S \times X)Hc$. We have that $\{(a, i)\} \in \mathrm{CON}_{A\times A'}(b, i)$ and $S \times X \in \mathrm{CON}_{A\times A'}(a, i)$. Hence, $S \times X \in \mathrm{CON}_{A\times A'}(b, i)$ and $((b, i), S \times X)Hc$. Therefore, if $(a, S)\Lambda(H)\langle\mathfrak{E}|\mathfrak{D}\rangle$, then also $(b, S)\Lambda(H)\langle\mathfrak{E}|\mathfrak{D}\rangle$. □

Lemma 6.7. $(\Lambda(H) \times \mathrm{Id}_{A'}) \circ \mathrm{EV} = H$. *for all* $H\colon A \times A' \Vdash A''$.

Proof. Let $((a, b), U) \in \mathrm{CON}_{A\times A'}$ and $c \in A''$ with $((a, b), U)((\Lambda(H) \times \mathrm{Id}_{A'}) \circ \mathrm{EV})c$. Because of Lemma 4.2, there exists $((a', b'), U') \in \mathrm{CON}_{A\times A'}$ such that

$$((a, b), U) \vdash_{A\times A'} ((a', b'), U'), \tag{67}$$

$$((a', b'), U')((\Lambda(H) \times \mathrm{Id}_{A'}) \circ \mathrm{EV})c. \tag{68}$$

In addition there is some $(((\langle\mathfrak{B}|\mathfrak{A}\rangle, \bar{b}), \overline{U}) \in \mathrm{CON}_{(A'\to A'')\times A'}$ with

$$((a', b'), U')(\Lambda(H) \times \mathrm{Id}_{A'})(((\langle\mathfrak{B}|\mathfrak{A}\rangle, \bar{b}), \overline{U}), \tag{69}$$

$$(((\langle\mathfrak{B}|\mathfrak{A}\rangle, \bar{b}), \overline{U})\,\mathrm{EV}\,c, \tag{70}$$

where we obtain with Statement (69) that

$$(a', \mathrm{pr}_1(U'))\Lambda(H)(\langle\mathfrak{B}|\mathfrak{A}\rangle, \mathrm{pr}_1(\overline{U})), \tag{71}$$

$$(b', \mathrm{pr}_2(U')) \vdash' (\bar{b}, \mathrm{pr}_2(\overline{U})). \tag{72}$$

As a consequence of Statement (71), we gain that for $((i, X), e) \in \bigcup \mathrm{pr}_{2,1}(\overline{U})$,

$$((a', i), \mathrm{pr}_1(U') \times X)He, \tag{73}$$

which implies that

$$\left((a', \bar{b}), \mathrm{pr}_1(U') \times \mathrm{pr}_2(\overline{U})\right)H\,\mathrm{AP}((\bar{b}, \mathrm{pr}_2(\overline{U})), \bigcup \mathrm{pr}_{2,1}(\overline{U})\Big). \tag{74}$$

With Statements (67) and (72), we therefore have that

$$\Big((a, b), U)H\,\mathrm{AP}((\bar{b}, \mathrm{pr}_2(\overline{U})), \bigcup \mathrm{pr}_{2,1}(\overline{U})\Big). \tag{75}$$

By Condition 6.2(2), there is now some $j \in A''$ so that

$$((a, b), U)H\left(j, \mathrm{AP}\left((\bar{b}, \mathrm{pr}_2(\overline{U})), \bigcup \mathrm{pr}_{2,1}(\overline{U})\right)\right), \tag{76}$$

$$j \sim \mathfrak{B}_2(\bar{b}, \mathrm{pr}_2(\overline{U}))\Big[\mathrm{AP}((\bar{b}, \mathrm{pr}_2(\overline{U})), \bigcup \mathrm{pr}_{2,1}(\overline{U}))\Big]. \tag{77}$$

Since by Statement (70)

$$\left(\mathfrak{B}_2(\bar{b}, \mathrm{pr}_2(\overline{U})), \mathrm{AP}\left((\bar{b}, \mathrm{pr}_2(\overline{U})), \bigcup \mathrm{pr}_{2,1}(\overline{U})\right)\right) \vdash'' c,$$

it follows that also

$$\left(j, \mathrm{AP}\left((\bar{b}, \mathrm{pr}_2(\overline{U})), \bigcup \mathrm{pr}_{2,1}(\overline{U})\right)\right) \vdash'' c.$$

Together with Statement (76), we thus obtain that $((a,b),U)Hc$.

Next assume conversely that $((a,b),U)Hc$. Because of Lemmas 4.2, 4.3 and 4.4, there exist $((a',b'),U')$, $((a'',b''),U'') \in \mathrm{CON}_{A\times A'}$ and $(j,V) \in \mathrm{CON}''$ such that

$$((a,b),U) \vdash_{A\times A'} ((a',b'),U'), \tag{78}$$
$$((a',b'),U') \vdash_{A\times A'} ((a'',b''),U''), \tag{79}$$
$$((a'',b''),U'')H(j, V \cup \{\Delta''\}), \tag{80}$$
$$(j,V) \vdash'' c. \tag{81}$$

Set

$$\mathfrak{A} = \{\, ((b'', \mathrm{pr}_2(U'')), d) \mid d \in V \,\},$$

and let $\mathfrak{B} \in \mathrm{AC}(\mathfrak{A})$ be as in Lemma 5.8. Then we have for $(i,X) \in \mathrm{CON}'$ that

$$((a'',i), \mathrm{pr}_1(U'') \times X)H(\mathfrak{B}_2(i,X), \mathrm{AP}((i,X),\mathfrak{A})).$$

This shows that $(a'', \mathrm{pr}_1(U''))\Lambda(H)\langle\mathfrak{B}|\mathfrak{A}\rangle$. Moreover, we obtain with Statement (78) that

$$((a,b),U)(\Lambda(H) \times \mathrm{Id}_{A'})(((\langle\mathfrak{B}|\mathfrak{A}\rangle, b'), \{\langle\mathfrak{B}|\mathfrak{A}\rangle\} \times \mathrm{pr}_2(U')). \tag{82}$$

Now, note that $\mathrm{AP}((b', \mathrm{pr}_2(U')),\mathfrak{A}) = V$ and $\mathfrak{B}_2(b', \mathrm{pr}_2(U')) = j$. Therefore, it follows with Statement (81) that

$$(\mathfrak{B}_2(b', \mathrm{pr}_2(U')), \mathrm{AP}((b', \mathrm{pr}_2(U')),\mathfrak{A})) \vdash'' c.$$

Let $\mathbb{Z} = \{\langle\mathfrak{B}|\mathfrak{A}\rangle\} \times \mathrm{pr}_2(U')$. Then $\mathbb{Z} \in \mathrm{CON}_{(A'\to A'')\times A'}(\langle\mathfrak{B}|\mathfrak{A}\rangle, b')$. In addition, we have that $((\langle\mathfrak{B}|\mathfrak{A}\rangle, b'), \mathbb{Z})\,\mathrm{EV}\,c$. With Statement (82) we now gain that

$$((a,b),U)((\Lambda(H) \times \mathrm{Id}_{A'}) \circ \mathrm{EV})c,$$

as was to be shown. □

Lemma 6.8. $\Lambda((G \times \mathrm{Id}_{A'}) \circ \mathrm{EV}) = G$, *for all* $G\colon A \Vdash A' \to A''$.

Proof. Let $(a, V) \in \mathrm{CON}$ and $\langle \mathfrak{W} | \mathfrak{V} \rangle \in A' \to A''$ so that

$$(a, V)\Lambda((G \times \mathrm{Id}_{A'}) \circ \mathrm{EV})\langle \mathfrak{W} | \mathfrak{V} \rangle. \tag{83}$$

Moreover, assume that $\mathfrak{V} = \{\, ((i_\nu, X_\nu), e_\nu) \mid 1 \leq \nu \leq n \,\}$. Then we have for every $1 \leq \nu \leq n$ that

$$((a, i_\nu), V \times X_\nu)((G \times \mathrm{Id}_{A'}) \circ \mathrm{EV})e_\nu.$$

Thus, there are $(((\langle \mathfrak{B}^{(\nu)} | \mathfrak{A}^{(\nu)} \rangle, b_\nu), \mathbb{Z}_\nu) \in \mathrm{CON}_{(A' \to A'') \times A'}$ with

$$((a, i_\nu), V \times X_\nu)(G \times \mathrm{Id}_{A'})(((\langle \mathfrak{B}^{(\nu)} | \mathfrak{A}^{(\nu)} \rangle, b_\nu), \mathbb{Z}_\nu), \tag{84}$$

$$(((\langle \mathfrak{B}^{(\nu)} | \mathfrak{A}^{(\nu)} \rangle, b_\nu), \mathbb{Z}_\nu) \,\mathrm{EV}\, e_\nu. \tag{85}$$

So, we obtain

$$(a, V)G(\langle \mathfrak{B}^{(\nu)} | \mathfrak{A}^{(\nu)} \rangle, \mathrm{pr}_1(\mathbb{Z}_\nu)), \tag{86}$$

$$(i_\nu, X_\nu) \vdash' (b_\nu, \mathrm{pr}_2(\mathbb{Z}_\nu)), \tag{87}$$

$$\left(\mathfrak{B}_2^{(\nu)}(b_\nu, \mathrm{pr}_2(\mathbb{Z}_\nu)), \mathrm{AP}((b_\nu, \mathrm{pr}_2(\mathbb{Z}_\nu)), \bigcup \mathrm{pr}_{2,1}(\mathbb{Z}_\nu)\right) \vdash'' e_\nu. \tag{88}$$

Because of Statement (87), we moreover have that

$$\{\mathfrak{B}_2^{(\nu)}(b_\nu, \mathrm{pr}_2(\mathbb{Z}_\nu))\} \in \mathrm{CON}'(\mathfrak{B}_2^{(\nu)}(i_\nu, X_\nu)), \tag{89}$$

$$\mathrm{AP}\left((b_\nu, \mathrm{pr}_2(\mathbb{Z}_\nu)), \bigcup \mathrm{pr}_{2,1}(\mathbb{Z}_\nu)\right) \subseteq \mathrm{AP}\left((i_\nu, X_\nu), \bigcup \mathrm{pr}_{2,1}(\mathbb{Z}_\nu)\right), \tag{90}$$

and, hence, that

$$\left(\mathfrak{B}_2^{(\nu)}(i_\nu, X_\nu), \mathrm{AP}\left((i_\nu, X_\nu), \bigcup \mathrm{pr}_{2,1}(\mathbb{Z}_\nu)\right)\right) \vdash'' e_\nu. \tag{91}$$

By Statement (86) and Lemma 4.3, there exists $(\langle \mathfrak{C} | \mathfrak{D} \rangle, \langle \mathcal{E} | \mathcal{D} \rangle) \in \mathrm{CON}_{A' \to A''}$ with

$$(a, V)G(\langle \mathfrak{C} | \mathfrak{D} \rangle, \langle \mathcal{E} | \mathcal{D} \rangle), \tag{92}$$

$$(\langle \mathfrak{C} | \mathfrak{D} \rangle, \langle \mathcal{E} | \mathcal{D} \rangle) \vdash_{A' \to A''} (\langle \mathfrak{B}^{(\nu)} | \mathfrak{A}^{(\nu)} \rangle, \mathrm{pr}_1(\mathbb{Z}_\nu)), \tag{93}$$

for all $1 \leq \nu \leq n$. Thus, $\{\langle\mathfrak{B}^{(\nu)}|\mathfrak{A}^{(\nu)}\rangle\} \in \mathrm{CON}_{A'\to A''}(\langle\mathfrak{E}|\mathfrak{D}\rangle)$. By Condition 5.10(2a), it follows that

$$\mathfrak{E}_2(i_\nu, X_\nu) \sim \mathfrak{B}_2^{(\nu)}(i_\nu, X_\nu)\,[\mathrm{AP}((i_\nu, X_\nu), \mathfrak{A}^{(\nu)})].$$

Since $\mathrm{pr}_1(\mathbb{Z}_\nu) \in \mathrm{CON}_{A'\to A''}(\langle\mathfrak{B}^{(\nu)}|\mathfrak{A}^{(\nu)}\rangle)$, we hence obtain with Condition 5.10(2b) that moreover

$$\mathfrak{E}_2(i_\nu, X_\nu) \sim \mathfrak{B}_2^{(\nu)}(i_\nu, X_\nu)\left[\mathrm{AP}((i_\nu, X_\nu), \bigcup \mathrm{pr}_{2,1}(\mathbb{Z}_\nu))\right].$$

As a consequence of Statement (91), we therefore gain that

$$\left(\mathfrak{E}_2(i_\nu, X_\nu), \mathrm{AP}\left((i_\nu, X_\nu), \bigcup \mathrm{pr}_{2,1}(\mathbb{Z}_\nu)\right)\right) \vdash'' e_\nu.$$

Because it follows with Statement (93) that

$$\left(\mathfrak{E}_2(i_\nu, X_\nu), \mathrm{AP}\left((i_\nu, X_\nu), \bigcup \mathcal{D}\right)\right) \vdash'' \mathrm{AP}\left((i_\nu, X_\nu), \bigcup \mathrm{pr}_{2,1}(\mathbb{Z}_\nu)\right),$$

we conclude that

$$\left(\mathfrak{E}_2(i_\nu, X_\nu), \mathrm{AP}\left((i_\nu, X_\nu), \bigcup \mathcal{D}\right)\right) \vdash'' e_\nu,$$

from which we additionally gain that for $((t, S), c) \in \mathfrak{W}$,

$$\left(\mathfrak{E}_2(t, S), \bigcup \left\{ \mathrm{AP}((d, U), \bigcup \mathcal{D}\right) \mid (d, U) \in \mathrm{DS}((t, S), \mathfrak{V})\right\})$$
$$\vdash'' \mathrm{RS}((t, S), \mathfrak{V}).$$

Because of Axiom 3.1(8) there is thus some $j \in A''$ so that

$$\mathrm{RS}((t, S), \mathfrak{V}) \in \mathrm{CON}''(j),$$
$$\left(\mathfrak{E}_2(t, S), \bigcup \{ \mathrm{AP}((d, U), \bigcup \mathcal{D}\right) \mid (d, U) \in \mathrm{DS}((t, S), \mathfrak{V}) \}) \vdash'' j.$$

It follows that also

$$\left(\mathfrak{E}_2(t, S), \mathrm{AP}\left((t, S), \bigcup \mathcal{D}\right)\right) \vdash'' (j, \mathrm{RS}((t, S), \mathfrak{V})).$$

By definition, $(t, S) \vdash' \overline{\mathrm{CL}}((t, S), \mathrm{pr}_1(\bigcup \mathcal{D} \cup \mathfrak{D}))$. Hence, there is some $(b, T) \in \mathrm{CON}'$ with

$$(t, S) \vdash' (b, T),$$
$$(b, T) \vdash' \overline{\mathrm{CL}}\Big((t, S), \mathrm{pr}_1\left(\bigcup \mathcal{D} \cup \mathfrak{D}\right)\Big).$$

Then

$$\mathrm{AP}\left((b,T),\bigcup\mathcal{D}\right)=\mathrm{AP}\left((t,S),\bigcup\mathcal{D}\right)\quad\text{and}\quad\mathfrak{E}_2(b,T)=\mathfrak{E}_2(t,S).$$

Thus, we also have that

$$\left(\mathfrak{E}_2(b,t),\mathrm{AP}\left((b,T),\bigcup\mathcal{D}\right)\right)\vdash''(j,\mathrm{RS}((t,S),\mathfrak{V})).$$

So far we have shown that

$$\begin{gathered}((a,t),V\times S)(G\times\mathrm{Id}_{A'})(((\langle\mathfrak{E}|\mathfrak{D}\rangle,b),\langle\mathcal{E}|\mathcal{D}\rangle\times T),\\ (((\langle\mathfrak{E}|\mathfrak{D}\rangle,b),\langle\mathcal{E}|\mathcal{D}\rangle\times T)\,\mathrm{EV}(j,\mathrm{RS}((t,S),\mathfrak{V})),\end{gathered}$$

from which we obtain that

$$((a,t),V\times S)((G\times\mathrm{Id}_{A'})\circ\mathrm{EV})(j,\mathrm{RS}((t,S),\mathfrak{V})).$$

Because of Statement (83) and Condition 6.2(2), there is also some $\bar{\jmath}\in A''$ so that

$$((a,t),V\times S)((G\times\mathrm{Id}_{A'})\circ\mathrm{EV})\bar{\jmath}\quad\text{and}\quad\bar{\jmath}\sim c\,[\mathrm{RS}((t,S),\mathfrak{V})].$$

With Condition 4.1(9) there then some $k\in A''$ such that $\{j,\bar{\jmath}\}\in\mathrm{CON}''(k)$ and $((a,t),V\times S)((G\times\mathrm{Id}_{A'})\circ\mathrm{EV})k$. It follows that also $j\sim c\,[\mathrm{RS}((t,S),\mathfrak{V})]$. So, we have shown that

$$(\langle\mathfrak{E}|\mathfrak{D}\rangle,\langle\mathcal{E}|\mathcal{D}\rangle)\vdash_{A'\to A''}\langle\mathfrak{W}|\mathfrak{V}\rangle,$$

from which we obtain with Statement (92) that $(a,V)G\langle\mathfrak{W}|\mathfrak{V}\rangle$.

Next, conversely, let $(a,V)\in\mathrm{CON}$ and $\langle\mathfrak{W}|\mathfrak{V}\rangle\in A'\to A''$ with

$$(a,V)G\langle\mathfrak{W}|\mathfrak{V}\rangle.$$

Then there is some $(\langle\mathfrak{B}|\mathfrak{A}\rangle,\langle\mathcal{G}|\mathcal{F}\rangle)\in\mathrm{CON}_{A'\to A''}$ with

$$(a,V)G(\langle\mathfrak{B}|\mathfrak{A}\rangle,\langle\mathcal{G}|\mathcal{F}\rangle),\tag{94}$$

$$(\langle\mathfrak{B}|\mathfrak{A}\rangle,\langle\mathcal{G}|\mathcal{F}\rangle)\vdash_{A'\to A''}\langle\mathfrak{W}|\mathfrak{V}\rangle.\tag{95}$$

Let $\mathfrak{V}=\{\,(i_\nu,X_\nu),e_\nu)\mid 1\le\nu\le m\,\}$. Then we obtain for $1\le\nu\le m$ that

$$\left(\mathfrak{B}_2(i_\nu,X_\nu),\mathrm{AP}\left((i_\nu,X_\nu),\bigcup\mathcal{F}\right)\right)\vdash''e_\nu.$$

Moreover, there are $(b_\nu, Y_\nu) \in \mathrm{CON}'$, for $1 \le \nu \le m$, with

$$(i_\nu, X_\nu) \vdash' (b_\nu, Y_\nu),$$
$$(b_\nu, Y_\nu) \vdash' \overline{\mathrm{CL}}\Big((i_\nu, X_\nu), \mathrm{pr}_1\left(\bigcup \mathcal{F} \cup \mathcal{A}\right)\Big).$$

It follows that

$$\mathrm{AP}\left((i_\nu, X_\nu), \bigcup \mathcal{F}\right) = \mathrm{AP}\left((b_\nu, Y_\nu), \bigcup \mathcal{F}\right) \text{ and } \mathfrak{B}_2(i_\nu, X_\nu) = \mathfrak{B}_2(b_\nu, Y_\nu).$$

Thus,

$$\left(\mathfrak{B}_2(b_\nu, Y_\nu), \mathrm{AP}\left((b_\nu, Y_\nu), \bigcup \mathcal{F}\right)\right) \vdash'' e_\nu.$$

Set $\mathbb{Z}_\nu = \langle \mathcal{G} | \mathcal{F} \rangle \times Y_\nu$. Then $\mathbb{Z}_\nu \in \mathrm{CON}_{(A' \to A'') \times A'}(\langle \mathfrak{B} | \mathfrak{A} \rangle, b_\nu)$ and

$$((a, i_\nu), V \times X_\nu)(G \times \mathrm{Id}_{A'})(((\langle \mathfrak{B} | \mathfrak{A} \rangle, b_\nu), \mathbb{Z}_\nu),$$
$$(((\langle \mathfrak{B} | \mathfrak{A} \rangle, b_\nu), \mathbb{Z}_\nu)\, \mathrm{EV}\, e_\nu,$$

which means we have that

$$((a, i_\nu), V \times X_\nu)((G \times \mathrm{Id}_{A'}) \circ \mathrm{EV}) e_\nu.$$

It remains to verify Condition 6.2(2). Let to this end $((c, T), d) \in \mathfrak{W}$. Because of Statement (95) and Condition 5.10(3b), there is some $k \in A''$ with

$$k \sim d\,[\mathrm{RS}((c, T), \mathfrak{V})], \tag{96}$$
$$\left(\mathfrak{B}_2(c, T), \bigcup \left\{ \mathrm{AP}\left((e, U), \bigcup \mathcal{F}\right) \mid (e, U) \in \mathrm{DS}((c, T), \mathfrak{V}) \right\}\right) \vdash'' k. \tag{97}$$

It follows that also

$$\left(\mathfrak{B}_2(c, T), \mathrm{AP}\left((c, T), \bigcup \mathcal{F}\right)\right) \vdash'' (k, \mathrm{RS}((c, T), \mathfrak{V})).$$

Let $(\bar{b}, \overline{Y}) \in \mathrm{CON}'$ such that

$$(c, T) \vdash' (\bar{b}, \overline{Y}),$$
$$(\bar{b}, \overline{Y}) \vdash' \overline{\mathrm{CL}}\Big((c, T), \mathrm{pr}_1\left(\bigcup \mathcal{F} \cup \mathfrak{A}\right)\Big).$$

Then

$$\mathrm{AP}\left((c,T),\bigcup\mathcal{F}\right)=\mathrm{AP}\left((\bar{b},\overline{Y}),\bigcup\mathcal{F}\right)\quad\text{and}\quad\mathfrak{B}_2(c,T)=\mathfrak{B}_2(\bar{b},\overline{Y}),$$

and hence,

$$\left(\mathfrak{B}_2(\bar{b},\overline{Y}),\mathrm{AP}\left((\bar{b},\overline{Y}),\bigcup\mathcal{F}\right)\right)\vdash''(k,\mathrm{RS}((c,T),\mathfrak{V})).$$

Set $\mathbb{Z}=\langle\mathcal{G}|\mathcal{F}\rangle\times\overline{Y}$. Then $\mathbb{Z}\in\mathrm{CON}_{(A'\to A'')\times A'}(\langle\mathfrak{B}|\mathfrak{A}\rangle,\bar{b}))$ and

$$\begin{gathered}((a,c),V\times T)(G\times\mathrm{Id}_{A'})(((\langle\mathfrak{B}|\mathfrak{A}\rangle,\bar{b}),\mathbb{Z}),\\ (((\langle\mathfrak{B}|\mathfrak{A}\rangle,\bar{b}),\mathbb{Z})\,\mathrm{EV}(k,\mathrm{RS}((c,T),\mathfrak{V})),\end{gathered}$$

which implies that

$$((a,c),V\times T)((G\times\mathrm{Id}_{A'})\circ\mathrm{EV})k.$$

Because of Statement (96) this shows that

$$(a,V)\Lambda((G\times\mathrm{Id}_{A'})\circ\mathrm{EV})\langle\mathfrak{W}|\mathfrak{V}\rangle.$$

□

Proposition 6.9. *Let A and A' be an information system with witnesses. Then $(A\to A',\mathrm{EV})$ is their exponent in* **ISW**.

With Propositions 5.24 and 5.25, we moreover have that if both A and A' satisfy Condition (ALG), (BC) or both of them, then $(A\to A',\mathrm{EV})$ is their exponent in **aISW**, **bcISW** and **abcISW**, respectively.

As we have already seen, ISW as well as **aISW**, **bcISW** and **abcISW** contain a terminal object. Moreover, we have shown how to construct the categorical product of information systems with witnesses.

Theorem 6.10. *The category **ISW** of information systems with witnesses and approximable mappings as well as its full subcategories **aISW**, **bcISW** and **abcISW**, respectively, of information systems with witnesses satisfying Condition (ALG), (BC) or both of them are Cartesian closed.*

7. Final Remarks

This chapter is a continuation of Ref. [1] where information systems with witnesses were introduced as a logic-style representation of L-domains: the category of these information systems with approximable mappings as morphisms was shown to be equivalent to the category of L-domains with Scott continuous functions. As demonstrated in Ref. [3], the latter category is one of the two Cartesian closed full subcategories of the category of continuous domains. It follows in particular that the category of information systems with witnesses is Cartesian closed as well. In the present chapter, a direct construction of exponentiation in this category is presented.

Logic-style representations of domains — in particular those forming a Cartesian closed category — allow incorporating the domains into proof assistants, as done in the Minlog system, developed by the Munich logic group (cf. Ref. [12]).

Minlog is an interactive proof system based on first-order natural deduction calculus. It is intended to reason about higher-type computable functionals, using minimal rather than classical or intuitionistic logic. Minlog implements a *theory of computable functionals.* The underlying semantics is the Scott–Ershov model of partial continuous functionals, with *free algebras* as base types. These algebras are viewed as domains represented by Scott's information systems, whose tokens are constructor trees possibly involving the symbol $*$ ("no information") (Ref. [13]).

By using information systems with witnesses instead of Scott's information systems, a larger class of data structures and computable functionals can be considered. Examples of topological hyperspaces are known that are L-domains with respect to superset inclusion, but are not bounded-complete (cf., e.g., Ref. [3, p. 58]).

Acknowledgement

The research leading to these results has received funding from the People Programme (Marie Curie Actions) of the European Union's Seventh Framework Programme FP7/2007-2013/ under REA grant agreement no. PIRSES-GA-2013-612638-CORCON.

References

[1] D. Spreen, Generalised information systems capture L-domains. *Theoret. Comput. Sci.* **869**, 1–28 (2021). doi.org/10.1016/j.tcs.2020.12.044.

[2] T. Coquand, Categories of embeddings, *Theoret. Comput. Sci.* **68**, 221–238 (1989).

[3] A. Jung, *Cartesian Closed Categories of Domains. CWI Tracts*, vol. 66. Centrum voor Wiskunde en Informatica, Amsterdam, 1989.

[4] A. Jung, The classification of continuous domains. In *Proc., Fifth Annual IEEE Symposium on Logic in Computer Science.* IEEE Computer Society Press, New York, 1990, pp. 35–40.

[5] D. Scott, Domains for denotational semantics. In M. Nielsen *et al.* (eds.), *Automata, Languages and Programming*, Aarhus (1982). *Lecture Notes in Computer Science*, vol. 140. Springer, Berlin, 1982, pp. 577–613.

[6] S. Abramsky and A. Jung, Domain theory. In S. Abramsky *et al.* (eds.), *Handbook of Logic in Computer Science*, vol. 3. Clarendon Press, Oxford, 1994, pp. 1–168.

[7] R.M. Amadio and P.-L. Curien, *Domains and Lambda-Calculi.* Cambridge University Press, Cambridge, 1998.

[8] G. Gierz, K.H. Hoffmann, K. Keimel, J.D. Lawson, M.W. Mislove, and D.S. Scott, *Continuous Lattices and Domains.* Cambridge University Press, Cambridge, 2003.

[9] C.A. Gunter, *Semantics of Programming Languages.* MIT-Press, Cambridge, MA, 1992.

[10] C.A. Gunter and D.S. Scott, Semantic domains. In J. van Leeuwen (ed.), *Handbook of Theoretical Computer Science*, vol. B, Formal Models and Semantics. Elsevier, Amsterdam, 1990, pp. 633–674.

[11] V. Stoltenberg-Hansen, I. Lindström, and E. R. Griffor, *Mathematical Theory of Domains.* Cambridge University Press, Cambridge, 1994.

[12] The Minlog System. http://www.minlog-system.de.

[13] U. Berger, K. Miyamoto, M. Seisenberger, and H. Schwichtenberg, Minlog — A tool for program extraction supporting algebras and coalgebras. In A. Corradini *et al.* (eds.), *Algebra and Coalgebra in Computer Science*, Winchester, UK, 2011. *Lecture Notes in Computer Science*, vol. 6859. Springer, Berlin 2011, pp. 393–399.

[14] R. Hoofman, Continuous information systems, *Inform. Computation.* 105, 42–71 (1993).

https://doi.org/10.1142/9789811245220_0005

Chapter 5

A Constructive Version of Carathéodory's Convexity Theorem

Josef Berger[*,§] and Gregor Svindland[†,‡]

[*] *Mathematisches Institut Ludwig-Maximilians-Universität München, München, Germany*

[†] *Institut für Mathematische Stochastik & House of Insurance, Leibniz Universität Hannover, Hannover, Germany*

[‡] *jberger@math.lmu.de*

[§] *svindland@insurance.uni-hannover.de*

Carathéodory's Convexity Theorem states that each element in the convex hull of a subset A of $\mathbb{R}^m$ can be written as the convex combination of $m+1$ elements of A. We prove an approximate constructive version of Carathéodory's Convexity Theorem for totally bounded sets.

For each $n \in \mathbb{N}$, define

$$I_n := \{1, \ldots, n\}.$$

For any $\lambda \in \mathbb{R}^n$, we denote by λ_i the ith coordinate of λ, that is $\lambda = (\lambda_1, \ldots, \lambda_n)$. Let

$$S_n := \left\{ \lambda \in \mathbb{R}^n \mid \forall i \in I_n \, (0 \leq \lambda_i) \wedge \sum_{i \in I_n} \lambda_i = 1 \right\}.$$

The linear space generated by $x^1, \ldots, x^n \in \mathbb{R}^m$ is denoted by

$$\operatorname{span}\{x^1, \ldots, x^n\} := \left\{ \sum_{i \in I_n} \lambda_i x^i \mid \lambda \in \mathbb{R}^n \right\}.$$

A subset A of $\mathbb{R}^m$ is inhabited if there exists $x \in A$. The *convex hull* of an inhabited subset A of $\mathbb{R}^m$ is

$$\operatorname{co} A = \left\{ \sum_{i \in I_n} \lambda_i x^i \mid \lambda \in S_n,\ x_i \in A(i \in I_n), n \in \mathbb{N} \right\}.$$

A set $U \subseteq \mathbb{R}^n$ is *located* if it is inhabited and if for all $x \in \mathbb{R}^n$ the distance

$$d(x, U) = \inf\{\|x - y\| \mid y \in U\}$$

exists, where throughout this chapter $\| \cdot \|$ denotes the Euclidean norm on $\mathbb{R}^n$. U is said to be *totally bounded* if U is inhabited and if for every $\varepsilon > 0$ there exists a finite subset $F \subseteq U$ such that

$$\forall x \in U\ \exists y \in F\ \|x - y\| < \varepsilon.$$

Note that any totally bounded set is located [1, Proposition 2.2.9]. Let $x^1, \ldots, x^n \in \mathbb{R}^m$ and recall that

(1) The set $\{x^i \mid i \in I_n\}$ is *linearly independent* if

$$\forall\, \lambda \in \mathbb{R}^n \left(\|\lambda\| > 0\ \Rightarrow\ \|\sum_{i \in I_n} \lambda_i x^i\| > 0 \right),$$

(2) $\{x^i \mid i \in I_n\}$ is *linearly dependent* if

$$\exists\, \lambda \in \mathbb{R}^n \left(\|\lambda\| > 0\ \wedge\ \sum_{i \in I_n} \lambda_i x^i = 0 \right).$$

The following lemma seems to be folklore, but we could not find a proof in the constructive mathematics literature. As we will need it later on, we provide a proof for the sake of completeness.

Lemma 0.1. *Let $x^1, \ldots, x^n \in \mathbb{R}^m$. If $n > m$, then $(x^i)_{i \in I_n}$ are not linearly independent.*

Proof. It suffices to prove the assertion for $n = m + 1$. Assume that $\{x^i \mid i \in I_{m+1}\}$ is linearly independent.

Case $m = 1$: By linear independence we have $|x^1| > 0$ and $|x^2| > 0$. Set $\lambda_1 := x^2$ and $\lambda_2 := -x^1$. Then $\lambda = (\lambda_1, \lambda_2) \in \mathbb{R}^2$ satisfies $\|\lambda\| > 0$ and

$$\lambda_1 x^1 + \lambda_2 x^2 = x^2 x^1 - x^1 x^2 = 0,$$

which is a contradiction.

Case $m \geq 2$: As $\|x^{m+1}\| > 0$, we have that $|x_j^{m+1}| > 0$ for some $j \in I_m$. Without loss of generality we assume that $j = m$. Consider the vectors

$$v^i := x_m^{m+1} x^i - x_m^i x^{m+1}, \quad i \in I_m.$$

We have $v_m^i = 0$ for all $i \in I_m$, so we may identify the vectors v^i with elements of $\mathbb{R}^{m-1}$. Moreover, $\{v^i \mid i \in I_m\}$ is linearly independent. Indeed, consider $\lambda \in \mathbb{R}^m$ with $\|\lambda\| > 0$. Then

$$\sum_{i \in I_m} \lambda_i v^i = \sum_{i \in I_m} (\lambda_i x_m^{m+1}) x^i + \left(- \sum_{i \in I_m} \lambda_i x_m^i \right) x^{m+1}.$$

Since $|\lambda_k| > 0$ for some $k \in I_m$ and as $|x_m^{m+1}| > 0$, we have $\|\tilde{\lambda}\| > 0$ where $\tilde{\lambda} \in \mathbb{R}^{m+1}$ is given by $\tilde{\lambda}_i := \lambda_i x_m^{m+1}$, $i \in I_m$ and $\tilde{\lambda}_{m+1} = -\sum_{i \in I_m} \lambda_i x_m^i$. Linear independence of $(x^i)_{i \in I_{m+1}}$ now implies that

$$\left\| \sum_{i \in I_m} \lambda_i v^i \right\| = \left\| \sum_{i \in I_{m+1}} \tilde{\lambda}_i x^i \right\| > 0.$$

Thus, by erasing the last coordinate of the v^i, we have constructed m linear independent vectors in $\mathbb{R}^{m-1}$. Continuing this reduction procedure, if necessary, will eventually produce two linearly independent vectors in $\mathbb{R}$ which is a contradiction according to the case $m = 1$ above. □

Corollary 0.2. *Suppose that* $\{x^1, \ldots, x^n\} \subseteq \mathbb{R}^m$ *is linearly independent. Then*

(1) $n \leq m$;
(2) *if* $n = m$, *then* $x^1, \ldots, x^n$ *is a basis of* $\mathbb{R}^m$, *that is*

$$\mathbb{R}^m = \mathrm{span}\{x^1, \ldots, x^n\}.$$

Proof. (1) is obvious by Lemma 0.1. As for (2), note that $V := \mathrm{span}\{x^1, \ldots, x^m\}$ is a closed located linear subspace of $\mathbb{R}^m$ ([1, Lemma 4.1.2 and Corollary 4.1.5]). We show that $\mathbb{R}^m \subseteq V$. To this end, let $x \in \mathbb{R}^m$. We have to show that $d(x, V) = 0$, that is $\neg d(x, V) > 0$. Assume that $d(x, V) > 0$. Then $\{x^1, \ldots, x^m, x\}$ is linearly independent, see [1, Lemma 4.1.10]. This is a contradiction to Lemma 0.1. □

Lemma 0.3. *Let* $x^1, \ldots, x^n \in \mathbb{R}^m$. *Then* $\mathrm{co}\{x^1, \ldots, x^n\}$ *is located. Moreover, if* $n \geq 2$ *and* $\{x^1 - x^n, x^2 - x^n, \ldots, x^{n-1} - x^n\}$ *is linearly independent, then* $\mathrm{co}\{x^1, \ldots, x^n\}$ *is closed.*

Proof. Locatedness follows from [1, Propositions 2.2.6 and 2.2.9]. Next we prove that $\mathrm{co}\{x^1, \ldots, x^n\}$ is closed. Let $(y^k)_{k \in \mathbb{N}}$ be a sequence in $\mathrm{co}\{x^1, \ldots, x^n\}$ that converges to $y \in \mathbb{R}^m$. Further, let $\lambda^k \in S_n$ be such that

$$y^k = \sum_{i=1}^{n} \lambda_i^k x^i = x^n + \sum_{i=1}^{n-1} \lambda_i^k (x^i - x^n).$$

Then

$$y^k - y^l = \sum_{i=1}^{n-1} (\lambda_i^k - \lambda_i^l)(x^i - x^n).$$

By linear independence of $x^1 - x^n, \ldots, x^{n-1} - x^n$, it follows that both the linear mapping

$$(\mu_1, \ldots, \mu_{n-1}) \mapsto \sum_{i=1}^{n-1} \mu_i (x^i - x^n),$$

on $\mathbb{R}^{n-1}$ and its inverse are bounded linear injections, see [1, Corollary 4.1.5]. Hence, the sequence $(\lambda_1^k, \ldots, \lambda_{n-1}^k)_{k \in \mathbb{N}} \subseteq \mathbb{R}^{n-1}$ is Cauchy and thus converges to a limit $(\lambda_1, \ldots, \lambda_{n-1}) \in \mathbb{R}^{n-1}$. One verifies that

$$\lambda := \left(\lambda_1, \ldots, \lambda_{n-1}, 1 - \sum_{i=1}^{n-1} \lambda_i\right) \in S_n,$$

and that $y = \sum_{i=1}^{n} \lambda_i x^i \in \mathrm{co}\{x^1, \ldots, x^n\}$. □

Lemma 0.4. *For $n \geq 2$, fix $x^1, \ldots, x^n \in \mathbb{R}^m$ such that $\{x^1 - x^n, \ldots, x^{n-1} - x^n\}$ is linearly dependent. Moreover, let $x \in \mathrm{co}\{x^1, \ldots, x^n\}$. Then for each $\varepsilon > 0$ there exists $j \in I_n$ and $y \in \mathrm{co}\{x^i \mid i \in I_n \setminus \{j\}\}$ such that $\|x - y\| < \varepsilon$.*

Proof. Let $\lambda \in S_n$ such that $x = \sum_{i \in I_n} \lambda_i x^i$, and let $M > 0$ be such that $M > \|x^i\|$ for all $i \in I_n$. For all $i \in I_n$ either $\lambda_i > 0$ or $\lambda_i < \frac{\varepsilon}{2M}$. Suppose that there exists $j \in I_n$ such that $\lambda_j < \frac{\varepsilon}{2M}$. For $i \in I_n \setminus \{j\}$, let

$$\mu_i := \lambda_i + \frac{\lambda_j}{n-1}.$$

Note that $\mu_i \geq 0$ for all $i \in I_n \setminus \{j\}$ and that

$$\sum_{i \in I_n \setminus \{j\}} \mu_i = \sum_{i \in I_n} \lambda_i = 1.$$

Set

$$y := \sum_{I_n \setminus \{j\}} \mu_i x^i \in \mathrm{co}\{x^i \mid i \in I_n \setminus \{j\}\}.$$

Then

$$\|x - y\| \leq \lambda_j \|x^j\| + \frac{\lambda_j}{n-1} \sum_{i \in I_n \setminus \{j\}} \|x^i\| \leq 2M\lambda_j < \varepsilon.$$

Hence, the assertion of the lemma is proved in this case. Thus, we may from now on assume that $\lambda_i > 0$ for all $i \in I_n$. In that case, as the set $\{x^1 - x^n, \ldots, x^{n-1} - x^n\}$ is linearly dependent, there exists $\tilde{\nu} \in \mathbb{R}^{n-1}$ with $\|\tilde{\nu}\| > 0$ such that

$$\sum_{i \in I_{n-1}} \tilde{\nu}_i (x^i - x^n) = 0.$$

Let $\nu_i := \tilde{\nu}_i$ for $i \in I_{n-1}$ and $\nu_n := -\sum_{i \in I_{n-1}} \tilde{\nu}_i$. Then $\nu = (\nu_1, \ldots, \nu_n) \in \mathbb{R}^n$ satisfies

$$\|\nu\| > 0, \quad \sum_{i \in I_n} \nu_i = 0 \quad \text{and} \quad \sum_{i \in I_n} \nu_i x^i = 0.$$

In particular, there exists $k \in I_n$ such that $\nu_k > 0$. Let

$$\beta := \max\left\{\frac{\nu_i}{\lambda_i} \mid i \in I_n\right\}.$$

Then $\beta > 0$ and for all $i \in I_n$ we have $\tilde{\mu}_i := \lambda_i - \frac{1}{\beta}\nu_i \geq 0$ and

$$\sum_{i \in I_n} \tilde{\mu}_i = \sum_{i \in I_n} \lambda_i = 1 \quad \text{and} \quad x = \sum_{i \in I_n} \tilde{\mu}_i x^i.$$

Pick $j \in I_n$ such that $\nu_j > 0$ and

$$\beta - \frac{\nu_j}{\lambda_j} < \frac{\varepsilon\beta}{2M}.$$

Then $\tilde{\mu}_j < \frac{\varepsilon}{2M}$, so we are in the situation we covered in the first part of this proof and may thus construct $y \in \mathrm{co}\{x^i \mid i \in I_n \setminus \{j\}\}$ such that $\|x - y\| < \varepsilon$. □

For the following lemma, we recall that a subset M of a set N is said to be *detachable* from N if

$$\forall x \in N\ (x \in M \vee x \notin M).$$

Moreover, $\mathcal{P}_\omega(I_n)$ denotes the set of all finite subsets of I_n.

Lemma 0.5. *Let $n \geq 2$ and $x^1, \ldots, x^n \in \mathbb{R}^m$. Denote by $\mathcal{L}$ the set of all $J \subseteq I_n$ such that $|J| \geq 2$ and there exists $i \in J$ such that $\{x^j - x^i \mid j \in J \setminus \{i\}\}$ is linearly independent. Suppose that $\mathcal{L}$ is detachable from $\mathcal{P}_\omega(I_n)$. Then for all inhabited $J \subseteq I_n$ with $|J| \geq 2$, either*

(1) *there exists* $i \in J$ *such that* $\{x^j - x^i \mid j \in J \setminus \{i\}\}$ *is linearly independent; or*
(2) *there exists* $i \in J$ *such that* $\{x^j - x^i \mid j \in J \setminus \{i\}\}$ *is linearly dependent.*

Proof. Let $J \subseteq I_n$ be inhabited with $|J| \geq 2$. Note that

$$\{i,j\} \in \mathcal{L} \Leftrightarrow \|x^i - x^j\| > 0 \quad \text{and} \quad \neg(\{i,j\} \in \mathcal{L}) \Leftrightarrow \|x^i - x^j\| = 0. \tag{0.1}$$

Hence, as $\mathcal{L}$ is detachable from $\mathcal{P}_\omega(I_n)$, for arbitrary $i, j \in J$ we have either $\|x^j - x^i\| > 0$ or $\|x^j - x^i\| = 0$, and thus we know either that there are $i, j \in J$ such that $\|x^j - x^i\| = 0$ or else that $\|x^j - x^i\| > 0$ for all $i, j \in J$. In the first case, (2) holds. In the second case, the set

$$\mathcal{L}(J) := \{J' \mid (J' \in \mathcal{L}) \wedge (J' \subseteq J)\},$$

which is detachable from $\mathcal{P}_\omega(I_n)$, is inhabited. Pick a set $\tilde{J} \in \mathcal{L}(J)$ of maximal cardinality. If $\tilde{J} = J$, then (1) holds. If $\tilde{J} \subsetneq J$, let $i \in \tilde{J}$ be such that $\{x^j - x^i \mid j \in \tilde{J} \setminus \{i\}\}$ is linearly independent. Note that

$$\operatorname{span}\{x^j - x^i \mid j \in \tilde{J} \setminus \{i\}\}$$

is located and closed [1, Lemma 4.1.2 and Corollary 4.1.5]. For $k \in J \setminus \tilde{J}$, suppose that

$$d(x^k - x^i, \operatorname{span}\{x^j - x^i \mid j \in \tilde{J} \setminus \{i\}\}) > 0.$$

Then $\{x^k - x^i\} \cup \{x^j - x^i \mid j \in \tilde{J} \setminus \{i\}\}$ is linearly independent [1, Lemma 4.1.10]. Thus, $\tilde{J} \cup \{k\} \in \mathcal{L}(J)$, which contradicts maximality of $\tilde{J}$. Hence,

$$d(x^k - x^i, \operatorname{span}\{x^j - x^i \mid j \in \tilde{J} \setminus \{i\}\}) = 0,$$

that is, (2) holds. □

Definition 0.6. A formula φ is *conditionally constructive* if there exist a $k \in \mathbb{N}$ and a subset M of I_k such that the detachability of M from I_k implies φ.

It is straightforward to verify that conditionally constructive formulas are closed under conjunction and implication and may be used unconditionally in the proof of falsum:

Lemma 0.7. *Let the formulas* φ *and* ψ *be conditionally constructive. Then*

(1) *if* $\varphi \Rightarrow \nu$, *then* ν *is conditionally constructive;*
(2) $\varphi \wedge \psi$ *is conditionally constructive;*
(3) $(\varphi \Rightarrow \neg\nu) \Rightarrow \neg\nu$.

Proof. See [2, Lemma 2 and Proposition 6]. □

The following proposition shows that Carathéordory's Convexity Theorem is conditionally constructive.

Proposition 0.8. *Fix an inhabited set* $A \subseteq \mathbb{R}^m$ *and* $x \in \mathrm{co}A$. *Then the following statement is conditionally constructive:*

CCT(A) *There are vectors* $z^1, \ldots, z^k \in A$ *with* $k \leq m+1$ *such that*

$$x \in \mathrm{co}\{z^1, \ldots, z^k\}.$$

Proof. Let $x^1, \ldots, x^n \in A$ and $\lambda \in S_n$ such that $x = \sum_{i \in I_n} \lambda_i x^i$, and define $\mathcal{L}$ as in Lemma 0.5. Furthermore, define subsets $\Omega_i \subseteq \mathcal{P}_\omega(I_n) \times I_3$ $(i \in I_3)$ by

$$\begin{aligned}(J,1) \in \Omega_1 &\Leftrightarrow J \in \mathcal{L},\\ (J,2) \in \Omega_2 &\Leftrightarrow |J| \geq 1 \wedge d(x, \mathrm{co}\{x^j \mid j \in J\}) = 0,\\ (J,3) \in \Omega_3 &\Leftrightarrow |J| \geq 1 \wedge d(x, \mathrm{co}\{x^j \mid j \in J\}) > 0.\end{aligned}$$

Suppose that $\bigcup_{i \in I_3} \Omega_i$ is detachable from $\mathcal{P}_\omega(I_n) \times I_3$, which in particular implies that $\mathcal{L}$ is detachable from $\mathcal{P}_\omega(I_n)$. We prove, under this assumption, that there is $\tilde{J} \in \mathcal{P}_\omega(I_n)$ with $|\tilde{J}| \leq m+1$ such that

$$x \in \mathrm{co}\{x^j \mid j \in \tilde{J}\}.$$

Suppose that $\Omega_3 = \emptyset$. Then for all $j \in I_n$ we have $\neg(\|x - x^j\| = 0)$, i.e., $\|x - x^j\| = 0$. Hence, $x = x^1 = \ldots = x^n$, and the assertion holds. Thus, $\bigcup_{i \in I_3} \Omega_i$ being detachable, we may assume that Ω_3 is inhabited. Let

$$0 < \varepsilon < \min\{d(x, \mathrm{co}\{x^j \mid j \in J\}) \mid (J,3) \in \Omega_3\}.$$

Note that Ω_2 is inhabited, because $(I_n, 2) \in \Omega_2$. Let $\tilde{J} \in \mathcal{P}_\omega(I_n)$ be of minimal cardinality among all $J \in \mathcal{P}_\omega(I_n)$ such that $(J,2) \in \Omega_2$. If $\tilde{J} = \{j\}$, then $\|x - x^j\| = 0$, that is, $x = x^j$, and the assertion is proved. Hence, we may assume that $|\tilde{J}| \geq 2$. Since $\mathcal{L}$ is detachable from $\mathcal{P}_\omega(I_n) \times I_3$, we may apply Lemma 0.5 and conclude that either

$\tilde{J} \in \mathcal{L}$ or there is $i \in \tilde{J}$ such that $\{x^j - x^i \mid j \in \tilde{J} \setminus \{i\}\}$ is linearly dependent. Suppose the latter, and let $y \in \mathrm{co}\{x^j \mid j \in \tilde{J}\}$ be such that $\|x - y\| < \varepsilon/2$. By Lemma 0.4, there is $k \in \tilde{J}$ such that

$$d(y, \mathrm{co}\{x^j \mid j \in \tilde{J} \setminus \{k\}\}) < \varepsilon/2,$$

which implies that

$$\begin{aligned} d(x, \mathrm{co}\{x^j \mid j \in \tilde{J} \setminus \{k\}\}) &\leq \|x - y\| + d(y, \mathrm{co}\{x^j \mid j \in \tilde{J} \setminus \{k\}\}) \\ &< \varepsilon. \end{aligned}$$

Thus, $\neg((\tilde{J}\setminus\{k\}, 3) \in \Omega_3)$, that is, $(\tilde{J}\setminus\{k\}, 2) \in \Omega_2$, which contradicts the minimality of $\tilde{J}$. Therefore, $\{x^j - x^i \mid j \in \tilde{J} \setminus \{i\}\}$ is not linearly dependent and we must have $\tilde{J} \in \mathcal{L}$. Then, according to Lemma 0.3, $\mathrm{co}\{x^j \mid j \in \tilde{J}\}$ is closed. Since $(\tilde{J}, 2) \in \Omega_2$, that is, $d(x, \mathrm{co}\{x^j \mid j \in \tilde{J}\}) = 0$, and by closedness of $\mathrm{co}\{x^j \mid j \in \tilde{J}\}$, we infer that $x \in \mathrm{co}\{x^j \mid j \in \tilde{J}\}$. Moreover, $|\tilde{J}| \leq m + 1$ by Corollary 0.2. □

As a consequence of Proposition 0.8, we obtain the already advertised approximate version of Carathéordory's Convexity Theorem for totally bounded sets, namely that the convex hull $\mathrm{co}A$ of a totally bounded set $A \subseteq \mathbb{R}^m$ is approximated up to arbitrary small error by the subset consisting of all convex combinations of degree $m + 1$:

$$\mathrm{co}^{m+1}A := \left\{ \sum_{i \in I_{m+1}} \lambda_i z^i \mid z^i \in A\,(i = 1, \ldots, m+1), \lambda \in S_{m+1} \right\}.$$

Clearly, $\mathrm{co}^{m+1}A \subseteq \mathrm{co}A$. Hence, if we could prove that $\mathrm{co}^{m+1}A$ is closed, which we in general cannot, then $\mathrm{co}A = \mathrm{co}^{m+1}A$ as is classically always the case. Indeed, classically $\mathrm{co}A = \mathrm{co}^{m+1}A$ is compact whenever A is compact.

In the following, $\overline{\mathrm{co}}A$ denotes the closure of $\mathrm{co}A$ in $\mathbb{R}^m$. Similarly, $\overline{\mathrm{co}^{m+1}}A$ denotes the closure of $\mathrm{co}^{m+1}A$.

Theorem 0.9. *Suppose that $A \subseteq \mathbb{R}^m$ is totally bounded. Then for every $x \in \mathrm{co}A$ and every $\varepsilon > 0$ there is $y \in \mathrm{co}^{m+1}A$ such that $\|x - y\| < \varepsilon$. In particular, $\overline{\mathrm{co}}A = \overline{\mathrm{co}^{m+1}}A$ and $\overline{\mathrm{co}}A$ is compact.*

Proof. Let

$$\kappa : S_{m+1} \times A^{m+1} \to \mathbb{R}^m,$$
$$(\lambda_1, \dots, \lambda_{m+1}, z^1, \dots, z^{m+1}) \mapsto \sum_{i \in I_{m+1}} \lambda_i z^i.$$

As κ is uniformly continuous and its domain is totally bounded, its range $\mathrm{co}^{m+1}A$ is totally bounded as well, see [1, Proposition 2.2.6], and, hence, $\overline{\mathrm{co}^{m+1}A}$ is compact. We show that $\mathrm{co}A \subseteq \overline{\mathrm{co}^{m+1}A}$. Fix $x \in \mathrm{co}A$. We have to show that

$$d(x, \mathrm{co}^{m+1}A) = 0,$$

that is,

$$\neg(d(x, \mathrm{co}^{m+1}A) > 0).$$

According to Lemma 0.7 and Proposition 0.8, it suffices to prove this under the assumption that CCT(A) holds. But obviously

$$d(x, \mathrm{co}^{m+1}A) > 0$$

contradicts CCT(A). □

References

[1] D.S. Bridges and L.S. Vîţă, *Techniques of Constructive Analysis.* Universitext, Springer-Verlag, 2006.

[2] J. Berger and G. Svindland, *On Farkas' Lemma and Related Propositions in BISH*, https://arxiv.org/abs/2101.03424 (2021).

https://doi.org/10.1142/9789811245220_0006

Chapter 6

Varieties of the Weak Kőnig Lemma and the Disjunctive Dependent Choice

Josef Berger[*,§], Hajime Ishihara[†,¶] and Takako Nemoto[‡,‖]

[*] *Mathematisches Institut der Universität München, Germany*

[†] *School of Information Science, Japan Advanced Institute of Science and Technology, Japan*

[‡] *Department of Architectural Design, Faculty of Environmental Study, Hiroshima Institute of Technology, Japan*

[§] *jberger@math.lmu.de*

[¶] *ishihara@jaist.ac.jp*

[‖] *nemototakako@gmail.com*

In this chapter, we investigate decompositions of varieties of the weak Kőnig lemma into the lesser limited principle of omniscience and versions of the disjunctive dependent choice.

1. Introduction

The *Weak Kőnig lemma*

WKL: Every infinite (binary) tree has a branch

(see Definition 2.1), which, along with its classical contraposition, the *fan theorem*, plays an important role in the Friedman–Simpson program, called (classical) reverse mathematics [1] as well as in intuitionistic mathematics [2, 4.7]. Recently, a couple of versions of WKL, that is, $\mathrm{WKL}_{\leq 2}$ for trees having at most two nodes at each level and WKL_c for convex trees, have found to play important roles in constructive reverse mathematics; see [3,4] for constructive reverse mathematics. Berger *et al.* [5] have proved that $\mathrm{WKL}_{\leq 2}$ is equivalent to the *binary expansion* of real numbers in the unit interval:

BE: Every real number in $[0, 1]$ has a binary expansion,

and that WKL_c is equivalent to the *intermediate value theorem*:

IVT: If $f : [0, 1] \to \mathbb{R}$ is a uniformly continuous function with $f(0) < 0 < f(1)$, then there exists $x \in [0, 1]$ such that $f(x) = 0$,

in a formal system $\mathbf{EL}_0$ for constructive reverse mathematics; see [5] for the system $\mathbf{EL}_0$, and Nemoto [6] for formal systems of constructive reverse mathematics. They have revealed the following implications and equivalences over the system $\mathbf{EL}_0$:

$$\begin{array}{ccccccc} & & \mathrm{BE} & & \mathrm{IVT} & & \\ & & \updownarrow & & \updownarrow & & \\ \mathrm{LLPO} & \longleftarrow & \mathrm{WKL}_{\leq 2} & \longleftarrow & \mathrm{WKL}_c & \longleftarrow & \mathrm{WKL}; \end{array}$$

see Kihara [7] for separations among the implications.

On the other hand, it is well known that WKL implies the *lesser limited principle of omniscience* (LLPO or Σ_1^0-DML):

$$\neg(\exists n(\sigma(n) \neq 0) \wedge \exists n(\tau(n) \neq 0)) \to \neg\exists n(\sigma(n) \neq 0) \vee \neg\exists n(\tau(n) \neq 0)$$

for all infinite binary sequences σ and τ, which is an instance of De Morgan's law (DML):

$$\neg(A \wedge B) \to \neg A \vee \neg B$$

for Σ^0_1 formulae $A \equiv \exists n(\sigma(n) \neq 0)$ and $B \equiv \exists n(\tau(n) \neq 0)$. Berger *et al.* [8] showed that WKL can be decomposed into the logical principle LLPO and a function existence axiom, the *disjunctive dependent choice* ($\mathrm{DC}^\vee$):

$$\forall a(A(a) \vee B(a)) \rightarrow \exists \gamma \forall n[(\gamma(n) = 0 \rightarrow A(\overline{\gamma}(n))) \wedge (\gamma(n) = 1 \rightarrow B(\overline{\gamma}(n)))],$$

for Π^0_1 formulae A and B, where a ranges over the set of finite binary sequences, γ over the set of infinite binary sequences, and $\overline{\gamma}(n)$ denotes the initial segments of γ of length n.

The aim of this chapter is to give similar decompositions of $\mathrm{WKL}_{\leq 2}$ and WKL_c into the logical principle LLPO and, as function existence axioms, two versions of the disjunctive dependent choice $\mathrm{DC}^\vee$ in the system $\mathbf{EL}_0$ are presented.

In Section 2, we give characterisations, respectively, of trees having at most two nodes at each level and of convex trees, and also propose two conditions on (the Π^0_1 formulae of) $\mathrm{DC}^\vee$. In Section 3, we introduce the notion of a separated tree, and show the equivalence between LLPO and the statement that *every tree is separated*. In Section 4, we show that WKL, $\mathrm{WKL}_{\leq 2}$ and WKL_c for separated trees imply $\mathrm{DC}^\vee$, $\mathrm{DC}^\vee$ with the first condition and $\mathrm{DC}^\vee$ with the second condition, respectively. In Section 5, we show their converses, that is, $\mathrm{DC}^\vee$, $\mathrm{DC}^\vee$ with the first condition and $\mathrm{DC}^\vee$ with the second condition imply WKL, $\mathrm{WKL}_{\leq 2}$ and WKL_c for separated trees, respectively. We conclude the chapter with consequences, as corollaries, of what we have obtained, in Section 6.

2. Preliminaries

In this section, we introduce the basic notions and the notations which will be used throughout the chapter, and characterise trees having at most two nodes at each level and convex trees.

Throughout the chapter, k, m and n are supposed to range over the set N of natural numbers, and a, b, c, d and e over the set $\{0,1\}^*$ of finite binary sequences; $|a|$ denotes the length of a, and $a(k)$ the $k+1$th element of a, where $k < |a|$; $a * b$ denotes the concatenation of a and b, and $\overline{a}(n)$ the initial segments of a of length n, where $n \leq |a|$. We write $\{0,1\}^n$ and $\{0,1\}^{\leq n}$ for $\{a \in \{0,1\}^* \mid |a| = n\}$ and $\{a \in \{0,1\}^* \mid |a| \leq n\}$, respectively, and 0^n and 1^n

for $\langle 0,\dots,0\rangle$ and $\langle 1,\dots,1\rangle$ in $\{0,1\}^n$, respectively. Also γ, σ and τ are supposed to range over the set $N \to \{0,1\}$ of infinite binary sequences; $\overline{\gamma}(n)$ denotes the initial segments of γ of length n, that is, $\overline{\gamma}(n) = \langle \gamma(0),\dots,\gamma(n-1)\rangle$.

Let $a \preceq b$ denote that a is an *initial segment* of b, that is,

$$a \preceq b \Leftrightarrow \exists c \in \{0,1\}^{\leq |b|}(a * c = b),$$

and let $a \sqsubset b$ denote that a is *on the left of* b (b is *on the right of* a), that is,

$$a \sqsubset b \Leftrightarrow \exists c \in \{0,1\}^{\leq |a|}(c * \langle 0\rangle \preceq a \wedge c * \langle 1\rangle \preceq b).$$

A subset T of $\{0,1\}^*$ is *detachable* if there exists $\delta \in \{0,1\}^* \to \{0,1\}$ such that $\delta(a) \neq 0 \leftrightarrow a \in T$ for all a.

Definition 2.1. A *tree* is a detachable subset T of $\{0,1\}^*$ such that $\langle\rangle \in T$ and if $b \in T$ and $a \preceq b$, then $a \in T$ for all a and b. An infinite binary sequence γ is a *branch* of T if $\overline{\gamma}(n) \in T$ for all n. A tree T

- is *infinite* if for each n there exists $a \in T$ with $|a| = n$;
- *has at most two nodes at each level* if

$$|a| = |b| = |c| \to a = b \vee b = c \vee c = a,$$

  for all $a, b, c \in T$;
- is *convex* if

$$|a| = |b| = |c| \wedge a \sqsubset c \sqsubset b \to c \in T,$$

  for all $a, b \in T$ and c.

For a tree T, let C_0^T and C_1^T be predicates on $\{0,1\}^* \times N$ given by

$$C_0^T(a,n) \equiv \exists c \in \{0,1\}^n(a * \langle 0\rangle * c \in T),$$
$$C_1^T(a,n) \equiv \exists c \in \{0,1\}^n(a * \langle 1\rangle * c \in T),$$

for each a and n. Note that

$$C_0^T(a,n+1) \to C_0^T(a,n) \quad \text{and} \quad C_1^T(a,n+1) \to C_1^T(a,n),$$

for all a and n, and that

$$(C_0^T(b,n) \vee C_1^T(b,n)) \wedge a * \langle 0 \rangle \preceq b \rightarrow C_0^T(a, |b| - |a| + n),$$

and

$$(C_0^T(b,n) \vee C_1^T(b,n)) \wedge a * \langle 1 \rangle \preceq b \rightarrow C_1^T(a, |b| - |a| + n),$$

for all a, b and n.

The following is a characterisation of trees having at most two nodes at each level using the predicates C_0^T and C_1^T.

Proposition 2.2. *A tree T has at most two nodes at each level if and only if*

$$(a * \langle 0 \rangle \preceq b \rightarrow \neg(C_1^T(a, |b| - |a|) \wedge C_0^T(b,0) \wedge C_1^T(b,0)))$$
$$\wedge (a * \langle 1 \rangle \preceq b \rightarrow \neg(C_0^T(a, |b| - |a|) \wedge C_0^T(b,0) \wedge C_1^T(b,0))),$$

for all a and b.

Proof. Suppose that T is a tree having at most two nodes at each level, and consider a and b. If $a * \langle 0 \rangle \preceq b$, $C_1^T(a, |b| - |a|)$, $C_0^T(b,0)$ and $C_1^T(b,0)$, then there exists $c \in \{0,1\}^{|b|-|a|}$ such that $a * \langle 1 \rangle * c \in T$, and, since $b * \langle 0 \rangle \in T$, $b * \langle 1 \rangle \in T$ and $|a * \langle 1 \rangle * c| = |b * \langle 0 \rangle| = |b * \langle 1 \rangle|$, we have $a * \langle 1 \rangle * c = b * \langle 0 \rangle$, $b * \langle 0 \rangle = b * \langle 1 \rangle$ or $b * \langle 1 \rangle = a * \langle 1 \rangle * c$, a contradiction. Similarly, if $a * \langle 1 \rangle \preceq b$, $C_0^T(a, |b| - |a|)$, $C_0^T(b,0)$ and $C_1^T(b,0)$, then we have a contradiction.

Conversely, suppose that T is a tree such that

$$(a * \langle 0 \rangle \preceq b \rightarrow \neg(C_1^T(a, |b| - |a|) \wedge C_0^T(b,0) \wedge C_1^T(b,0)))$$
$$\wedge (a * \langle 1 \rangle \preceq b \rightarrow \neg(C_0^T(a, |b| - |a|) \wedge C_0^T(b,0) \wedge C_1^T(b,0))),$$

for all a and b. Consider $a, b, c \in T$ with $|a| = |b| = |c|$, and assume that $a \neq b$, $b \neq c$ and $c \neq a$. Furthermore, we may assume without loss of generality that $a \sqsubset b \sqsubset c$. Then there exist d and e such that either $d * \langle 0 \rangle \preceq e$, $e * \langle 0 \rangle \preceq a$, $e * \langle 1 \rangle \preceq b$ and $d * \langle 1 \rangle \preceq c$; or $d * \langle 1 \rangle \preceq e$, $e * \langle 1 \rangle \preceq c$, $e * \langle 0 \rangle \preceq b$ and $d * \langle 0 \rangle \preceq a$. In the first case, since $e * \langle 0 \rangle \in T$, $e * \langle 1 \rangle \in T$ and $d * \langle 1 \rangle \preceq \overline{c}(|e| + 1) \in T$, we have $C_1^T(d, |e| - |d|)$, $C_0^T(e,0)$ and $C_1^T(e,0)$, a contradiction. In the second case, similarly, we have a contradiction. Therefore, $a = b$, $b = c$ or $c = a$. $\square$

We need the following technical lemma before characterising convex trees.

Lemma 2.3. *Let T be a tree such that*

$$
\begin{aligned}
(a * \langle 0\rangle \preceq b \rightarrow \forall n \neg(C_1^T(a, |b|-|a|+n) \wedge C_0^T(b, n) \wedge \neg C_1^T(b, n))) \\
\wedge (a * \langle 1\rangle \preceq b \rightarrow \forall n \neg(C_0^T(a, |b|-|a|+n) \wedge \neg C_0^T(b, n) \\
\wedge C_1^T(b, n))),
\end{aligned}
$$

for all a and b. Then for each n, k, each $a, b \in \{0,1\}^{\leq n}$ and each $c \in \{0,1\}^k$,

$$
\begin{aligned}
&(|b|+k+1 \leq n \wedge a * \langle 0\rangle \preceq b \wedge C_1^T(a, |b|-|a|+k) \wedge C_0^T(b, k) \\
&\quad \rightarrow b * \langle 1\rangle * c \in T) \\
&\wedge (|b|+k+1 \leq n \wedge a * \langle 1\rangle \preceq b \wedge C_0^T(a, |b|-|a|+k) \wedge C_1^T(b, k) \\
&\quad \rightarrow b * \langle 0\rangle * c \in T).
\end{aligned}
$$

Proof. Suppose that

$$
\begin{aligned}
(a * \langle 0\rangle \preceq b \rightarrow \forall n \neg(C_1^T(a, |b|-|a|+n) \wedge C_0^T(b, n) \wedge \neg C_1^T(b, n))) \\
\wedge (a * \langle 1\rangle \preceq b \rightarrow \forall n \neg(C_0^T(a, |b|-|a|+n) \wedge \neg C_0^T(b, n) \\
\wedge C_1^T(b, n))),
\end{aligned}
$$

for all a and b. Then, given an n, we proceed by induction on k. Consider $a, b \in \{0,1\}^{\leq n}$ such that $|b|+1 \leq n$, $a * \langle 0\rangle \preceq b$, $C_1^T(a, |b|-|a|)$ and $C_0^T(b, 0)$. If $b * \langle 1\rangle \notin T$, then $\neg C_1^T(b, 0)$, a contradiction to our assumption. Therefore, $b * \langle 1\rangle \in T$. Similarly, for $a, b \in \{0,1\}^{\leq n}$ with $|b|+1 \leq n$, $a * \langle 1\rangle \preceq b$, $C_0^T(a, |b|-|a|)$ and $C_1^T(b, 0)$, we have $b * \langle 0\rangle \in T$.

Consider $a, b \in \{0,1\}^{\leq n}$ and $c \in \{0,1\}^{k+1}$ such that $|b|+(k+1)+1 \leq n$, $a * \langle 0\rangle \preceq b$, $C_1^T(a, |b|-|a|+k+1)$ and $C_0^T(b, k+1)$, and assume that $b * \langle 1\rangle * c \notin T$. Note that $|b * \langle 1\rangle * c| = |b|+1+(k+1) \leq n$, and assume further that $C_1^T(b, k+1)$. Then there exists $d \in \{0,1\}^{k+1}$ such that $b * \langle 1\rangle * d \in T$, and, since $c \neq d$, there exist $e \in \{0,1\}^{\leq n}$ with $b * \langle 1\rangle \preceq e$ and $c', d' \in \{0,1\}^{\leq k}$ with $|c'| = |d'|$ such that either $b * \langle 1\rangle * c = e * \langle 0\rangle * c'$ and $b * \langle 1\rangle * d = e * \langle 1\rangle * d'$, or

$b * \langle 1 \rangle * c = e * \langle 1 \rangle * c'$ and $b * \langle 1 \rangle * d = e * \langle 0 \rangle * d'$. Note that, since $|b * \langle 1 \rangle * c| = |e * \langle 0 \rangle * c'| = |e * \langle 1 \rangle * c'|$, we have $|b| + k + 1 = |e| + |c'|$ in both cases. In the first case, since $|e| - |b| + |c'| = k + 1$, we have $C_0^T(b, |e| - |b| + |c'|)$, $C_1^T(e, |c'|)$ and $e * \langle 0 \rangle * c' \notin T$, a contradiction to the induction hypothesis. In the second case, since $a * \langle 0 \rangle \preceq e$ and $|e| - |a| + |c'| = |b| - |a| + (k+1)$, we have $C_1^T(a, |e| - |a| + |c'|)$, $C_0^T(e, |c'|)$ and $e * \langle 1 \rangle * c' \notin T$, a contradiction to the induction hypothesis. Therefore, $\neg C_1^T(b, k+1)$, and this, together with $C_1^T(a, |b| - |a| + k + 1)$ and $C_0^T(b, k + 1)$, contradicts our assumption. Thus, $b * \langle 1 \rangle * c \in T$. Similarly, for $a, b \in \{0,1\}^{\leq n}$ and $c \in \{0,1\}^{k+1}$ with $|b| + (k+1) + 1 \leq n$, $a * \langle 1 \rangle \preceq b$, $C_0^T(a, |b| - |a| + k + 1)$ and $C_1^T(b, k + 1)$, we have $b * \langle 0 \rangle * c \in T$. □

The following is a characterisation of convex trees using the predicates C_0^T and C_1^T.

Proposition 2.4. *A tree T is convex if and only if*

$$(a * \langle 0 \rangle \preceq b \rightarrow \forall n \neg (C_1^T(a, |b| - |a| + n) \wedge C_0^T(b, n) \wedge \neg C_1^T(b, n)))$$
$$\wedge (a * \langle 1 \rangle \preceq b \rightarrow \forall n \neg (C_0^T(a, |b| - |a| + n) \wedge \neg C_0^T(b, n) \wedge C_1^T(b, n))),$$

for all a and b.

Proof. Suppose that T is a convex tree, and consider a and b. Given an n, if $a * \langle 0 \rangle \preceq b$, $C_1^T(a, |b| - |a| + n)$, $C_0^T(b, n)$ and $\neg C_1^T(b, n)$, then there exist $c \in \{0,1\}^n$ and $d \in \{0,1\}^{|b|-|a|+n}$ such that $b * \langle 0 \rangle * c \in T$, $a * \langle 1 \rangle * d \in T$ and $|b * \langle 0 \rangle * c| = |a * \langle 1 \rangle * d|$, and hence for any $e \in \{0,1\}^n$, since $b * \langle 0 \rangle * c \sqsubset b * \langle 1 \rangle * e \sqsubset a * \langle 1 \rangle * d$, we have $b * \langle 1 \rangle * e \in T$; whence $C_1^T(b, n)$, a contradiction. Similarly, given an n, if $a * \langle 1 \rangle \preceq b$, $C_0^T(a, |b| - |a| + n)$, $\neg C_0^T(b, n)$ and $C_1^T(b, n)$, then we have a contradiction.

Conversely, suppose that

$$(a * \langle 0 \rangle \preceq b \rightarrow \forall n \neg (C_1^T(a, |b| - |a| + n) \wedge C_0^T(b, n) \wedge \neg C_1^T(b, n)))$$
$$\wedge (a * \langle 1 \rangle \preceq b \rightarrow \forall n \neg (C_0^T(a, |b| - |a| + n) \wedge \neg C_0^T(b, n) \wedge C_1^T(b, n))),$$

for all a and b. Now consider $a, b \in T$ and c with $|a| = |b| = |c|$ and $a \sqsubset c \sqsubset b$. Then there exist d and e such that either $d * \langle 0 \rangle \preceq e$,

$e * \langle 0 \rangle \preceq a$, $e * \langle 1 \rangle \preceq c$ and $d * \langle 1 \rangle \preceq b$; or $d * \langle 1 \rangle \preceq e$, $e * \langle 1 \rangle \preceq b$, $e * \langle 0 \rangle \preceq c$ and $d * \langle 0 \rangle \preceq a$. Note that setting $n = |a|$ $(= |b| = |c|)$ and $k = n - (|e| + 1)$, we have $d, e \in \{0,1\}^{\leq n}$ and $|e| + k + 1 \leq n$. In the first case, since $d * \langle 0 \rangle \preceq e$, $C_1^T(d, |e| - |d| + k)$, $C_0^T(e, k)$ and $c = e * \langle 1 \rangle * c'$ for some $c' \in \{0,1\}^k$, we have $c \in T$, by Lemma 2.3. In the second case, similarly, we have $c \in T$. □

In the following, α and β are supposed to range over $\{0,1\}^* \times N \to \{0,1\}$. Furthermore, since we are interested in the Σ_1^0 formulae

$$\exists n(\alpha(a,n) \neq 0) \quad \text{and} \quad \exists n(\beta(a,n) \neq 0)$$

on $\{0,1\}^*$, or their negations as Π_1^0 formulae, replacing α and β by

$$\alpha'(a,n) = \max\{\alpha(a,k) \mid k \leq n\} \quad \text{and} \quad \beta'(a,n) = \max\{\beta(a,k) \mid k \leq n\},$$

respectively, we may assume without loss of generality that for each a, $\alpha(a,n)$ and $\beta(a,n)$ are nondecreasing in n, in the sense that

$$\alpha(a,n+1) = 0 \to \alpha(a,n) = 0 \quad \text{and} \quad \beta(a,n+1) = 0 \to \beta(a,n) = 0,$$

for all n.

Using such α and β, the axiom of *disjunctive dependent choice* for Π_1^0 formulae (Π_1^0-DC$^\vee$) is formulated by

$$\begin{aligned}
&\forall a[\forall n(\alpha(a,n) = 0) \vee \forall n(\beta(a,n) = 0)] \\
&\quad \to \exists \gamma \forall n[(\gamma(n) = 0 \to \forall m(\alpha(\overline{\gamma}(n), m) = 0)) \\
&\quad \wedge (\gamma(n) = 1 \to \forall m(\beta(\overline{\gamma}(n), m) = 0))],
\end{aligned}$$

or

$$\begin{aligned}
&\forall a[\neg\exists n(\alpha(a,n) \neq 0) \vee \neg\exists n(\beta(a,n) \neq 0)] \\
&\quad \to \exists \gamma \forall n[(\gamma(n) = 0 \to \neg\exists m(\alpha(\overline{\gamma}(n), m) \neq 0)) \\
&\quad \wedge (\gamma(n) = 1 \to \neg\exists m(\beta(\overline{\gamma}(n), m) \neq 0))].
\end{aligned}$$

Various conditions on α and β, hence on the Π_1^0 formulae, yield various versions of Π_1^0-DC$^\vee$, and we consider the following conditions on α and β.

Definition 2.5. Let $\alpha, \beta \in \{0,1\}^* \times N \to \{0,1\}$. Then α and β *satisfy*

- *the condition* (†) if

$$
\begin{aligned}
(a * \langle 0\rangle \preceq b \to \beta(a, |b|+1) \neq 0 \vee \alpha(b, |b|+1) \neq 0 \\
\vee\, \beta(b, |b|+1) \neq 0) \wedge (a * \langle 1\rangle \preceq b \to \alpha(a, |b|+1) \neq 0 \\
\vee\, \alpha(b, |b|+1) \neq 0 \vee \beta(b, |b|+1) \neq 0),
\end{aligned}
$$

  for all a and b;
- *the condition* (‡) if

$$
\begin{aligned}
(a * \langle 0\rangle \preceq b \to \forall n(\beta(a, |b|+n+1) \neq 0 \vee \beta(b, |b|+n+1) = 0)) \\
\wedge\, (a * \langle 1\rangle \preceq b \to \forall n(\alpha(a, |b|+n+1) \neq 0 \\
\vee\, \alpha(b, |b|+n+1) = 0)),
\end{aligned}
$$

  for all a and b.

In the following, we examine the following axioms of disjunctive dependent choice for Π^0_1 formulae corresponding to the above conditions.

Π^0_1-$\mathrm{DC}^\vee$(†): The disjunctive dependent choice for Π^0_1 formulae with α and β satisfying the condition (†).

Π^0_1-$\mathrm{DC}^\vee$(‡): The disjunctive dependent choice for Π^0_1 formulae with α and β satisfying the condition (‡).

Remark 2.6. In Propositions 4.6, 4.8, 5.5 and 5.6, we will observe, by Lemma 4.7 and the notion of a T-faithfulness, that the conditions (†) and (‡) correspond to the conditions in Propositions 2.2 and 2.4, respectively.

3. Separated Trees

In this section, we introduce the notion of a separated tree, and show the equivalence between LLPO and the statement that *every tree is separated*, from which we infer that WKL for separated trees is purely function existential without the logical principle LLPO.

Definition 3.1. A tree T is *separated* if

$$\neg\exists n(C_0^T(a,n) \wedge \neg C_1^T(a,n)) \vee \neg\exists n(C_1^T(a,n) \wedge \neg C_0^T(a,n)),$$

or

$$\forall n(C_0^T(a,n) \to C_1^T(a,n)) \vee \forall n(C_1^T(a,n) \to C_0^T(a,n)),$$

for all a.

Remark 3.2. For a tree T, let A_T and B_T be Σ_1^0 subsets of $\{0,1\}^*$ given by

$$A_T = \{a \in \{0,1\}^* \mid \exists n(C_0^T(a,n) \wedge \neg C_1^T(a,n))\},$$
$$B_T = \{a \in \{0,1\}^* \mid \exists n(C_1^T(a,n) \wedge \neg C_0^T(a,n))\}.$$

Then A_T and B_T are disjoint if

$$\neg[\exists n(C_0^T(a,n) \wedge \neg C_1^T(a,n)) \wedge \exists n(C_1^T(a,n) \wedge \neg C_0^T(a,n))],$$

for all a, and we may say that A_T and B_T are *separated* if

$$\neg\exists n(C_0^T(a,n) \wedge \neg C_1^T(a,n)) \vee \neg\exists n(C_1^T(a,n) \wedge \neg C_0^T(a,n)),$$

for all a. Note that if A_T and B_T are separated, then they are disjoint, and the converse holds in the presence of LLPO. A tree T is separated if A_T and B_T are separated.

Lemma 3.3. *For each σ and each τ, there exists a convex tree T having at most two nodes at each level such that*

(1) *if T has a branch, then $\neg\exists n(\sigma(n) \neq 0)$ or $\neg\exists n(\tau(n) \neq 0)$;*
(2) *if $\neg(\exists n(\sigma(n) \neq 0) \wedge \exists n(\tau(n) \neq 0))$, then T is infinite;*
(3) *if $\neg(\exists n(\sigma(n) \neq 0) \wedge \exists n(\tau(n) \neq 0))$ and T is separated, then $\neg\exists n(\sigma(n) \neq 0)$ or $\neg\exists n(\tau(n) \neq 0)$.*

Proof. Let T be a subset of $\{0,1\}^*$ given by

$$T = \{\langle\rangle\} \cup \{\langle 0\rangle * 1^n \mid \forall k < n(\sigma(k) = 0)\} \cup \{\langle 1\rangle * 0^n \mid \forall k < n(\tau(k) = 0)\}.$$

Then it is straightforward to show that T is a convex tree having at most two nodes at each level.

If T has a branch γ, then either $\gamma(0) = 0$ or $\gamma(0) = 1$; in the first case, we have $\neg\exists n(\sigma(n) \neq 0)$; in the second case, we have $\neg\exists n(\tau(n) \neq 0)$.

Assume that $\neg(\exists n(\sigma(n) \neq 0) \wedge \exists n(\tau(n) \neq 0))$, and note that it is equivalent to that $\forall k < n(\sigma(k) = 0)$ or $\forall k < n(\tau(k) = 0)$ for all n. Then for each $n > 0$, either $\forall k < n - 1(\sigma(k) = 0)$ or $\forall k < n - 1(\tau(k) = 0)$; in the former case, we have $\langle 0 \rangle * 1^{n-1} \in T$ and $|\langle 0 \rangle * 1^{n-1}| = n$; in the latter case, we have $\langle 1 \rangle * 0^{n-1} \in T$ and $|\langle 1 \rangle * 0^{n-1}| = n$. Therefore, T is infinite.

Assume that $\neg(\exists n(\sigma(n) \neq 0) \wedge \exists n(\tau(n) \neq 0))$ and T is separated. Then for each a, either $C_0^T(a, n) \to C_1^T(a, n)$ for all n, or $C_1^T(a, n) \to C_0^T(a, n)$ for all n. In the first case, if there exists n such that $\tau(n) \neq 0$, then $\neg\exists m(\sigma(m) \neq 0)$; hence $C_0^T(\langle\rangle, n + 1)$ and $\neg C_1^T(\langle\rangle, n + 1)$, a contradiction. Therefore, $\neg\exists n(\tau(n) \neq 0)$. In the second case, similarly, we have $\neg\exists n(\sigma(n) \neq 0)$. □

For the sake of completeness, we give a proof of the following proposition using Lemma 3.3.

Proposition 3.4.

(1) WKL *implies* LLPO,
(2) $\mathrm{WKL}_{\leq 2}$ *implies* LLPO,
(3) WKL_c *implies* LLPO.

Proof. Let σ and τ be such that $\neg(\exists n(\sigma(n) \neq 0) \wedge \exists n(\tau(n) \neq 0))$. Then there exists an infinite, convex tree T having at most two nodes at each level, by Lemma 3.3(2). Therefore, by WKL, $\mathrm{WKL}_{\leq 2}$ or WKL_c, there exists a branch of T, and so $\neg\exists n(\sigma(n) \neq 0)$ or $\neg\exists n(\tau(n) \neq 0)$, by Lemma 3.3(1). □

Theorem 3.5. *The following are equivalent.*

(1) *Every tree is separated.*
(2) LLPO.

Proof. (1) $\Rightarrow$ (2): Suppose that every tree is separated, and let σ and τ be such that $\neg(\exists n(\sigma(n) \neq 0) \wedge \exists n(\tau(n) \neq 0))$. Then there exists an infinite, convex tree T having at most two nodes at each level, by Lemma 3.3(2). Therefore, since T is separated, we have either $\neg\exists n(\sigma(n) \neq 0)$ or $\neg\exists n(\tau(n) \neq 0)$, by Lemma 3.3(3).

(2) $\Rightarrow$ (1): Let T be a tree. Given an a, define σ and τ by

$$\sigma(n) = 1 \leftrightarrow \exists k \leq n(\neg C_0^T(a,k) \wedge C_1^T(a,k)),$$

$$\tau(n) = 1 \leftrightarrow \exists k \leq n(C_0^T(a,k) \wedge \neg C_1^T(a,k)),$$

for each n. Then, trivially, $\neg(\exists n(\sigma(n) \neq 0) \wedge \exists n(\tau(n) \neq 0))$, and so, by LLPO, either $\neg\exists n(\sigma(n) \neq 0)$ or $\neg\exists n(\tau(n) \neq 0)$; in the first case, we have $C_1^T(a,n) \rightarrow C_0^T(a,n)$ for all n; in the second case, we have $C_0^T(a,n) \rightarrow C_1^T(a,n)$ for all n. □

In the rest of the chapter, we investigate the following varieties of the weak Kőnig lemma, which are purely function existential without LLPO.

- WKL^-: Every infinite, separated tree has a branch.
- $\mathrm{WKL}^-_{\leq 2}$: Every infinite, separated tree having at most two nodes at each level has a branch.
- WKL^-_c: Every infinite, separated, convex tree has a branch.

4. From WKL to DC$^\vee$

In this section, we show that WKL^-, $\mathrm{WKL}^-_{\leq 2}$ and WKL^-_c imply $\mathrm{DC}^\vee$, $\mathrm{DC}^\vee(\dagger)$ and $\mathrm{DC}^\vee(\ddagger)$, respectively.

Let

$$S_{\alpha,\beta}(a,n) \equiv \forall k < |a|[(\alpha(\overline{a}k,n) \neq 0 \rightarrow a(k) = 1) \\ \wedge (\beta(\overline{a}k,n) \neq 0 \rightarrow a(k) = 0)],$$

for each a and n. Then $S_{\alpha,\beta}(\langle\rangle,n)$, $S_{\alpha,\beta}(b,n) \wedge a \preceq b \rightarrow S_{\alpha,\beta}(a,n)$ and, since α and β are nondecreasing, $S_{\alpha,\beta}(a,n+1) \rightarrow S_{\alpha,\beta}(a,n)$ for all a, b and n. Therefore, the subset

$$T_{\alpha,\beta} = \{a \mid S_{\alpha,\beta}(a,|a|)\},$$

of $\{0,1\}^*$ is a tree.

Lemma 4.1. *For each γ, the following are equivalent.*

(1) *γ is a branch of $T_{\alpha,\beta}$;*

(2) $(\gamma(n) = 0 \rightarrow \neg\exists m(\alpha(\overline{\gamma}(n),m) \neq 0)) \wedge (\gamma(n) = 1 \rightarrow \neg\exists m(\beta(\overline{\gamma}(n),m) \neq 0))$ *for all n.*

Proof. (1) $\Rightarrow$ (2): Suppose that γ is a branch of $T_{\alpha,\beta}$, that is, $\overline{\gamma}(n) \in T_{\alpha,\beta}$, or $S_{\alpha,\beta}(\overline{\gamma}(n), n)$, for all n. Given n and m, if $\gamma(n) = 0$ and $\alpha(\overline{\gamma}(n), m) \neq 0$, then, since $S_{\alpha,\beta}(\overline{\gamma}(N), N)$ and $\alpha(\overline{\gamma}(n), N) \neq 0$ for $N = \max\{n+1, m\}$, we have $\gamma(n) = 1$, a contradiction. Therefore, $\gamma(n) = 0 \to \neg\exists m(\alpha(\overline{\gamma}(n), m) \neq 0)$ for all n. Similarly, we have $\gamma(n) = 1 \to \neg\exists m(\beta(\overline{\gamma}(n), m) \neq 0)$ for all n.

(2) $\Rightarrow$ (1): Suppose that

$$(\gamma(n) = 0 \to \neg\exists m(\alpha(\overline{\gamma}(n), m) \neq 0)) \wedge (\gamma(n) \\ = 1 \to \neg\exists m(\beta(\overline{\gamma}(n), m) \neq 0)),$$

for all n. Then for each n and each $k < n$, if $\alpha(\overline{\gamma}(k), n) \neq 0$, then $\alpha(\overline{\gamma}(k), m) \neq 0$ for some $m \leq n$, and so $\gamma(k) = 1$; similarly, if $\beta(\overline{\gamma}(k), n) \neq 0$, then $\gamma(k) = 0$. Therefore, $S_{\alpha,\beta}(\overline{\gamma}(n), n)$, or $\overline{\gamma}(n) \in T_{\alpha,\beta}$ for all n. □

Definition 4.2. Let $\alpha, \beta \in \{0,1\}^* \times N \to \{0,1\}$. Then α and β are *separated* if

$$\neg\exists n(\alpha(a, n) \neq 0) \vee \neg\exists n(\beta(a, n) \neq 0),$$

for each a.

Remark 4.3. α and β are separated if and only if Σ^0_1 subsets A and B of $\{0,1\}^*$, given by

$$A = \{a \in \{0,1\}^* \mid \exists n(\alpha(a, n) \neq 0)\},$$
$$B = \{a \in \{0,1\}^* \mid \exists n(\beta(a, n) \neq 0)\},$$

are separated.

Lemma 4.4. *Let α and β be separated. For each a and n, if $S_{\alpha,\beta}(a, n)$, then for each m there exists $c \in \{0,1\}^m$ such that $S_{\alpha,\beta}(a * c, n)$.*

Proof. Suppose that α and β are separated. Then $\alpha(a, n) = 0$ or $\beta(a, n) = 0$ for all a and n. Consider a and n with $S_{\alpha,\beta}(a, n)$. Then we show that

$$\forall m \exists c \in \{0,1\}^m S_{\alpha,\beta}(a * c, n)$$

by induction on m. It is trivial for $m = 0$. For induction step, by the induction hypothesis, there exists $c \in \{0,1\}^m$ such that $S_{\alpha,\beta}(a * c, n)$. If $\alpha(a * c, n) = 0$, then set $c' = c * \langle 0 \rangle$; if $\beta(a * c, n) = 0$, and then set $c' = c * \langle 1 \rangle$. Then we have $c' \in \{0,1\}^{m+1}$ and $S_{\alpha,\beta}(a * c', n)$. □

Proposition 4.5. *If α and β are separated, then $T_{\alpha,\beta}$ is infinite and separated.*

Proof. Suppose that α and β are separated. Then for each n, since $S_{\alpha,\beta}(\langle\rangle, n)$, there exists $c \in \{0,1\}^n$ such that $S_{\alpha,\beta}(c, n)$, by Lemma 4.4; hence $c \in T_{\alpha,\beta}$. Therefore, $T_{\alpha,\beta}$ is an infinite tree. For each a, either $\neg\exists n(\alpha(a,n) \neq 0)$ or $\neg\exists n(\beta(a,n) \neq 0)$. In the first case, given an n, assume that $C_1^{T_{\alpha,\beta}}(a, n)$. Then there exists $c \in \{0,1\}^n$ such that $a * \langle 1\rangle * c \in T_{\alpha,\beta}$, that is, $S_{\alpha,\beta}(a * \langle 1\rangle * c, |a| + n + 1)$; hence $S_{\alpha,\beta}(a, |a| + n + 1)$. Therefore, since $\alpha(a, |a| + n + 1) = 0$, we have $S_{\alpha,\beta}(a * \langle 0\rangle, |a| + n + 1)$, and so there exists $c \in \{0,1\}^n$ such that $S_{\alpha,\beta}(a * \langle 0\rangle * c, |a| + n + 1)$, by Lemma 4.4. Since $a * \langle 0\rangle * c \in T_{\alpha,\beta}$, we have $C_0^{T_{\alpha,\beta}}(a, n)$. Thus, $C_1^{T_{\alpha,\beta}}(a, n) \to C_0^{T_{\alpha,\beta}}(a, n)$ for all n. In the second case, similarly, we have $C_0^{T_{\alpha,\beta}}(a, n) \to C_1^{T_{\alpha,\beta}}(a, n)$ for all n. □

Proposition 4.6. *If α and β satisfy the condition (†), then $T_{\alpha,\beta}$ has at most two nodes at each level.*

Proof. Suppose that

$$
\begin{aligned}
&(a * \langle 0\rangle \preceq b \to \beta(a, |b| + 1) \neq 0 \vee \alpha(b, |b| + 1) \neq 0 \vee \beta(b, |b| + 1) \neq 0) \\
&\quad \wedge (a * \langle 1\rangle \preceq b \to \alpha(a, |b| + 1) \neq 0 \vee \alpha(b, |b| + 1) \neq 0 \\
&\quad \vee \beta(b, |b| + 1) \neq 0),
\end{aligned}
$$

for all a and b, and consider a and b with $a * \langle 0\rangle \preceq b$. Then $\beta(a, |b| + 1) \neq 0$, $\alpha(b, |b| + 1) \neq 0$ or $\beta(b, |b| + 1) \neq 0$. If $C_1^{T_{\alpha,\beta}}(a, |b| - |a|)$, $C_0^{T_{\alpha,\beta}}(b, 0)$ and $C_1^{T_{\alpha,\beta}}(b, 0)$, then, since there exist $c \in \{0,1\}^{|b|-|a|}$ such that $a*\langle 1\rangle*c \in T_{\alpha,\beta}$, we have $S_{\alpha,\beta}(a*\langle 1\rangle*c, |b|+1)$, $S_{\alpha,\beta}(b*\langle 0\rangle, |b|+1)$ and $S_{\alpha,\beta}(b * \langle 1\rangle, |b| + 1)$; hence $\beta(a, |b| + 1) = 0$, $\alpha(b, |b| + 1) = 0$ and $\beta(b, |b| + 1) = 0$, a contradiction. Therefore,

$$\neg(C_1^{T_{\alpha,\beta}}(a, |b| - |a|) \wedge C_0^{T_{\alpha,\beta}}(b, 0) \wedge C_1^{T_{\alpha,\beta}}(b, 0)).$$

For a and b with $a * \langle 1\rangle \preceq b$, similarly, we have

$$\neg(C_0^{T_{\alpha,\beta}}(a, |b| - |a|) \wedge C_0^{T_{\alpha,\beta}}(b, 0) \wedge C_1^{T_{\alpha,\beta}}(b, 0)).$$

Thus,

$$
\begin{aligned}
&(a * \langle 0\rangle \preceq b \to \neg(C_1^T(a, |b| - |a|) \wedge C_0^T(b, 0) \wedge C_1^T(b, 0))) \\
&\quad \wedge (a * \langle 1\rangle \preceq b \to \neg(C_0^T(a, |b| - |a|) \wedge C_0^T(b, 0) \wedge C_1^T(b, 0))),
\end{aligned}
$$

for all a and b, and so $T_{\alpha,\beta}$ has at most two nodes at each level, by Proposition 2.2. □

Lemma 4.7. *If α and β are separated, then*

$$\neg C_0^{T_{\alpha,\beta}}(a,n) \wedge C_1^{T_{\alpha,\beta}}(a,n) \rightarrow \alpha(a,|a|+n+1) \neq 0,$$
$$C_0^{T_{\alpha,\beta}}(a,n) \wedge \neg C_1^{T_{\alpha,\beta}}(a,n) \rightarrow \beta(a,|a|+n+1) \neq 0,$$

for all a and n.

Proof. Suppose that α and β are separated. Given a and n, assume that $\neg C_0^{T_{\alpha,\beta}}(a,n)$ and $C_1^{T_{\alpha,\beta}}(a,n)$. Then, since $a * \langle 1\rangle * c \in T_{\alpha,\beta}$ for some $c \in \{0,1\}^n$, we have $S_{\alpha,\beta}(a * \langle 1\rangle * c, |a|+n+1)$; hence $S_{\alpha,\beta}(a, |a|+n+1)$. If $\alpha(a,|a|+n+1) = 0$, then $S_{\alpha,\beta}(a*\langle 0\rangle, |a|+n+1)$, and so $S_{\alpha,\beta}(a*\langle 0\rangle*d, |a|+n+1)$ for some $d \in \{0,1\}^n$, by Lemma 4.4, that is, $C_0^{T_{\alpha,\beta}}(a,n)$, a contradiction. Therefore, $\alpha(a,|a|+n+1) \neq 0$. Similarly, if $C_0^{T_{\alpha,\beta}}(a,n) \wedge \neg C_1^{T_{\alpha,\beta}}(a,n)$, then $\beta(a,|a|+n+1) \neq 0$. □

Proposition 4.8. *If α and β are separated and satisfy the condition (‡), then $T_{\alpha,\beta}$ is convex.*

Proof. Suppose that α and β are separated such that

$$(a * \langle 0\rangle \preceq b \rightarrow \forall n(\beta(a,|b|+n+1) \neq 0 \vee \beta(b,|b|+n+1) = 0))$$
$$\wedge\ (a * \langle 1\rangle \preceq b \rightarrow \forall n(\alpha(a,|b|+n+1) \neq 0 \vee \alpha(b,|b|+n+1) = 0)),$$

for all a and b, and consider a and b with $a*\langle 0\rangle \preceq b$. Then $\beta(a,|b|+n+1) \neq 0$ or $\beta(b,|b|+n+1) = 0$ for all n. Given an n, if $C_1^{T_{\alpha,\beta}}(a,|b|-|a|+n)$, $C_0^{T_{\alpha,\beta}}(b,n)$ and $\neg C_1^{T_{\alpha,\beta}}(b,n)$, then, since there exists $c \in \{0,1\}^{|b|-|a|+n}$ such that $a * \langle 1\rangle * c \in T_{\alpha,\beta}$, that is, $S_{\alpha,\beta}(a * \langle 1\rangle * c, |b|+n+1)$, we have $\beta(a,|b|+n+1) = 0$, and, by Lemma 4.7, $\beta(b,|b|+n+1) \neq 0$; a contradiction. Therefore,

$$\neg(C_1^{T_{\alpha,\beta}}(a,|b|-|a|+n) \wedge C_0^{T_{\alpha,\beta}}(b,n) \wedge \neg C_1^{T_{\alpha,\beta}}(b,n)),$$

for all n. Similarly, for n, a and b with $a * \langle 1\rangle \preceq b$, we have

$$\neg(C_0^{T_{\alpha,\beta}}(a,|b|-|a|+n) \wedge \neg C_0^{T_{\alpha,\beta}}(b,n) \wedge C_1^{T_{\alpha,\beta}}(b,n)),$$

for all n. Thus,

$$(a * \langle 0\rangle \preceq b \to \forall n \neg (C_1^T(a, |b| - |a| + n) \wedge C_0^T(b, n) \wedge \neg C_1^T(b, n)))$$
$$\wedge\, (a * \langle 1\rangle \preceq b \to \forall n \neg (C_0^T(a, |b| - |a| + n) \wedge \neg C_0^T(b, n) \wedge C_1^T(b, n))),$$

for all a and b, and so $T_{\alpha,\beta}$ is convex, by Proposition 2.4. □

Theorem 4.9.

(1) WKL$^-$ *implies* Π_1^0-DC$^\vee$;
(2) WKL$^-_{\leq 2}$ *implies* Π_1^0-DC$^\vee$(†);
(3) WKL^-_c *implies* Π_1^0-DC$^\vee$(‡).

Proof. (1) Suppose that $\neg\exists n(\alpha(a, n) \neq 0) \vee \neg\exists n(\beta(a, n) \neq 0)$ for all a, that is, α and β are separated. Then $T_{\alpha,\beta}$ is infinite and separated, by Proposition 4.5; hence, by WKL$^-$, there exists a branch γ of $T_{\alpha,\beta}$. Therefore,

$$(\gamma(n) = 0 \to \neg\exists m(\alpha(\overline{\gamma}(n), m) \neq 0)) \wedge (\gamma(n) = 1$$
$$\to \neg\exists m(\beta(\overline{\gamma}(n), m) \neq 0)),$$

for all n, by Lemma 4.1.

(2) Suppose that α and β are separated and satisfy the condition (†). Then $T_{\alpha,\beta}$ is infinite, separated and having at most two nodes at each level, by Propositions 4.5 and 4.6; hence, by WKL$^-_{\leq 2}$, there exists a branch γ of $T_{\alpha,\beta}$. Therefore,

$$(\gamma(n) = 0 \to \neg\exists m(\alpha(\overline{\gamma}(n), m) \neq 0)) \wedge (\gamma(n) = 1$$
$$\to \neg\exists m(\beta(\overline{\gamma}(n), m) \neq 0))$$

for all n, by Lemma 4.1.

(3) Suppose that α and β are separated and satisfy the condition (‡). Then $T_{\alpha,\beta}$ is infinite, separated and convex, by Propositions 4.5 and 4.8; hence, by WKL^-_c, there exists a branch γ of $T_{\alpha,\beta}$. Therefore,

$$(\gamma(n) = 0 \to \neg\exists m(\alpha(\overline{\gamma}(n), m) \neq 0)) \wedge (\gamma(n) = 1$$
$$\to \neg\exists m(\beta(\overline{\gamma}(n), m) \neq 0)),$$

for all n, by Lemma 4.1. □

5. From $\mathrm{DC}^\vee$ to WKL

In this section, we show that $\mathrm{DC}^\vee$, $\mathrm{DC}^\vee(\dagger)$ and $\mathrm{DC}^\vee(\ddagger)$ imply WKL^-, $\mathrm{WKL}^-_{\leq 2}$ and WKL^-_c, respectively.

Definition 5.1. Let T be a tree, and let $\alpha, \beta \in \{0,1\}^* \times N \to \{0,1\}$. Then α and β are *T-faithful* if

$$\exists k \leq n(\neg C_0^T(a,k) \wedge C_1^T(a,k)) \to \alpha(a,n) \neq 0,$$

$$\exists k \leq n(C_0^T(a,k) \wedge \neg C_1^T(a,k)) \to \beta(a,n) \neq 0,$$

for each a and n.

Lemma 5.2. *Let T be a tree, and let α and β be T-faithful. For each a and n, if $S_{\alpha,\beta}(a,n)$, then*

$$C_0^T(\langle\rangle, |a|+k) \vee C_1^T(\langle\rangle, |a|+k) \to C_0^T(a,k) \vee C_1^T(a,k),$$

for all k with $|a|+k \leq n$.

Proof. Given an n, we proceed by induction on a. It is trivial for $a = \langle\rangle$. Suppose that $S_{\alpha,\beta}(a * \langle 0\rangle, n)$ and $|a * \langle 0\rangle| + k \leq n$. Since $S_{\alpha,\beta}(a,n)$ and $|a|+1+k \leq n$, we have

$$C_0^T(\langle\rangle, |a|+1+k) \vee C_1^T(\langle\rangle, |a|+1+k) \to C_0^T(a, 1+k) \vee C_1^T(a, 1+k),$$

by the induction hypothesis. Since $S_{\alpha,\beta}(a * \langle 0\rangle, n)$, we have $\alpha(a,n) = 0$; hence $\neg\exists i \leq n(\neg C_0^T(a,i) \wedge C_1^T(a,i))$, that is, $\forall i \leq n(C_1^T(a,i) \to C_0^T(a,i))$. Therefore,

$$C_0^T(\langle\rangle, |a * \langle 0\rangle| + k) \vee C_1^T(\langle\rangle, |a * \langle 0\rangle| + k) \to C_0^T(a, 1+k),$$

and, since $C_0^T(a, 1+k) \to C_0^T(a * \langle 0\rangle, k) \vee C_1^T(a * \langle 0\rangle, k)$, we have

$$C_0^T(\langle\rangle, |a*\langle 0\rangle|+k) \vee C_1^T(\langle\rangle, |a*\langle 0\rangle|+k) \to C_0^T(a*\langle 0\rangle, k) \vee C_1^T(a*\langle 0\rangle, k).$$

Similarly, if $S_{\alpha,\beta}(a * \langle 1\rangle, n)$ and $|a * \langle 1\rangle| + k \leq n$, then

$$C_0^T(\langle\rangle, |a*\langle 1\rangle|+k) \vee C_1^T(\langle\rangle, |a*\langle 1\rangle|+k) \to C_0^T(a*\langle 1\rangle, k) \vee C_1^T(a*\langle 1\rangle, k).$$

□

Proposition 5.3. *Let T be a tree, and let α and β be T-faithful. If T is infinite, then $T_{\alpha,\beta}$ is a subtree of T.*

Proof. Suppose that T is infinite. Then $C_0^T(\langle\rangle, |a|) \vee C_1^T(\langle\rangle, |a|)$ for all a. For each a, if $S_{\alpha,\beta}(a, |a|)$, then $C_0^T(a, 0) \vee C_1^T(a, 0)$, by Lemma 5.2; hence $a \in T$. □

Proposition 5.4. *If T is a separated tree, then there exist separated, T-faithful α and β.*

Proof. Let T be a separated tree, and define $\alpha, \beta \in \{0,1\}^* \times N \to \{0,1\}$ by

$$\alpha(a, n) = 1 \leftrightarrow \exists k \leq n(\neg C_0^T(a, k) \wedge C_1^T(a, k)),$$
$$\beta(a, n) = 1 \leftrightarrow \exists k \leq n(C_0^T(a, k) \wedge \neg C_1^T(a, k)),$$

for each a and n. Then, trivially, α and β are T-faithful. Since T is separated, either $\forall n(C_0^T(a, n) \to C_1^T(a, n))$ or $\forall n(C_1^T(a, n) \to C_0^T(a, n))$; in the former case, we have $\neg\exists n(\beta(a, n) \neq 0)$; in the latter case, we have $\neg\exists n(\alpha(a, n) \neq 0)$. □

Proposition 5.5. *If T is a separated tree having at most two nodes at each level, then there exist separated, T-faithful α and β satisfying the condition (†).*

Proof. Suppose that T is a tree having at most two nodes at each level. Then

$$(a * \langle 0\rangle \preceq b \to \neg(C_1^T(a, |b| - |a|) \wedge C_0^T(b, 0) \wedge C_1^T(b, 0)))$$
$$\wedge\ (a * \langle 1\rangle \preceq b \to \neg(C_0^T(a, |b| - |a|) \wedge C_0^T(b, 0) \wedge C_1^T(b, 0))),$$

for all a and b, by Proposition 2.2. Define $\alpha, \beta \in \{0,1\}^* \times N \to \{0,1\}$ by

$$\alpha(a, n) = 1 \leftrightarrow \exists k \leq n(\neg C_0^T(a, k) \wedge C_1^T(a, k)) \vee \neg C_0^T(a, 0),$$
$$\beta(a, n) = 1 \leftrightarrow \exists k \leq n(C_0^T(a, k) \wedge \neg C_1^T(a, k)),$$

for each a and n. Then, trivially, α and β are T-faithful. Given an a, if $\neg C_0^T(a, 0)$, then $\neg\exists n(\beta(a, n) \neq 0)$; if $C_0^T(a, 0)$, then, since T is separated, either $\neg\exists n(\beta(a, n) \neq 0)$ or $\neg\exists n(\alpha(a, n) \neq 0)$. Therefore,

α and β are separated. Consider a and b with $a * \langle 0 \rangle \preceq b$. Then

$$\neg(C_1^T(a, |b| - |a|) \wedge C_0^T(b, 0) \wedge C_1^T(b, 0))).$$

If $\beta(a, |b| + 1) = 0$, $\alpha(b, |b| + 1) = 0$ and $\beta(b, |b| + 1) = 0$, then

$$\forall k \leq |b| + 1(C_0^T(a, k) \rightarrow C_1^T(a, k)) \quad \text{and}$$
$$\forall k \leq |b| + 1(C_0^T(b, k) \leftrightarrow C_1^T(b, k)),$$

and, since $C_0^T(b, 0)$, we have $C_0^T(a, |b| - |a|)$ and $C_1^T(b, 0)$; hence $C_1^T(a, |b| - |a|)$, a contradiction. Therefore, $\beta(a, |b|+1) \neq 0$, $\alpha(b, |b|+1) \neq 0$ or $\beta(b, |b| + 1) \neq 0$. Similarly, for a and b with $a * \langle 1 \rangle \preceq b$, we have $\alpha(a, |b| + 1) \neq 0$, $\alpha(b, |b| + 1) \neq 0$ or $\beta(b, |b| + 1) \neq 0$. □

Proposition 5.6. *If T is a separated, convex tree, then there exist separated, T-faithful α and β satisfying the condition (‡).*

Proof. Suppose that T is a convex tree. Then

$$(a * \langle 0 \rangle \preceq b \rightarrow \forall n \neg(C_1^T(a, |b| - |a| + n) \wedge C_0^T(b, n) \wedge \neg C_1^T(b, n)))$$
$$\wedge (a * \langle 1 \rangle \preceq b \rightarrow \forall n \neg(C_0^T(a, |b| - |a| + n) \wedge \neg C_0^T(b, n) \wedge C_1^T(b, n))),$$

for all a and b, by Proposition 2.4. Define $\alpha, \beta \in \{0, 1\}^* \times N \rightarrow \{0, 1\}$ by

$$\alpha(a, n) = 1 \leftrightarrow \exists k \leq n(\neg C_0^T(a, k) \wedge C_1^T(a, k)),$$
$$\beta(a, n) = 1 \leftrightarrow \exists k \leq n(C_0^T(a, k) \wedge \neg C_1^T(a, k)),$$

for each a and n. Then, trivially, α and β are T-faithful, and, since T is separated, α and β are separated. Consider a and b with $a * \langle 0 \rangle \preceq b$. Then

$$\neg(C_1^T(a, |b| - |a| + k) \wedge C_0^T(b, k) \wedge \neg C_1^T(b, k)),$$

for all k. Given an n, if $\beta(a, |b| + n + 1) = 0$ and $\beta(b, |b| + n + 1) \neq 0$, then $C_0^T(a, k) \rightarrow C_1^T(a, k)$ for all $k \leq |b| + n + 1$ and there exists $k \leq |b| + n + 1$ such that $C_0^T(b, k) \wedge \neg C_1^T(b, k)$, and, since $C_0^T(a, |b| - |a| + k)$, we have $C_1^T(a, |b| - |a| + k)$; a contradiction. Therefore, $\beta(a, |b| + n + 1) \neq 0$ or $\beta(b, |b| + n + 1) = 0$ for all n. For a and b with $a * \langle 1 \rangle \preceq b$, similarly, we have $\alpha(a, |b| + n + 1) \neq 0$ or $\alpha(b, |b| + n + 1) = 0$ for all n. □

Theorem 5.7.

(1) Π^0_1-$\mathrm{DC}^\vee$ *implies* WKL^-;
(2) Π^0_1-$\mathrm{DC}^\vee(\dagger)$ *implies* $\mathrm{WKL}^-_{\leq 2}$;
(3) Π^0_1-$\mathrm{DC}^\vee(\ddagger)$ *implies* WKL^-_c.

Proof.

(1) Let T be an infinite, separated tree. Then there exist separated, T-faithful α and β, by Proposition 5.4. Therefore, by Π^0_1-$\mathrm{DC}^\vee$, there exists γ such that

$$(\gamma(n) = 0 \to \neg\exists m(\alpha(\overline{\gamma}(n), m) \neq 0)) \wedge (\gamma(n) = 1 \\ \to \neg\exists m(\beta(\overline{\gamma}(n), m) \neq 0))$$

for all n, and so γ is a branch of $T_{\alpha,\beta}$, by Lemma 4.1. Since T is infinite, $T_{\alpha,\beta}$ is a subtree of T, by Proposition 5.3, and thus γ is a branch of T.

(2) Let T be an infinite, separated tree having at most two nodes at each level. Then there exist separated, T-faithful α and β satisfying the condition ($\dagger$), by Proposition 5.5. Therefore, by Π^0_1-$\mathrm{DC}^\vee(\dagger)$, there exists γ such that

$$(\gamma(n) = 0 \to \neg\exists m(\alpha(\overline{\gamma}(n), m) \neq 0)) \wedge (\gamma(n) = 1 \\ \to \neg\exists m(\beta(\overline{\gamma}(n), m) \neq 0)),$$

for all n, γ is a branch of $T_{\alpha,\beta}$, by Lemma 4.1. Since T is infinite, $T_{\alpha,\beta}$ is a subtree of T, by Proposition 5.3, and thus γ is a branch of T.

(3) Let T be an infinite, separated, convex tree. Then there exist separated, T-faithful α and β satisfying the condition ($\ddagger$), by Proposition 5.6. Therefore, by Π^0_1-$\mathrm{DC}^\vee(\ddagger)$, there exists γ such that

$$(\gamma(n) = 0 \to \neg\exists m(\alpha(\overline{\gamma}(n), m) \neq 0)) \wedge (\gamma(n) = 1 \\ \to \neg\exists m(\beta(\overline{\gamma}(n), m) \neq 0)),$$

for all n, and so γ is a branch of $T_{\alpha,\beta}$, by Lemma 4.1. Since T is infinite, $T_{\alpha,\beta}$ is a subtree of T, by Proposition 5.3, and thus γ is a branch of T. □

6. Corollaries

We conclude the chapter by giving corollaries of what we have obtained in the previous sections.

Corollary 6.1. *The following are equivalent.*

(1) WKL^-,
(2) Π^0_1-$\mathrm{DC}^\vee$.

Proof. By Theorems 4.9(1) and 5.7(1). □

Corollary 6.2. *The following are equivalent.*

(1) WKL,
(2) $\mathrm{WKL}^- + \mathrm{LLPO}$,
(3) Π^0_1-$\mathrm{DC}^\vee + \mathrm{LLPO}$.

Proof. By Proposition 3.4(1), Theorem 3.5 and Corollary 6.1. □

Corollary 6.3. *The following are equivalent.*

(1) $\mathrm{WKL}^-_{\leq 2}$,
(2) Π^0_1-$\mathrm{DC}^\vee(\dagger)$.

Proof. By Theorems 4.9(2) and 5.7(2). □

Corollary 6.4. *The following are equivalent.*

(1) $\mathrm{WKL}_{\leq 2}$,
(2) $\mathrm{WKL}^-_{\leq 2} + \mathrm{LLPO}$,
(3) Π^0_1-$\mathrm{DC}^\vee(\dagger) + \mathrm{LLPO}$.

Proof. By Proposition 3.4(2), Theorem 3.5 and Corollary 6.3. □

Corollary 6.5. *The following are equivalent.*

(1) WKL^-_c,
(2) Π^0_1-$\mathrm{DC}^\vee(\ddagger)$.

Proof. By Theorems 4.9(3) and 5.7(3). □

Corollary 6.6. *The following are equivalent.*

(1) WKL_c,
(2) $\mathrm{WKL}^-_c + \mathrm{LLPO}$,
(3) Π^0_1-$\mathrm{DC}^\vee(\ddagger) + \mathrm{LLPO}$.

Proof. By Proposition 3.4(3), Theorem 3.5 and Corollary 6.5. □

Acknowledgement

The second and the last authors thank the Japan Society for the Promotion of Science (JSPS), Core-to-Core Program (A. Advanced Research Networks) for supporting the research.

References

[1] S.G. Simpson, Subsystems of second order arithmetic. *Perspectives in Mathematical Logic*. Springer-Verlag, Berlin, 1999. doi: 10.1007/978-3-642-59971-2.
[2] A.S. Troelstra and D. van Dalen, *Constructivism in Mathematics,* vols. I, 121, Studies in Logic and the Foundations of Mathematics. North-Holland Publishing Co., Amsterdam, 1988.
[3] H. Ishihara, Constructive reverse mathematics: compactness properties. In *From Sets and Types to Topology and Analysis*, vol. 48, Oxford Logic Guides. Oxford University Press, Oxford, 2005, pp. 245–267. doi: 10.1093/acprof:oso/9780198566519.003.0016.
[4] H. Ishihara, An introduction to constructive reverse mathematics. In D. Bridges, H. Ishihara, M. Rathjen, and H. Schwichtenberg (eds.), *Handbook of Constructive Mathematics.* Cambridge University Press, forthcoming.
[5] J. Berger, H. Ishihara, T. Kihara, and T. Nemoto, The binary expansion and the intermediate value theorem in constructive reverse mathematics, *Arch. Math. Logic.* **58**(1–2), 203–217 (2019). doi: 10.1007/s00153-018-0627-2.
[6] T. Nemoto, Systems for constructive reverse mathematics. In D. Bridges, H. Ishihara, M. Rathjen, and H. Schwichtenberg, (eds.), *Handbook of Constructive Mathematics.* Cambridge University Press, forthcoming.
[7] T. Kihara, Degrees of incomputability, realizability and constructive reverse mathematics (February 2020).
[8] J. Berger, H. Ishihara, and P. Schuster, The weak Kőnig lemma, Brouwer's Fan theorem, de Morgan's law, and dependent choice, *Rep. Math. Logic.* **47**, 63–86 (2012).

https://doi.org/10.1142/9789811245220_0007

Chapter 7

Intermediate Goodstein Principles

David Fernández-Duque*, Oriola Gjetaj† and Andreas Weiermann‡
Department of Mathematics: Analysis, Logic and Discrete Mathematics
Ghent University, Belgium
**David.FernandezDuque@UGent.be*
†Oriola.Gjetaj@UGent.be
‡Andreas.Weiermann@UGent.be

The original Goodstein process proceeds by writing natural numbers in nested exponential k-normal form, then successively raising the base to $k+1$ and subtracting one from the end result. Such sequences always reach zero, but this fact is unprovable in Peano arithmetic. In this chapter, we instead consider notations for natural numbers based on the Ackermann function. We define three new Goodstein processes, obtaining new independence results for ACA_0, ACA_0' and ACA_0^+, theories of second-order arithmetic related to the existence of Turing jumps.

1. Introduction

Goodstein's principle [1] is arguably the oldest example of a purely number-theoretic statement known to be independent of PA, as it does not require the coding of metamathematical notions such as Gödel's provability predicate [2]. The proof proceeds by transfinite induction up to the ordinal ε_0 [3]. PA does not prove such transfinite induction, and indeed Kirby and Paris later showed that Goodstein's principle is unprovable in PA [4].

Goodstein's original principle involves the termination of certain sequences of numbers. Say that m is in *nested (exponential) base-k normal form* if it is written in standard exponential base k, with each exponent written in turn in base k. Thus, for example, 20 would become $2^{2^2}+2^2$ in nested base-2 normal form. Then, define a sequence $(g_k(0))_{m\in\mathbb{N}}$ by setting $g_0(m) = m$ and defining $g_{k+1}(m)$ recursively by writing $g_k(m)$ in nested base-$(k+2)$ normal form, replacing every occurrence of $k+2$ by $k+3$, then subtracting one (unless $g_k(m) = 0$, in which case $g_{k+1}(m) = 0$).

In the case that $m = 20$, we obtain

$$g_0(20) = 20 = 2^{2^2} + 2^2,$$

$$g_1(20) = 3^{3^3} + 3^3 - 1 = 3^{3^3} + 3^2 \cdot 2 + 3 \cdot 2 + 2,$$

$$g_2(20) = 4^{4^4} + 4^2 \cdot 2 + 4 \cdot 2 + 2 - 1 = 4^{4^4} + 4^2 \cdot 2 + 4 \cdot 2 + 1,$$

and so forth. At first glance, these numbers seem to grow superexponentially. It should thus be a surprise that, as Goodstein showed, for every m there is k^* for which $g_{k^*}(m) = 0$.

By coding finite Goodstein sequences as natural numbers in a standard way, Goodstein's principle can be formalised in the language of arithmetic, but this formalised statement is unprovable in PA. Independence can be shown by proving that the Goodstein process takes at least as long as stepping down the *fundamental sequences* below ε_0; these are canonical sequences $(\xi[n])_{n<\omega}$ such that $\xi[n] < \xi$ for all ξ and for limit ξ, $\xi[n] \to \xi$ as $n \to \infty$. For standard fundamental sequences below ε_0, PA does not prove that the sequence $\xi > \xi[1] > \xi[1][2] > \xi[1][2][3], \ldots$, is finite.

Exponential notation is not suitable for writing very big numbers (e.g., Graham's number [5]), in which case it may be convenient to use systems of notation which employ faster-growing functions. In [6], T. Arai, S. Wainer and the authors have shown that the Ackermann function may be used to write natural numbers, giving rise to a new Goodstein process which is independent of the theory ATR_0 of *arithmetical transfinite recursion*; this is a theory in the language of second-order arithmetic, which is much more powerful than PA.

The main axiom of ATR_0 states that for any set X and ordinal α, the α-Turing jump of X exists; we refer the reader to [7] for details.

The idea is, for each $k \geq 2$, to define a notion of Ackermannian normal form for each $m \in \mathbb{N}$. Having done this, we can define Ackermannian Goodstein sequences analogously to Goodstein's original version. The normal forms used in [6] are defined using an elaborate "sandwiching" procedure first introduced in [8], approximating a number m by successive branches of the Ackermann function. In this chapter, we consider simpler, and arguably more intuitive, normal forms, also based on the Ackermann function. By making variations of how these normal forms are treated, we show that these give rise to three different Goodstein-like processes, independent of ACA_0, ACA_0' and ACA_0^+, respectively. As was the case for ATR_0, these are theories of second-order arithmetic, which state that certain Turing jumps exist. Recall that ACA_0 is a theory of second-order arithmetic whose characteristic axiom is that if X is a set of natural numbers, then the Turing jump of X also exists as a set. The more powerful theory ACA_0' asserts that, for all $n \in \mathbb{N}$ and $X \subseteq \mathbb{N}$, the n-Turing jump of X exists, while ACA_0^+ asserts that its ω-jump exists; see [7] for details. The theory ACA_0 is a conservative extension of Peano arithmetic, hence has proof-theoretic ordinal ε_0 [9]. The proof-theoretic ordinal of ACA_0' is ε_ω [10], and that of ACA_0^+ is $\varphi_2(0)$ [11]; we will briefly review these ordinals later in the text, but refer the reader to standard texts such as [9,12] for a more detailed treatment of proof-theoretic ordinals. Preliminary versions of some of the results reported here appeared originally in [13].

2. Basic Definitions

Our Goodstein processes will be based on a parametrised version of the Ackermann function, as given by the following definition.

Definition 2.1. For $a, b \in \mathbb{N}$ and $k \geq 2$, we define $A_a(k, b)$ by the following recursion. As a shorthand, we write $A_a b$ instead of $A_a(k, b)$. Then, we define

(1) $A_0 b := k^b$;
(2) $A_{a+1} 0 := A_a^k 0$;
(3) $A_{a+1}(b+1) := A_a^k A_{a+1} b$.

Throughout the text we maintain the convention of writing $A_a b$ instead of $A_a(k, b)$ when $k \geq 2$ is clear from context. It is well known that for every fixed a, the function $b \mapsto A_a b$ is primitive recursive and the function $a \mapsto A_a 0$ is not primitive recursive. It will be convenient to establish some simple lower bounds for the Ackermann function.

Lemma 2.2. *Fix $k \geq 2$ and write $A_x y$ for $A_x(k, y)$. Let $a, b \in \mathbb{N}$. Then,*

(1) *If $a < a'$, then $A_a b \leq A_{a'} b$ and if $b < b'$, then $A_a b \leq A_a b'$.*
(2) $kA_a b \leq A_a(b+1) \leq A_{a+1} b$.
(3) *If $k \geq 3$, $A_a b \geq b^2 + b$.*

Proof. Most inequalities follow by induction on a with a secondary induction on b. For 2.2, we note that $A_a b \geq A_0 b \geq 3^b \geq b^2 + b$. □

We use the Ackermann function to define k normal forms for natural numbers. These normal forms emerged from discussions with Toshiyasu Arai and Stan Wainer, which finally led to the definition of a more powerful normal form defined in [8] and used to prove termination in [6].

Lemma 2.3. *Fix $k \geq 2$. For all $m > 0$, there exist unique $a, b, c \in \mathbb{N}$ such that*

(1) $m = A_a b + c$.
(2) $A_a 0 \leq m < A_{a+1} 0$.
(3) $A_a b \leq m < A_a(b+1)$.

We write $m = {}_k A_a b + c$ in this case. This means that we have in mind an underlying context fixed by k and that for the number m we have uniquely associated the numbers a, b, c. Note that it could be possible that $A_{a+1} 0 = A_a b$ for some a, b, so that we have to choose the right representation for the context; in this case, Item 2.3 guarantees that a is chosen to take the maximal possible value.

By rewriting all of a, b, c in such a normal form, we obtain the *nested Ackermann k-normal form* of m. If we only rewrite a and c in such a normal form (but not b), we arrive at the *index-nested Ackermann k-normal form* of m; note that in this case, b should be regarded as a constant. If we instead rewrite b and c iteratively, we arrive at the *argument-nested Ackermann k-normal form* of m.

The following properties of normal forms are not hard to prove from the definitions.

Lemma 2.4. *Fix $k \geq 2$.*

(1) *If $m = A_a b + c + 1$ is in k-normal form, then $A_a b + c$ is in k-normal form as well.*
(2) *$A_a^\ell 0$ is in k-normal form for every ℓ such that $0 < \ell < k$.*
(3) *If $A_a b$ is in k-normal form, then for every $\ell < b$, the number $A_a \ell$ is also in k-normal form.*

Note that if $m = {}_k A_a b + c$, it may still be the case that $c > A_a b$. In this case, it will sometimes be convenient to count the number of occurrences of $A_a b$. We thus write $m \equiv_k A_a b \cdot q + c'$ if $m = {}_k A_a b + c$ for $c = A_a b \cdot (q-1) + c'$ and $c' < A_a b$, and say that $A_a b \cdot q + c'$ is in **extended normal form.**

3. Proof-theoretic Ordinals

We work with standard notations for ordinals. We use the function $\xi \mapsto \varepsilon_\xi$ to enumerate the fixed points of $\xi \mapsto \omega^\xi$. With $\alpha, \beta \mapsto \varphi_\alpha(\beta)$ we denote the binary Veblen function, where $\beta \mapsto \varphi_\alpha(\beta)$ enumerates the common fixed points of all $\varphi_{\alpha'}$ with $\alpha' < \alpha$. We often omit parentheses and simply write $\varphi_\alpha \beta$. Then $\varphi_0 \xi = \omega^\xi$, $\varphi_1 \xi = \varepsilon_\xi$, $\varphi_2 0$ is the first fixed point of the function $\xi \mapsto \varphi_1 \xi$, $\varphi_\omega 0$ is the first common fixed point of the function $\xi \mapsto \varphi_n \xi$, and Γ_0 is the first ordinal closed under $\alpha, \beta \mapsto \varphi_\alpha \beta$. In fact, not much ordinal theory is presumed in this chapter; we almost exclusively work with ordinals less than $\varphi_2 0$, which can be written in terms of addition and the functions $\xi \mapsto \omega^\xi$, $\xi \mapsto \varepsilon_\xi$. It will be convenient for us to adopt the convention that $\varepsilon_{-1} = 0$. We assume familiarity with ordinal addition and multiplication. We also use partial subtraction: recall that if $\alpha > \beta$, there is a unique ordinal η, such that $\beta + \eta = \alpha$. We will denote this unique η by $-\beta + \alpha$. For more details, we refer the reader to standard texts such as [9,12].

Definition 3.1. Let Λ be an ordinal. A *system of fundamental sequences on* Λ is a function $\cdot[\cdot] \colon \Lambda \times \mathbb{N} \to \Lambda$ such that $\alpha[n] \leq \alpha$ with equality holding if and only if $\alpha = 0$, and $\alpha[n] \leq \alpha[m]$ whenever $n \leq m$. The system of fundamental sequences is *convergent* if

$\lambda = \lim_{n\to\infty} \lambda[n]$ whenever λ is a limit, and has the *Bachmann property* if whenever $\alpha[n] < \beta < \alpha$, it follows that $\alpha[n] \leq \beta[1]$.

Define $\lambda[\![n]\!] = \lambda[0][1], \ldots, [n]$. It is clear that if Λ is an ordinal, then for every $\alpha < \Lambda$ and every m there is n such that $\alpha[\![n]\!] = 0$, but this fact is not always provable in weak theories. The ordinal $\varphi_2 0$ enjoys a natural system of fundamental sequences satisfying the Bachmann property [14].

Definition 3.2. Let $\omega_0(\alpha) := \alpha$ and $\omega_{k+1}(\alpha) = \omega^{\omega_k(\alpha)}$. The standard fundamental sequences for ordinals up to $\varphi_2 0$ are defined as follows:

(1) If $\alpha = \omega^\beta + \gamma$ with $0 < \gamma < \alpha$, then $\alpha[k] := \omega^\beta + \gamma[k]$.
(2) If $\alpha = \omega^\beta > \beta$, then we set $\alpha[k] := 0$ if $\beta = 0$, $\alpha[k] := \omega^\gamma \cdot k$ if $\beta = \gamma + 1$, and $\alpha[k] := \omega^{\beta[k]}$ if $\beta \in \mathrm{Lim}$.
(3) If $\alpha = \varepsilon_\beta > \beta$, then $\alpha[k] := \omega_k(1)$ if $\beta = 0$, $\alpha[k] := \omega_k(\varepsilon_\gamma + 1)$ if $\beta = \gamma + 1$, and $\alpha[k] := \varepsilon_{\beta[k]}$ if $\beta \in \mathrm{Lim}$.
(4) If $\alpha = \varphi_2 0$, then $\alpha[0] = 1$ and $\alpha[n+1] = \varepsilon_{\alpha[n]}$.

We will use these fundamental sequences to establish new independence results by appealing to proof-theoretic ordinals. Such ordinals are defined in terms of a given ordinal notation system with fundamental sequences, and in this case, we use our notation system for $\varphi_2 0$. Each of the theories T we are interested in can be assigned a proof-theoretic ordinal $\|T\| \leq \varphi_2 0$, which gives us a wealth of information about what is (un)provable in T. First, it characterises the amount of transfinite induction available. Recall that transfinite induction for an ordinal α is the scheme

$$\mathrm{TI}(\alpha) := \forall \xi < \bar{\alpha}\Big(\big(\forall \zeta < \xi\ \varphi(\zeta)\big) \to \varphi(\xi)\Big) \to \forall \xi < \bar{\alpha}\ \varphi(\xi),$$

where for our purposes φ is a formula in the language of Peano arithmetic and $\bar{\alpha}$ is the numeral coding α. Then $\|T\|$ can be characterised as the least ordinal ξ such that $T \nvdash \mathrm{TI}(\xi)$.

The ordinal $\|T\|$ can also be used to bound the provably total computable functions in T.[a] Given $\xi \leq \varphi_2 0$, let $H(\xi)$ be the least

[a]It should be remarked that proof-theoretic ordinals defined in terms of transfinite induction or in terms of computable functions are not necessarily equivalent, but they tend to coincide for 'natural' theories, including those considered in this text.

n so that $\xi[\![n]\!] = 0$, and for $n \in \mathbb{N}$, let $F_\xi(n) = H(\xi[n])$. Then, the proof-theoretic ordinal of T is also the least $\xi \leq \varphi_2(0)$ such that $T \nvdash \forall\zeta < \xi\ \forall n\ \exists m\ (m = F_\xi(n))$, if it exists (otherwise, we may define it to be ∞, or just leave it undefined).

Say that a partial function $f\colon \mathbb{N} \to \mathbb{N}$ is *computable* if there is a Σ_1 formula $\varphi_f(x, y)$ in the language of first-order arithmetic (with no other free variables) such that for every m, n, $f(m) = n$ if and only if $\varphi_f(m, n)$ holds. The function f is *provably total* in a theory T if $T \vdash \forall x \exists y \varphi_f(x, y)$ (more precisely, f is provably total if there is at least one such choice of φ_f).

Theorem 3.3. *Define*

- $\|\mathsf{ACA}_0\| = \varepsilon_0$ [9],
- $\|\mathsf{ACA}_0'\| = \varepsilon_\omega$ [10], *and*
- $\|\mathsf{ACA}_0^+\| = \varphi_2(0)$ [11].

Then, for $T \in \{\mathsf{ACA}_0, \mathsf{ACA}_0', \mathsf{ACA}_0^+\}$ *and* $\alpha < \varphi_2(0)$, *the following are equivalent* [15]:

(1) $\alpha < \|T\|$.
(2) $T \vdash \mathrm{TI}(\alpha)$.
(3) F_α *is provably total in* T.
(4) *There exists a provably total computable function* $f\colon \mathbb{N} \to \mathbb{N}$ *in* T *such that* $f(n) > F_\alpha(n)$ *for all* n.

The Bachmann property will be useful in transferring unprovability results stemming from proof-theoretic ordinals to the setting of Goodstein processes, in view of the following.

Proposition 3.4. *Let* Λ *be an ordinal with a system of fundamental sequences satisfying the Bachmann property, and let* $(\xi_n)_{n\in\mathbb{N}}$ *be a sequence of elements of* Λ *such that, for all* n, $\xi_n[n+1] \leq \xi_{n+1} \leq \xi_n$. *Then, for all* n, $\xi_n \geq \xi_0[\![n]\!]$.

Proof. Let $\preceq_k$ be the reflexive transitive closure of $\{(\alpha[k], \alpha) : \alpha < \varphi_2(0)\}$. We need a few properties of these orderings. Clearly, if $\alpha \preceq_k \beta$, then $\alpha \leq \beta$. It can be checked by a simple induction and the Bachmann property that, if $\alpha[n] \leq \beta < \alpha$, then $\alpha[n] \preceq_1 \beta$. Moreover, $\preceq_k$ is monotonic in the sense that if $\alpha \preceq_k \beta$, then $\alpha \preceq_{k+1} \beta$, and if $\alpha \preceq_k \beta$, then $\alpha[k] \preceq_k \beta[k]$ (see, e.g., [14] for details).

We claim that for all n, $\xi_n \succeq_n \xi_0[1], \ldots, [n]$, from which the desired inequality immediately follows. For the base case, we use the fact that $\succeq_0$ is transitive by definition. For the successor, note that the induction hypothesis yields $\xi_0[1], \ldots, [n] \preceq_n \xi_n$, hence $\xi_0[1], \ldots, [n+1] \preceq_{n+1} \xi_n[n+1]$. Then consider three cases.

Case 1 ($\xi_{n+1} = \xi_n$). By transitivity and monotonicity, $\xi_0[1], \ldots, [n+1] \preceq_{n+1} \xi_0[1], \ldots, [n] \preceq_n \xi_n = \xi_{n+1}$ yields $\xi_0[1], \ldots, [n+1] \preceq_{n+1} \xi_{n+1}$.

Case 2 ($\xi_{n+1} = \xi_n[n+1]$). Then $\xi_0[1], \ldots, [n+1] \preceq_{n+1} \xi_n[n+1] = \xi_{n+1}$.

Case 3 ($\xi_n[n+1] < \xi_{n+1} < \xi_n$). The Bachmann property yields $\xi_n[n+1] \preceq_1 \xi_{n+1}$, and since $\xi_0[1], \ldots, [n+1] \preceq_{n+1} \xi_n[n+1]$, monotonicity and transitivity yield $\xi_0[1], \ldots, [n+1] \preceq_{n+1} \xi_{n+1}$. □

As an immediate corollary, we obtain that for such a sequence of ordinals $(\xi_n)_{n\in\mathbb{N}}$, if m is least so that $\xi_m = 0$, then $m \geq F_{\xi_0}(n)$. So, our strategy will be to show that the respective Goodstein process for T grows at least as quickly as $F_{\|T\|}$, from which we obtain that the theorem is unprovable in T.

4. Goodstein Sequences for ACA_0

Each of the three notions of nested normal form we have discussed (nested index-nested and argument-nested) leads to a Goodstein principle of differing proof-theoretical strength. The key here is that each notion of normal form leads to a different base change operation, and each base change operation leads to a faster- or slower-terminating Goodstein process. In this section, we study argument-nested normal forms, and show that they lead to a Goodstein principle of the same proof-theoretic strength of the original.

Definition 4.1. For $2 \leq k \leq \ell$ and $m \in \mathbb{N}$, define $m\left(\begin{smallmatrix}\ell\\k\end{smallmatrix}\right)$ by

(1) $0\left(\begin{smallmatrix}\ell\\k\end{smallmatrix}\right) := 0$.
(2) $m\left(\begin{smallmatrix}\ell\\k\end{smallmatrix}\right) := A_{a\left(\begin{smallmatrix}\ell\\k\end{smallmatrix}\right)}(\ell, b) + c\left(\begin{smallmatrix}\ell\\k\end{smallmatrix}\right)$ if $m = {}_kA_a(k, b) + c$.

We will write $m(k^+)$ instead of $m\left(\begin{smallmatrix}k+1\\k\end{smallmatrix}\right)$.

It is not hard to check that if $m \equiv_k A_a(k,b) \cdot d + e$, then

$$m\left(\tfrac{\ell}{k}\right) = A_{a\left(\frac{\ell}{k}\right)}(\ell, b) \cdot d + e\left(\tfrac{\ell}{k}\right).$$

We then define a new Goodstein process based on this new base change operator.

Definition 4.2. Let $\ell < \omega$. Put $\mathrm{G}_0^{\mathrm{A}}(\ell) := \ell$. Assume recursively that $\mathrm{G}_k^{\mathrm{A}}(\ell)$ is defined and $\mathrm{G}_k^{\mathrm{A}}(\ell) > 0$. Then, $\mathrm{G}_{k+1}^{\mathrm{A}}(\ell) = \mathrm{G}_k^{\mathrm{A}}(\ell)\left(\tfrac{k+3}{k+2}\right) - 1$. If $\mathrm{G}_k^{\mathrm{A}}(\ell) = 0$, then $\mathrm{G}_{k+1}^{\mathrm{A}}(\ell) := 0$.

We will show that for every ℓ there is i with $\mathrm{G}_i^{\mathrm{A}}(\ell) = 0$. In order to prove this, we first establish some natural properties of the base-change operation.

Lemma 4.3. *Fix $k \geq 2$ and let $m, n \in \mathbb{N}$.*

(1) $m \leq m(k^+)$.
(2) *If $m < n$, then $m(k^+) < n(k^+)$.*

Proof. Write $A_a b$ for $A_a(k,b)$ and $B_a b$ for $A_a(k+1,b)$. The first assertion is proved by induction on m. It clearly holds for $m = 0$. If $m = {}_k A_a b + c$, then the induction hypothesis yields $m = A_a b + c \leq B_{a(k^+)}(b) + c(k^+) = m(k^+)$.

For the second, we proceed by induction on n with a secondary induction on m. The assertion is clear if $m = 0$. Let $m = {}_k A_a b + c$ and $n = {}_k A_{a'} b' + c'$. First we establish some useful inequalities. Note that $A_a b \leq m < A_{a+1} 0 = A_a^k 0 = A_a A_a^{k-1} 0$. So $b < A_a^{k-1} 0 \leq B_{a(k^+)}^{k-1} 0$. Moreover $c < A_a(b+1)$ yields

$$c(k^+) < B_{a(k^+)}(b+1) \leq B_{a(k^+)} A_{a(k^+)}^{k-1} 0 \leq B_{a(k^+)}^k 0.$$

Now, consider several cases.

Case 1 ($a < a'$). By the induction hypothesis, $a < a'$ yields $(a+1)(k^+) \leq a'(k^+)$. This yields

$$\begin{aligned} m(k^+) &= B_{a(k^+)} b + c(k^+) < B_{a(k^+)} B_{a(k^+)}^{k-1} 0 + B_{a(k^+)}^k 0 \\ &= 2 B_{a(k^+)}^k 0 \leq B_{a(k^+)}^{k+1} 0 = B_{a(k^+)+1} 0 \leq B_{(a+1)(k^+)} 0 \\ &\leq B_{a'(k^+)}(b') + c'(k^+) = n(k^+), \end{aligned}$$

where the second equality uses Lemma 2.2.

Case 2 ($a = a'$ and $b < b'$). We consider two sub-cases:

Case 2.1 ($A_a(b+1) < n$). Since $m < A_a(b+1)$ and $b+1 \leq b'$, then by the induction hypothesis $m(k^+) < B_{a(k^+)}(b+1) \leq B_{a'(k^+)}b' \leq n(k^+)$.

Case 2.2 ($A_a(b+1) = n$). Write $m = A_a b \cdot d + e$ and we consider two further sub-cases depending on the value of n.

Case 2.2.1 ($a = 0$). In this case, we may write $m \equiv_k k^b \cdot d + e$, with $d < k$ and $e < k^b$, and n has k-normal form k^{b+1}. The induction hypothesis applied to $e < k^b$ yields $e(k^+) < (k+1)^b$. We then have that $m(k^+) = (k+1)^b \cdot d + e(k^+) < (k+1)^b \cdot k + (k+1)^b = (k+1)^{(b+1)} = n(k^+)$.

Case 2.2.2 ($a > 0$). Write $m \equiv_k A_a b \cdot d + e$. Note that $d < A_a(b+1)$ and $e < A_a b$, which by the induction hypothesis yields $e(k^+) < B_{a(k^+)}(b)$. Let $r = B^k_{a(k^+)-1} B_{a(k^+)}(b)$, so that $B_{a(k^+)-1} r = B_{a(k^+)}(b+1)$. Then,

$$\begin{aligned} m(k^+) &= B_{a(k^+)}(b) \cdot d + e(k^+) \leq B_{a(k^+)}(b) \cdot A_a(b+1) + B_{a(k^+)}(b) \\ &< r^2 + r \leq B_{a(k^+)-1} r = B_{a(k^+)}(b+1) \\ &\leq B_{a(k^+)}((b+1)) = n(k^+), \end{aligned}$$

where the second inequality follows by

$$A_a(b+1) = A^k_{a-1} A_a b \leq B^k_{a-1} B_{a(k^+)}(b) = r,$$

and the third inequality uses Lemma 2.2.

Case 3 ($a = a'$ and $b' = b$). Since $m < n$, then we must have that $c < c'$, and

$$m(k^+) = B_{a(k^+)} b + c(k^+) < B_{a(k^+)} b + c'(k^+) = n(k^+).$$

□

Thus, the base-change operation is monotonic. Next we see that it also preserves normal forms.

Lemma 4.4. *Fix $k \geq 2$. If $m = A_a(k, b) + c$ is in k-normal form, then $m(k^+) = A_{a(k^+)}(k+1, b) + c(k^+)$ is in $k+1$ normal form.*

Proof. Write $A_x y$ for $A_x(k, y)$ and $B_x y$ for $A_x(k+1, y)$. Let $m =_k A_a b + c$. We have that $m < A_{a+1} 0$, $m < A_a(b+1)$, and $c < A_a b$. So, $A_a b < A_{a+1} 0 < A^k_a 0 = A_a A^{k-1}_a 0$. Hence, $b < A^{k-1}_a 0 < B^{k-1}_{a(k^+)} 0$;

Since $A_a b$ is in k-normal form, Lemma 4.3 yields $c(k^+) < B_{a(k^+)}b < B_{a(k^+)}B^{k-1}_{a(k^+)}0$. So

$$m(k^+) < B_{a(k^+)}B^{k-1}_{a(k^+)}0 + B_{a(k^+)}B^{k-1}_{a(k^+)}0$$
$$= 2B^{k}_{a(k^+)}0 \leq B^{k+1}_{a(k^+)}0 = B_{a(k^+)+1}0 \leq B_{(a+1)(k^+)}0.$$

Now, we check that $m(k^+) < B_{a(k^+)}(b+1)$.

If $a = 0$, then $m = {}_kA_0 b + c \equiv_k k^b \cdot d + e$ for some $d < k$ and $e < k^b$. Note that $k^b \cdot d' + e$ is in extended normal form if $1 \leq d' < d$, from which it readily follows that

$$m(k^+) = (k+1)^b \cdot d + e(k^+) < (k+1)^b \cdot k + (k+1)^b$$
$$= (k+1)^{b+1} = B_0(b+1).$$

In the remaining case, write $m \equiv_k A_a b \cdot d + e$, with $e < A_a b$, so that $d < A_a(b+1)$. By Lemma 2.2, $B_{a(k^+)}(b+1) \geq B_{a(k^+)}B_{a(k^+)}b \geq (B_{a(k^+)}b)^2 + B_{a(k^+)}b$, so

$$m(k^+) = B_{a(k^+)}b \cdot d + e(k^+) < B_{a(k^+)}b \cdot A_{a(k^+)}(b+1) + B_{a(k^+)}b$$
$$\leq (B_{a(k^+)}b)^2 + B_{a(k^+)}b \leq B_{a(k^+)}(b+1).$$

So, $A_{a(k^+)}(k+1, b) + c(k^+)$ is in $k+1$-normal form. □

Definition 4.5. For $k \geq 2$, define $\cdot\left(\begin{smallmatrix}\omega\\k\end{smallmatrix}\right) : \mathbb{N} \to \varepsilon_0$ as follows:

(1) $0\left(\begin{smallmatrix}\omega\\k\end{smallmatrix}\right) := 0$.
(2) If $m = {}_kA_a b + c$, then $m\left(\begin{smallmatrix}\omega\\k\end{smallmatrix}\right) := \omega^{\omega\cdot(a(\begin{smallmatrix}\omega\\k\end{smallmatrix}))+b} + c\left(\begin{smallmatrix}\omega\\k\end{smallmatrix}\right)$.

The function $\cdot\left(\begin{smallmatrix}\omega\\k\end{smallmatrix}\right)$ is the *base k ordinal assignment,* and provides a monotone mapping of the natural numbers into the ordinals.

Lemma 4.6. *If $m < n < \omega$, then $m\left(\begin{smallmatrix}\omega\\k\end{smallmatrix}\right) < n\left(\begin{smallmatrix}\omega\\k\end{smallmatrix}\right)$.*

Proof. Proof by induction on n with subsidiary induction on m. The assertion is clear if $m = 0$. Let $m = {}_kA_a b + c$ and $n = {}_kA_{a'}b' + c'$.

Case 1 ($a < a'$). By the induction hypothesis, $a\left(\begin{smallmatrix}\omega\\k\end{smallmatrix}\right) < a'\left(\begin{smallmatrix}\omega\\k\end{smallmatrix}\right)$, thus $\omega^{\omega\cdot a(\begin{smallmatrix}\omega\\k\end{smallmatrix})} < \omega^{\omega\cdot a'(\begin{smallmatrix}\omega\\k\end{smallmatrix})}$. We have $c < m < A_{a+1}0 \leq A_{a'}0 \leq n$, and the induction hypothesis yields $c\left(\begin{smallmatrix}\omega\\k\end{smallmatrix}\right) < (A_{a'}0)\left(\begin{smallmatrix}\omega\\k\end{smallmatrix}\right) = \omega^{\omega\cdot a'(\begin{smallmatrix}\omega\\k\end{smallmatrix})}$.

Since $\omega^{\omega\cdot a\left(\frac{\omega}{k}\right)+b} < \omega^{\omega\cdot a'\left(\frac{\omega}{k}\right)}$, then $m\left(\frac{\omega}{k}\right) = \omega^{\omega\cdot a\left(\frac{\omega}{k}\right)+b} + c\left(\frac{\omega}{k}\right) < \omega^{\omega\cdot a'\left(\frac{\omega}{k}\right)} \leq n\left(\frac{\omega}{k}\right)$.

Case 2 ($a = a'$). We consider several sub-cases.

Case 2.1 ($b<b'$). By the induction hypothesis, $\omega^{\omega\cdot a\left(\frac{\omega}{k}\right)+b} < \omega^{\omega\cdot a\left(\frac{\omega}{k}\right)+b'}$. We have $c < A_a b$, and the subsidiary induction hypothesis yields

$$c\left(\tfrac{\omega}{k}\right) < (A_a b)\left(\tfrac{\omega}{k}\right) = \omega^{\omega\cdot a\left(\frac{\omega}{k}\right)+b} < \omega^{\omega\cdot a\left(\frac{\omega}{k}\right)+b'}.$$

So, $m\left(\frac{\omega}{k}\right) = \omega^{\omega\cdot a\left(\frac{\omega}{k}\right)+b} + c\left(\frac{\omega}{k}\right) < \omega^{\omega\cdot a\left(\frac{\omega}{k}\right)+b'} \leq n\left(\frac{\omega}{k}\right)$.

Case 2.2 ($b = b'$). Then $c < c'$ and by the induction hypothesis, $c\left(\frac{\omega}{k}\right) < c'\left(\frac{\omega}{k}\right)$. Hence, $m\left(\frac{\omega}{k}\right) = \omega^{\omega\cdot a\left(\frac{\omega}{k}\right)+b} + c\left(\frac{\omega}{k}\right) < \omega^{\omega\cdot a'\left(\frac{\omega}{k}\right)+b'} + c'\left(\frac{\omega}{k}\right) = n\left(\frac{\omega}{k}\right)$. □

The general idea of the proof of termination is that the functions $\cdot\left(\frac{\omega}{k}\right)$ can be used to assign a decreasing sequence of ordinals to any Goodstein sequence. Since no decreasing sequence of ordinals can be infinite, the sequence must terminate at some point. For this to work, we must show that the ordinal assignment is a monotone, in the following sense.

Lemma 4.7. *For any $m \in \mathbb{N}$ and $k \geq 2$, $m\left(\frac{k+1}{k}\right)\left(\frac{\omega}{k+1}\right) = m\left(\frac{\omega}{k}\right)$.*

Proof. Proof by induction on m. The assertion is clear for $m = 0$. Assume $m = {}_kA_a(k,b)+c$. Then, $m(k^+) ={}_{k+1} A_{a(k^+)}(k+1,b)+c(k^+)$, and the induction hypothesis yields

$$\begin{aligned} m(k^+)\left(\tfrac{\omega}{k+1}\right) &= (A_{a(k^+)}(k+1,b) + c(k^+))\left(\tfrac{\omega}{k+1}\right) \\ &= \omega^{\omega\cdot a(k^+)\left(\frac{\omega}{k+1}\right)+b} + c(k^+)\left(\tfrac{\omega}{k+1}\right) \\ &= \omega^{\omega\cdot a\left(\frac{\omega}{k}\right)+b} + c\left(\tfrac{\omega}{k}\right) = m\left(\tfrac{\omega}{k}\right). \end{aligned}$$

□

Lemma 4.8. *Given $k, m < \omega$ with $k \geq 2$, $(m(k^+) - 1)\left(\frac{\omega}{k+1}\right) \geq m\left(\frac{\omega}{k}\right)[k]$.*

Proof. We prove the claim by induction on c. Write $A_x y$ for $A_x(k,y)$ and $B_x y$ for $A_x(k+1,y)$. Let $m = {}_kA_a b + c$.

Case 1 ($c > 0$). Then the induction hypothesis and Lemma 4.6 yields

$$\begin{aligned}
(m(k^+) - 1)\tbinom{\omega}{k+1} &= (B_{a(k^+)}b + c(k^+) - 1)\tbinom{\omega}{k+1} \\
&= \omega^{\omega \cdot a(k^+)\binom{\omega}{k+1} + b} + (c(k^+) - 1\tbinom{\omega}{k+1}) \\
&\geq \omega^{\omega \cdot a\binom{\omega}{k} + b} + c\tbinom{\omega}{k}[k] \\
&= (A_a b + c)\tbinom{\omega}{k}[k] = m\tbinom{\omega}{k}[k].
\end{aligned}$$

Case 2 ($c = 0$). We consider several sub-cases.

Case 2.1 ($a > 0$ and $b > 0$) We have $m = A_a b$. Then by the induction hypothesis and Lemma 4.6,

$$\begin{aligned}
(m(k^+) - 1)\tbinom{\omega}{k+1} &= (B_{a(k^+)}b - 1)\tbinom{\omega}{k+1} \\
&\geq B_{a(k^+)}(b - 1)\tbinom{\omega}{k+1} \\
&= \omega^{\omega \cdot a(k^+)\binom{\omega}{k+1} + (b-1)} \\
&\geq \omega^{\omega \cdot a\binom{\omega}{k} + (b-1)} \\
&= \omega^{\omega \cdot a\binom{\omega}{k} + b[k]} = m\tbinom{\omega}{k}[k].
\end{aligned}$$

Case 2.2 ($a > 0$ and $b = 0$). In this case $m = A_a 0$, and the induction hypothesis yields

$$\begin{aligned}
(m(k^+) - 1)\tbinom{\omega}{k+1} &= (B_a 0 - 1)\tbinom{\omega}{k+1} \\
&= (B^{k+1}_{a(k^+)-1}0 - 1)\tbinom{\omega}{k+1} \\
&\geq (B^{k}_{a(k^+)-1}0)\tbinom{\omega}{k+1} \\
&= \omega^{\omega \cdot (a(k^+)-1)\binom{\omega}{k+1} + B^{k-1}_{a(k^+)-1}0} \\
&\geq \omega^{\omega \cdot a\binom{\omega}{k}[k] + B^{k-1}_{a(k^+)-1}0} \\
&\geq \omega^{\omega \cdot (\binom{\omega}{k}a)[k]} = A_a 0\tbinom{\omega}{k}[k].
\end{aligned}$$

Case 2.3 ($a = 0$ and $b > 0$). Then $m = A_0 b$.

$$\begin{aligned}(m(k^+) - 1)\left(\tfrac{\omega}{k+1}\right) &= (B_0 b - 1)\left(\tfrac{\omega}{k+1}\right)\\ &\geq (B_0(b-1)\left(\tfrac{\omega}{k+1}\right) \cdot k)\\ &= \omega^{b-1} \cdot k\\ &= \omega^{b[k]} = (c\left(\tfrac{\omega}{k}\right))[k],\end{aligned}$$

since $B_0(b-1) \cdot k$ is in $k+1$ normal form.

Case 2.4 ($a = 0$ and $b = 0$) In this case $m\left(\tfrac{\omega}{k}\right)[k] = B_0 0\left(\tfrac{\omega}{k}\right)[k] = (\omega^{\omega \cdot 0 + 0})[k] = 1[k] = 0$, so the lemma follows. □

The proof of termination follows by observing that for every ℓ,

$$o_k^{\mathrm{A}}(\ell) := \mathrm{G}_k^{\mathrm{A}}(\ell)\left(\tfrac{\omega}{k}\right),$$

is a decreasing sequence of ordinals, hence must be finite. In the following, we make this precise. It is well-known that the so-called slow-growing hierarchy at level $\varphi_\omega 0$ matches up with the Ackermann function, so one might expect that the corresponding Goodstein process can be proved terminating in $\mathsf{PA} + \mathrm{TI}(\varphi_\omega 0)$. This is true but, somewhat surprisingly, much less is needed here. We can lower $\varphi_\omega 0$ to $\varepsilon_\omega = \varepsilon_0$.

Theorem 4.9. *For all $\ell < \omega$, there exists a $k < \omega$ such that $\mathrm{G}_k^{\mathrm{A}}(\ell) = 0$. This is provable in $\mathsf{PA} + \mathrm{TI}(\varepsilon_0)$.*

Proof. If $\mathrm{G}_k^{\mathrm{A}}(\ell)\left(\tfrac{k}{\omega}\right) > 0$, then, by the previous lemmata,

$$\begin{aligned}o_{k+1}^{\mathrm{A}}(\ell) = \mathrm{G}_{k+1}^{\mathrm{A}}(\ell)\left(\tfrac{\omega}{k+3}\right) &= (\mathrm{G}_k^{\mathrm{A}}(\ell)\left(\tfrac{k+3}{k+2}\right) - 1)\left(\tfrac{\omega}{k+3}\right)\\ &< \mathrm{G}_k^{\mathrm{A}}(\ell)\left(\tfrac{k+3}{k+2}\right)\left(\tfrac{\omega}{k+3}\right) = \mathrm{G}_k^{\mathrm{A}}(\ell)\left(\tfrac{\omega}{k+2}\right) = o_k^{\mathrm{A}}(\ell).\end{aligned}$$

Since $(o_k^{\mathrm{A}}(\ell))_{k<\omega}$ cannot be an infinite decreasing sequence of ordinals, there must be some k with $o_k^{\mathrm{A}}(\ell) = 0$, yielding $\mathrm{G}_k^{\mathrm{A}}(\ell) = 0$. □

Theorem 4.10. $\mathsf{ACA}_0 \nvdash \forall \ell \exists k\ \mathrm{G}_k^{\mathrm{A}}(\ell) = 0$.

Proof. Recall that $o_k^{\mathrm{A}}(A_\ell(2,0)) = \mathrm{G}_k^{\mathrm{A}}(A_\ell(2,0))\left(\tfrac{\omega}{k+2}\right)$. Let ℓ_n be given by $\ell_0 = 0$ and $\ell_{n+1} = A_{\ell_n}(2,0)$. Then, we have that $o_0^{\mathrm{A}}(\ell_{n+1}) = \omega_n(1)$. Lemmas 4.8 and 4.6 yield $o_k^{\mathrm{A}}(A_{\ell_{n+1}}(2,0))[k+1] \leq o_{k+1}^{\mathrm{A}}(A_{\ell_{n+1}}(2,0)) < o_k^{\mathrm{A}}(A_{\ell_{n+1}}(2,0))$, hence Proposition 3.4 yields

$o_k^{\mathrm{A}}(\ell_{n+1}) \geq o_0^{\mathrm{A}}(A_\ell(2,0))[\![k]\!]$. So the least k such that $\mathrm{G}_k^{\mathrm{A}}(\ell_{n+1}) = 0$ is at least as big as $H(\omega_n) = F_{\varepsilon_0}(n)$. By Theorem 3.3, $\|\mathsf{ACA}_0\| = \varepsilon_0$, and thus ACA_0 does not prove that this k is always defined as a function of ℓ_{n+1}, so that $\mathsf{ACA}_0 \nvdash \forall \ell \exists k\ \mathrm{G}_k^{\mathrm{A}}(A_\ell(2,0))) = 0$. □

5. Goodstein Sequences for ACA_0'

The base change operator of the previous section corresponds to the notion that $A_a b + c$ is written in such a way that a and c are also expressed using the Ackermann function, but b is treated as a parameter. But what if we instead treat a as a parameter and express b using the Ackermann function? One may expect this to lead to a *weaker* Goodstein result, since the function $a \mapsto A_a b$ grows much faster than $b \mapsto A_a b$. Surprisingly, this is not the case, as the new Goodstein process is independent of ACA_0'.

Definition 5.1. Given $k \geq 2$, $m \in \mathbb{N}$ and $2 \leq k < \ell$, define $m\langle {\ell \atop k} \rangle$ by

(1) $0\langle {\ell \atop k} \rangle := 0$.
(2) $m\langle {\ell \atop k} \rangle := A_a\left(\ell, b\langle {\ell \atop k} \rangle\right) + c\langle {\ell \atop k} \rangle$ if $m = {}_k A_a(k,b) + c$.

We write $\langle k^+ \rangle$ instead of $\langle {k+1 \atop k} \rangle$.

The idea is that now, a is being treated as a constant, while b should be regarded as being iteratively represented using the Ackermann function. As before, if $m \equiv_k A_a(k,b) \cdot d + e$, then

$$m\left({\ell \atop k} \right) = A_a(\ell, b\langle {\ell \atop k} \rangle) \cdot d + e\langle {\ell \atop k} \rangle.$$

We then define a new Goodstein process based on this new base change operator. This new base change operator leads to a second Goodstein process.

Definition 5.2. Let $\ell < \omega$. Put $\mathrm{G}_0^{\mathrm{B}}(\ell) := \ell$. Assume recursively that $\mathrm{G}_k^{\mathrm{B}}(\ell)$ is defined and $\mathrm{G}_k^{\mathrm{B}}(\ell) > 0$. Then $\mathrm{G}_{k+1}^{\mathrm{B}}(\ell) = \mathrm{G}_k^{\mathrm{B}}(\ell)\langle {k+3 \atop k+2} \rangle - 1$. If $\mathrm{G}_k^{\mathrm{B}}(\ell) = 0$, then $\mathrm{G}_{k+1}^{\mathrm{B}}(\ell) := 0$.

We will prove monotonicitiy using a similar method as in the previous section.

Lemma 5.3. *Fix $k \geq 2$ and let $m, n \in \mathbb{N}$. Then: If $m < n$, then $m\langle k^+ \rangle < n\langle k^+ \rangle$.*

Proof. Write $A_x y$ for $A_x(k,y)$ and $B_x y$ for $A_x(k+1,y)$.

The proof is by induction on n with a subsidiary induction on m. The assertion is clear if $m=0$. Let $m = {}_k A_a b + c$ and $n = {}_k A_{a'} b' + c'$. We distinguish cases according to the position of a relative to a', the position of b relative to b', etc. Note that from the choice of a, a' we must have $a \leq a'$.

Case 1 ($a < a'$). Write $m \equiv_k A_a b \cdot d + e$. We have $A_a b \leq m < A_{a+1}0 = A_a A_a^{k-1}0$. For $\ell < k$, we have that $A_a^\ell 0$ is in k-normal form by Lemma 2.4. Thus, the induction hypothesis yields $b\langle k^+\rangle < B_a^{k-1}0$. The number $A_a b$ is in k-normal form and so the induction hypothesis applied to $e < A_a b$ yields $e\langle k^+\rangle < B_a(b\langle k^+\rangle) \leq B_a^k 0$. Moreover, $d < m < A_{a+1}0$. This yields

$$
\begin{aligned}
m\langle k^+\rangle &= B_a(b\langle k^+\rangle)\cdot d + e\langle k^+\rangle \\
&\leq B_a B_a^{k-1}0 \cdot A_a^k 0 + B_a B_a^{k-1}0 \\
&\leq (B_a^k 0)^2 + B_a^k 0 < B_a^{k+1}0 = B_{a+1}0 \\
&\leq B_{a'}(b\langle k^+\rangle) + c'\langle k^+\rangle = n\langle k^+\rangle,
\end{aligned}
$$

where the second inequality follows from

$$A_{a+1}0 = A_a^k 0 \leq B_a^k 0,$$

and the second-to-last from Lemma 2.2.

Case 2 ($a = a'$). Note that in this case $b \leq b'$. Consider the following sub-cases.

Case 2.1 ($a = a'$ and $b < b'$). Consider two sub-cases.

Case 2.1.1 ($A_a(b+1) < n$). Since n is in k-normal form and $b+1 \leq b'$, we see that $A_a(b+1)$ is in k-normal form by Lemma 2.4. Then, the induction hypothesis yields $m\langle k^+\rangle < B_a((b+1)\langle k^+\rangle) \leq B_a(b'\langle k^+\rangle) \leq n\langle k^+\rangle$.

Case 2.1.2 ($A_a(b+1) = n$). We know that $m = A_a b + c < A_a(b+1) = n$. Consider two further sub-cases.

Case 2.1.2.1 ($a = 0$). This means that $m \equiv_k k^b \cdot d + e < k^{b+1} = n$ for some $d < k$ and $e < k^b$, where n has k-normal form k^{b+1}.

The induction hypothesis yields $b\langle k^+\rangle < (b+1)\langle k^+\rangle$ and $e\langle k^+\rangle < (k+1)^{b\langle k^+\rangle}$. We then have that

$$m\langle k^+\rangle = (k+1)^{b\langle k^+\rangle} \cdot d + e\langle k^+\rangle < (k+1)^{b\langle k^+\rangle} \cdot k + (k+1)^{b\langle k^+\rangle}$$
$$\leq (k+1)^{b\langle k^+\rangle+1} \leq (k+1)^{(b+1)\langle k^+\rangle} = n\langle k^+\rangle.$$

Case 2.1.2.2 ($a > 0$). Write $m \equiv_k A_a b \cdot d + e$. Note that $d < A_a(b+1)$ and $e < A_a b$, which by the induction hypothesis yields $e\langle k^+\rangle < B_a(b\langle k^+\rangle)$. Let $r = B^k_{a-1} B_a(b\langle k^+\rangle)$, so that $B_{a-1} r = B_a(b\langle k^+\rangle + 1)$. Then,

$$m\langle k^+\rangle = B_a(b\langle k^+\rangle) \cdot d + e\langle k^+\rangle \leq B_a(b\langle k^+\rangle) \cdot A_a(b+1) + B_a(b\langle k^+\rangle)$$
$$< r^2 + r \leq B_{a-1} r = B_a(b\langle k^+\rangle + 1)$$
$$\leq B_a((b+1)\langle k^+\rangle) = n\langle k^+\rangle,$$

where the second inequality follows by

$$A_a(b+1) = A^k_{a-1} A_a b \leq B^k_{a-1} B_a(b\langle k^+\rangle) = r,$$

and the third inequality uses Lemma 2.2.

Case 2.2 ($b' = b$). Then $c < c'$ and the induction hypothesis yields

$$m\langle k^+\rangle = B_a(b\langle k^+\rangle) + c\langle k^+\rangle$$
$$< B_a(b\langle k^+\rangle) + c'\langle k^+\rangle = n\langle k^+\rangle.$$

□

Thus, the base-change operation is monotone. Next we see that it also preserves normal forms.

Lemma 5.4. *If* $m = {}_k A_a(k, b) + c$, *then* $m\langle k^+\rangle =_{k+1} A_a(k+1, b\langle k^+\rangle) + c\langle k^+\rangle$.

Proof. Write $A_x y$ for $A_x(k, y)$ and $B_x y$ for $A_x(k+1, y)$. Assume that $m = {}_k A_a(k, b) + c$. Then, $m < A_{a+1} 0$, $m < A_a(b+1)$, and $c < A_a b$. Clearly, $B_a 0 \leq m\langle k^+\rangle$. By Lemma 2.4, $A_{a+1} 0$ is in k-normal form, so that by Lemma 5.3, $c < A_{a+1} 0$ yields $c\langle k^+\rangle < B_{a+1} 0$. Since $A_a b$ is in k-normal form, Lemma 5.3 yields $c\langle k^+\rangle < B_a(b\langle k^+\rangle)$. It remains to check that we also have $m\langle k^+\rangle < B_a(b\langle k^+\rangle + 1)$.

If $a = 0$, then write $m \equiv_k A_0 b \cdot d + e$ for some $d < k$ and $e < k^b$. Then, $m\langle k^+\rangle = (k+1)^{b\langle k^+\rangle} \cdot d + e\langle k^+\rangle < (k+1)^{b\langle k^+\rangle+1}$ and thus $m\langle k^+\rangle = {}_k(k+1)^{b\langle k^+\rangle} + c\langle k^+\rangle$. In the remaining case, we have for $a > 0$ that

$$\begin{aligned} m\langle k^+\rangle &= B_a(b\langle k^+\rangle) + c\langle k^+\rangle < B_a(b\langle k^+\rangle) + B_a(b\langle k^+\rangle) \\ &= 2B_a(b\langle k^+\rangle) \le B_a(b\langle k^+\rangle + 1), \end{aligned}$$

with the last equality following from Lemma 2.2. So $B_a(b\langle k^+\rangle)+c\langle k^+\rangle$ is in $k+1$-normal form. □

The Goodstein process arising from this base-change operator is also terminating. In order to prove this, we must assign ordinals to natural numbers, in such a way that the process gives rise to a decreasing (hence, finite) sequence. For each k, we define a function $\cdot\langle {\omega \atop k} \rangle \colon \mathbb{N} \to \Lambda$, where Λ is a suitable ordinal, in such a way that $m\langle {\omega \atop k} \rangle$ is computed from the k-normal form of m. Unnested Ackermannian normal forms correspond to ordinals below $\Lambda = \varepsilon_\omega$, as the following map shows. In the following, recall that as per our convention, $\varepsilon_{-1} = 0$.

Definition 5.5. For $k \ge 2$, define $\cdot\langle {\omega \atop k} \rangle \colon \mathbb{N} \to \varepsilon_\omega$ as follows:

(1) $0\langle {\omega \atop k} \rangle := 0$.
(2) $m\langle {\omega \atop k} \rangle := \omega^{\varepsilon_{a-1}+b\langle {\omega \atop k} \rangle} + c\langle {\omega \atop k} \rangle$ if $m = {}_kA_a(k,b) + c$.

The ordinal assignment in this case is once again monotone.

Lemma 5.6. *If $m < n < \omega$, then $m\langle {\omega \atop k} \rangle < n\langle {\omega \atop k} \rangle$.*

Proof. Proof by induction on n with subsidiary induction on m. The assertion is clear if $m = 0$. Write $A_x y$ for $A_x(k,y)$ and let $m = {}_kA_a b + c$ and $n = {}_kA_{a'} b' + c'$. We distinguish cases according to the position of a relative to a', the position of b relative to b', etc. Note that $a \le a'$.

Case 1 ($a < a'$). We have $c < m < A_{a+1}0 \le A_{a'}0$ and, since $A_{a'}0 \le n$, the induction hypothesis yields $c\langle {\omega \atop k} \rangle < \omega^{\varepsilon_{a'-1}+0\langle {\omega \atop k} \rangle} = \omega^{\varepsilon_{a'-1}+1}$. We have $b < m < A_{a+1}0 \le A_{a'}0$ and the induction hypothesis yields $b\langle {\omega \atop k} \rangle < \omega^{\varepsilon_{a'-1}+0\langle {\omega \atop k} \rangle} = \omega^{\varepsilon_{a'-1}+1}$. It follows that $\varepsilon_{a-1} + b\langle {\omega \atop k} \rangle < \omega^{\varepsilon_{a'-1}+1}$, hence, $m\langle {\omega \atop k} \rangle = \omega^{\varepsilon_{a-1}+b\langle {\omega \atop k} \rangle} + c\langle {\omega \atop k} \rangle < \omega^{\varepsilon_{a'-1}+1} \le n\langle {\omega \atop k} \rangle$.

Case 2 ($a = a'$). Note that then $b \leq b'$. We consider several sub-cases.

Case 2.1 ($b < b'$). The induction hypothesis yields $b\langle {\omega \atop k} \rangle < b'\langle {\omega \atop k} \rangle$. Hence, $\omega^{\varepsilon_{a-1}+b\langle {\omega \atop k} \rangle} < \omega^{\varepsilon_{a-1}+b'\langle {\omega \atop k} \rangle}$. We have $c < A_a b$, and the subsidiary induction hypothesis yields $c\langle {\omega \atop k} \rangle < \omega^{\varepsilon_{a-1}+b\langle {\omega \atop k} \rangle} < \omega^{\varepsilon_{a-1}+b'\langle {\omega \atop k} \rangle}$. Putting things together, we see $m\langle {\omega \atop k} \rangle = \omega^{\varepsilon_{a-1}+b\langle {\omega \atop k} \rangle} + c\langle {\omega \atop k} \rangle < \omega^{\varepsilon_{a-1}+b'\langle {\omega \atop k} \rangle} \leq n\langle {\omega \atop k} \rangle$.

Case 2.2 ($b = b'$). The inequality $m < n$ yields $c < c'$ and the induction hypothesis yields $c\langle {\omega \atop k} \rangle < c'\langle {\omega \atop k} \rangle$. Hence, $m\langle {\omega \atop k} \rangle = \omega^{\varepsilon_{a-1}+b\langle {\omega \atop k} \rangle} + c\langle {\omega \atop k} \rangle < \omega^{\varepsilon_{a-1}+b\langle {\omega \atop k} \rangle} + c'\langle {\omega \atop k} \rangle = n\langle {\omega \atop k} \rangle$. □

As before, our ordinal assignment is invariant under base change.

Lemma 5.7. $m\langle {k+1 \atop k} \rangle\langle {\omega \atop k+1} \rangle = m\langle {\omega \atop k} \rangle$.

Proof. Write $A_x y$ for $A_x(k, y)$ and $B_x y$ for $A_x(k+1, y)$ and proceed by induction on m. The assertion is clear for $m = 0$. Let $m = {}_k A_a b + c$. Then, $m\langle {k+1 \atop k} \rangle = {}_k B_a(b\langle {k+1 \atop k} \rangle) + c\langle {k+1 \atop k} \rangle$, and the induction hypothesis yields

$$\begin{aligned} m\langle {k+1 \atop k} \rangle\langle {\omega \atop k+1} \rangle &= \big(B_a(b\langle {k+1 \atop k} \rangle) + c\langle {k+1 \atop k} \rangle\big)\langle {\omega \atop k+1} \rangle \\ &= \omega^{\varepsilon_{a-1}+b\langle {k+1 \atop k} \rangle\langle {\omega \atop k+1} \rangle} + c\langle {k+1 \atop k} \rangle\langle {\omega \atop k+1} \rangle \\ &= \omega^{\varepsilon_{a-1}+b\langle {\omega \atop k} \rangle} + c\langle {\omega \atop k} \rangle = m\langle {\omega \atop k} \rangle. \end{aligned}$$

□

As we did for the first Goodstein process, we define

$$o^{\mathrm{B}}_k(\ell) := \mathrm{G}^{\mathrm{B}}_k(\ell)\langle {\omega \atop k+2} \rangle.$$

The ordinals $o^{\mathrm{B}}_k(\ell)$ are decreasing on k provided they are nonzero, from which the termination of the process follows.

Theorem 5.8. *For all $\ell < \omega$, there exists a $k < \omega$ such that $\mathrm{G}^{\mathrm{B}}_k(\ell) = 0$. This is provable in* $\mathsf{PA} + \mathrm{TI}(\varepsilon_\omega)$.

Proof. By the previous lemmata,

$$\begin{aligned} o^{\mathrm{B}}_{k+1}(\ell) &= \mathrm{G}^{\mathrm{B}}_{k+1}(\ell)\langle {\omega \atop k+3} \rangle = (\mathrm{G}^{\mathrm{B}}_k(\ell)\langle k^+ \rangle - 1)\langle {\omega \atop k+3} \rangle \\ &< \mathrm{G}^{\mathrm{B}}_k(\ell)\langle k^+ \rangle\langle {\omega \atop k+3} \rangle = \mathrm{G}^{\mathrm{B}}_k(\ell)\langle {\omega \atop k+2} \rangle = o^{\mathrm{B}}_k(\ell). \end{aligned}$$

Since $(o^{\mathrm{B}}_k(\ell))_{k<\omega}$ cannot be an infinite decreasing sequence of ordinals, there must be some k with $o^{\mathrm{B}}_k(\ell) = 0$, yielding $\mathrm{G}^{\mathrm{B}}_k(\ell) = 0$. □

Now, we are going to show that for every $\alpha < \varepsilon_\omega$, $\mathsf{PA} + \mathrm{TI}(\alpha) \nvdash \forall \ell \exists k\ b_k(\ell) = 0$. In view of Proposition 3.4, the following technical lemma will be crucial for proving our main independence result for ACA_0'.

Lemma 5.9. *Given $k, m < \omega$ with $k \geq 2$,*

$$(m\langle {k+1 \atop k} \rangle - 1)\langle {\omega \atop k+1} \rangle \geq m\langle {\omega \atop k} \rangle[k-1].$$

Proof. Write $A_x y$ for $A_x(k, y)$ and $B_x y$ for $A_a(k+1, y)$. We prove the claim by induction on m. Let $m = {}_k A_a b + c$.

Case 1 ($c > 0$). Then the induction hypothesis and Lemma 5.6 yield

$$\begin{aligned}
(m\langle k^+\rangle - 1)\langle {\omega \atop k+1} \rangle &= \omega^{\varepsilon_{a-1} + b\langle k^+\rangle\langle {\omega \atop k+1} \rangle} + (c\langle k^+\rangle - 1)\langle {\omega \atop k+1} \rangle \\
&\geq \omega^{\varepsilon_{a-1} + b\langle {\omega \atop k} \rangle} + c\langle {\omega \atop k} \rangle[k-1] \\
&= \left(\omega^{\varepsilon_{a-1} + b\langle {\omega \atop k} \rangle} + c\langle {\omega \atop k} \rangle\right)[k-1] \\
&= (A_a b + c)\langle {\omega \atop k} \rangle[k-1] = m\langle {\omega \atop k} \rangle[k-1].
\end{aligned}$$

Case 2 ($c = 0$). We consider several sub-cases.

Case 2.1 ($a > 0$ and $b > 0$). The induction hypothesis yields

$$\begin{aligned}
(m\langle k^+\rangle - 1)\langle {\omega \atop k+1} \rangle &= (B_a(b\langle k^+\rangle) - 1)\langle {\omega \atop k+1} \rangle \\
&\geq B_a(b\langle k^+\rangle - 1)\langle {\omega \atop k+1} \rangle \cdot k \\
&= \omega^{\varepsilon_{a-1} + (b\langle k^+\rangle - 1)\langle {\omega \atop k+1} \rangle} \cdot k \\
&\geq \omega^{\varepsilon_{a-1} + b\langle {\omega \atop k} \rangle[k-1]} \cdot k \geq \omega^{\varepsilon_{a-1} + b\langle {\omega \atop k} \rangle}[k-1] \\
&= m\langle {\omega \atop k} \rangle[k-1],
\end{aligned}$$

since $B_a(b\langle k^+\rangle - 1)$ is in $k+1$ normal form by Lemmas 2.4 and 5.4.

Case 2.2 ($a > 0$ and $b = 0$). Then, the induction hypothesis yields

$$\begin{aligned}
(m\langle k^+\rangle - 1)\langle {\omega \atop k+1} \rangle &= (B_a 0 - 1)\langle {\omega \atop k+1} \rangle = (B_{a-1}^{k+1} 0 - 1)\langle {\omega \atop k+1} \rangle \\
&= (B_{a-1}^{k-1} B_{a-1}^2 0 - 1)\langle {\omega \atop k+1} \rangle \\
&\geq (B_{a-1}^{k-1} 2 - 1)\langle {\omega \atop k+1} \rangle > B_{a-1}^{k-1} 1\langle {\omega \atop k+1} \rangle
\end{aligned}$$

$$\begin{aligned}&\geq \omega_{k-1}(\varepsilon_{a-1}+1) = \varepsilon_{a-1}[k-1]\\ &= A_a 0\langle {}^{\omega}_{k}\rangle[k-1] = m\langle {}^{\omega}_{k}\rangle[k-1],\end{aligned}$$

where in the last inequality we use that $B^{\ell}_{a-1}0$ is in $k+1$ normal form for $\ell \leq k$ by Lemma 2.4 and Lemma 5.4, and an easy induction on k.

Case 2.3 ($a = 0$ and $b > 0$). Then the induction hypothesis yields similarly as in Case 2.1:

$$\begin{aligned}(m\langle k^+\rangle - 1)\langle {}^{\omega}_{k+1}\rangle &= (B_0(b\langle k^+\rangle) - 1)\langle {}^{\omega}_{k+1}\rangle\\ &= ((k+1)^{b\langle k^+\rangle} - 1)\langle {}^{\omega}_{k+1}\rangle\\ &\geq ((k+1)^{b\langle k^+\rangle - 1}\cdot k)\langle {}^{\omega}_{k+1}\rangle\\ &= \omega^{(b\langle k^+\rangle - 1)\langle {}^{\omega}_{k+1}\rangle}\cdot k \geq \omega^{b\langle {}^{\omega}_{k}\rangle[k-1]}\\ &\quad\cdot k > m\langle {}^{\omega}_{k}\rangle[k-1],\end{aligned}$$

since $(k+1)^{b\langle k^+\rangle - 1}\cdot k$ is in extended $(k+1)$-normal form.

Case 2.4 ($a = 0$ and $b = 0$). Recall that by convention, $\varepsilon_{-1} = 0$. Then, $m\langle {}^{\omega}_{k}\rangle[k-1] = B_0 0\langle {}^{\omega}_{k}\rangle[k-1] = \omega^{\varepsilon_{-1}+0}[k-1] = \omega^0[k-1] = 0$, and the claim follows. □

Theorem 5.10. $\mathsf{ACA}_0' \nvdash \forall\ell\exists k\ \mathrm{G}^{\mathrm{B}}_{k}(\ell) = 0$.

Proof. The proof runs similarly to that of Theorem 4.9. This time, we define $\ell_n = A_{n+1}0$. Observe that ℓ_n is in normal form and $(\langle {}^{\omega}_{k}\rangle\ell_n) = \omega^{\varepsilon_n+0} = \varepsilon_n$. Moreover, $\varepsilon_n = \varepsilon_\omega[n]$. It follows in view of Lemma 5.9 that $\forall\ell\exists k\ \mathrm{G}^{\mathrm{B}}_{k}(\ell) = 0$ implies that F_{ε_ω} is total, which is not provable in $\mathsf{PA} + \mathrm{TI}(\alpha)$ for any $\alpha < \varepsilon_\omega$. □

6. Goodstein Sequences for ACA_0^+

Finally, we consider what should intuitively be the strongest version of our Goodstein process: that where the base change is simultaneously applied to a, b, and c, the resulting Goodstein principle will then be independent of ACA_0^+. The intuition should be that an expression $A_a b+c$ is written so that each of a, b, c is hereditarily represented using the Ackermann function.

Definition 6.1. If $2 \leq k < \ell < \omega$ and $m \in \mathbb{N}$, define $m\left\{\begin{smallmatrix}\ell\\k\end{smallmatrix}\right\}$ recursively by

(1) $0\left\{\begin{smallmatrix}\ell\\k\end{smallmatrix}\right\} := 0$
(2) $m\left\{\begin{smallmatrix}\ell\\k\end{smallmatrix}\right\} := A_{a\left\{\begin{smallmatrix}\ell\\k\end{smallmatrix}\right\}}(\ell, b\left\{\begin{smallmatrix}\ell\\k\end{smallmatrix}\right\}) + c\left\{\begin{smallmatrix}\ell\\k\end{smallmatrix}\right\}$ if $m = {}_kA_a(k,b)+c$.

We write $\{k^+\}$ instead of $\left\{\begin{smallmatrix}k+1\\k\end{smallmatrix}\right\}$.

As before, we observe that if $m \equiv_k A_a(k,b)\cdot d + e$, then

$$m\left\{\begin{smallmatrix}\ell\\k\end{smallmatrix}\right\} = A_{a\left\{\begin{smallmatrix}\ell\\k\end{smallmatrix}\right\}}(\ell, b\left\{\begin{smallmatrix}\ell\\k\end{smallmatrix}\right\})\cdot d + e\left\{\begin{smallmatrix}\ell\\k\end{smallmatrix}\right\}.$$

We can then define our final Goodstein process based on this new base change operator.

Definition 6.2. Let $\ell < \omega$. Put $\mathrm{G}_0^{\mathrm{C}}(\ell) := \ell$. Assume recursively that $\mathrm{G}_k^{\mathrm{C}}(\ell)$ is defined and $\mathrm{G}_k^{\mathrm{C}}(\ell) > 0$. Then, $\mathrm{G}_{k+1}^{\mathrm{C}}(\ell) = \mathrm{G}_k^{\mathrm{C}}(\ell)\left\{\begin{smallmatrix}k+3\\k+2\end{smallmatrix}\right\} - 1$. If $\mathrm{G}_k^{\mathrm{C}}(\ell) = 0$, then $\mathrm{G}_{k+1}^{\mathrm{C}}(\ell) := 0$.

Termination and independence results can then be obtained following the same general strategy as before.

Lemma 6.3. *If $m < n$ and $k \geq 2$, then $m\{k^+\} < n\{k^+\}$.*

Proof. The assertion is clear if $m = 0$. As before we write $A_x y$ for $A_x(k,y)$, $B_x y$ for $B_x(k+1,y)$, $m = {}_kA_a b + c$ and $n = {}_kA_{a'}b' + c'$.

Case 1 ($a < a'$). By the induction hypothesis $a < a'$ yields $a\{k^+\} \leq a'\{k^+\}$. Write $m \equiv_k A_a b \cdot d + e$. We have

$$m\{k^+\} = B_{a\{k^+\}}(b\{k^+\})\cdot d + e\{k^+\}.$$

There are a few things to consider: first, we have $A_a b \leq m < A_{a+1}0 = A_a^k 0 = A_a A_a^{k-1}0$. Then $b < A_a^{k-1}0$. By the induction hypothesis, $b\{k^+\} < B_{a\{k^+\}}^{k-1}0$. We have that $d < A_{a+1}0$. Moreover, $A_a b$ is in

k-normal form by Lemma 2.4 and $e < m \leq A_a b$ yields $e\{k^+\} < B_{a\{k^+\}}(b\{k^+\}) = B^k_{a\{k^+\}}0$ using our bound for $b\{k^+\}$. This yields

$$\begin{aligned} m\{k^+\} &= B_{a\{k^+\}}(b\{k^+\}) \cdot d + e\{k^+\} \\ &< B^k_{a\{k^+\}}0 \cdot A_{a+1}0 + B^k_{a\{k^+\}}0 \\ &\leq (B^k_{a\{k^+\}}0)^2 + B^k_{a\{k^+\}}0 \\ &\leq B^{k+1}_{a\{k^+\}}0 = B_{a\{k^+\}+1}0 \\ &\leq B_{a'\{k^+\}}(b'\{k^+\}) + c'\{k^+\} = n\{k^+\}, \end{aligned}$$

where we use Lemma 2.2 for the third inequality.

Case 2 ($a = a'$ and $b < b'$). Consider two sub-cases.

Case 2.1 ($A_a(b+1) < n$). Since $m < A_a(b+1)$ and $b+1 \leq b'$, then by the induction hypothesis $m\{k^+\} < B_{a\{k^+\}}((b+1)\{k^+\}) \leq B_{a\{k^+\}}(b'\{k^+\}) \leq n\{k^+\}$.

Case 2.2 ($A_a(b+1) = d$). Write $m \equiv_k A_a b \cdot d + e$. Here we consider two cases, depending on the value of a.

Case 2.2.1 ($a = 0$). In this case $m = A_0 b \cdot d + e = k^b \cdot d + e < k^{b+1} = n$, where $d < k$, $e < k^b$, and n has k-normal form k^{b+1}. The induction hypothesis yields $b\{k^+\} < (b+1)\{k^+\}$ and $e\{k^+\} < (k+1)^{b\{k^+\}}$. Then

$$\begin{aligned} m\{k^+\} &= (k+1)^{b\{k^+\}} \cdot d + e\{k^+\} < (k+1)^{b\{k^+\}} \cdot k + (k+1)^{b\{k^+\}} \\ &= (k+1)^{b\{k^+\}+1} \leq (k+1)^{(b+1)\{k^+\}} = n\{k^+\}. \end{aligned}$$

Case 2.2.2 ($a > 0$). We have $e < A_a b$ and by the induction hypothesis, $e\{k^+\} < B_{a\{k^+\}}(b\{k^+\})$. Moreover, $d < A_a(k, b+1)$. Let $r = B^k_{a\{k^+\}-1}B_{a\{k^+\}}(b\{k^+\})$, so that $B_{a\{k^+\}-1}r = B_{a\{k^+\}}(b\{k^+\}+1)$. Note that $r \geq A^k_{a-1}A_a b = A_a(b+1)$. Then,

$$\begin{aligned} m\{k^+\} &= B_{a\{k^+\}}(b\{k^+\}) \cdot d + e\{k^+\} \\ &< B_{a\{k^+\}}(b\{k^+\}) \cdot A_a(b+1) + B_{a\{k^+\}}(b\{k^+\}) \\ &\leq r^2 + r < B_{a\{k^+\}-1}r = B_{a\{k^+\}}(b\{k^+\}+1) \\ &\leq B_{a\{k^+\}}((b+1)\{k^+\}) = n\{k^+\}, \end{aligned}$$

where the third inequality uses Lemma 2.2.

Case 3 ($a = a'$ and $b' = b$). Since $m < n$, then it must be that $c < c'$, so

$$\begin{aligned} m\{k^+\} &= B_{a\{k+\}}(b\{k^+\}) + c\{k^+\} \\ &< B_{a\{k+\}}(b\{k^+\}) + c'\{k^+\} = n\{k^+\}. \end{aligned}$$

□

Thus, the base-change operation is monotone. Next we see that it also preserves normal forms.

Lemma 6.4. *If $m = A_a(k, b) + c$ is in k-normal form, then*

$$m\{k^+\} =_{k+1} A_{a\{k+\}}(k+1, b\{k^+\}) + c\{k^+\}.$$

Proof. As usual write $A_x y$ for $A_x(k, y)$ and $B_x y$ for $A_x(k+1, y)$. Let $m \equiv_k A_a b \cdot d + e$. We have that $A_a 0 \leq m < A_{a+1} 0$ and $A_a b \leq m < A_a(b+1)$. So, $A_a b < A_{a+1} 0 = A_a^k 0$. By Lemma 2.4, $A_a^\ell 0$ is in k-normal form for $\ell < k$. Hence, by Lemma 6.3, $b < A_a^{k-1} 0$ yields $b\{k^+\} < B_{a\{k+\}}^{k-1} 0$. We have that $d < m \leq A_a(b+1)$, and $A_a b$ is in k-normal form, Lemma 6.3 yields $e\{k^+\} < B_{a\{k+\}}(b\{k^+\}) < B_{a\{k+\}} B_{a\{k+\}}^{k-1} 0$. Let $r = B_{a\{k+\}}^k 0$. Then,

$$\begin{aligned} m\{k^+\} &= B_{a\{k+\}}(b\{k^+\}) \cdot d + e\{k^+\} \\ &\leq B_{a\{k+\}}^k 0 \cdot A_{a\{k+\}}^k 0 + B_{a\{k+\}}^k 0 \\ &\leq r^2 + r \leq B_{a\{k+\}+1} 0. \end{aligned}$$

Now, we check that $c\{k^+\} < B_{a\{k+\}}(b\{k^+\} + 1)$.

If $a = 0$, then $c \equiv_k A_0 b \cdot d + e = k^b \cdot d + e$ with $d < k$ and $e < k^b$. Then $e\{k^+\} < (k+1)^{b\{k^+\}}$. Thus,

$$\begin{aligned} m\{k^+\} &= (k+1)^{b\{k^+\}} \cdot d + e\{k^+\} \\ &\leq (k+1)^{b\{k^+\}} \cdot k + (k+1)^{b\{k^+\}} \\ &= (k+1)^{b\{k^+\}+1} = B_0(b\{k^+\} + 1), \end{aligned}$$

as needed.

In the remaining case, we have that $a > 0$. Let $r = B^k_{a\{k^+\}-1}B_{a\{k^+\}}(b\{k^+\})$. Then

$$\begin{aligned} m\{k^+\} &= B_{a\{k^+\}}(b\{k^+\}) \cdot d + e\{k^+\} \\ &< B_{a\{k^+\}}(b\{k^+\}) \cdot A_{a\{k^+\}}(b+1) + B_{a\{k^+\}}(b\{k^+\}) \\ &\le r^2 + r \le B_{a\{k^+\}}(b\{k^+\}+1), \end{aligned}$$

where the last inequality uses Lemma 2.2. So $B_{a\{k^+\}}(b\{k^+\}) + c\{k^+\}$ is in $(k+1)$-normal form. □

Finally, we provide an ordinal mapping to show that the Goodstein process terminates. Recall that $\varepsilon_{-1} = 0$ by convention, and $-1+\alpha$ is the unique η such that $1+\eta = \alpha$, provided $\alpha > 0$. For the sake of legibility, we write $\breve{\varepsilon}(\alpha)$ instead of $\varepsilon_{-1+\alpha}$.

Definition 6.5. Given $k \ge 2$, define a function $\cdot\{{}^{\omega}_{k}\} : \mathbb{N} \to \varphi_2(0)$ given by

(1) $0\{{}^{\omega}_{k}\} := 0$.
(2) $m\{{}^{\omega}_{k}\} := \omega^{\breve{\varepsilon}(a\{{}^{\omega}_{k}\})+b\{{}^{\omega}_{k}\}} + c\langle{}^{\omega}_{k}\rangle$ if $m = {}_kA_a(k,b) + c$.

As was the case for the previous mappings, the mappings in Definition 6.5 are strictly increasing and invariant under base change, as we show in the following lemmas.

Lemma 6.6. *If* $m < n$, *then* $m\{{}^{\omega}_{k}\} < n\{{}^{\omega}_{k}\}$.

Proof. Proof by induction on d with subsidiary induction on c. The assertion is clear if $c = 0$. Let $m = {}_kA_a b + c$ and $n = {}_kA_{a'}b' + c'$.

Case 1 ($a < a'$). By the induction hypothesis, $a\{{}^{\omega}_{k}\} < a'\{{}^{\omega}_{k}\}$, so $\breve{\varepsilon}(a\{{}^{\omega}_{k}\}) < \breve{\varepsilon}(a'\{{}^{\omega}_{k}\})$. We have $c < m < A_{a+1}0 \le A_{a'}0 \le n$, so the induction hypothesis yields

$$c\{{}^{\omega}_{k}\} < \omega^{\breve{\varepsilon}(a'\{{}^{\omega}_{k}\})+0\{{}^{\omega}_{k}\}} = \breve{\varepsilon}(a'\{{}^{\omega}_{k}\}).$$

We have $b < m < A_{a+1}0 \le A_{a'}0$ and the induction hypothesis also yields

$$b\{{}^{\omega}_{k}\} < \omega^{\breve{\varepsilon}(a'\{{}^{\omega}_{k}\})+0\{{}^{\omega}_{k}\}} = \breve{\varepsilon}(a'\{{}^{\omega}_{k}\}).$$

It follows that $\breve{\varepsilon}\,(a\,\{\genfrac{}{}{0pt}{}{\omega}{k}\}) + b\,\{\genfrac{}{}{0pt}{}{\omega}{k}\} < \breve{\varepsilon}\,(a'\,\{\genfrac{}{}{0pt}{}{\omega}{k}\})$, hence,

$$m\,\{\genfrac{}{}{0pt}{}{\omega}{k}\} = \omega^{\breve{\varepsilon}(a\{\genfrac{}{}{0pt}{}{\omega}{k}\})+b\{\genfrac{}{}{0pt}{}{\omega}{k}\}} + c\,\{\genfrac{}{}{0pt}{}{\omega}{k}\} < \breve{\varepsilon}\,(a'\,\{\genfrac{}{}{0pt}{}{\omega}{k}\}) \leq n\,\{\genfrac{}{}{0pt}{}{\omega}{k}\}.$$

Case 2 ($a = a'$). We consider several sub-cases.

Case 2.1 ($b < b'$). The induction hypothesis yields $b\,\{\genfrac{}{}{0pt}{}{\omega}{k}\} < b'\,\{\genfrac{}{}{0pt}{}{\omega}{k}\}$. Hence, $\omega^{\breve{\varepsilon}(a\{\genfrac{}{}{0pt}{}{\omega}{k}\})+b\{\genfrac{}{}{0pt}{}{\omega}{k}\}} < \omega^{\breve{\varepsilon}(a\{\genfrac{}{}{0pt}{}{\omega}{k}\})+b'\{\genfrac{}{}{0pt}{}{\omega}{k}\}}$. We have $c < A_a(k, b)$, and the subsidiary induction hypothesis yields

$$c\,\{\genfrac{}{}{0pt}{}{\omega}{k}\} < \omega^{\breve{\varepsilon}(a\{\genfrac{}{}{0pt}{}{\omega}{k}\})+b\{\genfrac{}{}{0pt}{}{\omega}{k}\}} < \omega^{\breve{\varepsilon}(a\{\genfrac{}{}{0pt}{}{\omega}{k}\})+b'\{\genfrac{}{}{0pt}{}{\omega}{k}\}},$$

so

$$m\,\{\genfrac{}{}{0pt}{}{\omega}{k}\} = \omega^{\breve{\varepsilon}(a\{\genfrac{}{}{0pt}{}{\omega}{k}\})+b\{\genfrac{}{}{0pt}{}{\omega}{k}\}} + c\,\{\genfrac{}{}{0pt}{}{\omega}{k}\} < \omega^{\breve{\varepsilon}(a\{\genfrac{}{}{0pt}{}{\omega}{k}\})+b'\{\genfrac{}{}{0pt}{}{\omega}{k}\}} \leq n\,\{\genfrac{}{}{0pt}{}{\omega}{k}\}.$$

Case 2.2 ($b = b'$). The inequality $m < n$ yields $c < c'$ and the induction hypothesis yields $c\,\{\genfrac{}{}{0pt}{}{\omega}{k}\} < c'\,\{\genfrac{}{}{0pt}{}{\omega}{k}\}$. Hence,

$$\begin{aligned} m\,\{\genfrac{}{}{0pt}{}{\omega}{k}\} &= \omega^{\breve{\varepsilon}(a\{\genfrac{}{}{0pt}{}{\omega}{k}\})+b\{\genfrac{}{}{0pt}{}{\omega}{k}\}} + c\,\{\genfrac{}{}{0pt}{}{\omega}{k}\} \\ &< \omega^{\breve{\varepsilon}(a\{\genfrac{}{}{0pt}{}{\omega}{k}\})+b\{\genfrac{}{}{0pt}{}{\omega}{k}\}} + c'\,\{\genfrac{}{}{0pt}{}{\omega}{k}\} = n\,\{\genfrac{}{}{0pt}{}{\omega}{k}\}. \end{aligned}$$

□

Next, we prove that the ordinal assignment is invariant under base change.

Lemma 6.7. *For all* $m < \omega$, $m\,\{\genfrac{}{}{0pt}{}{k+1}{k}\}\,\{\genfrac{}{}{0pt}{}{\omega}{k+1}\} = m\,\{\genfrac{}{}{0pt}{}{\omega}{k}\}$.

Proof. Proof by induction on m. The assertion is clear for $m = 0$. Assume $c = {}_kA_a b + c$. Then $c\{k^+\} = {}_kB_{a\{k^+\}}(b\{k^+\}) + c\{k^+\}$, and the induction hypothesis yields

$$\begin{aligned} m\{k^+\}\,\{\genfrac{}{}{0pt}{}{\omega}{k+1}\} &= (B_{a\{k^+\}}(b\{k^+\}) + c\{k^+\})\,\{\genfrac{}{}{0pt}{}{\omega}{k+1}\} \\ &= \omega^{\breve{\varepsilon}(a\{k^+\}\{\genfrac{}{}{0pt}{}{\omega}{k+1}\})+b\{k^+\}\{\genfrac{}{}{0pt}{}{\omega}{k+1}\}} + c\{k^+\}\,\{\genfrac{}{}{0pt}{}{\omega}{k+1}\} \\ &= \omega^{\breve{\varepsilon}(a\{\genfrac{}{}{0pt}{}{\omega}{k}\})+b\{\genfrac{}{}{0pt}{}{\omega}{k}\}} + c\,\{\genfrac{}{}{0pt}{}{\omega}{k}\} = m\,\{\genfrac{}{}{0pt}{}{\omega}{k}\}. \end{aligned}$$

□

With this, we define for $m, k \in \mathbb{N}$

$$o_k^{\mathrm{C}}(m) = \mathrm{G}_k^{\mathrm{C}}(m)\left\{\tfrac{\omega}{k+2}\right\}.$$

As in the proof of, e.g., Theorem 4.9, the sequence $(o_k^{\mathrm{C}}(m))_{k<\omega}$ is decreasing on k as long as $\mathrm{G}_k^{\mathrm{C}}(m) > 0$. Since $o_k^{\mathrm{C}}(m) \le \varphi_2 0$, we obtain the following.

Theorem 6.8. *For all* $\ell < \omega$, *there exists a* $k < \omega$ *such that* $c_k(\ell) = 0$. *This is provable in* $\mathsf{PA} + \mathrm{TI}(\varphi_2 0)$.

Finally, we show that for every $\alpha < \varphi_2 0$, $\mathsf{PA} + \mathrm{TI}(\alpha) \nvdash \forall \ell \exists k \; c_k(\ell) = 0$. For this, we need the following analogue of Lemma 4.8.

Lemma 6.9. *Given* $k, m < \omega$ *with* $k \ge 2$, $(m\{k^+\} - 1)\left\{\tfrac{\omega}{k+1}\right\} \ge m\left\{\tfrac{\omega}{k}\right\}[k-1]$.

Proof. We prove the claim by induction on m. Write $A_x y$ for $A_x(k, y)$ and $B_x y$ for $A_x(k+1, y)$. Let $m = {}_k A_a b + c \equiv_k A_a b \cdot d + e$.

Case 1 ($c > 0$). Then the induction hypothesis and Lemma 6.6 yield

$$\begin{aligned}
(m\{k^+\} - 1)\left\{\tfrac{\omega}{k+1}\right\} &= (B_{a\{k^+\}}(b\{k^+\}) + c\{k^+\} - 1)\left\{\tfrac{\omega}{k+1}\right\} \\
&= \omega^{\check{\xi}(a\{k^+\}\{\frac{\omega}{k+1}\}) + b\{k^+\}\{\frac{\omega}{k+1}\}} \\
&\quad + (c\{k^+\} - 1)\left\{\tfrac{\omega}{k+1}\right\} \\
&\ge \omega^{\check{\xi}(a\{\frac{\omega}{k}\}) + b\{\frac{\omega}{k}\}} + c\left\{\tfrac{\omega}{k}\right\}[k-1] \\
&= (\omega^{\check{\xi}(a\{\frac{\omega}{k}\}) + b\{\frac{\omega}{k}\}} + c\left\{\tfrac{\omega}{k}\right\})[k-1] \\
&= (A_a b + c)\left\{\tfrac{\omega}{k}\right\}[k-1] = c\left\{\tfrac{\omega}{k}\right\}[k-1].
\end{aligned}$$

Case 2 ($c = 0$). We consider several sub-cases.

Case 2.1 ($a > 0$ and $b > 0$). We have $m = A_a b$. The induction hypothesis yields

$$\begin{aligned}
(m\{k^+\} - 1)\left\{\tfrac{\omega}{k+1}\right\} &= (B_{a\{k^+\}}(b\{k^+\}) - 1)\left\{\tfrac{\omega}{k+1}\right\} \\
&\ge (B_{a\{k^+\}}(b\{k^+\} - 1) \cdot k)\left\{\tfrac{\omega}{k+1}\right\} \\
&= \left(\omega^{\check{\xi}(a\{\frac{\omega}{k+1}\}\{k^+\}) + (b\{k^+\} - 1)\{\frac{\omega}{k+1}\}}\right) \cdot k
\end{aligned}$$

$$\geq \left(\omega^{\check{\varepsilon}(a\{\frac{\omega}{k}\})+b\{\frac{\omega}{k}\}[k-1]}\right) \cdot k$$

$$> \left(\omega^{\check{\varepsilon}(a\{\frac{\omega}{k}\})+b\{\frac{\omega}{k}\}}\right)[k-1] = m\left\{\tfrac{\omega}{k}\right\}[k-1].$$

Case 2.2 ($a > 0$ and $b = 0$). In this case $m = A_a 0$. Then, the induction hypothesis yields

$$\begin{aligned}(m\{k^+\}-1)\left\{\tfrac{\omega}{k+1}\right\} &= (B_{a(k^+)}0 - 1)\left\{\tfrac{\omega}{k+1}\right\}\\ &= (B^{k+1}_{a\{k^+\}-1}0 - 1)\left\{\tfrac{\omega}{k+1}\right\}\\ &\geq (B^{k-1}_{a\{k^+\}-1}1)\left\{\tfrac{\omega}{k+1}\right\}\\ &\geq \omega_{k-1}\left(\check{\varepsilon}\left((a\{k^+\}-1)\left\{\tfrac{\omega}{k+1}\right\}\right)+1\right)\\ &\geq \omega_{k-1}\left(\check{\varepsilon}\left(a\left\{\tfrac{\omega}{k}\right\}[k-1]\right)+1\right)\\ &\geq \check{\varepsilon}\left(a\left\{\tfrac{\omega}{k}\right\}\right)[k-1] = A_a 0\left\{\tfrac{\omega}{k}\right\}[k-1],\end{aligned}$$

since $B^{\ell}_{a\{k^+\}-1}1$ is in $(k+1)$-normal form for $\ell < k+1$.

Case 2.3 ($a = 0$ and $b > 0$). Then the induction hypothesis yields

$$\begin{aligned}(m\{k^+\}-1)\left\{\tfrac{\omega}{k+1}\right\} &= (B_0(b\{k^+\})-1)\left\{\tfrac{\omega}{k+1}\right\}\\ &\geq (B_0(b\{k^+\}-1)\cdot k)\left\{\tfrac{\omega}{k+1}\right\}\\ &= (k+1)^{(b\{k^+\}-1)}\left\{\tfrac{\omega}{k+1}\right\}\cdot k\\ &= \omega^{(b\{k^+\}-1)\{\frac{\omega}{k+1}\}}\cdot k\\ &\geq \omega^{b\{\frac{\omega}{k}\}[k-1]}\cdot k \geq m\left\{\tfrac{\omega}{k}\right\}[k-1],\end{aligned}$$

since $(k+1)^{b\{k^+\}-1}\cdot k$ is in $(k+1)$-normal form.

Case 2.4 ($a = 0$ and $b = 0$). In this case $m\left\{\tfrac{\omega}{k}\right\}[k-1] = A_0 0\left\{\tfrac{\omega}{k}\right\}$ $[k-1] = \omega^0[k-1] = 0$, so the lemma follows. □

Theorem 6.10. $\mathsf{ACA}_0^+ \nvdash \forall\ell\exists k \mathrm{G}^{\mathrm{C}}_k(\ell) = 0$.

Proof. The proof once again runs similarly to that of Theorem 4.9 using in this case Lemma 6.9. In this case, we define ℓ_n recursively by $\ell_0 = 1$ and $\ell_{n+1} = A_{\ell_n}(2,0)$, and observe that $\ell_{n+1}\left\{\tfrac{\omega}{k}\right\} = \varphi_1^n 1 = \varphi_2 0[n]$. □

7. Concluding Remarks

This work follows that of Arai *et al.* [6], where it is shown that a more elaborate normal form based on the Ackermann function leads to an independence result of strength Γ_0. The simpler normal forms considered here, which are arguably more natural, lead to substantially weaker Goodstein principles. However, it is instructive to observe that changes in the notion of base change can lead to somewhat unpredictable modifications in proof-theoretic strength, naturally leading to independence results for ACA_0, ACA_0' and ACA_0^+, three prominent theories of reverse mathematics.

A natural question is whether the gap between ε_0 and Γ_0 can be bridged using further variants of the Ackermannian Goodstein process. One idea would be to perform the sandwiching of [6] only up to a fixed number n of steps. One may conjecture that such a process would lead to independence results for $\mathsf{TI}(\Gamma_0[n])$, where the fundamental sequence for Γ_0 is defined by iteration on the parameter α in $\varphi_\alpha\beta$. However, the results presented here should serve as evidence that the proof-theoretic strength of Goodstein principles is not easy to predict, so such conjectures should be taken with a grain of salt!

Acknowledgement

This work has been partially supported by the FWO-FWF Lead Agency Grant G030620N and BOF.DOC.2020.0052.01 Grant.

References

[1] R.L. Goodstein, Transfinite ordinals in recursive number theory, *J. Symbol. Logic.* **12**(4), 123–129, 12 (1947).

[2] K. Gödel, Über Formal Unentscheidbare Sätze der Principia Mathematica und Verwandter Systeme, I. *Monatshefte für Mathematik und Physik* **38**, 173–198 (1931).

[3] R.L. Goodstein, On the restricted ordinal theorem, *J. Symb. Logic.* **9**(2), 33–41 (1944).

[4] L. Kirby and J. Paris, Accessible independence results for Peano arithmetic, *Bull. London Math. Soc.* **14**(4), 285–293 (1982).

[5] R.L. Graham and B.L. Rothschild, Ramsey's theorem for n-dimensional arrays, *Bull. Am. Math. Soc.* **75**(2), 418–422, 03 (1969).
[6] T. Arai, D. Fernández-Duque, S. Wainer and A. Weiermann, Predicatively unprovable termination of the Ackermannian Goodstein principle. *Proceedings of the American Mathematical Society*, 2019. Accepted for publication.
[7] S.G. Simpson, *Subsystems of Second Order Arithmetic.* Cambridge University Press, New York, 2009.
[8] A. Weiermann, Ackermannian Goodstein principles for first order peano arithmetic. In *Sets and Computations, vol. 33 of Lecture Notes Series, Institute for Mathematical Sciences, National University of Singapore*, Hackensack, NJ, 2018, pp. 157–181. World Scientific. English summary.
[9] W. Pohlers, *Proof Theory, The First Step into Impredicativity.* Springer-Verlag, Berlin Heidelberg, 2009.
[10] B. Afshari and M. Rathjen, Ordinal analysis and the infinite Ramsey theorem. In *How the World Computes — Turing Centenary Conference and 8th Conference on Computability in Europe, CiE 2012*, Cambridge, UK, June 18-23, 2012. Proceedings, 2012, pp. 1–10.
[11] G. Leigh and M. Rathjen, An ordinal analysis for theories of self-referential truth, *Arch. Math. Log.* **49**(2), 213–247 (2010).
[12] K. Schütte, *Proof Theory.* Springer Verlag, 1977.
[13] F.D. David and A. Weiermann, Ackermannian goodstein sequences of intermediate growth. In Marcella Anselmo, Gianluca Della Vedova, Florin Manea, and Arno Pauly (eds.), *Beyond the Horizon of Computability — 16th Conference on Computability in Europe, CiE 2020, Fisciano, Italy, June 29–July 3, 2020, Proceedings, vol. 12098 of Lecture Notes in Computer Science*, Springer, 2020, pp. 163–174.
[14] Diana Schmidt. Built-up systems of fundamental sequences and hierarchies of number-theoretic functions, *Arch. Math. Logic.* **18**(1), 47–53 (1977).
[15] E.A. Cichon, W. Buchholz, and A. Weiermann, A uniform approach to fundamental sequences and hierarchies, *Math. Logic Quarterly* **40**, 273–286 (1994).

https://doi.org/10.1142/9789811245220_0008

Chapter 8

Infinite Horizon Extensive Form Games, Coalgebraically

Matteo Capucci*, Neil Ghani†, Clemens Kupke‡, Jérémy Ledent§ and Fredrik Nordvall Forsberg¶

Mathematically Structured Programming Group, Department of Computer and Information Sciences, University of Strathclyde, Scotland

*matteo.capucci@strath.ac.uk

†neil.ghani@strath.ac.uk

‡clemens.kupke@strath.ac.uk

§jeremy.ledent@strath.ac.uk

¶fredrik.nordvall-forsberg@strath.ac.uk

1. Introduction

Game theory is the study of how agents make decisions in order to maximise their outcomes [1,2]. A strategy profile describes how each agent will play the game, and is said to be a *Nash equilibrium* if no player has any incentive to deviate from their strategy; it is called *subgame perfect* if it is a Nash equilibrium in every subgame of the game. In a series of papers [3–6], Douglas Bridges investigated constructive aspects of the theory of games where players move simultaneously (so-called normal form games), and their preference relations. This chapter is concerned with a constructive treatment of games where players move sequentially.

A common way to model sequential games is using their *extensive form*: a game is represented as a tree, whose branching structure reflects the decisions available to the players, and whose leaves (corresponding to a complete "play" of the game) are decorated by payoffs for each player. When the number of rounds in the game is infinite (e.g., because a finite game is repeated an infinite number of times, or because the game may continue forever), the game tree needs to be infinitely deep. One way to handle such infinite trees is to consider them as the metric completion of finite trees, after equipping them with a suitable metric [17]. However, as a definitional principle, this only gives a method to construct functions into other complete metric spaces, and the explicit construction as a quotient of Cauchy sequences [8, Section 4.3] can be unwieldy to work with. Instead, we prefer to treat the infinite as the *dual* of the finite, in the spirit of category theory and especially the theory of coalgebras [9].

We are not the first to attack infinite extensive form games using coalgebraic methods. Lescanne [10,11], Lescanne and Perrinel [12] and Abramsky and Winschel [13] define infinite two-player games coalgebraically, and show that coinductive proof methods can be used to constructively prove properties of games. However, their definition only assigns utility to finite plays. For that reason, they restrict attention to *strongly convergent* strategies, i.e., strategy profiles that always lead to a leaf of the tree in a finite number of steps. This restriction rules out infinitely repeated games, where utility could be assigned using discounted sums or limiting averages — both methods crucially making use of the entire infinite history of the game. Building on our on work on infinitely repeated open games [14], we extend Lescanne's and Abramsky and Winschel's coalgebraic framework to not necessarily convergent strategies.

The *one-shot deviation principle* is a celebrated theorem of classical game theory. It asserts that a strategy is a subgame perfect equilibrium if and only if there is no profitable one-shot deviation in any subgame. While this principle holds for all finite games, in the case of infinite trees, it requires an extra assumption called *continuity at infinity* (see, e.g., Fudenberg and Tirole [15, Chapter 4.2]). Essentially, this property says that the actions taken in the distant future have a negligible impact on the current payoff. In the coalgebraic setting, Abramsky and Winschel [13] claim to prove the one-shot principle without continuity assumptions — we argue that this is not entirely the case. Indeed, they show that the natural coalgebraic equilibrium concept (which they call "SPE") satisfies the one-shot deviation principle. However, they do not discuss how this coalgebraic concept relates to the traditional notion of subgame perfect Nash equilibria. As we show in Theorem 4.12, these two notions are indeed equivalent, but only assuming continuity of the utility function. In that regard, the predicate "SPE" of Abramsky and Winschel (called $\Box\mathsf{Unimprov}$ in our work) is in fact closer to a coalgebraic version of the one-shot equilibrium.

Our proof of the one-shot deviation principle extends the previous ones in several ways. Compared to the one of Abramsky and Winschel [13], it applies to games where infinite plays are possible; and it relates to the more standard definition of subgame-perfect Nash equilibrium, $\Box\mathsf{Nash}$. Additionally, our theorem applies to *any* coalgebra of the extensive-form tree functor, whereas Abramsky and Winschel only work with the final coalgebra. Compared to the usual proofs found in the game theory literature, we carefully analyse the constructivity of the proof. The only extra assumption that we require is decidable equality on the set of players (which is typically finite), and decidability of the order relation on the set of payoffs (typically, the set of rational numbers). Moreover, continuity at infinity is usually expressed using uniform continuity; we remark that pointwise continuity suffices.

Structure of the Chapter We recall the basics of coalgebras for endofunctors in Section 2. In Section 3, we define infinite extensive form games as final coalgebras, and use properties of coalgebras to define notions such as strategies, moves, payoffs and equilibria in a game. We then relate our coalgebraic notions with the existing notions from the literature in Section 4. Throughout the chapter, we demonstrate how coinductive proof principles can be used to reason constructively about infinite games.

Notation We use $\mathcal{P} : \mathsf{Set} \to \mathsf{Set}$ for the covariant powerset functor mapping a set to its set of subsets. Given a set-indexed collection of sets $Y : I \to \mathsf{Set}$ and $i \in I$, we write interchangably Y_i or $Y(i)$ for the ith component of the collection. The dependent sum $\sum_{i \in I} Y_i$ is the disjoint union of all of the sets Y_i in the collection, with elements pairs (i, y) where $i \in I$ and $y \in Y_i$, while the dependent function space $\prod_{i \in I} Y_i$ is the set of functions mapping an input $i \in I$ to an element of Y_i. We write $A+B$ for the disjoint union of two sets A and B, with injections $\mathsf{inl} : A \to A+B$ and $\mathsf{inr} : B \to A+B$. We denote the set of natural numbers by $\mathbb{N}$, the set of positive natural numbers by $\mathbb{N}^+$, and write $[n] = \{0, \ldots, n-1\}$ for a canonical n-element set. We also write $\mathbf{1} = [1]$, $\mathbf{2} = [2]$ and so on for fixed small finite sets.

2. Coalgebraic Preliminaries

We assume familiarity with basic category theory.

2.1. *Final coalgebras*

Let $\mathcal{C}$ be a category and $F : \mathcal{C} \to \mathcal{C}$ an endofunctor. An F-coalgebra is a pair (A, α), where A is an object of $\mathcal{C}$, and $\alpha : A \to FA$ is a morphism. An F-coalgebra homomorphism from (A, α) to (B, β) is a morphism $f : A \to B$ preserving the coalgebra structure, i.e., such that the following diagram commutes:

$$\begin{array}{ccc} A & \xrightarrow{\alpha} & FA \\ {\scriptstyle f}\downarrow & & \downarrow{\scriptstyle F(f)} \\ B & \xrightarrow[\beta]{} & FB \end{array}$$

F-coalgebras and F-coalgebra homomorphisms form a category. If F is well behaved (e.g., finitary), this category will have a final object, called the *final F-coalgebra*, and denoted $(\nu F, \mathsf{out})$. Its universal property is a *corecursion* or *coinduction principle*: for every F-coalgebra (A, α), there exists a unique coalgebra homomorphism $\mathsf{unfold} : (A, \alpha) \to (\nu F, \mathsf{out})$.

We will make use of Lambek's Lemma, which says that for a final coalgebra $(\nu F, \mathsf{out})$, the map $\mathsf{out} : \nu F \to F(\nu F)$ is an isomorphism.

2.2. *Coinductive families and predicates*

Let I be a set. The category Set^I of *I-indexed families of sets* is the category whose objects are functors from I (viewed as a discrete category) to Set, and whose morphisms are natural transformations.

In other words, given two I-indexed families P and Q, a morphism $f : P \to Q$ in Set^I is a family of functions $f_i : P_i \to Q_i$ for each $i \in I$.

A *coinductive family* indexed by I is the final coalgebra νG of an endofunctor G on Set^I. Its corresponding "coinduction principle" says that for every G-coalgebra (P, g), there exists a unique morphism $\mathsf{unfold} : P \to \nu G$ making the following diagram in Set^I commute:

$$\begin{array}{ccc} P & \xrightarrow{g} & G(P) \\ {\scriptstyle \mathsf{unfold}}\downarrow & & \downarrow{\scriptstyle G(\mathsf{unfold})} \\ \nu G & \xrightarrow[\mathsf{out}]{} & G(\nu G) \end{array}$$

Let us spell out this principle. It says that for every I-indexed family P, if there is a family of functions $g_i : P_i \to G(P)_i$, then there is a unique family of morphisms $\mathsf{unfold}_i : P_i \to \nu G_i$ commuting with the coalgebra maps.

In particular, we will be interested in coinductive families indexed by the carrier X of a coalgebra (X, γ) of a functor $F : \mathsf{Set} \to \mathsf{Set}$, in which case there is a canonical way to obtain coinductive families via predicate liftings of F, as we now explain.

A *(proof-relevant) predicate lifting* of a functor $F : \mathsf{Set} \to \mathsf{Set}$ is a natural transformation $\{\varphi_X : \mathsf{Set}^X \to \mathsf{Set}^{FX}\}_{X \in \mathsf{Set}}$. Given an F-coalgebra $(X, \gamma : X \to FX)$ and a predicate lifting φ of F, we can define an endofunctor on Set^X by

$$\mathsf{Set}^X \xrightarrow{\varphi_X} \mathsf{Set}^{FX} \xrightarrow{-\circ\gamma} \mathsf{Set}^X,$$

and consider its final coalgebra. Similarly, in Section 3.3, we also use *proof-irrelevant* predicate liftings $\{\varphi_X : \mathcal{P}(X) \to \mathcal{P}(FX)\}_{X \in \mathsf{Set}}$.

3. Infinite Extensive Form Games

In the game theory literature [1,16,17], extensive form games are typically defined using a nonrecursive formulation. We take advantage of a more categorical presentation, as it is more compact, supports (co-)recursive function definitions and (co-)inductive reasoning, and smoothly generalises to richer semantic domains, e.g., metric, probabilistic and topological spaces. Throughout this section, let P be a finite set of players.

In this work we separate the description of an extensive form game into its "tree" part, representing the dynamical structure of the game, and its "payoff" part, classically given as a decoration on the leaves. The reason we do so stems from our goal of treating infinite games, where plays may be infinite. In that case, the correspondence between leaves and paths in the tree breaks down, making decorations on the leaves an ill-suited device to specify payoff functions.

In the following, we define first the tree part of a game (Definition 3.1), then the family of payoffs over a given extensive form tree (Definition 3.10), and finally an "extensive form game" as a combination of a given tree part with a payoff over it (Definition 3.13). During our journey to the latter definition, we also define the set of *moves* of a game as paths in its tree (Definition 3.7), and the set of *strategy profiles* as choices of a branch for each node of the tree (Definition 3.4). These are fundamental ingredients in the description and analysis of a game.

Definition 3.1. The set ETree^∞ of *infinite extensive form game trees with players P* is the final coalgebra $(\mathsf{ETree}^\infty, \mathsf{out}_{\mathsf{ETree}^\infty})$ of the functor $F_{\mathsf{ETree}} : \mathsf{Set} \to \mathsf{Set}$ defined by

$$F_{\mathsf{ETree}}(X) = \mathbf{1} + P \times \sum_{n \in \mathbb{N}^+} X^n.$$

This supports the Haskell-like data type

$$\text{data}\, \mathsf{ETree}^\infty = \mathsf{Leaf} \mid \mathsf{Node}\, P\, (n : \mathbb{N}^+)\, ([n] \to \mathsf{ETree}^\infty).$$

Concretely, a tree $T \in \mathsf{ETree}^\infty$ is either a leaf, indicating no further plays are possible, or an internal node labelled with a player $p \in P$ who is to play at that point in the game, and an *arity* $n \in \mathbb{N}^+$ representing the number of different moves available, followed by n subtrees. Crucially, being a final coalgebra, ETree^∞ includes paths of infinite depth.

Example 3.2 (Dollar Auction). The Dollar Auction is an infinite game introduced by Shubik [18] to exemplify a situation of "rational escalation". The game has two players, A and B, bidding over a dollar bill. Player A bids first and then players alternate turns. At each turn, a player chooses between two actions:

- *quit*, in which case the game ends and the other player wins the \$1.
- *bid*, which costs \$0.1, and yields the turn to the other player.

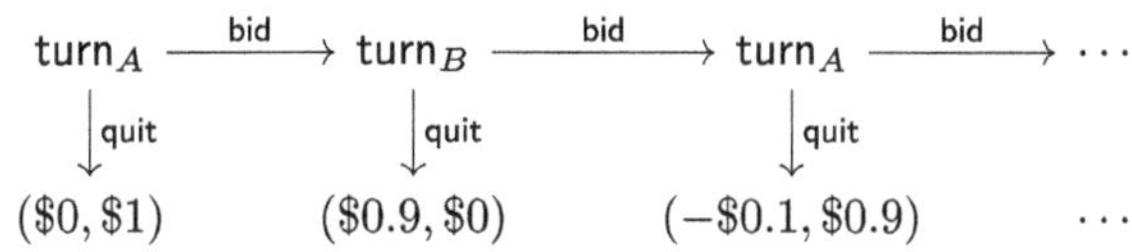

Note that when players bid, they immediately pay and are not refunded in case they lose the auction. The tree part of this game can be defined by mutual corecursion:

$$\mathsf{Dollar}_A = \mathsf{Node}\ A\ 2\ (\mathsf{Leaf}, \mathsf{Dollar}_B),$$
$$\mathsf{Dollar}_B = \mathsf{Node}\ B\ 2\ (\mathsf{Leaf}, \mathsf{Dollar}_A).$$

Then $\mathsf{Dollar} := \mathsf{Dollar}_A$, as A moves first. As an F_{ETree}-coalgebra, Dollar is defined starting from the coalgebra (D, δ), where $D = \{\mathsf{turn}_A, \mathsf{turn}_B, \mathsf{end}\}$, $P = \{A, B\}$ and δ is defined by

$$\delta : D \longrightarrow \mathbf{1} + P \times \sum_{n \in \mathbb{N}^+} D^n,$$
$$\delta\ \mathsf{turn}_A = \mathsf{inr}\ (A, 2, (\mathsf{end}, \mathsf{turn}_B)),$$
$$\delta\ \mathsf{turn}_B = \mathsf{inr}\ (B, 2, (\mathsf{end}, \mathsf{turn}_A)),$$
$$\delta\ \mathsf{end} = \mathsf{inl}\ *\,.$$

The coalgebra (D, δ) can be represented by the following automaton, where the two elements of $\mathbf{2}$ are named quit and bid:

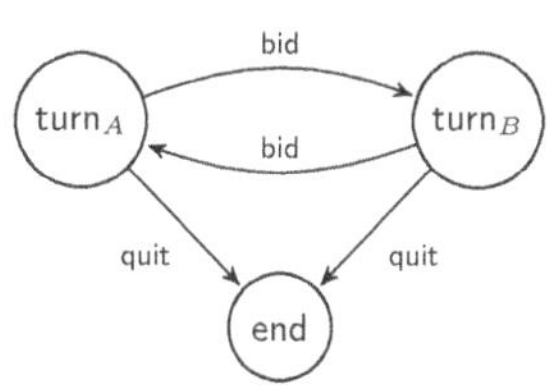

By terminality of $(\mathsf{ETree}^\infty, \mathsf{out}_{\mathsf{ETree}^\infty})$, there is a unique map $\mathsf{unfold}_{(D,\delta)} : (D, \delta) \to (\mathsf{ETree}^\infty, \mathsf{out}_{\mathsf{ETree}^\infty})$, and we define $\mathsf{Dollar} := \mathsf{unfold}_{(D,\delta)}(\mathsf{turn}_A)$. Thus, Dollar is the following infinite tree:

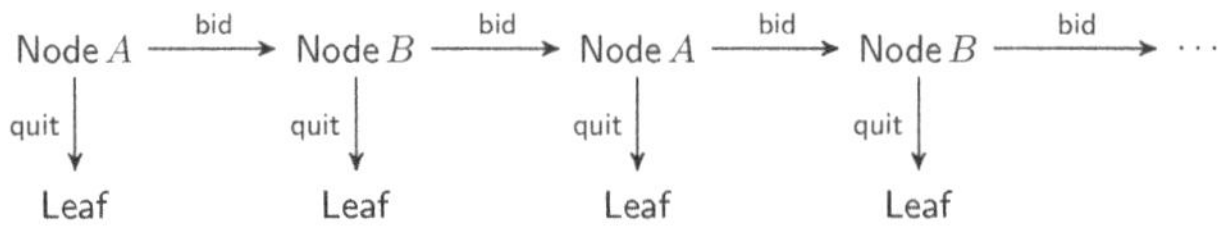

The payoff decoration making this into *the* Dollar Auction game will be discussed later in Example 3.11.

Example 3.3 (Repeated games). Let T be a finite, perfect-information, extensive-form game, with set of players P. Such games (without utility information) are represented as the elements of the initial algebra of F_{ETree} (see Capucci, Ghani, Ledent and Nordvall Forsberg [19, Section 2]). Any such tree can be converted to an F_{ETree}-coalgebra given by the automaton whose states and transitions correspond, respectively, to nodes and branches of T.

If we now identify the final states (given by leaves of T) with the initial state (given by the root of T) of the automaton, we get another F_{ETree}-coalgebra (Rep_T, ρ_T): the *repeated game coalgebra.*

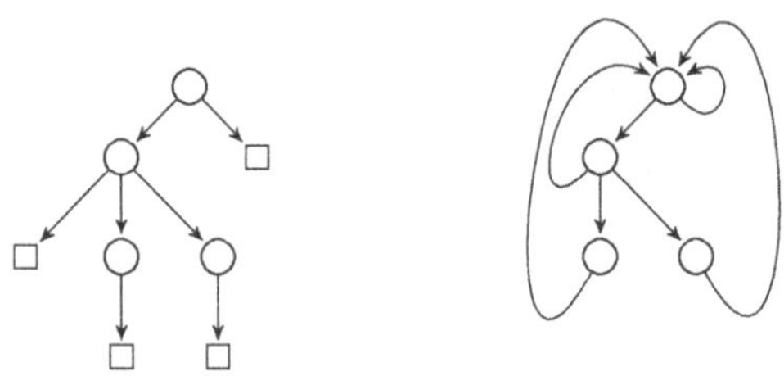

By terminality of $(\mathsf{ETree}^\infty, \mathsf{out}_{\mathsf{ETree}^\infty})$, there is a unique map $\mathsf{unfold}_{(\mathsf{Rep}_T, \rho_T)} : (\mathsf{Rep}_T, \rho_T) \to (\mathsf{ETree}^\infty, \mathsf{out}_{\mathsf{ETree}^\infty})$, and we define

$$T^\infty := \mathsf{unfold}_{(\mathsf{Rep}_T, \rho_T)}(\mathsf{root}).$$

One concrete example is the Market Entry game [20], a game with players $P = \{A, B\}$ described by the extensive-form tree M (here with payoff-labelled leaves):

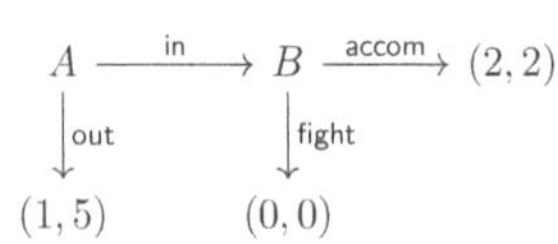

Player A decides whether to enter a new market or not. If staying out, the game ends, but if A enters, then player B has to decide whether to accommodate or fight the incumbent. In this case (Rep_T, ρ_T) corresponds to the automaton:

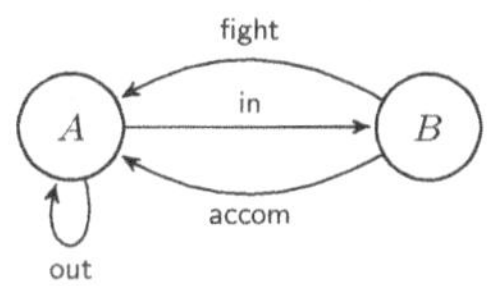

The induced infinite tree M^∞ is then given by

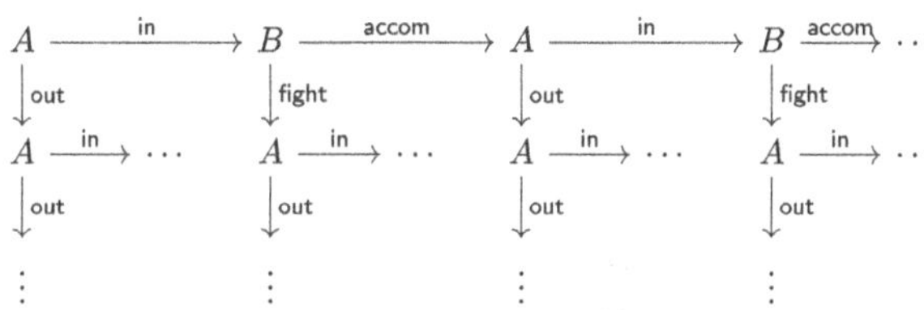

3.1. *Strategies and moves*

Throughout the section, let (X, γ) be an F_{ETree}-coalgebra.

3.1.1. *Strategy profiles*

A strategy profile for the coalgebra (X, γ) at state $x \in X$ consists of a choice of an action at each node in the game tree induced by (X, γ). Recall that an element of $F_{\mathsf{ETree}}(X)$ is either a leaf (of the form $\mathsf{inl}\, *$) or an internal node of the form $\mathsf{inr}\,(q, n, f)$, where q is a player, $n \in \mathbb{N}^+$ and $f : [n] \to X$.

Definition 3.4. We define the family of *strategy profiles* $\mathsf{prof}_{(X,\gamma)} : X \to \mathsf{Set}$ as the final coalgebra associated with the predicate lifting

$$\varphi_{\mathsf{prof},X} : (X \to \mathsf{Set}) \to F_{\mathsf{ETree}}(X) \to \mathsf{Set},$$
$$\varphi_{\mathsf{prof},X}\ P\ (\mathsf{inl}\, *) = \mathbf{1},$$
$$\varphi_{\mathsf{prof},X}\ P\ (\mathsf{inr}\,(q, n, f)) = [n] \times \prod_{a \in [n]} P_{f(a)}$$

i.e., we define $\mathsf{prof}_{(X,\gamma)}$ as the final coalgebra of the functor $F_{\mathsf{Prof}} : \mathsf{Set}^X \to \mathsf{Set}^X$ defined by $F_{\mathsf{Prof}}(P) = \varphi_{\mathsf{prof},X}(P) \circ \gamma$.

That $\mathsf{prof}_{(X,\gamma)}$ is the final coalgebra implies that for every $x \in X$, there is an isomorphism $s_x : \mathsf{prof}_{(X,\gamma)}(x) \to \varphi_{\mathsf{prof},X}(\mathsf{prof}_{(X,\gamma)})(\gamma(x))$. If $\gamma(x) = \mathsf{inr}\,(q, n, f)$, we thus have

$$s_x(\sigma) \in [n] \times \prod_{a\in[n]} \mathsf{prof}_{(X,\gamma)}(f\ a),$$

and we write $s_x(\sigma) = (\mathsf{now}\,\sigma, \mathsf{next}\,\sigma)$. Intuitively, $\mathsf{now}\,\sigma \in [n]$ is the branch chosen by player q in the current node x of the game tree; and $\mathsf{next}\,\sigma$ is a function assigning, for every branch $a \in [n]$ — not only the one chosen by player q — a strategy profile $\mathsf{next}\,\sigma\,a \in \mathsf{prof}_{(X,\gamma)}(f(a))$ starting at node $f(a)$.

Example 3.5. (Dollar Auction (continues from Example 3.2)). For the Dollar game of Example 3.2, we would expect the set of strategy profiles to be isomorphic to $\mathbf{2}^{\mathbb{N}}$, since a strategy profile selects, for every node of the game, an action in $\mathbf{2} \cong \{\mathsf{bid}, \mathsf{quit}\}$.

Formally, we check that $\mathsf{prof}_{(D,\delta)} : D \to \mathsf{Set}$ is the following family of sets, where (D, δ) is the coalgebra that defines the Dollar game in Example 3.2.

$$\mathsf{prof}_{(D,\delta)}(x) \cong \begin{cases} \mathbf{1} & \text{if } x = \mathsf{end} \\ \mathbf{2}^{\mathbb{N}} & \text{if } x = \mathsf{turn}_A, \mathsf{turn}_B. \end{cases}$$

Indeed, a function $\sigma \in \mathbf{2}^{\mathbb{N}}$ contains exactly the data of a strategy profile in $\mathsf{prof}_{(D,\delta)}(\mathsf{turn}_A)$, since we can define

$$\begin{aligned} &\mathsf{now}\,\sigma = \sigma(0) \in \mathbf{2} \\ &\mathsf{next}\,\sigma\ \mathsf{quit} = * \in \mathbf{1} \cong \mathsf{prof}_{(D,\delta)}(\mathsf{end}) \\ &\mathsf{next}\,\sigma\ \mathsf{bid} = (n \mapsto \sigma(n+1)) \in \mathbf{2}^{\mathbb{N}} \cong \mathsf{prof}_{(D,\delta)}(\mathsf{turn}_B) \end{aligned}$$

and similarly for a profile $\sigma \in \mathsf{prof}_{(D,\delta)}(\mathsf{turn}_B)$. It is straightforward (but cumbersome) to check that this satisfies the universal property of the final coalgebra of F_{Prof}.

Example 3.6. (Repeated games (continues from Example 3.3)). For a finite game T, we have defined in Example 3.3 its repeated game coalgebra (Rep_T, ρ_T), whose unfolding is the infinitely repeated game T^∞. A strategy profile for (Rep_T, ρ_T) with initial state x is the greatest solution to

$$\mathsf{prof}_{(\mathsf{Rep}_T,\rho_T)}(x) \cong \{\text{strategy profiles of } T|_x\} \times \prod_{\ell \in \mathsf{leaves}\ x} \mathsf{prof}_{(\mathsf{Rep}_T,\rho_T)}(\mathsf{root}),$$

where $T|_x$ is the subtree of T starting at $x \in \mathsf{Rep}_T$, root is the state corresponding to the root of the tree T, and $\mathsf{leaves}\ x$ denotes the set of leaves in the subtree $T|_x$. In the concrete case of the market entry game, this becomes (where we put $\mathsf{prof}_M = \mathsf{prof}_{(\mathsf{Rep}_M,\rho_M)}$ to ease notation):

$$\mathsf{prof}_M(A) \cong \{\mathsf{in}, \mathsf{out}\} \times \{\mathsf{fight}, \mathsf{accom}\} \times \underbrace{\mathsf{prof}_M(A) \times \mathsf{prof}_M(A) \times \mathsf{prof}_M(A)}_{\text{3 leaves accessible from } A},$$

$$\mathsf{prof}_M(B) \cong \{\mathsf{fight}, \mathsf{accom}\} \times \underbrace{\mathsf{prof}_M(A) \times \mathsf{prof}_M(A)}_{\text{2 leaves accessible from } B}.$$

Therefore, $\mathsf{prof}_M(A)$ is the final coalgebra of the functor

$$X \mapsto \{\mathsf{in}, \mathsf{out}\} \times \{\mathsf{fight}, \mathsf{accom}\} \times X^3.$$

3.1.2. *Moves*

The set of moves in the game is the set of paths in the tree, which is another coinductive family.

Definition 3.7. We define the family of *moves* $\mathsf{moves} : X \to \mathsf{Set}$ as the final coalgebra associated with the predicate lifting

$$\varphi_{\mathsf{moves},X} : (X \to \mathsf{Set}) \to F_{\mathsf{ETree}}(X) \to \mathsf{Set},$$
$$\varphi_{\mathsf{moves},X}\ P\ (\mathsf{inl}\ *) = \mathbf{1},$$
$$\varphi_{\mathsf{moves},X}\ P\ (\mathsf{inr}\ (q, n, f)) = \sum_{a \in [n]} P_{f(a)},$$

i.e., we define $\mathsf{moves}_{(X,\gamma)}$ as the final coalgebra of the functor $F_{\mathsf{moves}} : \mathsf{Set}^X \to \mathsf{Set}^X$ defined by $F_{\mathsf{moves}}(P) = \varphi_{\mathsf{moves},X}(P) \circ \gamma$.

Again, for every $x \in X$, we have an isomorphism $m_x : \mathsf{moves}_{(X,\gamma)}(x) \to \varphi_{\mathsf{moves},X}(\mathsf{moves}_{(X,\gamma)})(\gamma(x))$. If $\gamma(x) = \mathsf{inr}\ (q, n, f)$, we have

$$m_x(\pi) \in \sum_{a \in [n]} \mathsf{moves}_{(X,\gamma)}(f\,a).$$

Note that, as m_x is iso, if $\gamma(x) = \mathsf{inr}\,(q, n, f)$, then for each $a \in [n]$ and $\pi' \in \mathsf{moves}_{(X,\gamma)}(f\,a)$ there is a unique element $\mathsf{cons}(a, \pi') \in \mathsf{moves}_{(X,\gamma)}(x)$ such that $m_x(\mathsf{cons}(a, \pi')) = (a, \pi')$. Intuitively, $\mathsf{cons}(a, \pi')$ represents a path in the game tree starting at node x, where the first chosen branch is $a \in [n]$, and the following branches are chosen according to the path π'.

Example 3.8. (Dollar Auction (continues from Example 3.5)). The moves of Dollar are given by the final coalgebra of $X \mapsto \mathbf{1} + X$, i.e.,

$$\mathsf{moves}(\mathsf{Dollar}) \cong \mathbf{1} + \mathsf{moves}(\mathsf{Dollar}).$$

The final coalgebra of this functor is known as the *conatural numbers* $(\mathbb{N}^\infty, \mathsf{pred})$, which include all finite natural numbers and an "infinite" number ω. The map pred sends $0 \in \mathbb{N}^\infty$ to $\mathsf{inl}\ *$ and every other natural number to the right injection of its predecessor. The predecessor of ω is itself, $\mathsf{pred}\ \omega = \mathsf{inr}\,\omega$. Note that it is not decidable if a given conatural number x is finite or infinite; however, by applying pred a finite number of times, we can decide if $x \geq n$ for any finite natural number n.

We interpret $n \in \mathsf{moves}(\mathsf{Dollar})$ as the path starting from the root and ending at the n-th leaf, i.e., the play where players bid n times before one of them[a] decides to quit. The unique infinite play ω corresponds to infinite escalation, with players never quitting.

Similarly, $\mathsf{moves}_{(D,\delta)}$ is given by

$$\mathsf{moves}_{(D,\delta)}(x) \cong \begin{cases} \mathbf{1} & \text{if } x = \mathsf{end} \\ \mathbb{N}^\infty & \text{if } x = \mathsf{turn}_A, \mathsf{turn}_B. \end{cases}$$

[a]The player who quits can be determined from the parity of n.

Example 3.9. (Repeated games (continues from Example 3.6)). For any finite extensive-form game tree T, one has

$$\mathsf{moves}_{(\mathsf{Rep}_T, \rho_T)}(x) \cong (\mathsf{leaves}\ x) \times (\mathsf{leaves\ root})^{\mathbb{N}}.$$

In the specific instance of the market entry game M, moves are three: $1 : A \xrightarrow{\mathsf{out}} *$, $2 : A \xrightarrow{\mathsf{in}} B \xrightarrow{\mathsf{accom}} *$ and $3 : A \xrightarrow{\mathsf{in}} B \xrightarrow{\mathsf{fight}} *$, forming the set $\mathbf{3}$. These are all accessible from A, therefore M^∞ has set of moves specified by the final coalgebra of

$$X \mapsto \mathbf{3} \times X,$$

which is readily seen to be $\mathbf{3}^{\mathbb{N}}$. Indeed, a move of the repeated game is a move for every stage game. On the other hand, only moves 2 and 3 are accessible from B, therefore, we get

$$\mathsf{moves}_{(\mathsf{Rep}_M, \rho_M)}(x) \cong \begin{cases} \mathbf{3} \times \mathbf{3}^{\mathbb{N}} & \text{if } x = A \\ \mathbf{2} \times \mathbf{3}^{\mathbb{N}} & \text{if } x = B. \end{cases}$$

3.2. *Evaluating strategies*

In order to compare strategies, we need a way to assign a payoff to them. This is done in two steps: the *play function* turns a strategy profile into a sequence of moves; and the *payoff function* explains how outcomes turn into rewards for the players. This will allow us, in Section 3.3, to define several equilibrium concepts, i.e., predicates on strategy profiles that express when all players are happy with their given rewards.

3.2.1. *The play function*

We can use the universal property of final coalgebras to define a play function $\mathsf{play}_{(X,\gamma)} : \mathsf{prof}_{(X,\gamma)} \to \mathsf{moves}_{(X,\gamma)}$, which computes the sequence of moves generated by playing according to a strategy profile.

To define $\mathsf{play}_{(X,\gamma)} : \mathsf{prof}_{(X,\gamma)} \to \mathsf{moves}_{(X,\gamma)}$, we use the finality of $\mathsf{moves}_{(X,\gamma)}$, i.e., we want to give $\mathsf{prof}_{(X,\gamma)}$ a F_{moves}-coalgebra structure. It is sufficient to give, for $Q \in \mathsf{Set}^X$, a natural transformation p_Q :

$\varphi_{\mathsf{prof},X}(Q) \to \varphi_{\mathsf{moves},X}(Q)$ in $\mathsf{Set}^{F_{\mathsf{ETree}}(X)}$, which we can do as follows:

$$\begin{aligned} p_Q\ (\mathsf{inl}\ *)\ * &= * \\ p_Q\ (\mathsf{inr}\ (q,n,f))\ (a,\sigma) &= (a,\sigma(a)). \end{aligned}$$

Instantiating at component $Q = \mathsf{prof}_{(X,\gamma)}$ and composing with the isomorphism s_x gives $\mathsf{prof}_{(X,\gamma)}$ a F_{moves}-coalgebra structure, as required. Hence, there is a unique family of functions $\mathsf{play}_{(X,\gamma)}(x) : \mathsf{prof}_{(X,\gamma)}(x) \to \mathsf{moves}_{(X,\gamma)}(x)$, which, up to the isomorphisms s_x and m_x, satisfies the following definition.

$$\begin{aligned} &\mathsf{play} \in \prod_{x\in X} \left(\mathsf{prof}_{(X,\gamma)}(x) \to \mathsf{moves}_{(X,\gamma)}(x)\right) \\ &\mathsf{play}(x,*) = * \quad \text{when } \gamma(x) = (\mathsf{inl}\,*) \\ &\mathsf{play}(x,(a,\sigma)) = (a, \mathsf{play}(f(a),\sigma(a))) \quad \text{when } \gamma(x) = \mathsf{inr}\,(q,n,f). \end{aligned}$$

3.2.2. *Payoff functions and the game coalgebra*

From now on, let R be a set of "rewards" or "payoffs". Common choices are $R = \mathbb{Q}$, the rational numbers or $R = \mathbb{R}$, the real numbers, but dealing with infinite plays quickly leads to contemplating infinite rewards (negative and positive) as well. We assume R is ordered, and eventually, we will need to assume that this order is trichotomous. Recall that we denote by P the set of players. We write R^P for the set of functions $R \to P$, representing choices of payoffs for each of the players.

Definition 3.10. A payoff function for an F_{ETree}-coalgebra (X,γ) at state $x \in X$ is a function

$$u : \mathsf{moves}_{(X,\gamma)}(x) \to R^P.$$

The set of payoff functions for each $x \in X$ is denoted by $\mathsf{pay}_{(X,\gamma)}(x)$.

Example 3.11. (Dollar Auction (continues from Example 3.8)). Recall that $\mathsf{moves}_{(D,\delta)}(\mathsf{turn}_p)$ is given by the conatural numbers. The payoff function for the Dollar Auction game (where

$R = [-\infty, +\infty)$[b]) can be defined in two steps. First, we coinductively define a map into colists[c] of payoffs $\mathsf{List}^\infty\ R^P$ (which we think of as "ledgers"):

$$\mathsf{led} \in \prod_{x \in D} \left(\mathsf{moves}_{(D,\delta)}(x) \to \mathsf{List}^\infty\ R^P\right)$$

$$\mathsf{led\ end} * = \mathsf{Empty}$$

$$\mathsf{led\ turn}_A\ m = \begin{cases} (A \mapsto 0, B \mapsto 1) :: \mathsf{Empty} & \mathsf{pred}\ m = \mathsf{inl}\ * \\ (A \mapsto -0.1, B \mapsto 0) :: (\mathsf{led\ turn}_B\ n) & \mathsf{pred}\ m = \mathsf{inr}\ n \end{cases}$$

$$\mathsf{led\ turn}_B\ m = \begin{cases} (A \mapsto 1, B \mapsto 0) :: \mathsf{Empty} & \mathsf{pred}\ m = \mathsf{inl}\ * \\ (A \mapsto 0, B \mapsto -0.1) :: (\mathsf{led\ turn}_A\ n) & \mathsf{pred}\ m = \mathsf{inr}\ n. \end{cases}$$

Then the actual utility function is given by summing up componentwise all the payoffs collected by the players during the game:

$$u_{\mathsf{Dollar}}\ m = \sum_{n=0}^{+\infty} p_i, \quad \text{where } p = \mathsf{led\ turn}_A\ m.$$

Here p_i is defined to be zero when i is greater than the length of p.

In the case of $m = \omega$, this will unfold into an infinite sum where the summands alternate between $(A \mapsto -0.1, B \mapsto 0)$ and $(A \mapsto 0, B \mapsto -0.1)$, therefore, yielding the payoff vector $(A \mapsto -\infty, B \mapsto -\infty)$.

Example 3.12. (Repeated games (continues from Example 3.9)). The payoff function of an infinitely repeated game is obtained similarly to the previous example: "partial" payoffs are summed at each iteration of the stage game. Unlike the Dollar Auction, however, in an infinitely repeated game all plays are infinite, therefore, *discounting* has to be adopted. This means that at each stage of the game, payoff vectors are uniformly scaled by a discount factor $0 < \delta < 1$. Discounting reflects the real-world tendency to value future payoffs less than present ones.

---

[b]We assume that $-\infty + m = -\infty$ for every $m \in R$.

[c]Colists of A are "possibly infinite lists", i.e., terms of the final coalgebra of $X \mapsto 1 + A \times X$ for a given A. We denote $\mathsf{inl}\ *$ as Empty and $\mathsf{inr}\ (a, x)$ as $a :: x$.

Thus, let $v_T \in \prod_{x \in \mathsf{Rep}_T} \left(\mathsf{leaves}(x) \to \mathbb{R}^P\right)$ be the utility function of the (finite) stage game T (such as the one represented by the diagram of M in Example 3.3). For a given discount factor δ, we get a payoff function $u^{\delta}_{T^\infty} \in \prod_{x \in \mathsf{Rep}_T} \left(\mathsf{moves}_{(\mathsf{Rep}_T, \rho_T)}(x) \to \mathbb{R}^P\right)$ by setting (recall that $\mathsf{moves}_{(\mathsf{Rep}_T, \rho_T)}(x) = (\mathsf{leaves}\ x) \times (\mathsf{leaves\ root})^{\mathbb{N}}$)

$$u^{\delta}_{T^\infty}\ x\ (m_0, ms) := (v_T\ x\ m_0) + \sum_{i=0}^{+\infty} \delta^{i+1} \cdot (v_T\ x\ (ms\ i)).$$

The assumption of $|\delta| < 1$ guarantees the convergence of such a sum.

We are now ready to define the "game coalgebra" — an F_{ETree}-coalgebra which will enable us to study equilibria. We do this by collecting all the information needed: the current state of the game, a strategy profile and a payoff function.

Definition 3.13. Let (X, γ) be an F_{ETree}-coalgebra. The *game coalgebra* $(Z_{(X,\gamma)}, \Gamma)$ is the F_{ETree}-coalgebra with carrier set

$$Z_{(X,\gamma)} = \sum_{x \in X} \mathsf{prof}_{(X,\gamma)}(x) \times \mathsf{pay}_{(X,\gamma)}(x),$$

and dynamics given by the map Γ defined by

$$\Gamma(x, \sigma, u) = \begin{cases} \mathsf{inl} * & \text{if } \gamma(x) = \mathsf{inl} * \\ \mathsf{inr}\ (q, n, a \mapsto (f\ a, \mathsf{next}\,\sigma\,a, u_a)) & \text{if } \gamma(x) = \mathsf{inr}\ (q, n, f), \end{cases}$$

where $u_a(\pi') = u(\mathsf{cons}(a, \pi'))$, for $a \in [n]$ and $\pi' \in \mathsf{moves}_{(X,\gamma)}(f\ a)$.

3.3. *Equilibrium concepts*

We are now going to introduce the equilibrium concepts for our games: a "one-shot" equilibrium concept that expresses that no player can improve their payoff by unilaterallty deviating from their strategy in exactly one state, and the usual Nash equilibrium that expresses that none of the players can change their strategy unilaterally (at possibly arbitrarily many places) such that their overall payoff increases. Before we are able to introduce the "one-shot" equilibrium concept, we will first introduce a modal operator that allows to specify properties that should hold everywhere in the game tree.

3.3.1. *The "Everywhere" modality*

Notions from game theory such as subgame perfection require a predicate to hold at every node of a tree (i.e., in every subgame). This is achieved by the so-called "everywhere" or "henceforth" modality. Using standard techniques from coalgebra, we can construct it using a proof-irrelevant predicate lifting as follows:

Definition 3.14. Let (X, γ) be an F_{ETree}-coalgebra. Consider the predicate lifting $\varphi_\Box : \mathcal{P}(X) \to \mathcal{P}(F_{\mathsf{ETree}}X)$ defined by

$$\varphi_\Box(Q) = \{\mathsf{inl}\, *\} \cup \{\mathsf{inr}(q, n, f) \mid (\forall a \in [n])\, f(a) \in Q\},$$

and for a predicate $P \in \mathcal{P}(X)$, define $\Box P$ to be the greatest fixpoint of

$$F_{\Box,P}(U) = P \cap \gamma^{-1}(\varphi_\Box(U)).$$

A detailed discussion of this operator can be found, e.g., in Jacobs [21] where $\Box$ is referred to as the "henceforth" operator. Intuitively, given a predicate P, $\Box P$ holds at a state x of a coalgebra if x itself and all its descendants satisfy P, i.e., P is invariably true in the game starting at x. The $\Box$-operator satisfies the properties one would expect from basic modal logic.

Lemma 3.15. *The modality $\Box$ is monotone, i.e., if P implies Q, then $\Box P$ implies $\Box Q$. Furthermore, $\Box P \subseteq P$ and $\Box P \subseteq \Box\Box P$.*

Proof. Assume $P \subseteq Q$. To show $\Box P \subseteq \Box Q$, we use the finality of $\Box Q$ to conclude $\Box P \subseteq \Box Q$ by showing that $\Box P \subseteq F_{\Box,Q}(\Box P)$. This follows since $\Box P \subseteq F_{\Box,P}(\Box P)$ and $P \subseteq Q$. In the same way, $\Box P \subseteq P$ and $\Box P \subseteq \Box\Box P$. $\Box$

3.3.2. *Unimprovability*

A very simple equilibrium concept is the following: at each node of the game, the current player cannot improve their payoff by changing their action at the current node. We call this the "one-shot" equilibrium concept. Formally, we can encode it as follows. First we define a predicate $\mathsf{Unimprov}$ which verifies that, at a node in $Z_{(X,\gamma)}$, the current strategy is unimprovable for the current player. We then ask that this predicate holds everywhere in the tree using the "everywhere" modality.

Definition 3.16. We define the predicate $\mathsf{Unimprov}$ on $Z_{(X,\gamma)}$ by

$$
\begin{aligned}
&(x, \sigma, u) \in \mathsf{Unimprov} \\
&\qquad \text{if} \qquad \gamma(x) = \mathsf{inl}\ * \\
&\qquad \text{or} \qquad \gamma(x) = \mathsf{inr}\ (q, n, f) \\
&\qquad\qquad \text{and } \mathsf{now}\,\sigma \in \mathsf{argmax}(a \mapsto \pi_q(u_a(\mathsf{play}\ (f\ a)\ (\mathsf{next}\,\sigma\,a)))).
\end{aligned}
$$

The "one-shot" equilibrium concept can now be defined as $\Box\mathsf{Unimprov}$.

In words, $(x, \sigma, u) \in \mathsf{Unimprov}$ holds if x is a leaf, or if x is an internal node and $\mathsf{now}\,\sigma$ gives the best payoff for the current player q at the node according to the utility function u, assuming that all players continue to play according to σ. This equilibrium concept also occurs in Lescanne and Perrinel [12] and Abramsky and Winschel [13], who call it "subgame perfect equilibria". We prefer to reserve that name for the predicate $\Box\mathsf{Nash}$ that we will define in the next section.

3.3.3. *Nash equilibria and subgame perfection*

The predicate $\mathsf{Unimprov}$ from the previous section says that a player cannot improve their payoff by changing their action at the current node only. In contrast, Nash equilibria are concerned with deviations where a player may change their action at several nodes simultaneously. The only restriction is that all such nodes must belong to the same player. So, we first define a predicate $\equiv_p$ which characterises when two strategy profiles are the same, except for deviations by one player p. Since we want to allow an infinite number of deviations by player p, we define this as a coinductive predicate.

Definition 3.17. For each player $p \in P$ and state $x \in X$, we define a family of relations

$$\equiv_p \subseteq \mathsf{prof}_{(X,\gamma)}(x) \times \mathsf{prof}_{(X,\gamma)}(x),$$

as the maximal family such that for all $\sigma, \sigma' \in \mathsf{prof}_{(X,\gamma)}(x)$ and $x \in X$, we have $\sigma \equiv_p \sigma'$ if and only if one of the following is satisfied:

(1) $\gamma(x) = \mathsf{inl}\ *$; or
(2) $\gamma(x) = \mathsf{inr}\ (p, n, f)$ and $\mathsf{next}\,\sigma\, a \equiv_p \mathsf{next}\,\sigma'\, a$ for all $a \in [n]$; or
(3) $\gamma(x) = \mathsf{inr}\ (q, n, f)$ with $q \neq p$, $\mathsf{now}\,\sigma = \mathsf{now}\,\sigma'$ and $\mathsf{next}\,\sigma\, a \equiv_p \mathsf{next}\,\sigma'\, a$ for all $a \in [n]$.

We can use the universal property of $\equiv_p$ to deduce the following:

Lemma 3.18. *Assume the set of players P has decidable equality. For each player $p \in P$, the relation $\equiv_p$ is reflexive.*

Using $\equiv_p$ to talk about deviations, we can now formulate the Nash equilibrium concept. Unlike the "one-step" equilibrium, the definition for a Nash equilibrium relies on a global quantification over all possible alternative strategies for a player rather than a local quantification over transitions in the game coalgebra. Consequently, Nash is neither an inductive nor a coinductive predicate.

Definition 3.19. In the game coalgebra $(Z_{(X,\gamma)}, \Gamma)$, we define

$$
\begin{aligned}
&(x, \sigma, u) \in \mathsf{Nash} \\
&\quad \text{if } \forall p \in P.\ \forall \sigma' \in \mathsf{prof}_{(X,\gamma)}(x), \\
&\qquad (\sigma \equiv_p \sigma') \to \left(\pi_p u(\mathsf{play}\ x\ \sigma) \geq \pi_p u(\mathsf{play}\ x\ \sigma')\right).
\end{aligned}
$$

In other words, given an initial node x and payoff function u, the strategy profile σ is a Nash equilibrium if every possible deviation σ' by player p yields a payoff for p which is smaller than the one of σ.

We can now succinctly define the solution concept of subgame perfect Nash equilibria simply as $\Box\mathsf{Nash}$ — a strategy profile is subgame perfect if it is a Nash equilibrium in every subgame of the tree.

4. Relating Unimprovability and Subgame Perfect Nash Equilibria

In this section, we relate the coalgebraic subgame perfect Nash equilibria $\Box\mathsf{Nash}$ and the one-deviation equilibrium $\Box\mathsf{Unimprov}$, thus connecting our coalgebraic treatment with the standard notions from game theory. One direction is almost immediate:

Lemma 4.1. *Assume the set of players P has decidable equality. Let (x, σ, u) be a state of the game coalgebra $(Z_{(X,\gamma)}, \Gamma)$. If $(x, \sigma, u) \in \Box\mathsf{Nash}$, then $(x, \sigma, u) \in \Box\mathsf{Unimprov}$.*

Proof. Since $\Box$ is monotone by Lemma 3.15, it is sufficient to show that $(x, \sigma, u) \in \mathsf{Nash}$ implies $(x, \sigma, u) \in \mathsf{Unimprov}$. If $\gamma(x) = \mathsf{inl} *$, this is trivial, so we concentrate on the case when $\gamma(x) = \mathsf{inr}\,(q, n, f)$. By definition, we have to show that

$$\begin{aligned} &\pi_q(u_{\mathsf{now}\,\sigma}(\mathsf{play}\ (f(\mathsf{now}\,\sigma))\ (\mathsf{next}\,\sigma\ (\mathsf{now}\,\sigma)))) \\ &\quad \geq \pi_q(u_a(\mathsf{play}\ (f\ a)\ (\mathsf{next}\,\sigma\ a))), \end{aligned}$$

for every $a \in [n]$. For each a, let σ_a be the strategy profile with $\mathsf{now}\,\sigma_a = a$ and $\mathsf{next}\,\sigma_a = \mathsf{next}\,\sigma$. By Lemma 3.18, $\sigma \equiv_q \sigma_a$, and the conclusion follows from the assumption that $(x, \sigma, u) \in \mathsf{Nash}$. $\Box$

For the other direction, we need to assume that the utility function is suitably well behaved; this is known as *continuity at infinity* in the game theory literature [15, Chapter 4.2]. We formulate it more generally for arbitrary F_{moves}-coalgebras.

4.1. *Continuity at infinity*

To formally define continuity at infinity, we assume that the set of payoffs R is a metric space and R^P is the P-fold product of this metric space obtained via taking the maximum. To obtain a metric on a F_{moves}-coalgebra, we use the projections into the terminal sequence of F_{moves}. This technique can be formulated for arbitrary functors on indexed families of sets.

Definition 4.2. Let $H : \mathsf{Set}^X \to \mathsf{Set}^X$ be a functor, and (A, γ) an H-coalgebra. Recall that $\top_X(x) = \mathbf{1}$ is the terminal object in Set^X.

We define a family of natural transformations $(\gamma^i : A \to H^i(\top_X))_{i\in\mathbb{N}}$ inductively by

$$\begin{array}{ccc} A & \quad & A \\ \downarrow{\scriptstyle \gamma^0} & & {\scriptstyle \gamma}\downarrow \searrow^{\gamma^{i+1}} \\ \top_X & & HA \xrightarrow[H\gamma^i]{} H^{i+1}\top_X \end{array}$$

where $\gamma^0 := !_A$ is the unique morphism from A into the terminal object. We call states $a, a' \in A(x)$ *n-step equivalent*, and we write $a \sim_n a'$, when $\gamma^n_x(a) = \gamma^n_x(a')$. This induces a pseudometric on $A(x)$ by putting $d^x_{(A,\gamma)}(a, a') = 2^{-m}$, where $m = \sup\{n \mid a \sim_n a'\}$.

If H is finitary, i.e. determined by its action on finitely presentable objects [22], then if two states in an H-coalgebra (A, γ) agree for all finite observations, they are equal. Hence, in this case $d^x_{(A,\gamma)}$ is actually a metric:

Lemma 4.3. *Let $H : \mathsf{Set}^X \to \mathsf{Set}^X$ be a finitary functor. For each H-coalgebra (A, γ) and $x \in X$, $d^x_{(A,\gamma)}$ is a metric on $A(x)$.*

The lemma is a straightforward consequence of a similar result for Set-functors [23,24]. We apply the above lemma to the functor $F_{\mathsf{moves}} : \mathsf{Set}^X \to \mathsf{Set}^X$ from Definition 3.7, which is finitary. As a result, we are now ready to define continuity at infinity coalgebraically.

Definition 4.4. Let (X, γ) be an F_{ETree}-coalgebra. We call $u \in \mathsf{pay}_{(X,\gamma)}(x)$ continuous at infinity if $u : \mathsf{moves}_{(X,\gamma)}(x) \to R^P$ is *uniformly continuous* as a map between metric spaces, i.e., if

$$\forall \varepsilon > 0.\ \exists \delta > 0.\ \forall m, m' \cdot d^x_{\mathsf{moves}_{(X,\gamma)}}(m, m') < \delta$$
$$\to d_{R^P}(u(m), (u\ m')) < \varepsilon.$$

Remark 4.5. This generalises the usual formulation of continuity at infinity from the game theory literature (see, e.g., Fudenberg and Tirole [15, Definition 4.1]) to coalgebras. We observe that the weaker assumption of *pointwise* continuity would be sufficient to prove Theorem 4.12 (or the corresponding traditional statement [15, Theorem 4.2]). Classically, $\mathsf{moves}_{(X,\gamma)}(x)$ is compact [25], and the distinction disappears, but this is of course not constructively valid.

Spelling out the definition of $d^x_{\mathsf{moves}(X,\gamma)}$ and d_{R^P}, we arrive at the following concrete definition of continuity at infinity:

Proposition 4.6. *Let (X,γ) be an F_{ETree}-coalgebra. A payoff function $u \in \mathsf{pay}_{(X,\gamma)}(x)$ is continuous at infinity if and only if*

$$\begin{aligned}&\forall \varepsilon > 0.\ \exists n \in \mathbb{N}.\ \forall m, m' \cdot (m \sim_n m')\\ &\to \forall p \in P \cdot d_R(\pi_p(u\ m), \pi_p(u\ m')) < \varepsilon.\end{aligned}$$

Example 4.7. (Dollar Auction (continues from Example 3.11)). We claim the payoff function for Dollar is continuous at infinity. It will suffice to focus on one component, say $u_A = \pi_A \circ u$, since $\pi_B \circ u$ is the same up to a shift. Let us begin by specifying a metric on $R = [-\infty, +\infty)$[d]:

$$d_R(r, r') = |\arctan\ r - \arctan\ r'|.$$

This choice of metric makes R into a bounded space, since evidently $\operatorname{diam}(R) = \pi$. In particular, $d_R((u_A\ m), (u_A\ m'))$ is finite for every $m, m' \in \mathsf{moves}_{(D,\delta)}(x)$.

By applying tan at both sides[e] of $d_R((u_A\ m), (u_A\ m')) < \varepsilon$, our thesis becomes

$$\forall \varepsilon > 0.\ \exists n \in \mathbb{N}.\ \forall m, m'.\ m \sim_n m' \ \to \ \frac{|(u_A\ m) - (u_A\ m')|}{1 + (u_A\ m)(u_A\ m')} < \tan\ \varepsilon. \tag{4.1}$$

Observe that, when $m, m' \to +\infty$,

$$g((u_A\ m), (u_A\ m')) \longrightarrow 0 \quad \text{where } g(x, y) = \frac{|x - y|}{1 + xy}.$$

Since $(u_A\ m), (u_A\ m') \to -\infty$ by the definition of u_A and $g(x, y) \to 0$ as $x, y \to -\infty$. This convergence gives an $n \in \mathbb{N}$ for each chosen

[d]We cannot use the traditional Euclidean metric on $R = \mathbb{R}$, since $|-\infty - r| = +\infty$ for finite $r \in \mathbb{R}$. Any well-defined metric that makes R into a bounded space is equivalent.

[e]The function tan, by virtue of being monotone on the domain $(-\pi/2, +\pi/2)$, preserves inequalities for small enough ε (and, by previous considerations on the diameter of (R, d_R), for every value of $d_R((u_A\ m), (u_A\ m'))$).

$\tan \varepsilon > 0$. Suppose now $m \sim_n m'$. In this particular example, this can happen if and only if either $m, m' < n$ and $m = m'$, or $m, m' \geq n$. In the first case, $d_{\mathbb{R}^P}((u_A\ m), (u_A\ m')) = 0 < \tan \varepsilon$. In the second case, we have chosen n to satisfy $d_{\mathbb{R}^P}((u_A\ m), (u_A\ m')) < \tan \varepsilon$. Thus, we conclude that (4.1) is satisfied.

Example 4.8. (Repeated games (continues from Example 3.12)). The utility function of a repeated game with discounting is almost immediately seen to be continuous. Setting $v = \pi_p \circ (v_T\ A)$ (using notation from Example 3.12), we see that we are tasked to prove:

$$\forall \varepsilon > 0.\ \exists n \in \mathbb{N}.\ \forall (m_0, ms), (m_0', ms').$$
$$(m_0, ms') \sim_n (m_0', ms') \rightarrow$$
$$\left| (v(m_0) - v(m_0')) + \sum_{i=0}^{+\infty} \delta^{i+1}(v(ms\ i) - v(ms'\ i)) \right| < \varepsilon$$

In this case, $(m_0, ms') \sim_n (m_0', ms')$ holds exactly when m_0 and m_0' agree and (if $n > 0$) if ms and ms' also agree on their first n entries. When this happens, the first $n+1$ terms of the series cancel out. By convergence of said series (easily obtainable by comparison with a geometric series), we can make that quantity as small as we need to by eliding enough leading terms.

4.2. *The one-shot deviation principle*

The one-shot equilibrium concept states that there is no profitable single-node deviation. As an intermediate step towards subgame perfect Nash equilibria, we can also consider profitable deviations in a finite number of nodes. Following Lescanne [10, Section 5], this concept can be formalised as an *inductive* definition as follows:

Definition 4.9. Let $p \in P$ be a player, $x \in X$ a state and (X, γ) an F_{ETree}-coalgebra. We define a family of relations $\equiv_p^{\text{fin}} \subseteq \mathsf{prof}_{(X,\gamma)}(x) \times \mathsf{prof}_{(X,\gamma)}(x)$ inductively as the least family such that for all $\sigma, \sigma' \in \mathsf{prof}_{(X,\gamma)}(x)$ and $x \in X$, we have $\sigma \equiv_p^{\text{fin}} \sigma'$ iff one of the following is satisfied:

(1) $\sigma = \sigma'$; or
(2) $\gamma(x) = \mathsf{inl}\ *$; or

(3) $\gamma(x) = \mathsf{inr}\, p\, n\, f$ and $\mathsf{next}\, \sigma\, a \equiv_p^{\mathrm{fin}} \mathsf{next}\, \sigma'\, a$ for all $a \in [n]$; or
(4) $\gamma(x) = \mathsf{inr}\, q\, n\, f$ with $q \neq p$, $\mathsf{now}\, \sigma = \mathsf{now}\, \sigma'$ and $\mathsf{next}\, \sigma\, a \equiv_p^{\mathrm{fin}} \mathsf{next}\, \sigma'\, a$ for all $a \in [n]$.

Thus, strategies with $\sigma \equiv_p^{\mathrm{fin}} \sigma'$ can differ in their choice of action $\mathsf{now}\, \sigma \neq \mathsf{now}\, \sigma'$ at p-nodes; since the definition is inductive, this can only happen a finite number of times before reaching a base case. Given two strategies σ and σ', we can "truncate" σ after n rounds by replacing it with σ', resulting in a new strategy $\lceil \sigma \rceil_n^{\sigma'}$.

Lemma 4.10. *If $\sigma \equiv_p \sigma'$, then $\sigma \equiv_p^{fin} \lceil \sigma' \rceil_n^{\sigma}$ for any $n \in \mathbb{N}$. Conversely, if the set of players P has decidable equality, then $\sigma \equiv_p^{fin} \sigma'$ implies $\sigma \equiv_p \sigma'$.*

Proof. The first statement follows by induction on n. Since $\lceil \sigma' \rceil_0^{\sigma} = \sigma$, we have by $\sigma \equiv_p^{\mathrm{fin}} \lceil \sigma' \rceil_0^{\sigma}$ by (4.9). For the step case, consider the form of the proof $\sigma \equiv_p \sigma'$: if $\sigma \equiv_p \sigma'$ because $\gamma(x) = \mathsf{inl}\, *$, we immediately have $\sigma \equiv_p^{\mathrm{fin}} \lceil \sigma' \rceil_{n+1}^{\sigma}$; otherwise $p = q$ or $p \neq q$, and in either case $\mathsf{next}\, \sigma\, a \equiv_p \mathsf{next}\, \sigma'\, a$ for all $a \in [m]$, hence, by the induction hypothesis $\mathsf{next}\, \sigma\, a \equiv_p^{\mathrm{fin}} \lceil \mathsf{next}\, \sigma'\, a \rceil_n^{\mathsf{next}\, \sigma\, a}$ for every a, and by either (4.9) or (4.9) and how $\lceil \sigma \rceil_n^{\sigma'}$ is defined, we have $\sigma \equiv_p^{\mathrm{fin}} \lceil \sigma' \rceil_{n+1}^{\sigma}$ as required.

The second statement follows easily by induction on the proof that $\sigma \equiv_p^{\mathrm{fin}} \sigma'$, using decidable equality on P to invoke Lemma 4.1 for the base case 1. □

Although allowing a finite number of deviations might seem like a stronger notion of equilibrium than allowing just one, they turn out to be equivalent. This is because the one-shot equilibrium concept is quantified on every subgame: assuming a player can improve their payoff with a finite number of deviations, we can find a single profitable deviation by restricting to the subgame starting at the last deviation. Recall that an order relation $<$ is *trichotomous* if, for every pair of elements x, y, it is decidable whether $x < y$ or $x = y$ or $x > y$.

Lemma 4.11. *Let (x, σ, u) be a state of the game coalgebra $(Z_{(X,\gamma)}, \Gamma)$. Assume that the order relation $<$ on R is trichotomous. If there is a player p and a strategy σ' such that $\sigma \equiv_p^{fin} \sigma'$ and $\pi_p(u(\mathsf{play}\, x\, \sigma')) > \pi_p(u(\mathsf{play}\, x\, \sigma))$, then $(x, \sigma, u) \in \neg\Box\mathsf{Unimprov}$.*

Proof. By induction on the proof that $\sigma \equiv_p^{\text{fin}} \sigma'$, we can find a strategy σ'' that differs from σ (and agrees with σ') a minimum number of times, while still being a profitable deviation. In addition, there is a deepest node where σ'' differs from σ; let σ''' be the strategy that agrees with σ everywhere but at this node, where it agrees with σ'. By trichotomy, we either have $\pi_p(u(\mathsf{play}\, x\, \sigma''')) > \pi_p(u(\mathsf{play}\, x\, \sigma))$ or $\pi_p(u(\mathsf{play}\, x\, \sigma''')) \leq \pi_p(u(\mathsf{play}\, x\, \sigma))$. If the former, then this contradicts $\mathsf{Unimprov}$ at this node as required, so we only need to show that the latter is impossible. This is so, because the latter case violates the assumption that σ'' is minimal. □

Armed with this lemma, we can now tackle the difficult direction of the one-shot deviation principle, assuming the payoff function is continuous. For simplicity, we only state the theorem for $R = \mathbb{Q}$ with the standard metric $d_{\mathbb{Q}}(x, y) = |x - y|$. Note that the order on $\mathbb{Q}$ certainly is trichotomous.

Theorem 4.12. *Assume the set of players P has decidable equality. Let (x, σ, u) be a state of the game coalgebra $(Z_{(X,\gamma)}, \Gamma)$ with rewards $R = \mathbb{Q}$. If u is continuous at infinity and $(x, \sigma, u) \in \Box\mathsf{Unimprov}$, then $(x, \sigma, u) \in \Box\mathsf{Nash}$.*

Proof. Since $\Box P \subseteq \Box\Box P$ and $\Box$ is monotone by Lemma 3.15, it is sufficient to show that $(x, \sigma, u) \in \Box\mathsf{Unimprov}$ implies $(x, \sigma, u) \in \mathsf{Nash}$. Hence, assume a player p and a strategy σ' with $\sigma \equiv_p \sigma'$ are given; by trichotomy of $<$, it is enough to show that

$$u_p(\mathsf{play}\ x\ \sigma) < u_p(\mathsf{play}\ x\ \sigma'),$$

where we write $u_p(m) = \pi_p(u(m))$, is impossible. By continuity of u with $\varepsilon = u_p(\mathsf{play}\ x\ \sigma') - u_p(\mathsf{play}\ x\ \sigma)$, we find n such that if $m \sim_n \mathsf{play}\ x\ \sigma'$, then $|u_p(\mathsf{play}\ x\ \sigma') - u_p(m)| < \varepsilon$. Consider the history $m = \mathsf{play}\ x\ \sigma''$, where $\sigma'' = \lceil \sigma' \rceil_n^\sigma$ is σ' up to depth n, and σ thereafter. By construction, $m \sim_n \mathsf{play}\ x\ \sigma'$. We claim that σ'' is still a strictly profitable deviation, i.e. $u_p(\mathsf{play}\ x\ \sigma) < u_p(\mathsf{play}\ x\ \sigma'')$, by trichotomy of $<$ again: if $u_p(\mathsf{play}\ x\ \sigma) \geq u_p(\mathsf{play}\ x\ \sigma'')$, then both $|u_p(\mathsf{play}\ x\ \sigma') - u_p(\mathsf{play}\ x\ \sigma'')| \geq \varepsilon$ and $|u_p(\mathsf{play}\ x\ \sigma') - u_p(\mathsf{play}\ x\ \sigma'')| < \varepsilon$, which is absurd. Hence, we have found a profitable finite deviation, which by Lemmas 4.10 and 4.11 contradicts the assumption that $(x, \sigma, u) \in \Box\mathsf{Unimprov}$. Therefore, we must instead have $u_p(\mathsf{play}\ x\ \sigma) \geq u_p(\mathsf{play}\ x\ \sigma')$ as required. □

5. Conclusion and Future Work

In this chapter, we have built on Lescanne's and Abramksy and Winschel's coalgebraic treatment of infinite extensive form games to also consider plays that go on forever, rather than just convergent strategies that eventually lead to a terminal node. We are thus able to treat games such as infinitely repeated games, in addition to the games previously considered. We also connected the coalgebraic and traditional notions of equilibria by proving them equal under the assumption that the payoff function is continuous — a well-known result in the game theory literature, here extended to more general coalgebras. In future work, we plan to exploit techniques from coalgebra and automata theory to use our framework to *solve* various infinite-horizon games, i.e., to compute the set of equilibria for some given notion of equilibrium — Nash, subgame perfect equilibria, etc. We also hope to extend our recent translation of finite extensive form games into the framework of open games [19] to infinite games; since open games are well-suited for software implementation, this might point to another approach for computing the equilibria of infinite extensive form games in practice.

Acknowledgements

We would like to thank Pierre Lescanne and Philipp Zahn for discussions about coinduction and the one-step deviation principle, respectively. Thanks also to the anonymous reviewer whose comments and suggestions helped improve the chapter.

References

[1] K. Leyton-Brown and Y. Shoham, *Essentials of Game Theory: A Concise Multidisciplinary Introduction.* Morgan & Claypool, 2008.
[2] M. Osbourne and A. Rubinstein, *A Course in Game Theory.* MIT Press, 1994.
[3] D.S. Bridges, Preference and utility: A constructive development, *J. Math. Econ.* **9**(1), 165–185 (1982). doi: 10.1016/0304-4068(82)90025-8.

[4] D.S. Bridges, The constructive theory of preference relations on a locally compact space, *Indag. Math.* **92**(2), 141–165 (1989). doi: 10.1016/S1385-7258(89)80023-X.

[5] D.S. Bridges, The constructive theory of preference relations on a locally compact space II, *Math. Soc. Sci.* **27**(1), 1–9 (1994). ISSN 0165-4896. doi: 10.1016/0165-4896(94)00718-7.

[6] D.S. Bridges, First steps in constructive game theory, *Math. Log. Q.* **50**(4-5), 501–506 (2004).

[7] J. Mycielski and W. Taylor, A compactification of the algebra of terms, *Algebra Universalis.* **6**, 159–163 (1976).

[8] E. Bishop and D. Bridges, *Constructive Analysis.* Springer, 1985.

[9] J. Rutten, Universal coalgebra: A theory of systems, *Theoret. Comput. Sci.* **249**(1), 3–80 (2000). doi: 10.1016/S0304-3975(00)00056-6.

[10] P. Lescanne, A simple case of rationality of escalation. In R. Heckel and S. Milius (eds.), *CALCO '13, vol. 8089, LNCS.* Springer, 2013, pp. 191–204. doi: 10.1007/978-3-642-40206-7_15.

[11] P. Lescanne, Dependent types for extensive games, *J. Formaliz. Reason.* **11**(1), 1–17 (2018). doi: 10.6092/issn.1972-5787/7517.

[12] P. Lescanne and M. Perrinel, "Backward" coinduction, Nash equilibrium and the rationality of escalation, *Acta Informatica.* **49**(3), 117–137 (2012). doi: 10.1007/s00236-012-0153-3.

[13] S. Abramsky and V. Winschel, Coalgebraic analysis of subgame-perfect equilibria in infinite games without discounting, *Math. Struct. Comput. Sci.* **27**(5), 751–761 (2017). doi: 10.1017/S0960129515000365.

[14] N. Ghani, C. Kupke, A. Lambert, and F.N. Forsberg, A compositional treatment of iterated open games, *Theor. Comput. Sci.* **741**, 48–57 (2018). doi: 10.1016/j.tcs.2018.05.026.

[15] D. Fudenberg and J. Tirole, *Game Theory.* MIT Press, Cambridge, MA, 1991.

[16] H.W. Kuhn, Extensive games and the problem of information. In H.W. Kuhn and A.W. Tucker (eds.), *Contributions to the Theory of Games, vol. II (Annals of Mathematics Studies, 28).* Princeton University Press, 1953, pp. 193–216.

[17] R. Selten, Reexamination of the perfectness concept for equilibrium points in extensive games, *Int. J. Game Theory.* **4**, 25–55 (1975). ISSN 0020-7276. doi: 10.1007/BF01766400.

[18] M. Shubik, The dollar auction game: A paradox in noncooperative behavior and escalation, *J. Conflict Resolut.* **15**(1), 109–111 (1971).

[19] M. Capucci, N. Ghani, J. Ledent, and F. Nordvall Forsberg, Translating extensive form games to open games with agency. In *Applied Category Theory '21* (2021).

[20] R. Selten, The chain store paradox, *Theory Decis.* **9**(2), 127–159 (1978). doi: 10.1007/BF00131770.

[21] B. Jacobs, *Introduction to Coalgebra: Towards Mathematics of States and Observation.* Cambridge Tracts in Theoretical Computer Science, Cambridge University Press, 2016. doi: 10.1017/CBO9781316823187.

[22] J. Adamek and J. Rosicky, *Locally Presentable and Accessible Categories.* London Mathematical Society Lecture Note Series, Cambridge University Press, 1994. doi: 10.1017/CBO9780511600579.

[23] M. Barr, Terminal coalgebras in well-founded set theory, *Theor. Comput. Sci.* **114**(2), 299–315 (1993). ISSN 0304-3975. doi: 10.1016/0304-3975(93)90076-6.

[24] J. Worrell, *On Coalgebras and Final Semantics.* PhD thesis, University of Oxford, 2000.

[25] A. Kurz and D. Pattinson, Definability, canonical models, compactness for finitary coalgebraic modal logic, *Electron. Notes Theor. Comput. Sci.* **65**(1), 135–155 (2002). ISSN 1571-0661. doi: 10.1016/S1571-0661(04)80363-3.

https://doi.org/10.1142/9789811245220_0009

Chapter 9

Concurrent Gaussian Elimination

Ulrich Berger[*,§], Monika Seisenberger[*,§], Dieter Spreen[†,¶] and Hideki Tsuiki[‡,||]

[*] *Department of Computer Science, Swansea University Bay Campus, Fabian Way, Crymlyn Burrows, UK*

[†] *Department of Mathematics, University of Siegen, Germany*

[‡] *Graduate School of Human and Environmental Studies, Kyoto University, Kyoto, Japan*

[§] *u.berger*
[§] *m.seisenberger*
[§] *swansea.ac.uk*
[¶] *spreen@math.uni-siegen.de*
[||] *tsuiki@i.h.kyoto-u.ac.jp*

Working in a semi-constructive logical system that supports the extraction of concurrent programmes, we extract a programme inverting non-singular real-valued matrices from a constructive proof based on Gaussian elimination. Concurrency is used for efficient pivoting, that is, for finding an entry that is apart from zero in a nonnull vector of real numbers.

1. Introduction

A salient feature of constructive mathematics [1–3], which attracts mathematicians, logicians, philosophers and computer scientists, independently of their foundational position, is the fact that its notions are computationally meaningful and its proofs give rise to concrete realisations of computations, that is, programmes. Manifestations of this ability to "extract programmes" can be found, for example, in constructive type theories, functional interpretation and theories of realisability. Using computer implementations of constructive type theory, such as Nuprl [4], Coq [5] and Agda [6], or systems supporting realisability, such as Isabelle/HOL [7], and Minlog [8], substantial case studies have been undertaken that demonstrate the usefulness and scalability of programme extraction.

The exploration of the scope and limits of programme extraction is an active field of research that has led to many interesting results regarding the partial inclusion of classical logic [9–11] and mathematical principles such as various forms of choice [12–16] and induction principles [13,17–19]. The benefit of programme extraction is not only that extracted programmes come automatically with formal correctness proofs but also that important additional properties of programmes concerning computational complexity, totality and typability can be obtained.

An important paradigm in modern computing technology, that has only recently been connected with programme extraction, is *concurrency*, that is, the ability to run several "threads" of computations independently and asynchronously but allowing them to interact in a number of ways. Concurrency poses big challenges to computer science, since, on the one hand, it is indispensable for running computer systems efficiently on a range of different tasks simultaneously, on the other hand, the (often nondeterministic) behaviour of such systems is very difficult to control.

In [20], a semi-constructive logical system (CFP, Concurrent Fixed Point Logic) for the extraction of concurrent programmes was introduced and applied to extract a programme that converts between two different exact representations of real numbers. Concurrency was needed to avoid nontermination when dealing with potentially partial inputs (infinite Gray code [21]). In the present chapter, we apply CFP to another problem in computable analysis, the inversion of real-valued matrices. Here, nontermination is not the issue, but the potential gain in efficiency through concurrent computation. In both applications the interaction between concurrent threads is limited to winning a race: The result of the thread terminating first will be selected and the other killed. Nevertheless, controlling this form of concurrency at the logical level poses serious challenges.

The chapter is organised as follows: In Section 2, we introduce the main elements of the system IFP [17] our method of programme extraction is based on, and in Section 3, we instantiate IFP to the setting of real numbers. In Section 4, we present the system CFP that extends IFP by logical operators permitting the extraction of concurrent programmes. To be able to carry out our case study, we adjust CFP to allow for more than two concurrent threads and iterated "monadic" concurrency. Finally, Section 5 presents the proof of matrix inversion through Gaussian elimination where pivoting, that is, selecting a nonzero entry in a nonzero vector, is done concurrently.

2. Intuitionistic Fixed Point Logic

Our basic logical system for programme extraction is IFP, which is based on many-sorted intuitionistic first-order logic with equality and least and greatest fixed points. A detailed definition of its syntax, semantics and proof calculus, as well as applications to programme extraction, can be found in [17]. We highlight the most important points.

The *syntax* of IFP is parametric in a set of sorts (which we usually suppress), a set of constants and function symbols, and a set of predicate constants (all with fixed arities). *Formulas* A, B, *predicate* P, Q, and *operators* Φ, Ψ are defined simultaneously:

Formulas are $\bot, \top$ (falsity and truth), $s = t$ where s, t are first-order terms of the same sort, $P(\vec{t})$ where $\vec{t}$ is a tuple of first-order terms whose length is the arity of the predicate P, $A \wedge B$, $A \vee B$, $A \to B$ and $\forall x\, A$, $\exists x\, A$ where x is a first-order variable of a fixed sort.

Predicates are predicate variables, predicate constants, both of fixed arities, $\lambda\vec{x}\, A$ (abstraction) where $\vec{x}$ is a tuple of first-order variables and $\mu\,\Phi$, $\nu\,\Phi$. The common arity of the latter two constructs is the arity of the operator Φ.

Operators are of the form $\lambda X\, P$ where X is a predicate variable of the same arity as P which occurs freely in P only at strictly positive positions, that is, not in the left part of an implication. The arity of $\lambda X\, P$ is the common arity of X and P.

The application, $\Phi(Q)$, of an operator $\Phi = \lambda X\, P$ to a predicate Q of the same arity is defined as $P[Q/X]$. A definition $P \stackrel{\mathrm{Def}}{=} \mu\Phi$ will also be written as $P \stackrel{\mu}{=} \Phi(P)$. The notation $P \stackrel{\nu}{=} \Phi(P)$ has a similar meaning. If $\Phi = \lambda X \lambda\vec{x}\, A$, then we also write $P(\vec{x}) \stackrel{\mu}{=} A[P/X]$ and $P(\vec{x}) \stackrel{\nu}{=} A[P/X]$ instead of $P \stackrel{\mathrm{Def}}{=} \mu\Phi$ and $P \stackrel{\mathrm{Def}}{=} \nu\Phi$. Furthermore, we use the set-theoretic notations $P \subseteq Q$ for $\forall x\,(P(x) \to Q(x))$ and $P \cap Q$ for $\lambda x\,(P(x) \wedge Q(x))$, as well as the bounded quantifiers $\forall x \in P\, A$ and $\exists x \in P\, A$ for $\forall x\,(P(x) \to A)$ and $\exists x\,(P(x) \wedge A)$, respectively. In Sections 4 and 5, we will also use finitely iterated conjunctions and disjunctions which we define as

$$\bigvee_{1 \le i \le n} A(i) \stackrel{\mathrm{Def}}{=} A_1 \wedge (A_2 \wedge \ldots (A_n \wedge \top) \ldots),$$

$$\bigwedge_{1 \le i \le n} A(i) \stackrel{\mathrm{Def}}{=} A_1 \vee (A_2 \vee \ldots (A_n \vee \bot) \ldots).$$

The *semantics* of IFP has a classical and a constructive part.

The *classical semantics* is as usual for classical first-order predicate logic, that is, sorts are interpreted as sets (called carrier sets), function symbols as (set-theoretic) functions, and predicate constants as subsets of the corresponding cartesian products of carrier sets. Operators are interpreted as monotone predicate transformers and the predicates $\mu\,\Phi$ and $\nu\,\Phi$ as their least and greatest fixed points, also called inductive and coinductive definitions. No assumptions of computability or constructivity are made at this point.

The *constructive semantics* consist of a realisability interpretation of formulas and (co)inductive predicates. The space of potential realisers is an effective Scott-domain [22], D, defined by the recursive domain equation

$$\begin{aligned} D = (&\mathbf{Nil} + \mathbf{Left}(D) + \mathbf{Right}(D) + \mathbf{Pair}(D \times D) \\ &+ \mathbf{Fun}(D \to D))_\bot. \end{aligned} \tag{1}$$

This means that D is the disjoint union of the components labelled by the constructors $\mathbf{Nil}, \mathbf{Left}, \ldots$. D contains its own function space, $(D \to D)$, which would not be possible, set-theoretically, but is a standard construction in domain theory, since $(D \to D)$ consists of continuous functions only. Continuity refers to the Scott topology which is generated by a partial order $\sqsubseteq$ on D, which has $\bot$ as its least element (representing "undefined"). Thanks to the function space component, D carries the structure of a PCA (partial combinatory algebra) and can therefore interpret simple recursive types defined by the grammar

$$\rho, \sigma ::= \alpha \mid \mathbf{1} \mid \rho \times \sigma \mid \rho + \sigma \mid \rho \Rightarrow \sigma \mid \mathbf{fix}\, \alpha \,.\, \rho, \tag{2}$$

where α ranges over type variables and in $\mathbf{fix}\,\alpha\,.\,\rho$, ρ must be strictly positive in α. Hence, every type ρ denotes a subdomain $D(\rho)$. To define the semantics of a formula A, we first define a type $\tau(A)$ that represents the "propositional skeleton" of A. Then the semantics of A is defined as a set $\mathbf{R}(A) \subseteq D(\tau(A))$ of realisers of A, essentially following Kleene [23] and Kreisel [24] but, if possible, simplifying realisers in order to avoid junk in extracted programmes. We will usually write "$a\,\mathbf{r}\,A$" ("a realises A") for "$a \in \mathbf{R}(A)$" and omit the type information about a. For the simplification of realisers, *Harrop formulas* play a special role. These are formulas that do not contain disjunctions or free predicate variables at strictly positive positions. For example, $\bot$, equations and formulas of the form every $P(\vec{t})$ where P is a predicate constant are Harrop, and so is every negated formula since we view negation as implication to falsity. The realisability interpretation of an implication $A \to B$ where A and B are both nonHarrop is, as expected,

$$c\,\mathbf{r}\,(A \to B) \stackrel{\text{Def}}{=} \forall a\,(a\,\mathbf{r}\,A \to (c\,a)\,\mathbf{r}\,B),$$

while if A is Harrop, it is

$$b\,\mathbf{r}\,(A \to B) \stackrel{\text{Def}}{=} \mathbf{H}(A) \to b\,\mathbf{r}\,B.$$

Here $\mathbf{H}(A)$ means $\exists a\,(a\,\mathbf{r}\,A)$ or, equivalently, $\mathbf{Nil}\,\mathbf{r}\,A$ since, for Harrop formulas, $\mathbf{Nil}$ (an element of D signalling trivial computational content) is the only possible realiser. An even more restricted class of formulas are *noncomputational* (*nc*) formulas which do not contain free predicate variables or disjunctions at all. These formulas coincide with their realisability interpretation, that is, $\mathbf{H}(A)$ is identical to A for nc formulas A. Since first-order variables may range over abstract objects without computable structure, quantifiers have to be interpreted uniformly, that is,

$$a\,\mathbf{r}\,\forall x\, A(x) \stackrel{\text{Def}}{=} \forall x\,(a\,\mathbf{r}\,A(x)),$$

$$a\,\mathbf{r}\,\exists x\, A(x) \stackrel{\text{Def}}{=} \exists x\,(a\,\mathbf{r}\,A(x)).$$

Hence, the realiser of a universal formula does not depend on the quantified variable, and the realiser of an existential formula does not carry a witness.

The realisability interpretation of a least or greatest fixed point is the least or greatest fixed point of the realisability interpretation of the corresponding operator. For example, in the presence of a constant 0 and a successor function (+1), we can define the natural numbers as the least predicate containing 0 and being closed under (+1),

$$\mathbf{N}(x) \stackrel{\mu}{=} x = 0 \vee \exists y\,(x = y + 1 \wedge \mathbf{N}(y)). \tag{3}$$

The realisability interpretation of $\mathbf{N}$ is

$$\begin{aligned} n\,\mathbf{r}\,\mathbf{N}(x) \stackrel{\mu}{=} &(n = \mathbf{Left}(\mathbf{Nil}) \wedge x = 0) \vee \\ &(n = \mathbf{Right}(m) \wedge \exists y\,(m\,\mathbf{r}\,\mathbf{N}(y) \wedge x = y + 1)). \end{aligned} \tag{4}$$

Therefore, $n\,\mathbf{r}\,\mathbf{N}(x)$ means that x is a natural number and n is the unary representation of x where zero and successor are modelled by $\mathbf{Left}(\mathbf{Nil})$ and $\mathbf{Right}(\_)$. Examples of coinductive definitions will be given in Section 3.

The guiding principle for the *proof rules* of IFP is that they must be sound for the classical and the constructive semantics. This is achieved through the usual rules for intuitionistic logic, and rules expressing the least resp. greatest fixed point property of $\mu\Phi$ resp. $\nu\Phi$. For the natural numbers, these are equivalent to the closure under 0 and successor and the usual induction rule that infers, for an arbitrary predicate P, from $P(0)$ and $\forall x\,(P(x) \to P(x+1))$ that $\mathbf{N} \subseteq P$, that is, $\forall x\,(\mathbf{N}(x) \to P(x))$.

Regarding *axioms* one has a fair amount of flexibility. Any closed formula A that is true in the intended classical interpretation and for which one has a programme M that realises it can be taken as an axiom. In particular, every classically true nc formula can be taken as an axiom. More generally, any closed Harrop formula that is equivalent to its realisability interpretation may be taken as an axiom if we accept it as true in the intended interpretation. A large class of formulas enjoying this property is the class of *admissible Harrop formulas*, which is the least class that contains all nc formulas, is closed under conjunction and quantification and under implication, $A \to B$ where B is admissibly Harrop and the formula A is equivalent to its realisability interpretation (for example, that is the case if A is of the form $\mathbf{N}(t)$). An important and useful axiom is a generalisation of *Brouwer's thesis* (BT) [25], which can be stated as the following nc formula:

$$\mathbf{BT} \qquad \forall x\,(\neg\mathbf{Path}_{\prec}(x) \to \mathbf{Acc}_{\prec}(x)). \tag{5}$$

This refers to some nc relation $\prec$ and relates two ways of expressing that an element x is wellfounded with respect to $\prec$: The first way, $\neg\mathbf{Path}_{\prec}(x)$, says that there is no descending path starting with x where the existence of a path can be expressed coinductively as

$$\mathbf{Path}_{\prec}(x) \stackrel{\nu}{=} \exists y \prec x\,\mathbf{Path}_{\prec}(y). \tag{6}$$

The second way, $\mathbf{Acc}_{\prec}(x)$, expresses wellfoundedness inductively as the accessible part of $\prec$:

$$\mathbf{Acc}_{\prec}(x) \stackrel{\mu}{=} \forall y \prec x\,\mathbf{Acc}_{\prec}(y). \tag{7}$$

BT is stronger and more general than Brouwer's original thesis. The precise relationship is explained in [17]. As an example, we show the following:

Lemma 2.1. *Brouwer's thesis,* (5), *implies Markov's Principle, that is, the schema*

$$\forall x \in \mathbf{N}(A(x) \vee \neg A(x)) \wedge \neg\neg\exists x \in \mathbf{N}\,A(x) \to \exists x \in \mathbf{N}\,A(x). \tag{8}$$

Proof. To prove (8) set $y \prec x \stackrel{\text{Def}}{=} y = x + 1 \wedge \neg A(x)$. Then, clearly, $\neg\neg\exists x \in \mathbf{N}\,A(x)$ implies $\neg\mathbf{Path}_{\prec}(0)$ (if $\mathbf{Path}_{\prec}(0)$, then $\forall x \in \mathbf{N}\,\mathbf{Path}_{\prec}(x)$ follows by induction on $x \in \mathbf{N}$, hence, $\forall x \in \mathbf{N}\neg A(x)$). Therefore, by (5), $\mathbf{Acc}_{\prec}(0)$. We set $P(x) \stackrel{\text{Def}}{=} x \in \mathbf{N} \to \exists y \in \mathbf{N}\,A(y)$, with the goal to prove $\mathbf{Acc}_{\prec} \subseteq P$, by wellfounded induction on $\prec$, that is, strictly positive induction on $\mathbf{Acc}_{\prec}$. Therefore, we assume, as induction hypothesis, $\forall y \prec x\,P(y)$, and have to show $P(x)$. Hence, let

$x \in \mathbf{N}$. Since A is assumed to be decidable, we can do case analysis on $A(x)$. If $A(x)$ holds, then we are done. Otherwise, $x + 1 \prec x$ and therefore $P(x + 1)$ holds by the induction hypothesis. Since $x + 1 \in \mathbf{N}$, it follows $\exists y \in \mathbf{N}\, A(y)$, thus completing the inductive proof of $\mathbf{Acc}_\prec \subseteq P$. Hence, $P(0)$ which means $\exists x \in \mathbf{N}\, A(x)$. $\square$

Since Markov's Principle, and hence **BT**, is not generally accepted in constructive mathematics, we will mark uses of **BT** in the following.

Programme extraction is based on the soundness theorem for the constructive semantics:

Theorem 2.2 (Soundness). From an IFP proof of a formula A, one can extract a closed programme $M : \tau(A)$ such that $M\,\mathbf{r}\,A$ is provable.

A detailed proof can be found in [17].

The realisers extracted from proofs are given as programmes in a simple functional programming language, that is, an enriched lambda calculus given by the grammar

$$
\begin{aligned}
M, N, L ::= {} & a, b, c \quad \text{(variables)} \\
& \mid \lambda a.\, M \mid M\, N \mid \mathbf{rec}\, M \mid \bot \\
& \mid \mathbf{Nil} \mid \mathbf{Left}(M) \mid \mathbf{Right}(M) \mid \mathbf{Pair}(M, N) \\
& \mid \mathbf{case}\, M\, \mathbf{of}\, \{\, Cl_1, \ldots, Cl_n\}.
\end{aligned}
$$

The Cl_i are clauses of the form $C(a_1, \ldots, a_k) \to N$ where C is a constructor ($\in \{\mathbf{Nil}, \mathbf{Left}, \mathbf{Right}, \mathbf{Pair}\}$). A programme $\mathbf{rec}\, M$ denotes the least fixed point of the function defined by M. Details of the denotational semantics of programmes can be found in [17]. There, also a matching operational semantics is given.

A system similar to IFP is TCF (Theory of Computable Functionals), which is the formal system underlying the interactive proof system Minlog [8,26]. A comprehensive account of TCF and examples of programme extraction in Minlog can be found in [26]. The main difference between IFP and TCF is that while IFP concentrates on the integration of abstract and constructive mathematics, the focus in TCF is on higher type computable functionals as a standard model.

In the following, if we say of a formula that it "holds" or "implies" another formula, we always mean that the corresponding formulas are provable in the formal system in question, which will be IFP in the next two sections, and its concurrent variant, CFP, thereafter.

3. Real Numbers

We consider an instance of IFP with a sort ι for real numbers, the constants 0 and 1 and function symbols for addition, subtraction, multiplication, division, and absolute value, as well as relation constants $<$ and $\leq$. Furthermore, we assume the axioms of a real closed field as a set of nc formulas, as well as the *Archimedean axiom* that says that there are no infinite numbers, that is, by repeatedly subtracting 1 any real number will eventually become negative. Setting $y \prec x \stackrel{\text{Def}}{=} 0 \leq y \wedge y = x - 1$, this can be expressed as the nc formula

$$\forall x \, \neg \mathbf{Path}_{\prec}(x). \tag{9}$$

Hence, real numbers are given abstractly and axiomatically. In particular, no programmes for computing the field operations are available. However, we can use (3) to define the natural numbers as a subset $\mathbf{N}$ of the reals and prove, for example, that they are closed under addition,

$$\forall x, y \, (\mathbf{N}(x) \wedge \mathbf{N}(y) \to \mathbf{N}(x + y)). \tag{10}$$

The extracted programme will add (unary representations of) natural numbers. To extract programmes for (exact) real number computation, we need to be able to approximate real numbers by rational numbers. Therefore, we define the predicate

$$\mathbf{C}(x) \stackrel{\text{Def}}{=} \forall n \in \mathbf{N} \, \exists q \in \mathbf{Q} \, |x - q| \leq 2^{-n}, \tag{11}$$

where $\mathbf{Q}$ is a predicate characterising the rational numbers, for example,

$$\mathbf{Q}(x) \stackrel{\text{Def}}{=} \exists y, z, u \in \mathbf{N} \; x = \frac{y - z}{u + 1}. \tag{12}$$

We call a real number x satisfying $\mathbf{C}(x)$ *constructive*. A realiser of $\mathbf{C}(x)$ is a fast Cauchy sequence, that is, a sequence f of rational numbers converging quickly to x:

$$f\,\mathbf{r}\,\mathbf{C}(x) \;\leftrightarrow\; \forall n \in \mathbf{N}\,|x - f(n)| \le 2^{-n}. \tag{13}$$

The realisers of $\mathbf{C}(x)$ are sequences of rational numbers implemented as *functions* from the natural numbers to the rationals. To obtain as realisers sequences that are implemented as infinite *streams*, one can use the following coinductive predicate:

$$\mathbf{C}'(x) \overset{\text{Def}}{=} \mathbf{C}'(x,0), \text{ where } \mathbf{C}'(x,n) \overset{\nu}{=} \exists q \in \mathbf{Q}|x-q| \le 2^{-n} \wedge \mathbf{C}'(x,n+1). \tag{14}$$

Then a realiser of $\mathbf{C}'(x)$ is an infinite stream of rationals $q_0 : q_1 : \ldots$ such that $|x - q_n| \le 2^{-n}$ for all n. It is easy to prove that the predicates $\mathbf{C}$ and $\mathbf{C}'$ are equivalent. From the proof of this equivalence, one can extract programmes that translate between the function and the stream representation of Cauchy sequences.

In [17], it is shown that $\mathbf{C}$ is also equivalent to the coinductive predicate

$$\mathbf{S}(x) \overset{\nu}{=} \exists d \in \{-1,0,1\}\,(|d/2 - x| \le 1/2 \wedge \mathbf{S}(2x-d)). \tag{15}$$

A realiser of $\mathbf{S}(x)$ is a signed digit representations of x. The programmes extracted from the proof of the equivalence of $\mathbf{C}$ and $\mathbf{S}$ provide translations between the Cauchy and the signed digit representations. The coinductive characterisation of the signed digit representation has been generalised to continuous functions [27] and compact sets [28] and corresponding translations have been extracted.

A predicate of arity (ι) is called a *ring* if it contains 0 and 1 and is closed under addition, subtraction and multiplication. A ring that is closed under division by elements that are apart from 0 is called a *field*, where the *apartness relation* between real numbers is defined as

$$x \# y \overset{\text{Def}}{=} \exists k \in \mathbf{N}\,|x - y| \ge 2^{-k}. \tag{16}$$

Note that, in our setting, rings and fields are automatically ordered and satisfy the Archimedean axiom (9). A *constructive field* is a field F whose elements are all constructive, that is, $F \subseteq \mathbf{C}$.

Lemma 3.1. **Q**, **C**, **C′**, *and* **S** *are constructive fields.*

Recall that statements about validity of formulas are to be understood as provability in IFP. Therefore, a constructive field provides, via programme extraction, implementations of the field operations with respect to the real number representation defined by the field. Since closure under the multiplicative inverse is required only for numbers that are apart from 0, the extracted inversion programme has as extra argument a natural number k that guarantees that $|x| \geq 2^{-k}$. In the following, we show that this extra argument can be omitted if one is willing to accept the generalised Brouwers Thesis, **BT**.

Using **BT**, the Archimedean axiom (9) becomes $\forall x\,\mathbf{Acc}_{\prec}(x)$ (where, as in (9), $y \prec x \stackrel{\mathrm{Def}}{=} 0 \leq y \wedge y = x - 1$), and this can in turn be used to prove the following two induction rules, called *Archimedean Induction* [17]. Both rules are parametrised by a positive rational number q. In (17), P is an arbitrary predicate. In (18), R is a ring and B is arbitrary.

$$\frac{\forall x \neq 0\, P(x)}{\forall x \neq 0\,((|x| \leq q \rightarrow P(2x)) \rightarrow P(x))}, \tag{17}$$

$$\frac{\forall x \in R \setminus \{0\}\, B(x)}{\forall x \in R \setminus \{0\}\,(B(x) \vee (|x| \leq q \wedge (B(2x) \rightarrow B(x))))}. \tag{18}$$

Unlike (9), the Archimedean Induction rules do have computational content:

Lemma 3.2 (BT, [17]). (17) and (18) are provable.

If s realises the premise of (17), then the extracted programme realising the conclusion is $\mathbf{rec}\, s$.

If s realises the premise of (18), then the extracted programme realising the conclusion is

$$\chi\, a \stackrel{\mathrm{rec}}{=} \mathbf{case}\, s\, a\, \mathbf{of}\, \{\mathbf{Left}(b) \rightarrow b; \mathbf{Right}(f) \rightarrow f(\chi\,(\mathbf{mult}_2(a)))\},$$

where $\mathbf{mult}_2$ is extracted from a proof that R is closed under doubling.

Proof. (17) can be inferred from $\forall x\,\mathbf{Acc}_{\prec}(x)$ using properties of the function $x \mapsto 2^{-x}$.

(18) follows directly from (17), since with $P(x) \stackrel{\mathrm{Def}}{=} R(x) \rightarrow B(x)$, the premise of (18) implies the premise of (17). □

Lemma 3.3 (BT). Nonzero elements of a constructive field are apart from zero.

Proof. Let F be a constructive field. We have to show $\forall x \in F \setminus \{0\}\, x \# 0$. We use Archimedean Induction (18) with $q = 3$ and the fact that $2x \# 0 \to x \# 0$ to reduce this task to showing that for every $x \in F \setminus \{0\}$, $x \# 0$ or $|x| \leq 3$. Using the constructivity of F, we have $q \in \mathbf{Q}$ such that $|x - q| \leq 1$. If $|q| > 2$, then $|x| \geq 1$, hence $x \# 0$. If $|q| \leq 2$, then $|x| \leq 3$. □

Since the implication $2x \# 0 \to x \# 0$ is realised by the successor function, the programme extracted from the above proof is (after trivial simplifications)

$$\varphi_1(f) = \textbf{if}\, |\textbf{approx}(f, 0)| > 2\, \textbf{then}\, 0\, \textbf{else}\, 1 + \varphi_1(\textbf{mult}_2(f)),$$

where **approx** realises the constructivity of F, that is, if f realises $F(x)$ and $k \in \mathbf{N}$, then $\textbf{approx}(f, k)$ is a rational number whose distance to x is at most 2^{-k}.

Remark 3.4. It may seem strange that we introduce constructive analysis in an axiomatic way; however, this approach is not uncommon in constructive mathematics. For example, [3] introduces constructive axioms for the real line, [29,30] study an axiomatic theory of apartness and nearness, and [31] gives a constructive axiomatic presentation of limit spaces. In our case, the rationale for an axiomatic approach to the real numbers is that it allows us to separate computationally irrelevant properties of the real numbers such as algebraic identities and the Archimedean axiom from computationally relevant ones such as inclusion of the integers and approximability by rationals.

4. Concurrency

In [20], the system CFP (concurrent fixed point logic) is introduced that extends IFP by a logical operator $\downdownarrows$ permitting the extraction of programmes with two concurrent threads. However, to do Gaussian elimination concurrently, we need n threads where n varies through all positive numbers less or equal the dimension of the matrix to be inverted. We will first sketch the essential features of CFP and then describe an extension that caters for this more general form of concurrency.

4.1. CFP

CFP extends IFP by two logical operators, $A|_B$ (restriction) and $\downdownarrows(A)$ (two-threaded concurrency). Ultimately, we are interested only in the latter, but the former is needed to derive concurrency in a non-trivial way.

$A|_B$ (read 'A restricted to B') is a strengthening of the reverse implication $B \to A$. The two differ only in their realisability semantics (the following explanation is valid if B is noncomputational and A has only realisers that are defined ($\neq \bot$)): A realiser of $A|_B$ is a "conditional partial realiser" of A, that is, some $a \in \tau(A)$ such that

(i) if the restriction B holds, then a is defined; and
(ii) if a is defined, then a realises A.

In contrast, for a to realise $B \to A$, it suffices that if B holds, then a realises A; condition (ii) may fail, that is, if we only know that a is defined, we cannot be sure that a realises A. For example, if A is $B \vee C$, then $\mathbf{Left}(\mathbf{Nil})$ realises $B \to A$ but not $A|_B$ unless B holds. The technical definition permits arbitrary B but requires A to satisfy a syntactic condition, called "strictness", that ensures that A has only defined realisers.

Before defining strictness we need to extend the Harrop property to CFP formulas. Harropness is defined as for IFP but stipulating in addition that formulas of the form $A|_B$ or $\downdownarrows(A)$ are always non-Harrop. Now strictness is defined as follows: A nonHarrop implication is strict if the premise is nonHarrop. Formulas of the form $\diamond x\, A$ ($\diamond \in \{\forall, \exists\}$) or $\Box(\lambda X \lambda \vec{x}\, A)$ ($\Box \in \{\mu, \nu\}$) are *strict* if A is strict. Formulas of other forms (e.g., $B|_A$, $\downdownarrows(A)$, $X(\vec{t})$) are not strict.

For formulas of the $A|_B$ or $\downdownarrows(A)$ to be wellformed, we require A to be strict. In the following, it is always assumed that the strictness requirement is fulfilled.

Realisability for restriction is defined as

$$a\,\mathbf{r}\,(A|_B) \stackrel{\text{Def}}{=} (\mathbf{r}(B) \to a \neq \bot) \wedge (a \neq \bot \to a\,\mathbf{r}\,A),$$

where $\mathbf{r}(B) \stackrel{\text{Def}}{=} \exists b\, b\,\mathbf{r}\,B$ ("B is realisable").

$\Downarrow(A)$ (read "concurrently A") is not distinguished from A in the classical semantics but realisers of this formula can be computed by two processes which run concurrently, at least one of which has to terminate, and each terminating one has to deliver a realiser of A. To model this denotationally, the domain and the programming language are extended by a binary constructor **Amb** which (denotationally) is an exact copy of the constructor **Pair**. However, operationally, **Amb** is interpreted as a version of McCarthy's ambiguity operator amb [32] that corresponds to globally angelic choice [33]. Realisability of $\Downarrow(A)$ is defined as

$$c \,\mathbf{r}\, \Downarrow(A) \stackrel{\text{Def}}{=} c = \mathbf{Amb}(a, b) \wedge a, b : \tau(A) \wedge (a \neq \bot \vee b \neq \bot) \\ \wedge (a \neq \bot \to a \,\mathbf{r}\, A) \wedge (b \neq \bot \to b \,\mathbf{r}\, A).$$

We also set $\tau(\Downarrow(A)) \stackrel{\text{Def}}{=} \mathbf{A}(\tau(A))$, where $\mathbf{A}$ is a new type constructor with $D(\mathbf{A}(\rho)) \stackrel{\text{Def}}{=} \{\mathbf{Amb}(a, b) \mid a, b \in D(\rho)\} \cup \{\bot\}$.

The proof rules for restriction and concurrency are as follows:

$$\frac{B \to A_0 \vee A_1 \quad \neg B \to A_0 \wedge A_1}{(A_0 \vee A_1)|_B} \text{ Rest-intro } (A_0, A_1, B \text{ Harrop})$$

$$\frac{A|_B \quad A \to (A'|_B)}{A'|_B} \text{ Rest-bind} \qquad \frac{A}{A|_B} \text{ Rest-return}$$

$$\frac{A|_B \quad B' \to B}{A|_{B'}} \text{ Rest-antimon} \qquad \frac{A|_B \quad B}{A} \text{ Rest-mp}$$

$$\frac{}{A|_{\mathbf{False}}} \text{ Rest-efq} \qquad \frac{A|_B}{A|_{\neg\neg B}} \text{ Rest-stab}$$

$$\frac{A|_C \quad A|_{\neg C}}{\Downarrow(A)} \text{ Conc-lem} \qquad \frac{A}{\Downarrow(A)} \text{ Conc-return}$$

$$\frac{A \to B \quad \Downarrow(A)}{\Downarrow(B)} \text{ Conc-mp.}$$

Using the rules (Rest-stab) and (Rest-antimon), one can show that the rule (Conc-lem) is equivalent to

$$\frac{\neg\neg(B \vee C) \quad A|_B \quad A|_C}{\Downarrow(A)} \text{ Conc-class-or.} \tag{19}$$

We also have a variant of the Archimedean Induction which we call *Archimedean Induction with restriction*. It applies to rings R, strict predicates B and rational numbers $q > 0$:

$$\frac{\forall x \in R\,(B(x) \vee (|x| \leq q \wedge (B(2x) \to B(x))))}{\forall x \in R\,(B(x)|_{x\neq 0})}. \tag{20}$$

To extract programmes from CFP proofs, the programming language for IFP needs, in addition to the new constructor **Amb**, a strict version of application $M{\downarrow}N$ which is undefined if the argument N is. As indicated earlier, denotationally, **Amb** is just a constructor (like **Pair**). However, the operational semantics interprets **Amb** as a globally angelic choice, which matches the realisability interpretation of the concurrency operator $\downdownarrows$, as shown in [20].

Theorem 4.1 (Soundness of CFP). From a CFP proof of a formula A, one can extract a closed programme $M : \tau(A)$ such that $M\,\mathbf{r}\,A$ is provable.

Proof. The realisability of the new version (20) of Archimedean Induction is shown in the following Lemma. All other rules have simple realisers that can be explicitly defined [20]. For example, (Conc-lem) and (Conc-class-or), rules which permit to derive concurrency with a form of the law of excluded middle, are realised by the constructor **Amb**. □

Lemma 4.2. *Archimedean Induction with restriction is realisable. If s realises the premise of* (20)*, then χ, defined recursively by*

$$\chi\, a = \mathbf{case}\, s\, a\, \mathbf{of}\, \{\mathbf{Left}(b) \to b; \mathbf{Right}(f) \to f{\downarrow}(\chi\,(\mathbf{mult}_2\, a))\},$$

realises the conclusion.

Proof. Assuming $a\,\mathbf{r}\,R(x)$, we have to show

(1) $x \neq 0 \to \chi\, a \neq \bot$.
(2) $\chi\, a \neq \bot \to (\chi\, a)\,\mathbf{r}\,B(x)$.

For (1), it suffices to show (1') $\forall x \neq 0\, \forall a\, (a\,\mathbf{r}\,R(x) \to (\chi\, a)\,\mathbf{r}\,B(x))$ since $(\chi\, a)\,\mathbf{r}\,B(x)$ implies $\chi\, a \neq \bot$, by the strictness of $B(x)$.

We show (1') by Archimedean Induction (17). Let $x \neq 0$ and assume, as i.h.,

$$|x| \leq q \to \forall a'\, (a'\,\mathbf{r}\,R(2x) \to (\chi\, a')\,\mathbf{r}\,B(2x)).$$

We have to show $\forall a\, (a\,\mathbf{r}\,R(x) \to (\chi\, a)\,\mathbf{r}\,B(x))$. Assume $a\,\mathbf{r}\,R(x)$. Then $(s\, a)\,\mathbf{r}\,(B(x) \vee (|x| \leq q \wedge (B(2x) \to B(x))))$. If $s\, a = \mathbf{Left}(b)$ where

$b\,\mathbf{r}\,B(x)$, then $\chi\,a = b$ and we are done. If $s\,a = \mathbf{Right}(f)$ where $|x| \leq q$ and $f\,\mathbf{r}\,(B(2x) \to B(x))$, then $a'\,\mathbf{r}\,R(2x)$ for $a' = \mathbf{mult}_2\,a$. Thus, by i.h., $(\chi\,a')\,\mathbf{r}\,B(2x)$ and therefore $(f(\chi\,a'))\,\mathbf{r}\,B(x)$. Since $\chi\,a' \neq \bot$ by the strictness of $B(2x)$, we have $\chi\,a = f(\chi\,a')$ and we are done.

To prove (2), we consider the approximations of χ,

$$\chi_0\,a = \bot$$

$$\chi_{n+1}\,a = \mathbf{case}\,s\,a\,\mathbf{of}\,\{\mathbf{Left}(b) \to b; \mathbf{Right}(f) \to f{\downarrow}(\chi_n\,(\mathbf{mult}_2\,a))\}.$$

By continuity, if $\chi\,a \neq \bot$, then $\chi_n\,a \neq \bot$ for some $n \in \mathbf{N}$. Therefore, it suffices to show

$$\forall n \in \mathbf{N}\,\forall x, a\,(a\,\mathbf{r}\,R(x) \wedge \chi_n\,a \neq \bot \to (\chi\,a)\,\mathbf{r}\,B(x)).$$

We induce on n. The induction base is trivial since $\chi_0\,a = \bot$. For the step, assume $a\,\mathbf{r}\,R(x)$ and $\chi_{n+1}\,a \neq \bot$. Then $(s\,a)\,\mathbf{r}\,(B(x) \vee (|x| \leq q \wedge (B(2x) \to B(x))))$. If $s\,a = \mathbf{Left}(b)$ where $b\,\mathbf{r}\,B(x)$, then $\chi\,a = b$ and we are done. If $s\,a = \mathbf{Right}(f)$ where $|x| \leq q$ and $f\,\mathbf{r}\,(B(2x) \to B(x))$, then $\chi_{n+1}\,a = f{\downarrow}(\chi_n\,a')$ for $a' = \mathbf{mult}_2\,a$. It follows $\chi_n\,a' \neq \bot$. By i.h., $(\chi\,a')\,\mathbf{r}\,B(2x)$ and therefore $(f\,(\chi\,a'))\,\mathbf{r}\,B(x)$. But $f\,(\chi\,a') = f{\downarrow}(\chi\,a') = \chi\,a$ since $\chi\,a' \neq \bot$. □

The proof above can also be viewed as an instance of Scott-induction with the admissible predicate $\lambda d\,.\,\forall a\,(d\,a \neq \bot \to (\chi\,a)\,\mathbf{r}\,B(x))$ applied to $d = \chi$.

Remark 4.3. Thanks to the rule (Rest-stab), one can use classical logic for the right argument of a restriction. The rules (Rest-bind) and (Rest-return) show that restriction behaves like a monad in its left argument. On the other hand, the concurrency operator does not enjoy this property since a corresponding bind-rule is missing. Instead, one has only the weaker rule (Conc-mp), which can also be seen as a monotonicity rule. This shortcoming will be addressed in Section 4.3.

Lemma 4.4 (BT). Elements x of a constructive field satisfy $x \# 0|_{x \neq 0}$.

The realiser extracted from the proof is the programme φ_1 of Lemma 3.3.

Proof. The proof is the same as for Lemma 3.3, noting that for the proof of the premise, the assumption $x \neq 0$ is not used. □

Remark 4.5. In contrast to Lemma 4.4, the formula $\forall x\,(\mathbf{C}(x) \to (x = 0|_{x=0}))$ is not realisable (where $x = 0$ is again an atomic nc formula). This can be seen by a simple (domain-theoretic) continuity argument. If ψ were a realiser then $\psi(\lambda n\,.\,0) = \mathbf{Nil}$, since $\lambda n\,.\,0$ realises $\mathbf{C}(0)$ and $0 = 0$ is realisable, i.e., holds. Since ψ is continuous, there is some $k \in \mathbf{N}$ such that $\psi(f) = \mathbf{Nil}$ where $f(n) = 0$ if $n < k$ and $f(n) = 2^{-(k+1)}$ if $n \geq k$. Clearly, f realises $\mathbf{C}(2^{-(k+1)})$, but $\mathbf{Nil}$ does not realise the false equation $2^{-(k+1)} = 0$.

The following lemma is crucial for concurrent Gaussian elimination.

Lemma 4.6 (BT). If elements x, y of a constructive field are not both 0, then, concurrently, one of them is apart from 0, that is, $\Downarrow(x \# 0 \lor y \# 0)$.

The extracted programme is

$$\varphi_2(f, g) = \mathbf{Amb}(\mathbf{Left}{\downarrow}(\varphi_1(f)), \mathbf{Right}{\downarrow}(\varphi_1(g))).$$

Proof. By Lemma 4.4 and the fact that restriction is monotone in its left argument (which follows from the rule (Rest-bind)), we have $(x \# 0 \lor y \# 0)|_{x\neq 0}$ as well as $(x \# 0 \lor y \# 0)|_{y\neq 0}$. Since not both x and y are 0, we can apply rule (Rest-class-or). □

Remark 4.7. Intuitively, the extracted programme φ_2 consists of two processes, $\alpha = \mathbf{Left}{\downarrow}(\varphi_1(f))$ and $\beta = \mathbf{Right}{\downarrow}(\varphi_1(g))$, which search concurrently for approximations to x and y, respectively, until α finds a k where $|f(k)| > 2^{-k}$ or β finds an l where $|g(l)| > 2^{-l}$, returning the numbers with a corresponding flag **Left** or **Right**. Both, or only one of the searches, might be successful and which of the successful search results is taken is decided nondeterministically.

In this particular situation, concurrency could be avoided (even without synchronisation or interleaving): Since not both x and y are 0, $x^2 + y^2$ is nonzero and, hence, apart from 0. Computing sufficiently good approximations to $x^2 + y^2$ and x^2 and comparing them, one can decide whether x or y is apart from 0. However, to compute approximations of $x^2 + y^2$, both, x and y need to be equally well approximated, thus the overall computation time is the *sum* of the computation times for x and y. However, using concurrency the computation time will be the *minimum*. Imagine x and y being very small positive reals and the realiser f of $F(x)$ providing very fast, but the realiser g of $F(y)$, very slow, approximations. Then the extracted concurrent programme will terminate fast with a result $\mathbf{Left}(k)$ (computed by the process α searching f). Hence, this example illustrates the fact that concurrent realisability not only supports a proper treatment of partiality, as in the case study on infinite Gray code [20], but also enables us to exploit the efficiency gain of concurrency at the logical level.

4.2. *Finitely threaded concurrency*

CFP can be easily generalised to concurrency with an arbitrary finite number of threads. We generalise the operator $\downdownarrows$ to $\downdownarrows_n$, where n is a positive natural number (so that $\downdownarrows$ corresponds to $\downdownarrows_2$), and allow the constructor **Amb** to take an arbitrary positive but finite number of arguments. We define that a realises $\downdownarrows_n(A)$ if a is of the form $\mathbf{Amb}(a_1, \ldots, a_m)$ with $1 \leq m \leq n$ such that at least one a_i is defined and all defined a_i realise A:

$$a\,\mathbf{r} \downdownarrows_n(A) \stackrel{\text{Def}}{=} \bigvee_{1\leq m\leq n} (a = \mathbf{Amb}(a_1, \ldots, a_m) \wedge \bigvee_{1\leq i\leq m} a_i \neq \bot \wedge \bigwedge_{1\leq i\leq m} (a_i \neq \bot \to a_i\,\mathbf{r}\,A)).$$

The proof rules for $\downdownarrows_n$ are similar to those for $\downdownarrows$:

$$\frac{\neg\neg \bigvee_{1\leq i\leq n} B(i) \qquad \bigwedge_{1\leq i\leq n} A|_{B(i)}}{\downdownarrows_n(A)}\ \text{Conc-class-or-n},$$

$$\frac{A}{\downdownarrows_n(A)}\ \text{Conc-return-n} \qquad \frac{A \to B \qquad \downdownarrows_n(A)}{\downdownarrows_n(B)}\ \text{Conc-mp-n}.$$

Lemma 4.8 (BT). If elements $x_1, \ldots, x_n$ of a constructive field are not all 0, then, n-concurrently, one of them is apart from 0, that is, $\downdownarrows_n(x_1 \# 0 \vee \ldots \vee x_n \# 0)$.

Setting $\mathbf{In}_i \stackrel{\text{Def}}{=} \lambda a\,.\,\mathbf{Right}^{i-1}(\mathbf{Left}(a))$ for $i \in \{1, \ldots, n\}$, the extracted programme is

$$\varphi_3(f_1, \ldots, f_n) = \mathbf{Amb}(\mathbf{In}_1{\downarrow}(\varphi_1(f_1)), \ldots, \mathbf{In}_n{\downarrow}(\varphi_1(f_n))).$$

Proof. Similar to Lemma 4.6 using the rules for $\downdownarrows_n$. □

Remark 4.9. We treated the index n as a parameter on the meta-level, i.e., $\downdownarrows_n$ is an operator for every n. On the other hand, in a stringent formalisation, one has to treat n as a formal parameter, i.e., $\downdownarrows_n(A)$ is a formula with a free variable n. Finitely iterated conjunctions and disjunctions have to be formalised in a similar way. This can be easily done, using formal inductive definitions, however, we refrain from carrying this out, since it would obscure the presentation and impede the understanding of what follows.

4.3. *Monadic concurrency*

The concurrency operators $\downdownarrows$ and $\downdownarrows_n$ were sufficient for the examples studied so far, since these did not involve iterated concurrent computation, i.e., we did not use the result of one concurrent computation as a parameter for another one. In other words, we did not use concurrency "in sequence". However, such sequencing will be required in our concurrent modelling of Gaussian elimination (Section 5) since the choices of previous pivot elements influence the choice of the next one.

The problem of sequencing becomes apparent if we nest nondeterminism as defined by $\downdownarrows_n$ since this increases the index n bounding the number of processes. More precisely, the rule

$$\frac{\downdownarrows_n(\downdownarrows_n(A) \vee \mathbf{False})}{\downdownarrows_{n^2}(A)}$$

is realisable, but this is optimal — we cannot lower the number n^2 in the conclusion to n. Mathematically, the problem is that $\downdownarrows_n$ is not a monad. However, we can turn $\downdownarrows_n$ into a monad by an inductive definition:

$$\downdownarrows_n^*(A) \stackrel{\mu}{=} \downdownarrows_n(A \vee \downdownarrows_n^*(A)).$$

Clearly, $a\,\mathbf{r}\downdownarrows_n^*(A)$ holds iff a is of the form $\mathbf{Amb}(a_1, \ldots, a_m)$ for some $1 \leq m \leq n$ such that at least one a_i is defined and all defined a_i realise $A \vee \downdownarrows_n^*(A)$.

Note that $\downdownarrows_n^*(A)$ is wellformed for arbitrary formulas A, not only strict ones.

Lemma 4.10. *The following rules follow from the rules for* $\downdownarrows_n$ *and are hence realisable (for arbitrary formulas):*

$$\frac{A}{\downdownarrows_n^*(A)}\ \textit{Conc-return-n*} \qquad \frac{A \to \downdownarrows_n^*(B) \quad \downdownarrows_n(A)}{\downdownarrows_n^*(B)}\ \textit{Conc-weak-bind}$$

$$\frac{A \to \downdownarrows_n^*(B) \quad \downdownarrows_n^*(A)}{\downdownarrows_n^*(B)}\ \textit{Conc-bind.}$$

Proof. (Conc-return-n*) follows immediately from the definition of $\downdownarrows_n^*$.

To show (Conc-weak-bind), assume $A \to \downdownarrows_n^*(B)$ and $\downdownarrows_n(A)$. Then also $A \to B \vee \downdownarrows_n^*(B)$ and, hence, by (Conc-mp-n), $\downdownarrows_n(B \vee \downdownarrows_n^*(B))$. It follows $\downdownarrows_n^*(B)$, by the definition of $\downdownarrows_n^*$.

To prove (Conc-bind), we assume $A \to \downdownarrows_n^*(B)$ and show $\downdownarrows_n^*(A) \to \downdownarrows_n^*(B)$ by strictly positive induction on the definition of $\downdownarrows_n^*(A)$. Hence, it suffices to show $A \vee \downdownarrows_n^*(B) \to \downdownarrows_n^*(B)$, which follows from the assumption $A \to \downdownarrows_n^*(B)$. □

We characterise realisability of $\downdownarrows_n^*(A)$ by term rewriting. Call $\mathbf{Amb}(\vec{a})$ *sound* if $\forall i\, (a_i \neq \bot \to \exists b\, (a_i = \mathbf{Left}(b) \vee a_i = \mathbf{Right}(b))$. Now define

$$a \to_n a' \stackrel{\text{Def}}{=} \exists m \leq n \exists \vec{a} = a_1, \ldots, a_m \wedge a = \mathbf{Amb}(\vec{a}) \text{ sound}$$
$$\wedge \bigvee_i (\exists b\, a_i = a' = \mathbf{Left}(b) \vee a_i = \mathbf{Right}(a'))$$

Lemma 4.11. $a\, \mathbf{r} \downdownarrows_n^*(A)$ *iff* a *is reducible w.r.t.* $\to_n$ *and every reduction sequence* $a \to_n a' \to_n \ldots$ *terminates and ends with some* $\mathbf{Left}(b)$ *such that* $b\, \mathbf{r}\, A$.

Proof. "only if": Induction on $a\, \mathbf{r} \downdownarrows_n^*(A)$. Assume $a\, \mathbf{r} \downdownarrows_n^*(A)$. Then $a = \mathbf{Amb}(\vec{a})$ and a is reducible. Let $a \to_n a' \to_n \ldots$ be a reduction sequence. Case $a_i = a' = \mathbf{Left}(b)$. Then a' is not further reducible and $b\, \mathbf{r}\, A$, hence, we are done. Case $a_i = \mathbf{Right}(a')$ and $a'\, \mathbf{r} \downdownarrows_n^*(A)$. Then the induction hypothesis applies. "if": Assume that a is reducible and every reduction sequence $a \to_n a' \to_n \ldots$ terminates and ends with some $\mathbf{Left}(b)$ where $b\, \mathbf{r}\, A$. This means that the relation $\to_n$ restricted to iterated reducts of a is wellfounded.

We show $a\,\mathbf{r} \Downarrow_n^*(A)$ by induction on $\rightarrow_n$. Since a is reducible, $a \rightarrow_n a'$ for some a'. Hence, $a = \mathbf{Amb}(\vec{a})$ and either $a_i = a' = \mathbf{Left}(b)$ for some b, or else $a_i = \mathbf{Right}(a')$. Hence, the first part of the definition of $a\,\mathbf{r} \Downarrow_n^*(A)$ holds. For the second part, assume $a_i \neq \bot$. Since $\mathbf{Amb}(\vec{a})$ is sound, either $a_i = \mathbf{Left}(b)$ or $a_i = \mathbf{Right}(b)$. Assume $a_i = \mathbf{Left}(b)$. We have to show $b\,\mathbf{r}\,A$. Since $a \rightarrow_n \mathbf{Left}(b)$, it follows that $\mathbf{Left}(b)$ terminates a reductions sequence starting with a. Hence, $b\,\mathbf{r}\,A$, by assumption. Finally, assume $a_0 = \mathbf{Right}(a')$. We have to show $a'\,\mathbf{r} \Downarrow_n^*(A)$. Since $a \rightarrow_n a'$, this holds by induction hypothesis. □

The characterisation of $a\,\mathbf{r} \Downarrow_n^*(A)$ given in Lemma 4.11 shows that a can be viewed as a nondeterministic process requiring at most n processes running concurrently (or in parallel). Each step $a \rightarrow_n a'$ represents such a nondeterministic computation. The fact that *every* reduction sequence will lead to a realiser of A means that after each step all other computations running in parallel may safely be abandoned since no backtracking will be necessary.

5. Gaussian Elimination

By a *matrix* we mean a quadratic matrix with real coefficients. A matrix is *nonsingular* if its columns are linearly independent. Let F be a constructive field of reals (see Section 3). A matrix A is called F *matrix* if all the coefficients of A are in F. We prove that every nonsingular F matrix has an F left inverse, and extract from the proof a concurrent version of the well-known matrix inversion algorithm based on Gaussian elimination. The proof is concurrent, because the nonzero element of a column (the pivot) is computed concurrently.

We fix a dimension $n \in \mathbf{N}$ and let k, l range over $\mathbf{N}_n \stackrel{\text{Def}}{=} \{1, \ldots, n\}$, the set of valid matrix indices. For a matrix A, we let $A(k,l)$ be its entry at the k-th row and l-th column. We denote matrix multiplication by $A * B$, i.e.,

$$(A * B)(k,l) = \sum_{i=1}^{n} A(k,i) \cdot B(i,l).$$

By E, we denote the unit matrix, i.e.,

$$E(k,l) \stackrel{\text{Def}}{=} \begin{cases} 1 & \text{if } k = l \\ 0 & \text{otherwise.} \end{cases}$$

We call an F matrix *invertible* if there exists an F matrix B such that $B * A = E$.

To formalise the above, one may add sorts for vectors and matrices and a function for accessing their entries. Furthermore, a function symbol for matrix multiplication which needs three arguments, two for the matrices to be multiplied and one for the dimension. The formula for matrix multiplication can then be stated as an axiom since it is admissibly Harrop (see Section 2). Furthermore, one postulates admissible Harrop axioms for the existence of explicitly defined matrices, for example,

$$\begin{aligned} &\forall n \in \mathbf{N} \exists E \forall i, j \in \{1, \dots, n\}((i = j \to E(i,j) = 1) \\ &\quad \wedge (i \neq j \to E(i,j) = 0). \end{aligned}$$

The notion of linear independence can be expressed as a Harrop formula as well. There are many alternative formalisations which, essentially, lead to the same result, in particular, the extracted programme will be the same.

Theorem 5.1 (Gaussian elimination). Every nonsingular F matrix is invertible.

Proof. For matrices A, B and $i \in \{0, \dots, n\}$, we set

$$A \stackrel{i}{=} B \stackrel{\text{Def}}{=} \forall k, l\, (l > i \to A(k,l) = B(k,l)).$$

(A and B coincide on the columns $i+1, \dots, n$). Hence, $A \stackrel{0}{=} B$, means $A = B$ whereas $A \stackrel{n}{=} B$ always holds. To prove the theorem, it suffices to show:

Claim 5.2. *For all $i \in \{0, \dots, n\}$, if A is a nonsingular F matrix such that $A \stackrel{i}{=} E$, then A is invertible.*

We prove the Claim by induction on i.

In the proof, we need to bear in mind that, since we want to use concurrent pivoting (Lemma 4.8), and we iterate the argument, we can only hope to prove invertibility of A ω-concurrently, that is, we prove in fact

$$\forall i \in \{0,\dots,n\} \forall A \in F^{n\times n}_{\mathrm{ns}} (A \stackrel{i}{=} E \to \Downarrow^*_n(\exists B \in F^{n\times n}_{\mathrm{ns}}\, B * A = E)),$$

where $A \in F^{n\times n}_{(\mathrm{ns})}$ *means that* A *is an* n*-dimensional (nonsingular)* F *matrix.*

The base case, $i = 0$, is easy, since the hypothesis $A \stackrel{0}{=} E$ means $A = E$. Therefore, we can take $B \stackrel{\mathrm{Def}}{=} E$.

For the step, assume $i > 0$ and let $A \in F^{n\times n}_{\mathrm{ns}}$ such that $A \stackrel{i}{=} E$. It suffices to show that there is $C \in F^{n\times n}_{\mathrm{ns}}$ such that $C * A \stackrel{i-1}{=} E$. Since then $C * A \in F^{n\times n}_{\mathrm{ns}}$, and therefore, by induction hypothesis, we find $B \in F^{n\times n}_{\mathrm{ns}}$, with $B * (C * A) = E$. Hence, $(B * C) * A = E$, by the associativity of matrix multiplication.

Thanks to the rule (Conc-weak-bind), it suffices to find the matrix C *concurrently, that is, it suffices to prove* $\Downarrow_n(\exists C \in F^{n\times n}_{ns}\, C * A \stackrel{i-1}{=} E)$*, since the induction hypothesis implies*

$$(\exists C \in F^{n\times n}_{ns}\, C * A \stackrel{i-1}{=} E) \to \Downarrow^*_n(\exists B \in F^{n\times n}_{ns}\, B * A = E).$$

Since A is nonsingular and $A \stackrel{i}{=} E$, it is not the case that $A(k,i) = 0$ for all $k \le i$. Because otherwise the n-tuple $(\alpha_1,\dots,\alpha_n)$ with $\alpha_l = 0$ for $l < i$, $\alpha_i = 1$ and $\alpha_l = -A(l,i)$ for $l > i$ would linearly combine the columns of A to the zero-vector, that is, $\sum_{l=1}^n \alpha_l \cdot A(k,l) = 0$ for all k.

Therefore, by Lemma 4.8, we find, concurrently, $k_0 \le i$ such that $A(k_0,i) \# 0$. Hence, there exists $\alpha \in F$ such that $A(k_0,i)\cdot\alpha = 1$. Define the matrix C_1 by

$$C_1(k,l) \stackrel{\mathrm{Def}}{=} \begin{cases} \alpha & \text{if } k = l = k_0 \\ 1 & \text{if } k = l \neq k_0 \\ 0 & \text{otherwise,} \end{cases}$$

and set $A_1 \stackrel{\mathrm{Def}}{=} C_1 * A$ (multiplying the k_0-th row by α). Clearly, $A_1 \stackrel{i}{=} E$ and $A_1(k_0,i) = 1$. Further, define

$$C_2(k,l) \stackrel{\mathrm{Def}}{=} \begin{cases} 1 & \text{if } k = l \notin \{k_0,i\} \text{ or } (k,l) \in \{(k_0,i),(i,k_0)\} \\ 0 & \text{otherwise,} \end{cases}$$

and set $A_2 \stackrel{\text{Def}}{=} C_2 * A_1$ (swapping rows k_0 and i). Clearly, $A_2 \stackrel{i}{=} E$ and $A_2(i,i) = 1$. Finally, define

$$C_3(k,l) \stackrel{\text{Def}}{=} \begin{cases} -A_2(k,i) & \text{if } k \neq i \quad \text{and} \quad l = i \\ 1 & \text{if } k = l \\ 0 & \text{otherwise,} \end{cases}$$

and set $A_3 \stackrel{\text{Def}}{=} C_3 * A_2$ (subtracting the $A_2(k,i)$ multiple of row i from row k, for each $k \neq i$). Clearly, $A_3 \stackrel{i-1}{=} E$, and $C \stackrel{\text{Def}}{=} C_3 * C_2 * C_1$ is a nonsingular F matrix since C_1, C_2, C_3 are. By associativity of matrix multiplication we have $C * A = A_3 \stackrel{i-1}{=} E$. This completes the proof of the claim and hence the theorem. □

The above proof is written in such detail that its formalisation is straightforward and a programme can be extracted. We did this (by hand since CFP has yet to be implemented) with the lazy functional programming language Haskell as target language. In [20] it is shown how **Amb** can be interpreted as concurrency using the Haskell library `Control.Concurrent`.

Since Theorem 5.1 is stated w.r.t. an arbitrary constructive field F, the extracted programme is polymorphic in (realisers of) F. We tested the extracted matrix inversion programme with respect to the constructive fields $\mathbf{Q}$ (rational numbers (12)), $\mathbf{C}$ (fast rational Cauchy sequences implemented as functions (11)), and $\mathbf{C}'$ (fast rational Cauchy sequences implemented as streams (14)), running it with matrices some of whose entries are nonzero but very close to zero and hard to compute [34].

The result was that $\mathbf{Q}$ is hopeless since exact rational numbers become huge, in terms of the size of their representation as fractions, and thus computationally unmanageable. $\mathbf{C}$ and $\mathbf{C}'$ performed similarly with $\mathbf{C}'$ being slightly faster since the stream representation can exploit memoisation effects and Haskell's laziness.

We also compared the extracted programmes with a variant where concurrency is explicitly modelled through interleaving Cauchy sequences or digit streams. Here, one notes the effect that the slowest Cauchy sequence (the one whose computation consumes most time) determines the overall run time, instead of the fastest as it is the case with true concurrency.

6. Conclusion

This chapter aimed to demonstrate the potential of constructive mathematics regarding programme extraction, not only as a possibility in principle, but as a viable method that is able to capture computational paradigms that are crucial in current programming practice. We used CFP, an extension of a logic for programme extraction by primitives for partiality and concurrency, to extract a concurrent programme for matrix inversion based on Gaussian elimination. Although the proof of the correctness of the logic is quite involved, the extension is very simple and consists of just two new logical operators, restriction, a strengthening of implication that is able to control partiality, and an operator that permits computation with a fixed number of concurrent threads. The case study on Gaussian elimination showed that proofs in this system are very close to usual mathematical practice.

The system CFP uses the axiom **BT**, a generalisation of Brouwer's Thesis, which can also be viewed as an abstract form of bar induction. **BT** was used to turn the Archimedean axiom for real numbers into an induction schema which in turn allowed us to prove that nonzero elements of a constructive field are apart from zero. The usual proof of the latter result requires Markov's principle and the axiom of countable choice. Moreover, the introduction rule for the concurrency operator is a form of the law of excluded middle. For these reasons, CFP has to be regarded as a semi-constructive system, which begs the question to which extent our results can be regarded constructively valid. Classical logic could be partially avoided by negative translation, however, its use has computational significance since it allows us to undertake searches without having a constructive bound for the search. Replacing these unbounded searches by algorithms that use constructively obtained (worst-case) bounds may result in a dramatic loss of efficiency since the computation of these bounds may be expensive and useless since the search usually stops far earlier than predicted by the bound (see [35] for examples). It is an interesting problem to find a formal system that is fully constructive but, at the same time, allows for efficient programme extraction as in our semi-constructive system.

Acknowledgements

This work has received funding from the European Union's Horizon 2020 research and innovation programme under the Marie Sklodowska-Curie grant agreement No 731143 (CID). We also acknowledge support by the JSPS Core-to-Core Program (A. Advanced Research Networks) and JSPS KAKENHI grant number 15K00015.

References

[1] E. Bishop and D. Bridges, *Constructive Analysis.* Grundlehren der mathematischen Wissenschaften 279, Springer, 1985.

[2] D. Bridges, Constructive mathematics: a foundation for computable analysis, *Theor. Comput. Sci.* **219**, 95–109 (1999).

[3] D. Bridges and S. Reeves, Constructive mathematics in theory and programming practice, *Philosophia Mathe.* **7**(3), 65–104 (1999).

[4] R.L. Constable, *Implementing Mathematics with the Nuprl Proof Development System.* Prentice–Hall, New Jersey, 1986.

[5] Coq. The Coq Proof Assistant.

[6] Agda. Official website.

[7] S. Berghofer, Program Extraction in simply-typed Higher Order Logic. In *Types for Proofs and Programs (TYPES '02), vol. 2646, Lecture Notes in Computer Science*, Springer, 2003, pp. 21–38.

[8] U. Berger, K. Miyamoto, H. Schwichtenberg, and M. Seisenberger, Minlog — A tool for programme extraction for supporting algebra and coalgebra. In *CALCO-Tools, vol. 6859, LNCS.* Springer, 2011, pp. 393–399. doi: Doi:10.1007/978-3-642-22944-2_29.

[9] U. Berger, H. Schwichtenberg, and W. Buchholz, Refined programme extraction from classical proofs, *Ann. Pure Appl. Logic.* **114**, 3–25 (2002).

[10] R. Constable and C. Murthy, Finding computational content in classical proofs. In G. Huet and G. Plotkin, (eds.), *Logical Frameworks*, Cambridge University Press, 1991, pp. 341–362.

[11] M. Parigot, $\lambda\mu$–calculus: an algorithmic interpretation of classical natural deduction. In *Proceedings of Logic Programming and Automatic Reasoning, St. Petersburg, vol. 624, LNCS.* Springer, 1992, pp. 190–201.

[12] S. Berardi, M. Bezem, and T. Coquand, On the computational content of the axiom of choice, *J. Symbol. Log.* **63**(2), 600–622 (1998).

[13] U. Berger, A computational interpretation of open induction. In *Proc. of the Nineteenth Annual IEEE Symposium on Logic in Computer Science*. IEEE Computer Society, 2004, pp. 326–334. ISBN 0-7695-2192-4.

[14] U. Berger and P. Oliva, Modified bar recursion and classical dependent choice. In *Logic Colloquium 2001*. Springer, 2001.

[15] J.-L. Krivine, Dependent choice, 'quote' and the clock, *Theor. Comput. Sci.* **308**, 259–276 (2003).

[16] M. Seisenberger, Programs from proofs using classical dependent choice, *Ann. Pure Appl. Log.* **153**(1–3), 97–110 (2008).

[17] U. Berger and H. Tsuiki, Intuitionistic fixed point logic, *Ann. Pure Appl. Log.* **172**(3), 1–56 (2021).

[18] T. Powell, P. Schuster, and F. Wiesnet. A universal algorithm for Krull's theorem. *Information and Computation* (2021), to appear.

[19] P. Schuster, Induction in algebra: A first case study, *Logical Meth. Comput. Sci.* **9**(3:20), 1–19 (2013).

[20] U. Berger and H. Tsuiki. Extracting total Amb programmes from proofs. In *Proceeding of the European Symposium on Programming (ESOP 2022)* (2022). To appear: LNCS. Full version at https://arxiv.org/abs/2104.14669.

[21] H. Tsuiki, Real number computation through gray code embedding, *Theoret. Comput. Sci.* **284**(2), 467–485 (2002).

[22] G. Gierz, K.H. Hofmann, K. Keimel, J.D. Lawson, M. Mislove, and D.S. Scott, *Continuous Lattices and Domains, vol. 93. Encyclopedia of Mathematics and its Applications.* Cambridge University Press, 2003.

[23] S.C. Kleene, On the interpretation of intuitionistic number theory, *J. Symbol. Log.* **10**, 109–124 (1945).

[24] G. Kreisel, Interpretation of analysis by means of constructive functionals of finite types, *Constructivity Math.* 101–128 (1959).

[25] V. Veldman, Brouwer's real thesis on bars, *Philos. Scientiæ, Constructivism: Math. Log. Philos. Linguist.* **6**, 21–42 (2001).

[26] H. Schwichtenberg and S.S. Wainer, *Proofs and Computations.* Cambridge University Press, 2012.

[27] U. Berger, From coinductive proofs to exact real arithmetic: Theory and applications, *Logi. Meth. Comput. Sci.* **7**(1), 1–24 (2011).

[28] U. Berger and D. Spreen, A coinductive approach to computing with compact sets, *J. Logic Analysis.* **8** (2016).

[29] D. Bridges, H. Ishihara, P. Schuster, and L. Vita, Apartness, compactness and nearness, *Theoret. Comput. Sci.* **405**, 3–10 (2008).

[30] D. Bridges and L. Vita, Apartness spaces as a framework for constructive topology, *Ann. Pure Appl. Log.* **119**, 61–83 (2003).

[31] I. Petrakis, Limit spaces with approximations, *Ann. Pure Appl. Log.* **167**, 737–752 (2016).

[32] J. McCarthy, A basis for a mathematical theory of computation. In P. Braffort and D. Hirschberg (eds.), *Computer Programming and Formal Systems, vol. 35, Studies in Logic and the Foundations of Mathematics*. Elsevier, 1963, pp. 33–70. doi: https://doi.org/10.1016/S0049-237X(08)72018-4.
[33] W. Clinger and C. Halpern, Alternative semantics for McCarthy's amb. In W.G. Brookes S.D., A.W. Roscoe (eds.), *Seminar on Concurrency. CONCURRENCY 1984, vol. 197, Lecture Notes in Computer Science*, Springer, 1985. doi: 10.1007/3-540-15670-4_22.
[34] U. Berger, CFP (Concurrent fixed point logic) repository.
[35] D. Pattinson, Constructive domains with classical witnesses, *Logical Meth. Comput. Sci.* **17**(1), 19:1–19:30 (2021).

https://doi.org/10.1142/9789811245220_0010

Chapter 10

A Herbrandised Interpretation of Semi-Intuitionistic Second-Order Arithmetic with Function Variables

João Enes* and Fernando Ferreira†

Departamento de Matemática,
Faculdade de Ciências da Universidade de Lisboa,
Campo Grande, Portugal
*joaofenes@gmail.com
†fjferreira@fc.ul.pt

We present a functional interpretation of a semi-intutionistic second-order system with function variables. The semi-intuitionism includes principles like Markov's principle, the lesser limited principle of omniscience, weak König's lemma or the FAN principle. The interpretation is herbrandised in the sense that it collects witnesses into finite sets. As a consequence, a Herbrand-like witnessing result holds for our system.

1. Introduction

In a recent paper [1], the second author introduced a herbrandised interpretation of a semi-intuitionistic system of second-order arithmetic. It is semi-intuitionistic in two senses. Firstly, it contains principles which are not intuitionistically valid, as it is the case with Markov's principle, the lesser limited principle of omniscience or weak König's lemma. Secondly, the system not only does not validate classical logic but it is even incompatible with the limited principle of omniscience. This is due to the presence of a classically false FAN principle in the system. The interpretation is dubbed herbrandised because it collects and carries witnesses along a derivation in nonempty finite sets. The following is true. If $\exists x A(x)$ is a sentence provable in the system, where A is a first-order formula, then from one such proof one can find effectively a finite set of natural numbers $n_1, n_2, \ldots, n_k$ such that the disjunction

$$A(n_1) \vee A(n_2) \vee \cdots \vee A(n_k)$$

is true in the structure of the natural numbers. Note, in particular, that our classically false system only proves true first-order sentences. One should compare the above witnessing result with well-known results in both the classical case and the intuitionistic case. In the classical case, A must be restricted to bounded formulas (this is essentially Herbrand's theorem). In the intuitionistic case, the finite disjunction can be replaced by a single instance. The semi-intuitionistic setting is a very natural mid-point.

The second-order part of [1] used set-variables (for subsets of the natural numbers). It has been pointed out that, as opposed to classical systems (as they arise, for instance, in reverse mathematics [2]), the intuitionistic setting can become unduly restricted when using set variables instead of function variables. This is of no consequence in the classical case of reverse mathematics because we can treat functions as sets of ordered pairs. This move is not in general available in an intuitionistic setting. As it happens, we believe that the classical treatment can still be carried out within the semi-intuitionistic setting of [1]. However, it is indeed more natural to have primitive function variables in the language. For one, choice principles can be stated easily. The semi-intuitionistic system of this chapter contains both the following forms of the axiom of choice

$$\forall x \exists y A(x, y) \to \exists f \forall x \exists y \leq f(x)\, A(x, y),$$

where A is any formula of the language, and

$$\forall x \exists y A_{\mathrm{b}}(x, y) \to \exists f \forall x A_{\mathrm{b}}(x, f(x)),$$

where A_{b} is restricted to bounded formulas. This is the Δ_0-AC principle.

Second-order systems with function variables were briefly discussed in the last comment of [1], only to be dismissed because it was thought that the interpretation of such systems would be too complicated. Instead, it was suggested to use the bounded functional interpretation (cf. [3]) for a system with function symbols. Nonetheless, as was then observed, the latter interpretation suffers from the fact that the transformation of first-order formulas is not set theoretically correct. As a consequence, a Herbrand-like witnessing result, like the one mentioned above, does not seem to follow. In this chapter, we present a functional interpretation of a semi-intuitionistic second-order arithmetical system with function variables. It has some delicate points, but it is not unduly complicated. The interpretation is, in the end, made possible because it relies crucially on the observation that the second-order setting is automatically extensional. In a nutshell, the herbrandised interpretation works because we are restricting ourselves to a second-order system (as opposed to higher order systems).

The chapter is organised as follows. In the next section, we present the semi-intuitionistic second-order setting with function variables that we are going to study. Section 3 introduces an auxiliary theory to which we apply a herbrandised interpretation. This interpretation is described and discussed in Section 4, but the proof of the associated soundness theorem is deferred to an appendix. The fact that extensionality is automatic in second-order theories is used in Section 5 in order to finalise the interpretation of our original formal theory.

2. The Semi-intuitionistic System

Our language of second-order arithmetic $\mathcal{L}_2$ is based on the first-order language of Heyting Arithmetic (HA), as described in [4]. We include function symbols for all the primitive recursive functions, and we also have primitive bounded arithmetical quantifications $\forall x \leq t\,(\dots)$ and $\exists x \leq t\,(\dots)$, where x is a variable not occurring in the term t. The extension to the second-order part is obtained by allowing (unary) function variables f, g, h, etc., together with associated second-order quantifications $\forall f$ and $\exists f$. There are new first-order terms of the form $f(t)$, where f is a second-order variable and t is a first-order term. Formulas are built as usual, including the new second-order quantifications, starting from the atomic formulas $t = q$ and $t \leq q$, where t and q are first-order terms.

The only second-order terms of $\mathcal{L}_2$ are the function variables. There is no equality primitive symbol infixing between them. If, by chance, we write $f = g$, this is to be regarded as an abbreviation of $\forall x(f(x) = g(x))$. Similarly, we often write $f \leq g$ as an abbreviation of $\forall x(f(x) \leq g(x))$. The reader may worry that restricting second-order terms to function variables is too restrictive since we do not include function symbols nor have lambda terms in our syntax. Our restriction is nevertheless in tune with the language of second-order arithmetic of reverse mathematics, where second-order set variables are the only second-order terms. The present treatment has the advantage of keeping the syntax simple and, mathematically, it does not constitute any restriction. The reason why this is so is that we accept Δ_0-AC in our theories. As a consequence, whenever we have a first-order term $t(x)$, with a distinguished free variable x, we can prove $\exists f \forall x\,(f(x) = t(x))$. This fact, together with the claim of extensionality as in Proposition 5.3, shows that there is no restriction.

The base theory HA_2 is intuitionistic and it is the natural outgrowth of first-order Heyting Arithmetic to the second-order setting. Induction is kept unrestricted, i.e., we allow induction also for the new formulas of the language. The notion of bounded formula, or Δ_0-formula, is as usual: one just remarks that now there are more atomic formulas to start with because there are new first-order terms (in other words, Δ_0-formulas may have second-order parameters). It is well known that HA_2 proves the law of excluded middle for Δ_0-formulas. We now list the principles that we add to HA_2:

(1) Markov's principle MP. This is the scheme $\neg\forall x A_{\mathrm{b}}(x) \to \exists x \neg A_{\mathrm{b}}(x)$ where A_{b} is a bounded formula.
(2) The bounded independence of premises principle $\mathsf{bIP}_{\Pi^0_1}$. This is the scheme

$$(\forall x A_{\mathrm{b}}(x) \to \exists y B(y)) \to \exists w(\forall x A_{\mathrm{b}}(x) \to \exists y \leq w B(y)),$$

where A_{b} is a bounded formula and B is arbitrary.
(3) The principle of bounded contra-collection bCColl. This is the scheme

$$\forall w \exists x \leq z \forall y \leq w A_{\mathrm{b}}(x, y) \to \exists x \leq z \forall y A_{\mathrm{b}}(x, y),$$

where A_{b} is a bounded formula. (If this looks unfamiliar, do consider the classical contrapositive.)

(4) The first form of choice bAC mentioned in the introduction. This is the scheme

$$\forall x \exists y A(x,y) \rightarrow \exists f \forall x \exists y \leq f(x)\, A(x,y),$$

where A is an arbitrary formula.

(5) The Δ_0-AC principle already described in the introduction.

(6) The principle FAN. This is the scheme

$$\forall f \leq h \exists y\, A(f,y) \rightarrow \exists w \forall f \leq h \exists y \leq w\, A(f,y),$$

where A is an arbitrary formula.

(7) Weak König's lemma WKL. This is the scheme

$$\forall w \exists f \leq h \forall y \leq w A_{\mathrm{b}}(f,y) \rightarrow \exists f \leq h \forall y A_{\mathrm{b}}(f,y),$$

where A_{b} is a Δ_0-formula.

In the above, the second-order "bounded" quantifications $\exists f \leq h(\dots)$ and $\forall f \leq h(\dots)$ are, of course, abbreviations of $\exists f\, (\forall x (f(x) \leq h(x)) \wedge \dots)$ and of $\forall f\, (\forall x (f(x) \leq h(x)) \rightarrow \dots)$, respectively.

The main aim of this chapter is to give an interpretation of the theory HA_2 together with the above seven principles. This extended theory has some very interesting consequences. Using the law of excluded middle for Δ_0-formulas, we have

$$\forall x \exists y\, ((y = 0 \wedge A_{\mathrm{b}}(x)) \vee (y = 1 \wedge \neg A_{\mathrm{b}}(x))),$$

for bounded formulas A_{b}. Therefore, by Δ_0-AC, we can derive comprehension for Δ_0-formulas as follows:

$$\exists f \forall x\, (f(x) = 0 \leftrightarrow A_{\mathrm{b}}(x)).$$

Now, as explained in [1], with the help of WKL, it is possible to derive the following recursive comprehension scheme:

$$\forall x\, (\exists y A_{\mathrm{b}}(x,y) \leftrightarrow \forall z B_{\mathrm{b}}(x,z)) \rightarrow \exists f \forall x\, (f(x) = 0 \leftrightarrow \exists y A_{\mathrm{b}}(x,y)),$$

where A_{b} and B_{b} are Δ_0-formulas. Note that this implies the law of excluded middle for recursive predicates (i.e., predicates that can be put simultaneously in Σ_1^0 and Π_1^0 form, as in the antecedent of the above scheme).

The bounded contra-collection principle bCColl entails the lesser limited principle of omniscience LLPO:

$$\forall x \forall y\, (A_{\mathrm{b}}(x) \vee B_{\mathrm{b}}(y)) \to \forall x A_{\mathrm{b}}(x) \vee \forall y B_{\mathrm{b}}(y),$$

where A_{b} and B_{b} are bounded formulas. This is shown in [1].

Our formulation of WKL entails (even intuitionistically) the usual standard binary tree formulation, generally used in reverse mathematics. The formulation is in the form of a so-called bounded contra-collection principle. It is, in fact, a second-order version of bCColl. The principle FAN is, on the other hand, a second-order collection principle (this terminology of collection and contra-collection was introduced in [3]). As we have already said, FAN is not a set-theoretical correct principle. It even contradicts Errett Bishop's limited principle of omniscience LPO of [5]. This principle is $\forall x A_{\mathrm{b}}(x) \vee \exists x \neg A_{\mathrm{b}}(x)$, where A_{b} is a bounded formula. We can clearly use this principle to prove

$$\forall f \leq 1 \exists y\, (\exists x (f(x) = 0) \to f(y) = 0),$$

where $f \leq 1$ abbreviates $\forall z (f(z) \leq 1)$. Now, FAN would entail

$$\exists w \forall f \leq 1\, (\exists x (f(x) = 0) \leftrightarrow \exists y \leq w (f(y) = 0)),$$

which obviously leads to a contradiction (this argument is taken from Section 12.1 of [6]). What about a first-order version of the collection? That would be

$$\forall x \leq z \exists y\, A(x, y) \to \exists w \forall x \leq z \exists y \leq w\, A(x, y),$$

where A can be any formula. We did not include this principle because it can be proved by induction on z. For a brief discussion of this issue, the interested reader is referred to [1].

3. An Auxiliary Theory

In this section, we consider a certain auxiliary theory. As we shall see in Section 5, this theory contains HA_2 together with the seven principles of the last section. The supertheory is framed in a language $\mathcal{L}_2^{\min,\trianglelefteq}$ that extends $\mathcal{L}_2$. The extended language has two new primitive symbols. A symbol min that accepts a pair of second-order terms τ and ρ and yields a second-order term $\min(\tau, \rho)$. Observe that the second-order terms of the extended language are now not only the second-order variables, but rather all terms built from such variables using the min operator. For instance, if f, g and h are second-order variables, then $\min(f, g)$ or $\min(h, \min(f, g))$ are second-order terms.

The other new primitive symbol of $\mathcal{L}_2^{\min,\trianglelefteq}$ is a binary relation symbol $\trianglelefteq$ that infixes between second-order terms, thus obtaining new atomic formulas of the form $\tau \trianglelefteq \rho$, where τ and ρ are second-order terms.

Within the language $\mathcal{L}_2^{\min,\trianglelefteq}$, it is convenient to have primitive second-order bounded quantifications of the form $\forall f \trianglelefteq \tau\, A$ and $\exists f \trianglelefteq \tau\, A$, where τ is a second-order term in which the variable f does not occur. These two quantifiers are ruled by the axiom schemes $\forall f \trianglelefteq \tau A \leftrightarrow \forall f (f \trianglelefteq \tau \to A)$ and $\exists f \trianglelefteq \tau A \leftrightarrow \exists f (f \trianglelefteq \tau \wedge A)$, respectively. The notion of bounded second-order formula naturally emerges: it is the smallest class of formulas of $\mathcal{L}_2^{\min,\trianglelefteq}$ that contains the atomic formulas and is closed under Boolean connectives and first-order and second-order bounded quantifications.

The formal theory $\mathsf{HA}_2^{\min,\trianglelefteq}$ in the new extended language consists of HA_2 (it is, of course, understood that the induction scheme extends to the new formulas of the language) together with the axiom schemes regulating the second-order bounded quantifiers and the following universal axioms:

(i) $\min(f, h)(x) = \min(f(x), h(x))$
(ii) $f \trianglelefteq h \to \forall x (f(x) \leq h(x))$
(iii) $\min(f, h) \trianglelefteq h$

The second min symbol in (i) is the binary function symbol corresponding to the primitive recursive function that gives the minimum of two natural numbers.

The principles MP, $\mathsf{bIP}_{\Pi^0_1}$, bCColl and bAC are naturally extended to the new language. Caveats are the following. On the one hand, arbitrary formulas are now arbitrary in the sense of the new extended language. On the other hand, the notion of bounded formula is extended as explained above, with the new second-order bounded quantifications and the new atomic formulas to start with. Instead of introducing new nomenclature, we tolerate some ambiguity: the notion of bounded formula changes according to the language that we are considering. However, we do reserve the name of Δ_0-formula for formulas in the original language $\mathcal{L}_2$.

We now consider the choice principle tameAC, the tame axiom of choice:

$$\forall x \exists y \leq f(x) A_{\mathrm{b}}(x, y) \to \exists g \trianglelefteq f \forall x A_{\mathrm{b}}(x, g(x)),$$

where A_{b} is a Δ_0-formula. It is clear that bAC and tameAC entail Δ_0-AC. (This version of choice was first considered in the doctoral dissertation [7] of Patrícia Engrácia.)

Finally, we state modifications of the FAN principle and of weak König's lemma. Firstly, there is the principle $\mathsf{FAN}_{\trianglelefteq}$:

$$\forall f \trianglelefteq h \exists y A(f, k) \to \exists w \forall f \trianglelefteq h \exists y \leq w A(f, y),$$

where A is an arbitrary formula. Secondly, we have the principle $\mathsf{WKL}_{\trianglelefteq}$:

$$\forall w \exists f \trianglelefteq h \forall y \leq x A_b(f, y) \to \exists f \trianglelefteq h \forall y A_b(f, y),$$

where A_b is a bounded formula.

4. The Herbrandised Functional Interpretation

The functional interpretation of this chapter works in tandem with a combinatory calculus which is an extension of the usual Schönfinkel calculus (see [8], for instance) within the arithmetic framework. The types of this calculus consist of the ground type N of the natural numbers, arrow types $\sigma \to \tau$, for types σ and τ, and star types σ^*, for types σ. The intended meaning of σ^* is to give the type of all nonempty finite subsets of elements of type σ. This so-called star combinatory calculus was introduced in [9] for logic and considered in [1] for arithmetic. The interpretation is dubbed herbrandised because it collects witnesses into finite sets (in our case, by means of terms of star type). The first herbrandised functional interpretations were introduced in [10] in order to analyze nonstandard theories of arithmetic.

Functional interpretations are based on a translation of formulas of the interpreted language into formulas of an interpreting (target) language. The translations are formulas that have an existential import and, from proofs in the interpreted theory, one can effectively extract terms of the combinatory calculus and verify that these terms fulfill the existential claim of the translation. In this chapter, we proceed as follows. Firstly, we do not describe formally the target language. This could be done, but it is not really necessary for our purposes. Secondly, the verification is done semantically, that is, the witnessing terms are seen to work by checking that they do their job in a certain set-theoretic structure.

Let us start by describing the set theoretic structure. We can associate to each type σ a set-theoretic domain of variation S_σ, where $S_{\mathrm{N}} = \mathbb{N}$, $S_{\sigma\to\tau} = {}^{S_\sigma}S_\tau$ (the set of functions from S_σ to S_τ) and

$$S_{\sigma^*} = \{X \subseteq S_\sigma : X \text{ is finite and nonempty}\}.$$

The underlying domain of the full set-theoretic interpretation of the star combinatory calculus is the family $S_*^\omega = \langle S_\sigma \rangle_{\sigma\,\text{a type}}$.

The next order of business is to describe the herbrandised translation. We associate to each formula $A(x_1, \ldots, x_n, f_1, \ldots, f_r)$ of $\mathcal{L}_2^{\min,\trianglelefteq}$ (free first- and second-order variables as displayed) a relation

$$A_{\mathrm{H}}(a_1, \ldots, a_k, b_1, \ldots, b_m, x_1, \ldots, x_n, f_1, \ldots, f_r),$$

with the free variables as shown (the *a*s and the *b*s are typed variables ranging over appropriate fibers of S_*^ω). As is usual in a functional interpretation, we also associate to $A(x_1, \ldots, x_n, f_1, \ldots, f_r)$ the following relation $A^{\mathrm{H}}(x_1, \ldots, x_n, f_1, \ldots, f_r)$:

$$\exists a_1 \ldots \exists a_k \forall b_1 \ldots \forall b_m A_{\mathrm{H}}(a_1, \ldots, a_k, b_1, \ldots, b_m, x_1, \ldots, x_n, f_1, \ldots, f_r).$$

For notational simplicity, we write $A(x, f)$, $A^{\mathrm{H}}(x, f)$ and $A_{\mathrm{H}}(a, b, x, f)$, instead of carrying on with the tuple notation. Sometimes, we also omit the parameters x and f.

The next definition gives the herbrandised translation. In what follows, the infixed relation symbol in $f \preceq h$ between f of type $\mathrm{N} \to \mathrm{N}$ and h of type $\mathrm{N} \to \mathrm{N}^*$ is interpreted by $\forall x \exists y \in h(x)\,(f(x) \leq y)$. Note that we are using the same notation for second-order variables of the second-order theory and variables of type $\mathrm{N} \to \mathrm{N}$ in the target language.

Definition 4.1 (Herbrandised interpretation). To each formula A of the language $\mathcal{L}_2^{\min,\trianglelefteq}$, possibly with first- and second-order parameters, we assign relations A^{H} and A_{H} so that A^{H} is of the form $\exists a \forall b A_{\mathrm{H}}(a, b)$, according to the following clauses:

1. For bounded formulas A of $\mathcal{L}_2^{\min,\trianglelefteq}$, the relations A^{H} and A_{H} are simply given by the flatenings of A. The *flatening* of a formula A is the relation obtained from A by interpreting the binary relation symbol $\trianglelefteq$ by the relation that holds between two functions f and g if $\forall x(f(x) \leq g(x))$. In particular, the flatening of $f \trianglelefteq g$ is simply $\forall x(f(x) \leq g(x))$.

For the remaining cases, if we have already interpretations of A and B given by $\exists a \forall b A_{\mathrm{H}}(a,b)$ and $\exists d \forall e B_{\mathrm{H}}(d,e)$, respectively, then we define:

2. $(A \wedge B)^{\mathrm{H}}$ is given by $\exists a, d \forall b, e\, [A_{\mathrm{H}}(a,b) \wedge B_{\mathrm{H}}(d,e)]$
3. $(A \vee B)^{\mathrm{H}}$ is given by $\exists a, d \forall b', e'\, [\forall b \in b' A_{\mathrm{H}}(a,b) \vee \forall e \in e' B_{\mathrm{H}}(d,e)]$
4. $(A \rightarrow B)^{\mathrm{H}}$ is given by $\exists \phi, \chi \forall a, e\, [\forall b \in \phi a e A_{\mathrm{H}}(a,b) \rightarrow B_{\mathrm{H}}(\chi a, e)]$
5. $(\forall x A(x))^{\mathrm{H}}$ is given by $\exists \phi \forall x, b\, [A_{\mathrm{H}}(\phi x, b, x)]$
6. $(\exists x A(x))^{\mathrm{H}}$ is given by $\exists x'^{\mathrm{N}^*}, a \forall b'\, [\exists x \in x' \forall b \in b' A_{\mathrm{H}}(a,b,x)]$
7. $(\forall x \leq z A(x,z))^{\mathrm{H}}$ is given by $\exists a \forall b\, [\forall x \leq z A_{\mathrm{H}}(a,b,x,z)]$
8. $(\exists x \leq z A(x,z))^{\mathrm{H}}$ is given by $\exists a \forall b'\, [\exists x \leq z \forall b \in b' A_{\mathrm{H}}(a,b,x,z)]$
9. $(\forall f A(f))^{\mathrm{H}}$ is given by $\exists \phi \forall h^{\mathrm{N} \rightarrow \mathrm{N}^*}, b\, [\forall f \preceq h A_{\mathrm{H}}(\phi h, b, f)]$
10. $(\exists f A(f))^{\mathrm{H}}$ is given by $\exists h^{\mathrm{N} \rightarrow \mathrm{N}^*} \exists a \forall b'\, [\exists f \preceq h \forall b \in b' A_{\mathrm{H}}(a,b,f)]$
11. $(\forall f \trianglelefteq h A(f,h))^{\mathrm{H}}$ is given by $\exists a \forall b\, [\forall f \leq h A_{\mathrm{H}}(a,b,f,h)]$
12. $(\exists f \trianglelefteq h A(f,h))^{\mathrm{H}}$ is given by $\exists a \forall b'\, [\exists f \leq h \forall b \in b' A_{\mathrm{H}}(a,b,f,h)]$.

The lower H-translations are displayed inside the square brackets. Negation of a formula A is, as it is customary, rendered by $A \rightarrow \perp$. The above interpretation is a cumulative interpretation in the sense discussed in the introduction of [9]. Related matters are also discussed in [1], namely, the fact that the existential variables a appearing in the translations $\exists a \forall b A_{\mathrm{H}}(a,b)$ are of end star type, and that a crucial monotonicity property holds. These matters are briefly recalled in the appendix. One important feature of the translation is that first-order formulas and their translations are equivalent in S^{ω}_{*}. The only clauses in real need of checking are 6, 7 and 8 (clauses 9, 10, 11 and 12 do not apply because we are restricting ourselves to first-order formulas). Note that one must use the monotonicity property to check clause 7. These matters are discussed in [1] and briefly recalled in the appendix.

In order to state the soundness theorem of this section, we also need to describe the terms of the star combinatory calculus. As usual, terms are built from variables and certain constants by means of application. The constants include the combinators and the arithmetical constants (0, successor and the recursors). Both the combinators and the recursors are defined for the new types as well. The novel constants are dubbed star constants. There is a star constant $\mathfrak{s}_\sigma$ of type $\sigma \to \sigma^*$ for each type σ. Its intended meaning is to map elements a^σ to the singleton set $\{a\}$. There is a star constant $\bigcup_\sigma$ of type $\sigma^* \to (\sigma^* \to \sigma^*)$ for each type σ, whose intended meaning is to map elements c and d of type σ^* to their union $c \cup d$. And, finally, there is a star constant $\bigcup_{\sigma,\tau}$ of type $\sigma^* \to (\sigma \to \tau^*) \to \tau^*$ for each pair of types σ, τ. The intended meaning of $\bigcup_{\sigma,\tau}$ is to map $c : \sigma^*$ and $f : \sigma \to \tau^*$ to the finite indexed union $\bigcup_{w \in c} fw$. Of course, the intended meanings provide the interpretation of the new constants in the set-theoretic structure S^ω_*.

The conversions of the star combinatory calculus are described in [9] and [1]. The calculus has the Church–Rosser property and enjoys the property of strong normalisation. It can be shown that a closed normal term of type N is a numeral and that, from a closed term of type N*, one can effectively read off a nonempty finite set of numbers (these finite sets are the set-theoretic interpretation of the closed terms). For more information, see [9] and [1].

Theorem 4.2 (Auxiliary soundness theorem). Let $A(x, f)$ be a formula of $\mathcal{L}_2^{\min,\trianglelefteq}$ with its first-order free variables among x and its second-order free variables among f. Let $A^{\mathrm{H}}(x, f)$ be $\exists a \forall b A_{\mathrm{H}}(a, b, x, f)$. Suppose that the theory $\mathsf{HA}_2^{\min,\trianglelefteq} + \mathsf{MP} + \mathsf{bIP}_{\Pi^0_1} + \mathsf{bCColl} + \mathsf{bAC} + \mathsf{tameAC} + \mathsf{FAN}_\trianglelefteq + \mathsf{WKL}_\trianglelefteq$ proves $A(x, f)$. Then there are closed terms t of the star combinatory calculus such that

$$S^\omega_* \models \forall h \forall x \forall b \forall f \preceq h\, A_{\mathrm{H}}(thx, b, x, f).$$

The proof of this theorem is long-winded. Apart from the formulation of the language and theory of the theorem, the main difficulty was to get the interpretation right (in this case, specially clauses 9 and 10). Our verification is semantic. The reason is that we wanted to focus on the interpretation itself and not on the details of what is exactly required for the verification. We have discussed similar matters in [1]: see note II and the proof of proposition 22 of that paper. Using the methods described there, it can be shown that the theory

$\mathsf{HA}_2^{\min,\trianglelefteq} + \mathsf{MP} + \mathsf{bIP}_{\Pi^0_1} + \mathsf{bCColl} + \mathsf{bAC} + \mathsf{tameAC} + \mathsf{FAN}_{\trianglelefteq} + \mathsf{WKL}_{\trianglelefteq}$ is conservative over HA with respect to Π^0_2-sentences.

For the sake of completeness, we provide a proof of this theorem in the appendix. It is clear from the proof that the terms t can be effectively extracted from a formal derivation of $A(x, f)$.

5. The Main Result

The aim of this section is to argue that the theory of Theorem 4.2 contains HA_2 plus the seven principles of Section 2. We have observed in Section 3 that Δ_0-AC is provable in the theory of Theorem 4.2. Therefore, all that needs to be checked is that FAN and WKL are also provable in this theory. In order to show this, we need an extensionality result.

We consider an intermediate language $\mathcal{L}_2^{\min}$ and an intermediate theory $\mathsf{HA}_2^{\min}$ stated in $\mathcal{L}_2^{\min}$. The latter language is simply $\mathcal{L}_2$ together with the second-order operator min (the languages $\mathcal{L}_2^{\min}$ and $\mathcal{L}_2^{\min,\trianglelefteq}$ have the same terms). The theory $\mathsf{HA}_2^{\min}$ is obtained from HA_2 by adding the axiom (i) of Section 3 (the induction scheme is unrestricted for formulas of $\mathcal{L}_2^{\min}$).

Definition 5.1. Given a first-order term t of the language $\mathcal{L}_2^{\min}$ with a distinguished second-order variable f, consider $r(x)$ another first-order term, this time with a distinguished first-order variable x. We define the first-order term $t[r/f]$ according to the following clauses:

(a) If t is a first-order constant or variable, then $t[r/f] = t$.
(b) If t is of the form $\mathrm{h}(q_1, \ldots, q_n)$, where h is an n-ary first-order function symbol and $q_1, \ldots, q_n$ are first-order terms, then $t[r/f] := \mathrm{h}(q_1[r/f], \ldots, q_n[r/f])$.
(c) If t is of the form $g(q)$, where g is a second-order variable different from f and q is a first-order term, then $t[r/f] := g(q[r/f])$.
(d) If t is of the form $f(q)$, where q is a first-order term, then $t[r/f] := r(q[r/f])$.
(e) If t is of the form $\min(\tau, \rho)(q)$, where τ and ρ are second-order terms and q is a first-order term, then $t[r/f] := \min(\tau(q)[r/f], \rho(q)[r/f])$. The latter min symbol is the binary function symbol for the primitive recursive function that gives the minimum of two numbers.

Lemma 5.2. *Given a first-order term t of the language $\mathcal{L}_2^{\min}$ with a distinguished second-order variable f, consider $r(x)$ another first-order term, with a distinguished first-order variable x. Then the theory $\mathsf{HA}_2^{\min}$ proves*

$$\forall x(f(x) = r(x)) \to t = t[r/f].$$

Proof. The proof is by structural induction on the term t. We only check two cases. Firstly, suppose that t is of the form $f(q)$, where q is a first-order term. Then

$$t[r/f] = r(q[r/f]) = f(q[r/f]) = f(q) = t.$$

The first equality is justified by clause (d) of the above definition. The second equality is because we are assuming that $\forall x(f(x) = r(x))$. The third equality is by induction hypothesis.

Now suppose that t is of the form $\min(\tau, \rho)(q)$, where τ and ρ are second-order terms and q is a first-order term. We have

$$\begin{aligned} t[r/f] &= \min(\tau(q)[r/f], \rho(q)[r/f]) = \min(\tau(q), \rho(q)) \\ &= \min(\tau, \rho)(q) = t. \end{aligned}$$

The first equality is justified by clause (e) of the above definition. The second equality is by induction hypothesis and the third equality is by axiom (i) of Section 3. □

Consider A an arbitrary formula of $\mathcal{L}_2^{\min}$ with a distinguished second-order variable f and let $r(x)$ be a first-order term with a distinguished first-order variable x. It is clear how to define the formula $A[r/f]$ in the usual recursive way, starting with the atomic case. Note that the atomic case can be straightforwardly defined using Definition 5.1. Of course, we must require that there are no clashes of variables (the terminology is sometimes to say that r is free for f in A).

By the above lemma, the following is clear for the atomic case. The remaining cases follow, as usual, by structural induction.

Proposition 5.3 (Extensionality). Let A be an arbitrary formula of $\mathcal{L}_2^{\min}$ with a distinguished second-order variable f. Consider $r(x)$ a first-order term with a distinguished first-order variable x. Then the theory $\mathsf{HA}_2^{\min}$ proves

$$\forall x(f(x) = r(x)) \wedge A \to A[r/f],$$

provided that r is free for f in A.

When there is no ambiguity, we often write $A(r)$ instead of $A(r/f)$.

Corollary 5.4. *For every formula $A(f)$ of the language of $\mathcal{L}_2^{\min}$,*

$$\mathsf{HA}_2^{\min} \vdash \forall x(f(x) \leq h(x)) \to (A(\min(f,h)) \leftrightarrow A(f)).$$

Proof. Assume $\forall x(f(x) \leq h(x))$. Under this assumption, we have

$$\min(f,h)(x) = \min(f(x),h(x)) = f(x),$$

for every x. Now apply the above proposition. □

The following two propositions are crucial.

Proposition 5.5. *The theory $\mathsf{HA}_2^{\min,\trianglelefteq} + \mathsf{FAN}_{\trianglelefteq}$ proves the principle* FAN, *as stated in the language $\mathcal{L}_2$.*

Proof. We reason inside the theory $\mathsf{HA}_2^{\min,\trianglelefteq}$. Assume $\mathsf{FAN}_{\trianglelefteq}$ and suppose that $\forall f \leq h \exists y A(f,y)$, where A is a formula of $\mathcal{L}_2$. By the axiom (ii), we get $\forall f \trianglelefteq h \exists y A(f,y)$. Now, by $\mathsf{FAN}_{\trianglelefteq}$, there is a natural number w_0 such that $\forall f \trianglelefteq h \exists y \leq w_0\, A(f,y)$. Since, by axiom (iii), $\min(f,h) \trianglelefteq h$, we get $\exists y \leq w_0\, A(\min(f,h),y)$. By extensionality, if $f \leq h$, we can conclude $\exists y \leq w_0\, A(f,y)$. In sum, given $\forall f \leq h \exists y A(f,y)$, we have shown that $\exists w \forall f \leq h \exists y \leq w A(f,y)$. □

The next proposition can be proved in a similar way. Note that WKL is stated with Δ_0-formulas A_{b} of the language $\mathcal{L}_2$.

Proposition 5.6. *The theory $\mathsf{HA}_2^{\min,\trianglelefteq} + \mathsf{WKL}_{\trianglelefteq}$ proves the principle* WKL, *as stated in the language $\mathcal{L}_2$.*

We have shown that the theory HA_2 plus the seven principles of Section 2 is a subtheory of the theory of Theorem 4.2. Therefore, we have

Theorem 5.7 (Main soundness theorem). Let $A(x,f)$ be a formula of $\mathcal{L}$ with all the free first-order variables among x and the free second-order variables among f. Let $A^{\mathrm{H}}(x,f)$ be

$\exists a \forall b A_{\mathrm{H}}(a, b, x, f)$. If

$$\mathsf{HA}_2 + \mathsf{MP} + \mathsf{bIP}_{\Pi^0_1} + \mathsf{bCColl} + \mathsf{bAC} + \Delta_0\text{-}\mathsf{AC} + \mathsf{FAN} + \mathsf{WKL} \vdash A(x, f),$$

then there are closed terms t of the star combinatory calculus such that

$$S^\omega_* \models \forall h \forall x \forall b \forall f \preceq h A_{\mathrm{H}}(thx, b, x, f).$$

Moreover, the terms t can be found effectively from a formal proof of $A(x, f)$ in the said theory.

As a consequence, we get the following Herbrand-like witnessing result:

Corollary 5.8. *Suppose that $A(x)$ is a first-order formula of the language of $\mathcal{L}$, with only a free first-order variable x. If*

$$\mathsf{HA}_2 + \mathsf{MP} + \mathsf{bIP}_{\Pi^0_1} + \mathsf{bCColl} + \mathsf{bAC} + \Delta_0\text{-}\mathsf{AC} + \mathsf{FAN} + \mathsf{WKL} \vdash \exists x A(x),$$

then there are natural numbers $n_1, n_2, \ldots, n_k$ such that

$$S^\omega_* \models A(n_1) \vee A(n_2) \vee \cdots \vee A(n_k).$$

Moreover, such numbers can be found effectively from a proof of $\exists x A(x)$.

Proof. Let $A^{\mathrm{H}}(x)$ be $\exists a \forall b A_{\mathrm{H}} A(a, b, x)$. Since $(\exists x A(x))^{\mathrm{H}}$ is

$$\exists x'^{\mathrm{N}^*} \exists a \forall b' \left[\exists x \in x' \forall b \in b' A_{\mathrm{H}}(a, b, x)\right],$$

the above soundness theorem guarantees that we can effectively find closed terms q and t such that

$$S^\omega_* \models \forall b' \exists x \in q \forall b \in b' A_{\mathrm{H}}(t, b, x).$$

Hence, $S^\omega_* \models \exists x \in q \forall b A_{\mathrm{H}}(t, b, x)$. To see this, just note that the contrapositive is clear. Since q is a closed term of type N^*, one can read off effectively from q a finite set of natural numbers n_1, n_2, ..., n_k such that $S^\omega_* \models q = \{n_1, n_2, \ldots, n_k\}$. Therefore,

$$S^\omega_* \models \forall b A_{\mathrm{H}}(t, b, n_1) \vee \forall b A_{\mathrm{H}}(t, b, n_2) \vee \cdots \vee \forall b A_{\mathrm{H}}(t, b, n_k),$$

and so, $S^\omega_* \models \exists a \forall b A_{\mathrm{H}}(a, b, n_1) \vee \exists a \forall b A_{\mathrm{H}}(a, b, n_2) \vee \cdots \vee \exists a \forall b A_{\mathrm{H}}(a, b, n_k)$. To finish the proof, one just has to observe that, for each natural number n, $S_*^\omega \vDash A(n) \leftrightarrow \exists a \forall b A_{\mathrm{H}}(a,b,n)$. It is at this juncture that we use the fact that the formula $A(x)$ is first-order. As noticed in Section 4, the herbrandised interpretation of first-order formulas preserves truth in S_*^ω. □

In particular, the above also shows that the provable first-order sentences of the classically false theory $\mathsf{HA}_2 + \mathsf{MP} + \mathsf{bIP}_{\Pi_1^0} + \mathsf{bCColl} + \mathsf{bAC} + \Delta_0\text{-}\mathsf{AC} + \mathsf{FAN} + \mathsf{WKL}$ must be set-theoretically true, i.e., they hold in S_*^ω.

6. Appendix

As usual, Theorem 4.2 is proved by induction on the number of steps in a derivation. We need to prepare some ground with two remarks. The first remark is plain enough: we will systematically use lambda terms. This is permissible since we have Schönfinkel's combinators in our term calculus. The second remark is not so well known because it concerns star types and star terms. It was made in [1] and, in the next few paragraphs, we briefly recall the most important facts concerning this issue.

A type σ of the star combinatory calculus is called an *end star type* if it is of the form $\tau_1 \to \tau_2 \to \cdots \tau_n \to \rho^*$, where n can be zero. One can easily see by inspection that, in Definition 4.1, the existential variables a appearing in $\exists a \forall b A_{\mathrm{H}}(a,b)$ are always of end star type.

Let σ be an end star type of the form $\tau_1 \to \tau_2 \to \cdots \tau_n \to \rho^*$ and a and a' be terms of type σ. We say that $a \sqsubseteq_\sigma a'$ if

$$S_*^\omega \models \forall x^{\tau_1} \ldots x^{\tau_n} (a x_1 \ldots x_n \subseteq_{\rho^*} a' x_1 \ldots x_n),$$

where $c \subseteq_{\rho^*} c'$ abbreviates $S_*^\omega \models \forall z^\rho (z \in c \to z \in c')$, for c and c' of type ρ^*. Our interpretation of Definition 4.1 is cumulative in the sense given by the following lemma:

Lemma 6.1 (Monotonicity property). For any formula A of the language $\mathcal{L}_2^{\min,\trianglelefteq}$, possibly with first- or second-order parameters, the following holds:

$$S_*^\omega \models a_1 \sqsubseteq a_1' \wedge \ldots a_k \sqsubseteq a_k' \wedge A_{\mathrm{H}}(a_1,\ldots,a_k,b_1,\ldots,b_m)$$
$$\to A_{\mathrm{H}}(a_1',\ldots a_k',b_1,\ldots,b_m).$$

The proof of the monotonicity property is easy, following [1]. Clause 10 is the only new case for which it is worth saying something. It preserves monotonicity because, for h and $\tilde{h}$ of type $\mathrm{N} \to \mathrm{N}^*$ and f of type $\mathrm{N} \to \mathrm{N}$, one has that $h \sqsubseteq \tilde{h} \wedge f \preceq h \to f \preceq \tilde{h}$ holds in S_*^ω. Monotonicity is a very important property because it is crucially used in certain verifications. Another crucial fact that we need is that we can define majorant terms which apply to values of end star type. Let σ be the end star type as above. We define, for a and b of type σ,

$$a \bigsqcup_\sigma b := \lambda x_1^{\tau_1} \lambda x_2^{\tau_2} \cdots \lambda x_n^{\tau_n} . \bigcup_\rho (a x_1 \ldots x_n)(b x_1 \ldots x_n).$$

It is clear that $a \sqsubseteq_\sigma a \sqcup b$ and $b \sqsubseteq_\sigma a \sqcup b$ hold in S_*^ω. In fact, we need more. Given c of type N^* and f of type $\mathrm{N} \to \sigma$, we define

$$\bigsqcup_{w \in c} fw := \lambda x_1^{\tau_1} \lambda x_2^{\tau_2} \cdots \lambda x_n^{\tau_n} . \bigcup_{\mathrm{N},\rho} c\,(\lambda w. f w x_1 \cdots x_n).$$

Informally, $\bigsqcup_{w \in c} fw := \lambda x_1^{\tau_1} \lambda x_2^{\tau_2} \cdots \lambda x_n^{\tau_n} . \bigcup_{w \in c} f w x_1 \ldots x_n$. It is clear that $\forall z \in c\,(fz \sqsubseteq_\sigma \bigsqcup_{w \in c} fw)$ holds in S_*^ω.

The proof of Theorem 4.1 is by induction on the number of steps of the formal derivation. For the logical part of the theory, we rely on the formalisation of intuitionistic logic given in [8]. Since the proof is modular, many of the verifications have already been made in [1]. So, we will only be concerned with the verifications that involve directly the clauses for second-order quantifications.

In the following, we identify terms of the language $\mathcal{L}_2^{\min,\trianglelefteq}$ with certain terms of the (star) combinatory calculus. This identification is well known for the terms of the first-order part of $\mathcal{L}_2$. On the other hand, as remarked just before Definition 4.1, second-order variables of $\mathcal{L}_2$ are identified with variables of type $\mathrm{N} \to \mathrm{N}$. We also identify recursively the second-order terms $\min(\sigma, \tau)$ of $\mathcal{L}_2^{\min,\trianglelefteq}$ with the terms $\lambda x. \min(\sigma x, \tau x)$ of type $\mathrm{N} \to \mathrm{N}$. Now, first-order terms of the form $\sigma(t)$ are recursively identified with σt.

For ease of reading, we ignore parameters that do not play an important role in the proof. Also, instead of writing $\mathfrak{s}t$, we write $\{t\}$, instead of $\cup tq$, we write $t \cup q$ and instead of $\bigcup tq$, we write $\bigcup_{w \in t} qw$. In the following, we take A and B as in Definition 4.1.

Let us check the second-order quantification rules.

$\boldsymbol{A \to B(f) \Rightarrow A \to \forall f B(f)}$**, where** $\boldsymbol{f}$ **is not free in** $\boldsymbol{A}$**:** By induction, there are terms t and q such that

$$S^{\omega}_{*} \models \forall h, a, e \forall f \preceq h(\forall b \in thaeA_{\mathrm{H}}(a,b) \to B_{\mathrm{H}}(qha, e, f)).$$

We need to find terms r and s such that

$$S^{\omega}_{*} \models \forall a, h, e(\forall b \in raheA_{\mathrm{H}}(a,b) \to \forall f \preceq hB_{\mathrm{H}}(sah, e, f)),$$

which is immediate.

$\boldsymbol{\forall f A(f) \to A(\sigma)}$ **where** $\boldsymbol{\sigma}$ **is a second-order term of** $\boldsymbol{\mathcal{L}_2^{\mathrm{min},\trianglelefteq}}$**:** We need terms q, r and s of the star combinatory calculus such that

$$S^{\omega}_{*} \models \forall \phi, b'(\forall h \in q\phi b' \forall b \in r\phi b' \forall f \preceq hA_{\mathrm{H}}(\phi h, b, f) \to A_{\mathrm{H}}(s\phi, b', \sigma)).$$

Since $\sigma \preceq \lambda x.\{\sigma x\}$, we can just take $q\phi b' := \{\lambda x.\{\sigma x\}\}$, $r\phi b' := \{b'\}$ and $s\phi := \phi(\lambda x.\{\sigma x\})$.

$\boldsymbol{A(\sigma) \to \exists f A(f)}$ **where** $\boldsymbol{\sigma}$ **is a second-order term of** $\boldsymbol{\mathcal{L}_2^{\mathrm{min},\trianglelefteq}}$**:** We need to exhibit terms q, r and s of the star combinatory calculus such that

$$S^{\omega}_{*} \models \forall a, b'(\forall b \in qab' A_{\mathrm{H}}(a, b, \sigma) \to \exists f \preceq ra \forall b \in b' A_{\mathrm{H}}(sa, b, f)).$$

Just take $qab' := b'$, $ra := \lambda x.\{\sigma x\}$ and $sa := a$.

$\boldsymbol{A(f) \to B \Rightarrow \exists f A(f) \to B}$ **where** $\boldsymbol{f}$ **is not free in** $\boldsymbol{B}$**:** By induction, there are terms t and q such that

$$S^{\omega}_{*} \models \forall h, a, e\left[\forall f \preceq h(\forall b \in thaeA_{\mathrm{H}}(a, b, f) \to B_{\mathrm{H}}(qha, e))\right].$$

To witness the conclusion of the rule, we need terms r and s of the star combinatory calculus such that

$$S^{\omega}_{*} \models \forall h, a, e(\forall b' \in rhae \exists f \preceq h \forall b \in b' A_{\mathrm{H}}(a, b, f) \to B_{\mathrm{H}}(sha, e)).$$

We can take $rhae := \{thae\}$ and $s = q$.

Except for the scheme of induction, the nonlogical axioms of $\mathsf{HA}_2^{\mathrm{min},\trianglelefteq}$ are universal and, hence, they pose no problem. Induction is checked as in [1] using, of course, the recursors.

Next, we check the axioms for the second-order primitive bounded quantifiers.

$\boldsymbol{\forall f \trianglelefteq h A(f,h) \to \forall f (f \trianglelefteq h \to A(f,h))}$**:** We must find terms t and q such that

$$S^{\omega}_{*} \vDash \forall i, a, g, b' (\forall h \preceq i (\forall b \in tiagb' \forall f \leq h A_{\mathrm{H}}(a,b,f,h) \\ \to \forall f \preceq g (f \leq h \to A(qiag, b', f, h)))).$$

Take $tiagb' := \{b'\}$ and $qiag := a$.

$\boldsymbol{\forall f (f \trianglelefteq h \to A(f,h)) \to \forall f \trianglelefteq h A(f,h)}$**:** We must find terms t, q and r such that

$$\forall i, \phi, b' (\forall h \preceq i (\forall g \in ti\phi b' \forall b \in qi\phi b' \forall f \preceq g (f \leq h \to A_{\mathrm{H}}(\phi g, b, f, h)) \\ \to \forall f \leq h A_{\mathrm{H}}(ri\phi, b', f, f)))$$

holds in S^{ω}_{*}. The terms that work are $ti\phi b' := \{i\}$, $qi\phi b' := \{b'\}$ and $ri\phi := \phi i$.

$\boldsymbol{\exists f \trianglelefteq h A(f,h) \to \exists f (f \trianglelefteq h \wedge A(f,h))}$**:** We want terms t, q and r such that

$$\forall i, a, b'' (\forall h \preceq i (\forall b' \in tiab'' \exists f \leq h \forall b \in b' A_{\mathrm{H}}(a,b,f,h) \\ \to \exists f \preceq qia \forall b \in b'' (f \leq h \wedge A_{\mathrm{H}}(ria, b, f, h))))$$

holds in S^{ω}_{*}. Take $tiab'' := \{b''\}$, $ria := a$ and $qia := i$.

$\boldsymbol{\exists f (f \trianglelefteq h \wedge A(f,h)) \to \exists f \trianglelefteq h A(f,h)}$**:** We need terms t and q such that

$$\forall i, g, a, b'' (\forall h \preceq i (\forall b' \in tigab'' \exists f \preceq g \forall b \in b' (f \leq h \wedge A_{\mathrm{H}}(a,b,f,h)) \\ \to \exists f \leq h \forall b \in b'' A_{\mathrm{H}}(qiga, b, f, h)))$$

holds in S^{ω}_{*}. Take $tigab'' := \{b''\}$ and $qiga := a$.

This concludes the verification of $\mathsf{HA}_2^{\min,\trianglelefteq}$. It remains to check the semi-intuitionistic principles. We only discuss the principles in the following:

$\boldsymbol{\forall x \exists y A(x,y) \to \exists f \forall x \exists y \leq f(x) A(x,y)}$**:** We need terms t, q, r and s such that

$$\forall \phi, \psi, x', b'' (\forall x \in t\phi\psi x' b'' \forall b' \in q\phi\psi x' b'' \exists y \in \psi x \forall b \in b' A_{\mathrm{H}}(\phi x, b, x, y) \\ \to \exists f \preceq r\phi\psi \forall x \in x' \forall b' \in b'' \exists y \leq fx \forall b \in b' A_{\mathrm{H}}(s\phi\psi x, b, x, y))$$

holds in S^{ω}_{*}. We can simply take $t\phi\psi x' b'' := x'$, $q\phi\psi x' b'' := b''$, $r\phi\psi := \psi$ and $s\phi\psi := \phi$.

$\forall x \exists y \leq f(x) A_b(x,y) \to \exists g \trianglelefteq f \forall x A_b(x, g(x))$ where A_b is a Δ_0-formula: We are looking for terms t and q such that

$$S_*^\omega \models \forall h, x'[\forall f \preceq h(\forall x \in thx' \exists y \leq fxA_b(x,y) \\ \to \exists g \leq f \forall x \in x' A_b(x, gx))].$$

The term $thx' := x'$ works. The function g clearly exists. Note that the values of the domain of g that really matter are the finite number of values in x' (the others can be 0).

$\forall f \trianglelefteq h \exists k A(f,k) \to \exists n \forall f \trianglelefteq h \exists k \leq n A(f,k)$: We must find terms t, q and r such that

$$\forall g, k'^{\mathrm{N}^*}, a, b''(\forall h \preceq g(\forall b' \in tgk'ab'' \forall f \leq h \exists k \in k' \forall b \in b' A_{\mathrm{H}}(a,b,f,k) \\ \to \exists n \in qgk'a \forall b' \in b'' \forall f \leq h \exists k \leq n \forall b \in b' A_{\mathrm{H}}(rgk'a, b, f, k)))$$

holds in S_*^ω. It is easy to see that $tgk'ab'' := b''$, $qgk'a := k'$ and $rgk'a := a$ work by taking for n the maximum of the finite nonempty set k'.

$\forall x \exists f \trianglelefteq h \forall y \leq x A_b(f,y) \to \exists f \trianglelefteq h \forall y A_b(f,y)$ where A_b is bounded: We must find a term t such that

$$\forall g, y'(\forall h \preceq g(\forall x \in tgy' \exists f \leq h \forall y \leq x A_b(f,y) \\ \to \exists f \leq h \forall y \in y' A_b(f,y)))$$

holds in S_*^ω. Just take $tgy' := y'$.

Acknowledgement

Both authors acknowledge the support of Fundação para a Ciência e Tecnologia (FCT). The support of João Enes was via the FCT Lismath fellowship PD/BD/52639/2014. The support of Fernando Ferreira was via the research centre CMAFcIO under the FCT grant UIDB/04561/2020.

References

[1] F. Ferreira, The FAN principle and weak König's lemma in herbrandised second-order arithmetic, *Ann. Pure Appl. Log.* **171**(9), 102843 (21 pages) (2020).

[2] S.G. Simpson, *Subsystems of Second Order Arithmetic.* Perspectives in Mathematical Logic, Springer-Verlag, Berlin, 1999.

[3] F. Ferreira and P. Oliva, Bounded functional interpretation, *Ann. Pure Appl. Log.* **135**, 73–112 (2005).

[4] A.S. Troelstra and D. van Dalen, *Constructivism in Mathematics. An Introduction,* (vol. I). Number 121 in Studies in Logic and the Foundations of Mathematics, North Holland, Amsterdam, 1988.

[5] E. Bishop, *Foundations of Constructive Analysis.* McGraw-Hill, New York, 1967.

[6] U. Kohlenbach, *Applied Proof Theory: Proof Interpretations and their Use in Mathematics.* Springer Monographs in Mathematics, Springer-Verlag, Berlin, 2008.

[7] P. Engrácia, *Proof-theoretic Studies on the Bounded Functional Interpretation.* PhD thesis, Universidade de Lisboa, 2009.

[8] J. Avigad and S. Feferman, Gödel's functional ("Dialectica") interpretation. In S.R. Buss (ed.), *Handbook of Proof Theory, vol. 137, Studies in Logic and the Foundations of Mathematics*, North Holland, Amsterdam, 1998, pp. 337–405.

[9] F. Ferreira and G. Ferreira, A herbrandised functional interpretation of classical first-order logic, *Arch. Math. Logic.* **56**(5), 523–539 (2017).

[10] B.v.d. Berg, E. Briseid, and P. Safarik, A functional interpretation for nonstandard arithmetic, *Ann. Pure Appl. Logic.* **163**(12), 1962–1994 (2012).

https://doi.org/10.1142/9789811245220_0011

Chapter 11

More or Less Uniform Convergence

Henry Towsner

Department of Mathematics, University of Pennsylvania,
Philadelphia, USA
htowsner@math.upenn.edu

Uniform metastable convergence is a weak form of uniform convergence for a family of sequences. In this chapter, we explore the way that metastable convergence stratifies into a family of notions indexed by countable ordinals.

We give two versions of this stratified family; loosely speaking, they correspond to the model theoretic and proof theoretic perspectives. For the model theoretic version, which we call abstract α-uniform convergence, we show that uniform metastable convergence is equivalent to abstract α-uniform convergence for some α, and that abstract ω-uniform convergence is equivalent to uniformly bounded oscillation of the family of sequences.

The proof theoretic version, which we call concrete α-uniform convergence, is less canonical (it depends on a choice of ordinal notation), but appears naturally when "proof mining" convergence proofs to obtain quantitative bounds.

We show that these hierarchies are strict by exhibiting a family which is concretely $\alpha + 1$-uniformly convergent but not abstractly α-uniformly convergent for each $\alpha < \omega_1$.

1. Introduction

There are many examples of sequences which can be shown to converge classically, but for which the rate of convergence is not computable, and therefore we cannot expect a constructive proof of convergence.

On the other hand, in many specific cases one can show that even though the rate of convergence is not computable, some weaker quantitative property holds. For instance, Bishop gave a constructive analysis of the ergodic theorem [1] showing how to compute bounds on the "upcrossings" — roughly speaking, the number of times the sequence can have large changes in value. One cannot expect such a result in general: for instance, it is possible to have families of convergent sequences where there is no uniform bound on the number of upcrossings.

More generally, considerations from proof mining led to the introduction of *metastable convergence* in [2,3] (and studied earlier in [4,5]). Any classical proof that a family of sequences converges leads to a uniformly computable bound on the metastable convergence of the sequence.

Our goal in this chapter is to show that these two notions are related. Metastable convergence can be seen as a family of notions indexed by ordinals, with full metastable convergence corresponding to the ω_1 level, and the notion of bounded upcrossings corresponding to the ω level, with other notions in between. While this idea has appeared implicitly in the literature [6,7], the details have not been made explicit.

We define the general family of notions of abstract α-uniform convergence and show that uniform metastable convergence is equivalent to abstract α-uniform convergence for some $\alpha < \omega_1$. We introduce another, slightly stronger notion, concrete α-uniform convergence, which has the benefit of being more explicit but the disadvantage of depending on explicit representations of ordinals (specifically, fixed sequences α_n so that $\alpha = \sup_n(\alpha_n + 1)$ for each α). Finally, we introduce families of sequences $\mathcal{S}_\alpha$ (the sequences which "change value α times") and show that each $\mathcal{S}_\alpha$ is concretely $\alpha + 1$-uniformly convergent but not abstractly α-uniformly convergent.

2. Uniform Metastable Convergence

Throughout this chapter, we will focus on $\{0,1\}$-valued sequences, and we write $\bar{a}$ for the sequence $(a_n)_{n\in\mathbb{N}}$. The ideas generalise to sequences valued in any complete metric space, and we will occasionally discuss this generalisation in remarks.

We are interested in sets $\mathcal{S}$ of sequences such that every $\bar{a} \in \mathcal{S}$ converges. In particular, we are interested in questions about the uniformity of this convergence. The classic notion of uniform convergence — that there is some fixed m so that, for every $\bar{a} \in \mathcal{S}$, if $m, m' \geq n$ then $a_m = a_{m'}$ — is quite strong. (Indeed, in our restricted setting of $\{0,1\}$-valued sequences, it is easy to see that any uniformly convergent set of sequences is finite).

The following, weaker notion, is in some sense the weakest reasonable notion of uniformity.

Definition 2.1. Let $\mathcal{S}$ be a set of sequences. We say $\mathcal{S}$ *converges uniformly metastably* if for every $F : \mathbb{N} \to \mathbb{N}$, such that $n < F(n)$ and $F(n) \leq F(n+1)$ for all n, there exists an M_F so that, for every $\bar{a} \in \mathcal{S}$, there is an $m \leq M_F$ and a $c \in \{0,1\}$ so that for all $n \in [m, F(m)]$, $a_n = c$.

This notion has also been called *local stability* [2]: it says that we can find arbitrarily long intervals on which the sequence $\bar{a}$ has stabilised, where the length of the interval can even depend on how large the starting point of the interval is (that is, the interval has the form $[m, F(m)]$, where F could grow very quickly as a function of m).

Numerous results [2,3,8–12] have shown that, under some assumptions on $\mathcal{S}$, if every sequence in $\mathcal{S}$ converges, then actually $\mathcal{S}$ must converge uniformly metastability. These hypotheses are usually given in terms of logic, but in this simple setting a direct formulation is possible.

Definition 2.2. Suppose that, for each i, $\bar{a}^i = (a^i_n)_{n\in\mathbb{N}}$ is a sequence. A *limit sequence* is a sequence $\bar{b} = (b_n)_{n\in\mathbb{N}}$ such that there is an infinite set S so that, for each n, $\{i \in S \mid a^i_n \neq b_n\}$ is finite.

For example, suppose that

$$a_n^i = \begin{cases} 1 & \text{if } i < n \\ 0 & \text{otherwise.} \end{cases}$$

Then the only limit sequence is the sequence which is constantly equal to 0.

More generally, if $\lim_{i\to\infty} a_n^i$ converges to a value c, then in any limit sequence (b_n), $b_n = c$. When $\lim_{i\to\infty} a_n^i$ fails to converge, there must be multiple limits, including at least one where $b_n = 0$ and one where $b_n = 1$; the requirement that there is a single set S is a coherence condition.

For example, if

$$a_n^i = \begin{cases} 1 & \text{if } 2^n \mid i \\ 0 & \text{otherwise,} \end{cases}$$

then the possible limits are the sequence that is all 1s, or any sequence consisting of a finite, positive number of 1s followed by 0s: the choice where there are $k \geq 1$ 1s followed by 0s corresponding to take S to be, for example, the numbers divisible by 2^{k-1} but not 2^k; the limit which is constantly 1 corresponds to taking S to be, for instance, $\{1, 2, 4, 8, 16, \ldots\}$.

Lemma 2.3. *For any collection of sequences $\bar{a}^i$, there exists a limit sequence.*

Remark 2.4. In the more general setting [8], the notion of a limit sequence is replaced by an *ultraproduct*. This not only allows consideration of sequences valued in arbitrary metric spaces, it includes the case where different sequences come from different metric spaces.

Definition 2.5. We say a sequence $\bar{a} = (a_n)_{n\in\mathbb{N}}$ *converges* if there is some m and some $c \in \{0, 1\}$ so that for all $n \geq m$, $a_n = c$.

Let $\mathcal{S}$ be a set of sequences. We say $\mathcal{S}$ *converges uniformly metastably* if for every $F : \mathbb{N} \to \mathbb{N}$, there exists an M_F so that, for every $\bar{a} \in \mathcal{S}$, there is an $m \leq M_F$ and a $c \in \{0, 1\}$ so that for all $n \in [m, F(m)]$, $a_n = c$.

Theorem 2.6. *Suppose $\mathcal{S}$ is a set of sequences such that*

- *every sequence in $\mathcal{S}$ converges; and*
- *whenever $\{\bar{a}^i\}_{i\in\mathbb{N}} \subseteq \mathcal{S}$, every limit sequence of $\bar{a}^i$ is also in $\mathcal{S}$.*

Then $\mathcal{S}$ converges uniformly metastably.

Proof. Suppose $\mathcal{S}$ does not converge uniformly metastably; then there is an $F : \mathbb{N} \to \mathbb{N}$ so that, for each i, there is a $\bar{a}^i \in \mathcal{S}$ such that, for every $m \leq i$, there are $n_0, n_1 \in [m, F(m)]$ so that $a_{n_0} = 0$ and $a_{n_1} = 1$.

Let $\bar{b}$ be some limit of the $\bar{a}^i$. Then $\bar{b} \in \mathcal{S}$, so $\bar{b}$ converges. So there is some m and some c so that, for all $n \geq m$, $b_n = c$.

Choose an infinite set S so that, for all n, $\{i \in S \mid a^i_n \neq b_n\}$ is finite. In particular, the set of $i \in S$ such that, for all $n \in [m, F(m)]$, $a^i_n = b_n$ must be infinite, so we can find some $i \geq m$ in S. Then for every $n \in [m, F(m)]$, $a^i_n = b_n = c$. But this contradicts the choice of $\bar{a}^i$. □

This is the sense in which uniform metastable convergence is the weakest notion of uniform convergence: if $\mathcal{S}$ fails to satisfy it, then $\mathcal{S}$ has limits which do not converge.

We wish to associate sets of convergent sequences $\mathcal{S}$ to ordinals.

Definition 2.7. Let $\mathcal{S}$ be a set of sequences. We define $T_{\mathcal{S}}$ to be the tree of finite increasing sequences $0 < r_1 < \cdots < r_M$ such that, taking $r_0 = 0$, there is some $\bar{a} \in \mathcal{S}$ so that, for every $i < M$, there are $n_0, n_1 \in [r_i, r_{i+1}]$ with $a_{n_0} = 0$ and $a_{n_1} = 1$.

When $\alpha < \omega_1$ is an ordinal, we say that $\mathcal{S}$ converges abstractly α-uniformly if $T_{\mathcal{S}}$ has height strictly less than α.

Remark 2.8. Note that, in our setting, $\mathcal{S}$ converges abstractly ω-uniformly iff $\mathcal{S}$ converges abstractly n-uniformly for some finite n (and similarly for other limit ordinals).

When studying more general sequences, one would want to consider countably many trees corresponding to fluctuations of size approaching 0: we could consider the tree $T_{\mathcal{S},k}$ of sequences $0 = r_0 < r_1 < \cdots < r_M$ such that there is some $\bar{a} \in \mathcal{S}$ so that, for every $i < M$, there are $n_0, n_1 \in [r_i, r_{i+1}]$ with $|a_{n_0} - a_{n_1}| > 1/k$. Then we would say $\mathcal{S}$ converges abstractly α-uniformly if for each k, $T_{\mathcal{S},k}$ has height strictly less than α.

To connect this with metastability, we need to relate these sequences $(r_i)_{1 \leq i \leq M}$ to functions $F : \mathbb{N} \to \mathbb{N}$.

Definition 2.9. Given a strictly increasing sequence $\bar{r}$ with $0 = r_0$, we define $F_{\bar{r}}(n)$ by taking i least so $n \leq r_i$ and setting $F_{\bar{r}}(n) = r_{i+1}$.

For any function $F : \mathbb{N} \to \mathbb{N}$ such that $n < F(n)$ and $F(n) \leq F(n+1)$ for all n, we define a function $\hat{F} = F_{(F^i(0))_{i\in\mathbb{N}}}$.

Lemma 2.10. *For any $F : \mathbb{N} \to \mathbb{N}$ such that $n < F(n)$ and $F(n) \leq F(n+1)$ for all n, $F(n) \leq \hat{F}(n)$ for all n.*

Proof. Let $r_i = F^i(0)$ for all i, so $\hat{F} = F_{\bar{r}}$.

Let i be least so that $n \leq r_i = F^i(0)$. Then $\hat{F}(n) = F^{i+1}(0) = F(F^i(0)) \geq F(n)$. □

Note that if $M_{\hat{F}}$ witnesses metastability for $\hat{F}$, then it also witnesses metastability for F; in particular, this means that if we can show metastability for all functions of the form $F_{\bar{r}}$, we have shown metastability.

Theorem 2.11. *The set $\mathcal{S}$ converges uniformly metastably iff $\mathcal{S}$ converges abstractly α-uniformly for some $\alpha < \omega_1$.*

Proof. Suppose $\mathcal{S}$ does not converge abstractly α-uniformly for any $\alpha < \omega_1$ — that is, suppose the tree $T_{\mathcal{S}}$ is ill-founded. Then it has an infinite path $\bar{r}$, and the function $F_{\bar{r}}$ witnesses a failure of uniformly metastable convergence: by the definition of $T_{\mathcal{S}}$, for every M there is a $\bar{a} \in \mathcal{S}$ so that, for each $m \leq M$, letting i be least so $m \leq r_i$, there are $n_0, n_1 \in [r_i, r_{i+1}] \subseteq [m, F_{\bar{r}}(m)]$ with $a_{n_0} = 0$ and $a_{n_1} = 1$. Therefore, the uniform bound M_F cannot exist.

Conversely, suppose F witnesses a failure of uniformly metastable convergence, so also $\hat{F}$ witnesses a failure of uniformly metastable convergence. Then for each K, there is a sequence $\bar{a}$ so that for each $k < K$ there are $n_0, n_1 \in [k, \hat{F}(k)]$ so that $a_{n_0} = 0$ and $a_{n_1} = 1$. In particular, taking $K = F^M(0)$, we may find $\bar{a}$ so that for each $i < M$ there are such $n_0, n_1 \in [F^i(0), F^{i+1}(0)]$. Therefore, the sequence $(F^i(0))_{i\in\mathbb{N}}$ is an infinite path through $T_{\mathcal{S}}$. □

We mention one other notion of uniform convergence, which has been intensely studied [1,13–15]: bounds on jumps (which, in this context, are essentially the same as bounds on "oscillations" or "upcrossings" as they are sometimes known in the literature).

Definition 2.12. $\mathcal{S}$ has *uniformly bounded jumps* if there is a k so that whenever $\bar{a} \in \mathcal{S}$ and $n_0 < \cdots < n_k$ is a sequence, there is an $i < k$ so that $a_{n_i} = a_{n_{i+1}}$.

Theorem 2.13. *The set $\mathcal{S}$ has uniformly bounded jumps iff $\mathcal{S}$ converges abstractly ω-uniformly.*

Proof. Suppose $\mathcal{S}$ does not converge abstractly ω-uniformly, so for each k there is a sequence $0 < r_0^k < \cdots < r_k^k$ and a $\bar{a} \in \mathcal{S}$ so that for each $i < k$ there are $n_0^i, n_1^i \in [r_i^k, r_{i+1}^k]$ with $a_{n_0^i} = 0$ and $a_{n_1^i} = 1$. Then the sequence $n_0^0 < n_1^1 < n_2^0 < n_3^1 < \cdots$ shows that $\mathcal{S}$ does not have uniformly bounded jumps with bound k. Since this holds for any k, $\mathcal{S}$ does not have uniformly bounded jumps.

Conversely, suppose $\mathcal{S}$ does not have uniformly bounded jumps, so for each k there is an $\bar{a}$ and a sequence $n_0 < \cdots < n_k$ so that $a_{n_i} \neq a_{n_{i+1}}$ for each $i < k$. Then the sequence $n_1, n_3, n_5, \ldots$ belongs to $T_{\mathcal{S}}$. Since $T_{\mathcal{S}}$ contains arbitrarily long finite sequences, $T_{\mathcal{S}}$ has height at least ω. □

3. Ordinal Iterations

3.1. *Concrete α-uniform convergence*

Most naturally occurring examples of abstract α-uniform convergence satisfy a stronger property. This stronger property is not quite canonical — we need to fix a family of ordinal notations.

Definition 3.1. A *fundamental sequence*[a] for a countable ordinal $\alpha > 0$ is a sequence of ordinals $\alpha[n]$ for $n \in \mathbb{N}$ such that

- $\alpha[n] < \alpha$;
- $\alpha[n] \leq \alpha[n+1]$; and
- for every $\beta < \alpha$, there is an n with $\beta \leq \alpha[n]$.

For convenience, we define $0[n] = 0$.

When α is a successor — $\alpha = \gamma + 1$ — these conditions imply that $\alpha[n] = \gamma$ for all but finitely many n. When α is a limit ordinal, these conditions imply that $\lim_{n\to\infty} \alpha[n] = \alpha$.

[a]Various definitions of this notion which are not exactly equivalent are found in the literature, but the differences are generally minor.

For small ordinals, there are conventional choices of fundamental sequences, like $\omega[n] = n$, $\omega^2[n] = \omega \cdot n$, $\epsilon_0[n] = \omega_n$ (where $\omega_0 = 0$ and $\omega_{n+1} = \omega^{\omega_n}$), and so on, arising out of ordinal notation schemes.

For the remainder of the chapter, assume we have picked a system of fundamental sequences, so that for each countable ordinal α, the ordinals $\alpha[n]$ are defined. For convenience, we assume that $(\gamma + 1)[n] = \gamma$ for all successor ordinals.

Definition 3.2. Let $F : \mathbb{N} \to \mathbb{N}$ be a function with $F(n) > n$ and $F(n) \leq F(n+1)$ for all n. We define the α-iteration of F by

- $F^0(n) = n$;
- when $\alpha > 0$, $F^\alpha(n) = F^{\alpha[F(n)]}(F(n))$.

Then F^1 is just F, F^k is the usual k-fold iteration of F, $F^\omega(n) = F^{F(0)+1}(0)$ (assuming the conventional fundamental sequence $\omega[n] = n$ for ω), and so on. Note that the definition of this iteration does depend on the choice of fundamental sequences.

Note that these functions are not quite increasing in the ordinal: if $\alpha < \beta$, then we have $F^\alpha(n) \leq F^\beta(n)$ for sufficiently large n, but not necessarily when n is small. (For instance, compare $F^{1000}(3)$ to $F^\omega(3)$ for F not growing too quickly.)

When calculating $F^\alpha(n)$, there is a canonical sequence of values and ordinals associated with its computation, given by

- $r_0 = n$, $\beta_0 = \alpha$;
- $r_{i+1} = F(r_i)$, $\beta_{i+1} = \beta_i[r_{i+1}]$.

Since the sequence $\beta_0, \beta_1, \ldots$ is strictly decreasing, it terminates at some value k with $\beta_k = 0$, and we have

$$F^\alpha(n) = F^{\beta_0}(r_0) = F^{\beta_1}(r_1) = \cdots = F^{\beta_k}(r_k) = r_k.$$

Definition 3.3. We say $\mathcal{S}$ converges *concretely α-uniformly* if there is a $\beta < \alpha$ so that, for every $F : \mathbb{N} \to \mathbb{N}$ such that $F(n) > n$ for all n, for each $\bar{a} \in \mathcal{S}$, there is an m with $F(m) \leq F^\beta(0)$ and a c so that, for all $n \in [m, F(m)]$, $a_n = c$.

Remark 3.4. Again, in our restricted setting this is only interesting at successor ordinals where concrete $\alpha+1$-uniform convergence means $F^\alpha(0)$ always suffices as a bound.

With more general sequences, we would say that for each k there is a $\beta < \alpha$ so that, for every $F : \mathbb{N} \to \mathbb{N}$ such that $F(n) > n$ for all n, for each $\bar{a} \in \mathcal{S}$, there is an m with $F(m) \leq F^{\beta}(0)$ and a c so that, for all $n, n' \in [m, F(m)]$, $|a_n - a_{n'}| < 1/k$.

Lemma 3.5. *If $\mathcal{S}$ converges concretely α-uniformly, then $\mathcal{S}$ converges abstractly α-uniformly.*

Proof. Suppose $\mathcal{S}$ fails to converge abstractly α-uniformly, so $T_{\mathcal{S}}$ has the height $\geq \alpha$ (possibly ill-founded). For each $\beta < \alpha$, we will construct a function F and find a $\bar{a} \in \mathcal{S}$ witnessing the failure of strong uniform convergence.

Fix some $\beta < \alpha$. By induction on n, we choose a decreasing sequence of ordinals $\beta_n \leq \beta$ and a sequence $(r_i)_{i \leq n}$. We begin with $r_0 = 0$ and $\beta_0 = \beta$. Given β_n, we take r_{n+1} so that the set of sequences in $T_{\mathcal{S}}$ extending $(r_i)_{1 \leq i \leq n+1}$ has height $\geq \beta_n$. We then set $\beta_{n+1} = \beta_n[r_{n+1}]$. We continue until we reach some k so that $\beta_k = 0$.

Chose $\bar{a} \in \mathcal{S}$ so that, for each $i < k$, there are $n_0, n_1 \in [r_i, r_{i+1}]$ with $a_{n_0} = 0$ and $a_{n_1} = 1$.

We extend the sequence $(r_i)_{1 \leq i \leq k}$ arbitrarily (say $r_i = r_k + i$ for $i > k$) and set $F = F_{\bar{r}}$. Then the computation sequence for $F^{\beta}(0)$ is precisely

$$F^{\beta_0}(r_0) = F^{\beta_1}(r_1) = \cdots = F^{\beta_k}(r_k) = r_k.$$

Suppose $F(m) \leq F^{\beta}(0) = r_k$, so $m \leq r_{k-1}$. Then, for some $i < k$, we have $[r_i, r_{i+1}] \subseteq [m, F(m)]$, and therefore, there are $n_0, n_1 \in [m, F(m)]$ with $a_{n_0} = 0$ and $a_{n_1} = 1$.

Since we can construct some such F for any $\beta < \alpha$, $\mathcal{S}$ is not concretely α-uniformly convergent. □

As a syntactic analogue to Theorem 2.11, we expect that if we prove that $\mathcal{S}$ converges uniformly metastably in some reasonable theory T with proof-theoretic ordinal λ, then there should be some $\alpha < \lambda$ such that we can prove that $\mathcal{S}$ converges concretely α-uniformly. See [16] for an explicit example of such an analysis in the context of differential algebra.

Indeed, given any proof that a family of sequences converges, we can always ask what the corresponding value of α is. While concrete ω-uniform convergence — that is, an explicit bound on upcrossings — is common, it does not always hold [17], and the question of whether it holds is an interesting (and in some cases open [13]) question.

3.2. *A proper hierarchy*

We show that our notions form a proper hierarchy by constructing, for each α, a family $\mathcal{S}_\alpha$ of sequences which are concretely α-uniformly convergent but not abstractly $\alpha+1$-uniformly convergent.

Definition 3.6. We define $\alpha[c_1,\ldots,c_m]$ inductively by $\alpha[c_1,\ldots,c_{m+1}] = (\alpha[c_1,\ldots,c_m])[c_{m+1}]$.

We define $\mathcal{S}_\alpha$ as consisting of those sequences $(a_n)_{n\in\mathbb{N}}$ such that whenever $c_1,\ldots,c_k$ is a sequence such that

- c_1 is least so that $a_{c_1} \neq a_0$,
- for each $i < k$, c_{i+1} is the smallest value greater than c_i so that $a_{c_{i+1}} \neq a_{c_i}$,

then $\alpha[c_1,\ldots,c_k] \neq 0$.

Roughly speaking, this measures how many times the sequence changes from being 0s to being 1s. For example, $\mathcal{S}_k$ is the sequence which changes at most k times. $\mathcal{S}_\omega$ is the sequence where, taking n to be the first place where the sequence changes, it changes at most n additional times. The condition is essentially that if we collect the "runs" of consecutive 1s or 0s, the starting points form an α-large set in the sense of [18].

Note that, for any $\bar{a} \in \mathcal{S}_\alpha$, the statement that $\bar{a}$ belongs to $\mathcal{S}_\alpha$ really concerns some maximal finite sequence $c_1,\ldots,c_k$. However, it is convenient to phrase the definition this way — where we also consider initial segments $c_1,\ldots,c_{k'}$ for some $k' < k$ — because when $\bar{a} \notin \mathcal{S}_\alpha$, the sequence may be infinite, but some finite initial segment is long enough to witness that α reduces to 0.

Lemma 3.7. *If $\{\bar{a}^i\}_{i\in\mathbb{N}} \subseteq \mathcal{S}_\alpha$ then every limit sequence of $\bar{a}^i$ is also in $\mathcal{S}_\alpha$.*

Proof. Let $\bar{a}^i$ be given and consider some limit $\bar{b}$ witnessed by an infinite set S such that, for all n, $\{i \in S \mid a^i_n \neq b_n\}$ is finite. Take any sequence $c_1,\ldots,c_k$ for $\bar{b}$ as in the definition of $\mathcal{S}_\alpha$. Then we may find an i so that, for all $n \leq c_k$, $a^i_n = b_n$. Then, since $\bar{a}^i \in \mathcal{S}_\alpha$, we have $\alpha[c_1,\ldots,c_k] \neq 0$ as needed. □

In order to prove the results we need, we require an additional property on our fundamental sequences.

Definition 3.8. Suppose that, for all $\beta \in (0, \alpha]$, we have a fundamental sequence for β. We say these sequences are *monotone* if, for each $\beta \leq \alpha$ and any sequences $r_1, \ldots, r_k$ and $r'_1, \ldots, r'_k$ with $r_i \leq r'_i$ for all i, $\beta[r_1, \ldots, r_k] \leq \beta[r'_1, \ldots, r'_k]$.

The usual fundamental sequences on small ordinals are monotone. With a nonmonotone fundamental sequence, we could have $\omega^2[1] = 10{,}000$ while $\omega^2[2] = \omega$, and then we would have $\omega^2[1,1] = 9{,}999$ while $\omega^2[2,1] = 1$; this is the sort of anomaly we need to avoid.

Lemma 3.9. *$\mathcal{S}_\alpha$ is concretely $\alpha + 1$-uniformly convergent.*

Proof. Let F be given and $\bar{a} \in \mathcal{S}_\alpha$ and suppose towards a contradiction that for each $m \leq F^\alpha(0)$ there are $n_{m_0}, n_{m_1} \in [m, F(m)]$ with $a_{n_{m_0}} = 0$ and $a_{n_{m_1}} = 1$.

Take the computation sequence for $F^\alpha(0)$, where $r_0 = 0$, $\beta_0 = \alpha$, $r_{i+1} = F(r_i)$ and $\beta_{i+1} = \beta_i[r_{i+1}]$. Let k be least so $\beta_k = 0$, so $F^\alpha(0) = r_k$.

Each interval $[r_i, r_{i+1}]$ (including $[0, r_1]$) must contain the start of at least one new run, so taking the sequence $c_1, \ldots, c_k$ corresponding to α, we have $c_i \leq r_i$ for all i. Since $0 \neq \alpha[c_1, \ldots, c_k] \leq \alpha[r_1, \ldots, r_k] = \beta_k = 0$, we have the desired contradiction. □

Lemma 3.10. *$\mathcal{S}_\alpha$ is not abstractly α-uniformly convergent.*

Proof. Given a sequence $(r_i)_{i \leq k}$ with $\alpha[r_1, \ldots, r_k] \neq 0$, we may naturally associate a sequence $\bar{a} \in \mathcal{S}_\alpha$ by taking, for each n, the unique i such that $n \in [r_i, r_{i+1})$ (where $r_0 = 0$) and setting $a_n = i$ mod 2. This witnesses that any such sequence belongs to $T_{\mathcal{S}_\alpha}$.

So it suffices to show that, for any α, the tree of sequences $(r_i)_{i \leq k}$ with $\alpha[r_1, \ldots, r_k] \neq 0$ has height $\geq \alpha$. We need to show slightly more, namely that for any α and any d, the tree of sequences $(r_i)_{i \leq k}$ with $d < r_1$ and $\alpha[r_1, \ldots, r_k] \neq 0$ has height $\geq \alpha$. We proceed by induction on α; of course when $\alpha = 0$, this tree is empty, and so has height 0.

Suppose the claim holds for all $\beta < \alpha$ and let d be given. Fix any $r_1 > d$. The tree of sequences whose first element is r_1 and $\alpha[r_1, \ldots, r_k]$ is precisely the tree of sequences $(r_i)_{2 \leq i \leq k}$ such that $r_1 < r_2$ and $(\alpha[r_1])[r_2, \ldots, r_k] \neq 0$, which has height $\geq \alpha[r_1]$. Since, $\sup_n \alpha[n] = \alpha$ (even after discarding those n with $n \leq d$), the tree of these sequences has height $\geq \alpha$. □

3.3. *Distinguishing abstract and concrete α-uniformity*

Finally, we note that concrete α-uniformity really is stronger than abstract α-uniformity.

To illustrate the gap at the ordinal $\omega+1$, take any sufficiently fast-growing function $f : \mathbb{N} \to \mathbb{N}$ (in fact, $f(n) = 2n$ suffices). Consider the family $\mathcal{S}_f$ consisting of

- The sequence which is all 0s.
- For infinitely many n, the sequence given by

$$a_i^n = \begin{cases} 1 & \text{if } i = n + 2j \quad \text{for some } j < f(n). \\ 0 & \text{otherwise} \end{cases}$$

That is, $\bar{a}^n$ is the sequence

$$00 \cdots 00101010 \cdots 010000 \cdots$$

where the first 1 occurs at the n-th position, there are $f(n)$ alternations, and then the sequence finishes with infinitely many 0s. By choosing the set of n sufficiently sparsely, we can ensure that if $\bar{a}, \bar{b} \in \mathcal{S}_f$ and $a_i = b_i = 1$, then $\bar{a} = \bar{b}$. This guarantees that any limit of $\mathcal{S}_f$ is also in $\mathcal{S}_f$.

For any n, consider the function F such that $F(i) = n$ for $i < n$ and $F(i) = i + 1$ for $i > n$. Then

$$F^{\omega}(0) = F^{\omega[n]}(n) = 2n.$$

But the sequence $\bar{a}^n$ has both a 0 and a 1 on every interval $[i, F(i)]$ with $F(i) \leq 2n$.

We could address this gap in an individual case by tweaking the definition of concrete α-uniformity, either by using a different fundamental sequence for ω, or by allowing concrete α-uniformity to use a bound like $F^{\alpha+k}(d)$ for constants k, d that depend on the family $\mathcal{S}_f$ (but not on the function F). But, by choosing f growing sufficiently fast, we can still find families $\mathcal{S}_f$ which outpace any fixed modification of this kind.

Acknowledgement

Partially supported by NSF grant DMS-1600263.

References

[1] E. Bishop, A constructive ergodic theorem, *J. Math. Mech.* **17**, 631–639 (1967/1968).

[2] J. Avigad, P. Gerhardy, and H. Towsner, Local stability of ergodic averages, *Trans. Amer. Math. Soc.* **362**(1), 261–288 (2010). ISSN 0002-9947. doi: 10.1090/S0002-9947-09-04814-4.

[3] T. Tao, Norm convergence of multiple ergodic averages for commuting transformations, *Ergodic Theory Dynam. Systems* **28**(2), 657–688 (2008). ISSN 0143-3857.

[4] U. Kohlenbach, Some computational aspects of metric fixed-point theory, *Nonlinear Anal.* **61**(5), 823–837 (2005). ISSN 0362-546X. doi: 10.1016/j.na.2005.01.075.

[5] U. Kohlenbach and B. Lambov, Bounds on iterations of asymptotically quasi-nonexpansive mappings. In J.G.a. Falset, E.L. Fuster, and B. Sims (eds.), *International Conference on Fixed Point Theory and Applications*, Yokohama Publ., Yokohama, 2004, pp. 143–172. ISBN 4-946552-13-8.

[6] J. Gaspar and U. Kohlenbach, On Tao's "finitary" infinite pigeonhole principle, *J. Symbolic Logic* **75**(1), 355–371 (2010). ISSN 0022-4812. doi: 10.2178/jsl/1264433926.

[7] H. Towsner, Towards an effective theory of absolutely continuous measures. In P. Cégielski, A. Enayat, and R. Kossak (eds.), *Studies in Weak Arithmetics, vols. 3, 217, CSLI Lecture Notes*, CSLI Publ., Stanford, CA, 2016, pp. 171–229. ISBN 978-1-57586-953-7; 1-57586-953-5. Papers based on the conference "Journées sur les Arithmétiques Faibles (Weak Arithmetics Days)" (JAF33) held at the University of Gothenburg, Gothenburg, June 16–18, 2014 and the conference (JAF34) held at CUNY, New York, July 7–9, 2015.

[8] J. Avigad and J. Iovino, Ultraproducts and metastability, *New York J. Math.* **19**, 713–727 (2013). ISSN 1076-9803.

[9] S. Cho, A variant of continuous logic and applications to fixed point theory, *ArXiv e-prints* (October, 2016).

[10] E. Dueñez and J. Iovino, Model theory and metric convergence i: Metastability and dominated convergence. In *Beyond First Order Model Theory*. Chapman and Hall/CRC, 2017.

[11] U. Kohlenbach and A. Koutsoukou-Argyraki, Rates of convergence and metastability for abstract Cauchy problems generated by accretive operators, *J. Math. Anal. Appl.* **423**(2), 1089–1112 (2015). ISSN 0022-247X. doi: 10.1016/j.jmaa.2014.10.035.

[12] U. Kohlenbach and L. Leuştean, A quantitative mean ergodic theorem for uniformly convex Banach spaces, *Ergodic Theory Dynam.*

Systems **29**(6), 1907–1915 (2009). ISSN 0143-3857. doi: 10.1017/S0143385708001004.

[13] J. Avigad and J. Rute, Oscillation and the mean ergodic theorem for uniformly convex Banach spaces, *Ergodic Theory Dynam. Syst.* **35**(4), 1009–1027 (2015). ISSN 0143-3857. doi: 10.1017/etds.2013.90.

[14] U. Kohlenbach and P. Safarik, Fluctuations, effective learnability and metastability in analysis, *Ann. Pure Appl. Logic.* **165**(1), 266–304 (2014). ISSN 0168-0072. doi: 10.1016/j.apal.2013.07.014.

[15] H. Towsner, Nonstandard convergence gives bounds on jumps, *New York J. Math.* **25**, 651–667 (August, 2019).

[16] W. Simmons and H. Towsner, Proof mining and effective bounds in differential polynomial rings, *Adv. Math.* **343**, 567–623 (2019). ISSN 0001-8708. doi: https://doi.org/10.1016/j.aim.2018.11.026.

[17] E. Neumann, Computational problems in metric fixed point theory and their Weihrauch degrees, *Log. Methods Comput. Sci.* **11**(4), 4:20, 44 (2015). ISSN 1860-5974.

[18] J.B. Paris, A hierarchy of cuts in models of arithmetic. In L. Pacholski and J. Wierzejewski (eds.), *Model Theory of Algebra and Arithmetic (Proc. Conf., Karpacz, 1979), vol. 834, Lecture Notes in Math.*, Springer, Berlin-New York, 1980, pp. 312–337, ISBN 3-540-10269-8.

https://doi.org/10.1142/9789811245220_0012

Chapter 12

Constructive Theory of Ordinals

Thierry Coquand*,‡, Henri Lombardi†,§ and Stefan Neuwirth†,¶

*Computer Science and Engineering Department, University of Gothenburg, Sweden

†Laboratoire de mathématiques de Besançon, Université Bourgogne Franche-Comté, France

‡coquand@chalmers.se

§henri.lombardi@univ-fcomte.fr

¶stefan.neuwirth@univ-fcomte.fr

Martin-Löf [1] describes recursively constructed ordinals. He gives a constructively acceptable version of Kleene's computable ordinals. In fact, the Turing definition of computable functions is not needed from a constructive point of view. We give in this chapter a constructive theory of ordinals that is similar to Martin-Löf's theory, but based only on the two relations "$x \leqslant y$" and "$x < y$", i.e., without considering sequents whose intuitive meaning is a classical disjunction. In our setting, the operation "supremum of ordinals" plays an important rôle through its interactions with the relations "$x \leqslant y$" and "$x < y$". This allows us to approach as much as we may the notion of linear order when the property "$\alpha \leqslant \beta$ or $\beta \leqslant \alpha$" is provable only within classical logic. Our aim is to give a formal definition corresponding to intuition, and to prove that our constructive ordinals satisfy constructively all desirable properties.

1. Introduction

This chapter is written in the framework of informal constructive mathematics. We use Bishop's constructive set theory enriched with generalised inductive definitions (Bishop used this kind of constructions for measure theory, Borel sets and Lebesgue integration.)

In classical mathematics, a natural definition for an ordinal is to be an order type of a well-ordered set (see, e.g., [2, III.2.Ex.14]). Nevertheless, it is more convenient to use von Neumann ordinals, for which many results can be proved without using choice (see, e.g., [3, Chapitre 2] and [4, Chapitre II]).

Let us now propose a constructive approach. A binary relation $<$ on a set X is said to be *well-founded* if for any family of sets $(E_x)_{x\in X}$ indexed by X it is possible to construct elements of $\prod_{x\in X} E_x$ by $<$-induction. Precisely, each time a construction γ is given which from an element $a \in X$ and an element $\varphi \in \prod_{x\in X, x<a} E_x$ constructs an element $\gamma(a, \varphi) \in E_a$, there exists a unique $\Phi \in \prod_{x\in X} E_x$ such that for all $a \in X$ we have $\Phi(a) = \gamma(a, \Phi|_{x\in X, x<a})$. This notion has a clear constructive meaning.

In particular, let us consider a property for elements in X. If the property is $<$-hereditary, i.e., if it is true for $a \in X$ as soon as it is true for all $x \in X$ with $x < a$, then this property is true for all elements in X.

In constructive mathematics, Mines, Richman and Ruitenburg [5, Section I.6] spell out well-foundedness in a different but equivalent way and define an ordinal as a linearly ordered set for which the order relation is well-founded. So all subsets of $\mathbb{N}$ are ordinals even if we don't know whether they have a smallest element.

The Univalent Foundations Program [6, Section 10.3] considers "Grayson ordinals" (see [5, Exercise I.6.12]) in the framework of univalent homotopy type theory; the ordinals of a given universe turn out to form a set (and not a groupoid). This theory of ordinals differs from ours with respect to Axioms 8 and 9 for $\mathfrak{F}$-orders in the following.

Among other constructive points of view there are descriptions of countable ordinals constructed by induction by Brouwer [7], Gentzen [8], Church [9], Kleene [10], Heyting [11] and Martin-Löf [1, Chapter 3]. A constructive treatment of von Neumann ordinals based on transfinite recursion is given by Aczel and Rathjen [12, Section 9.4].

Brouwer proposes an inductive construction based on the idea that when ordinals α_n are defined for all $n \in \mathbb{N}$ and are linearly ordered well-founded sets, then we can describe the ordinal α corresponding intuitively to α_1 followed by α_2 followed by α_3 followed by.... The ordered set α defined by Brouwer will again be a linearly ordered well-founded set. And if the order relation on each α_i is decidable, the same is true for α.

Two Brouwer ordinals are in general not comparable (within intuitionistic logic): there is no general criterion allowing us to decide whether two ordinals have the same order type, and, when this is not the case, which is isomorphic to an initial segment of the other.

The paper [13] compares three distinct constructive approaches to constructive ordinals, denoted by Cnf, Brw and Ord, which are available in the framework of univalent homotopy type theory. The approach Brw is directly inspired by Brouwer ordinals.

Martin-Löf describes recursively constructed ordinals. He gives a constructively acceptable version of Kleene's computable ordinals. Intuitively, an ordinal à la Martin-Löf is inductively defined using the following two basic constructions:

- there is a minimum ordinal $\underline{0}$;
- if (α_n) is an explicit sequence of ordinals (indexed by $\mathbb{N}$ or by an $\mathbb{N}_k = \{\, n \in \mathbb{N} \mid n < k \,\}$), the supremum of the successors of the α_ns is an ordinal.[a]

To say that the definition is inductive is to say that every ordinal is constructed using the indicated rules.

In a constructive framework, we can drop Turing machines and replace Turing computability by intuitive (undefined) computability. In this case, the main difference between Brouwer and Martin-Löf ordinals is that Martin-Löf ordinals, being defined in a "parallel" way rather than in a "sequential" way, are more general: it is possible for any sequence of well-defined ordinals (α_n) to construct the supremum

[a]Martin-Löf denotes this supremum by $\sup(\alpha_n)$. In his setting, $\underline{0}$ is in fact the supremum of the empty sequence. Except for this case, his $\sup(\alpha_n)$ is the supremum of the successors of the α_ns; we shall prefer the notation $\mathrm{s}(\alpha_n)$.

of the successors of the ordinals α_n. A drawback is that there is no way to associate to a Martin-Löf ordinal a linearly ordered well-founded set with the same order type. For example, if the α_n are all equal to $\underline{0}$ or $\underline{1}$, it is a priori impossible to decide whether the supremum of the successors of the α_ns equals $\underline{1}$ or $\underline{2}$.

1.1. *Ordinals as trees*

Martin-Löf proposes to visualise an ordinal α as a well-founded tree with finite or countable branchings. The ordinal α is given with an index set denoted by In_α; in the sequel, it will be an element of the set $\mathfrak{F}_2$ of index sets consisting of $\mathbb{N}$ and its finite subsets $\mathbb{N}_k$.

- The tree with only its root represents $\underline{0}$.
- If $(t_i)_{i\in\mathrm{In}_\alpha}$ is a family of ordinal trees for a family of ordinals $(\alpha_i)_{i\in\mathrm{In}_\alpha}$, the supremum $\alpha = \mathrm{s}(\alpha_i)_{i\in\mathrm{In}_\alpha}$ of the successors of the α_is is given by the ordinal tree for which there are $\#\mathrm{In}_\alpha$ branches above the root and a copy of t_i is attached to the branch indexed by $i \in \mathrm{In}_\alpha$.

Consider the trees in Figure 1.

If $n \in \mathbb{N}$, the ordinal $\underline{n}$ can be represented by the tree with n successive unary branchings at n nodes, so that it has $n+1$ nodes.

The first infinite ordinal ω can be represented by the tree that has a countable branching above the root, the branches bearing the preceding trees (representing $\underline{n}$, $n \in \mathbb{N}$).

Its successor, denoted by $\omega+\underline{1}$, can be represented by the tree with unary branching above the root, the branch bearing the preceding tree.

The ordinal $\omega + \underline{2}$ can be represented by the tree with unary branching above the root, the branch bearing the preceding tree.

The ordinal $\omega + \omega$ can be represented by the tree that has a countable branching above the root, the branches bearing the trees representing $\omega + \underline{n}$, $n \in \mathbb{N}$.

More formally, such a tree can be defined as the set of its nodes, or branching points, suitably named. We may consider the set $\mathrm{Lst}(\mathbb{N})$ of finite lists of elements of $\mathbb{N}$. Let $n \in \mathbb{N}$ and ℓ, $\ell' \in \mathrm{Lst}(\mathbb{N})$. We denote by $n \frown \ell$ the list $[n, \ell_1, \ldots, \ell_k]$, where $\ell = [\ell_1, \ldots, \ell_k]$, by $\ell \frown n$ the list $[\ell_1, \ldots, \ell_k, n]$, and by $\ell \frown \ell'$ the concatenation of the lists ℓ and ℓ'.

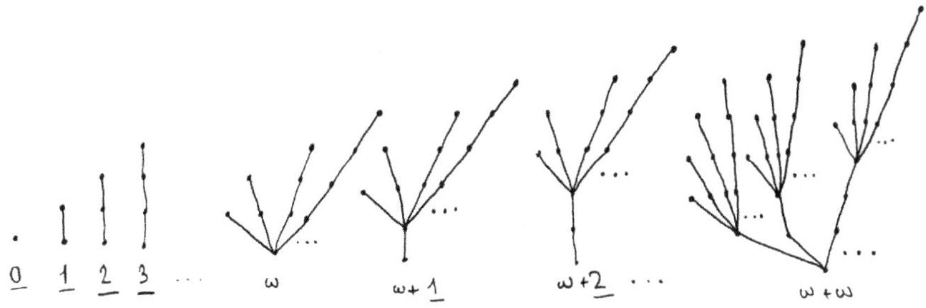

Fig. 1. Ordinal trees.

We remark that Lst(ℕ) can be enumerated in a natural way[b] and that the notion of an ℕ-indexed family in Lst(ℕ) corresponds, via such an enumeration, to the basic (undefined) notion of map from ℕ to ℕ.

A well-founded tree with finite or countable branchings can then be described as a detachable subset T of Lst(ℕ) which is inductively constructed according to the previously indicated process. T is closed by initial segments: if $\ell \in \mathrm{Lst}(\mathbb{N})$, $p \in \mathbb{N}$ and $\ell \frown p \in T$, then $\ell \in T$. Thus, to each ordinal α, we are associating a tree, defined as a suitable subset of Lst(ℕ), denoted by Tree(α).

If $n \in \mathbb{N}$, the ordinal $\underline{n}$ can be described by the finite sequence of $n+1$ lists $[\,]$, $[0]$, $[0,0]$, $\ldots$, $[0,\ldots,0]$.

The first infinite ordinal ω can be described by the subset of Lst(ℕ) enumerated by the infinite sequence $[\,]$, $[0]$, $[1]$, $[1,0]$, $[2]$, $[2,0]$, $[2,0,0]$, $[3]$, $[3,0]$, $[3,0,0]$, $[3,0,0,0]$, etc.

The ordinal $\omega+\underline{1}$ can be described by the infinite sequence $[\,]$, $[0]$, $[0,0]$, $[0,1]$, $[0,1,0]$, $[0,2]$, $[0,2,0]$, $[0,2,0,0]$, $[0,3]$, $[0,3,0]$, $[0,3,0,0]$, $[0,3,0,0,0]$, etc.

The ordinal $\omega+\underline{2}$ can be described by the infinite sequence $[\,]$, $[0]$, $[0,0]$, $[0,0,0]$, $[0,0,1]$, $[0,0,1,0]$, $[0,0,2]$, $[0,0,2,0]$, $[0,0,2,0,0]$, $[0,0,3]$, $[0,0,3,0]$, $[0,0,3,0,0]$, $[0,0,3,0,0,0]$, etc.

The ordinal $\omega+\omega$ can be described by the doubly infinite sequence $[\,]$, $[0]$, $[0,0]$, $[0,1]$, $[0,1,0]$, $[0,2]$, $[0,2,0]$, $[0,2,0,0]$, $[0,3]$, $[0,3,0]$, $[0,3,0,0]$, $[0,3,0,0,0]$, etc., $[1]$, $[1,0]$, $[1,0,0]$, $[1,0,1]$, $[1,0,1,0]$, $[1,0,2]$, $[1,0,2,0]$, $[1,0,2,0,0]$, $[1,0,3]$, $[1,0,3,0]$,

[b]For example, for $\ell = [\ell_1,\ldots,\ell_k] \in \mathrm{Lst}(\mathbb{N})$, we let $\mu(\ell) = \sum_{i=1}^{k}(\ell_i+1)$ and we enumerate the lists by increasing $\mu(\ell)$.

$[1,0,3,0,0]$, $[1,0,3,0,0,0]$, etc., $[2]$, $[2,0]$, $[2,0,0]$, $[2,0,0,0]$, $[2,0,0,1]$, $[2,0,0,1,0]$, $[2,0,0,2]$, $[2,0,0,2,0]$, $[2,0,0,2,0,0]$, $[2,0,0,3]$, $[2,0,0,3,0]$, $[2,0,0,3,0,0]$, $[2,0,0,3,0,0,0]$, etc..

* * *

We give in this chapter a constructive theory of ordinals that is similar to Martin-Löf's theory, but based only on the two relations "$x \leqslant y$" and "$x < y$", i.e., without considering sequents whose intuitive meaning is a classical disjunction.

In our setting, the operation "supremum of ordinals" plays an important rôle through its interactions with the relations "$x \leqslant y$" and "$x < y$". This allows us to approach as much as we may the notion of linear order when the property "$\alpha \leqslant \beta$ or $\beta \leqslant \alpha$" is provable only within classical logic. In the same way, the impossibility of constructively proving linear order for real numbers is circumvented by the introduction of $x < y$, $x \leqslant y$ and $\sup(x, y)$, which are all indispensable.

Our problem is to give a formal definition corresponding to intuition, and to prove that our constructive ordinals satisfy constructively all desirable properties.

* * *

The first step in Section 2 is to describe each of these desirable properties.

2. Linear Orders Associated to a Set of Index Sets

We define in this section the structure of linear orders associated to a set $\mathfrak{F}$ of index sets, $\mathfrak{F}$-orders for short.

2.1. *Index sets*

First, we need a set $\mathfrak{F}$ of index sets. An index set will be denoted by I, J, K, I', I'', J', I_a, I_b, etc.

An *index set* is simply a set that will be used as a set of indices for the families we shall consider. In the sequel, a finitely enumerated subset of A is always a subset of A defined à la Bishop by a map $\mathbb{N}_k \to A$. If A is discrete, a finitely enumerated subset of A is a detachable subset.

Properties of the set $\mathfrak{F}$ of index sets *We will assume that*

- $\mathbb{N}$ *and the finite sets* $\mathbb{N}_k = \{\, n \in \mathbb{N} \,|\, n < k \,\}$ $(k \geqslant 0)$ *are elements of* $\mathfrak{F}$;
- *any finitely enumerated subset*[c] *of an element of* $\mathfrak{F}$ *is isomorphic*[d] *to an element of* $\mathfrak{F}$;
- *if* $J \in \mathfrak{F}$, *the set of finitely enumerated subsets of* J *is isomorphic to an element of* $\mathfrak{F}$;
- $\mathfrak{F}$ *is closed by disjoint unions indexed by* $\mathfrak{F}$: *we will denote by* $I+J$ *a disjoint union of* I *and* J, *and by* $\sum_{i \in I} J_i$, *a disjoint union of the family* $(J_i)_{i \in I}$.

Disjoint unions are to be understood as direct sums in the category of sets. The disjoint union $J = \sum_{i \in I} J_i$ comes with a family $\iota_\ell \colon J_\ell \to J$ of injective maps realising J as the direct sum of the J_is in the category of sets.

If we restrict ourselves to countable ordinals, we can take for $\mathfrak{F}$ the set

$$\boxed{\mathfrak{F}_2 = \{\, \mathbb{N}_k \,|\, k \in \mathbb{N}, k \geqslant 0 \,\} \cup \{\mathbb{N}\}}$$

with convenient operations for the set of finite subsets of an $I \in \mathfrak{F}$ and for disjoint unions of elements of $\mathfrak{F}$ indexed by an element of $\mathfrak{F}$. Any other set $\mathfrak{F}$ of index sets will contain $\mathfrak{F}_2$.

An $\mathfrak{F}$-*indexed family of elements of* E is a family $(x_i)_{i \in I}$ where $I \in \mathfrak{F}$ and the x_is $\in E$. The set of $\mathfrak{F}$-indexed families of elements of E is denoted by $\mathrm{Fam}(\mathfrak{F}, E)$.

We shall restrict the use of subscripts for ordinal variables to this meaning, and use superscripts for all other uses.

2.2. *Axioms*

A structure of $\mathfrak{F}$-*order* on a set $(E, =)$ is given as $(E, <, \leqslant, 0_E, \sup, \mathrm{s})$, where

- $<$ and $\leqslant$ are binary relations defined on $(E, =)$;
- 0_E is an element of E and we let $E^* = \{\, \alpha \in E \,|\, 0_E < \alpha \,\}$;

---

[c]By definition, this is a subobject given by a function $\mathbb{N}_k \to \mathfrak{F}$.
[d]In the category of sets.

- sup is a map from $\mathrm{Fam}(\mathfrak{F}, E^*)$ to E^*: taking as input an element $(\alpha_i)_{i\in I}$ of $\mathrm{Fam}(\mathfrak{F}, E^*)$, it constructs an element of E^* denoted by $\alpha = \sup(\alpha_i)_{i\in I}$;
- s is a unary map from E to E^*: taking as input an element $\beta \in E$, it constructs an element of E^* denoted by $\mathrm{s}(\beta)$.

Definition 2.1. In order to write axioms with finite sup's, we define $\sup(\alpha, \beta)$ for $\alpha, \beta \in E$ in the following way (using implicitly Axiom 15): $\sup(0_E, \alpha) = \alpha = \sup(\alpha, 0_E)$; if $\alpha, \beta \in E^*$, $\sup(\alpha, \beta)$ is already defined.

These data are to satisfy the following axioms.

Axioms for $\mathfrak{F}$-orders

(1) $\alpha = \beta$ *if and only if* $\alpha \leqslant \beta$ *and* $\beta \leqslant \alpha$ *(reflexivity and anti-symmetry);*
(2) $0_E \leqslant \alpha$*;*
(3) *if* $\alpha < \alpha$*, then* $0_E = \beta$ *(irreflexivity);*
(4) *if* $\alpha < \beta$*, then* $\alpha \leqslant \beta$*;*
(5) *if* $\alpha \leqslant \beta$ *and* $\beta \leqslant \gamma$*, then* $\alpha \leqslant \gamma$ *(transitivity 1);*
(6) *if* $\alpha < \beta$ *and* $\beta \leqslant \gamma$*, then* $\alpha < \gamma$ *(transitivity 2);*
(7) *if* $\alpha \leqslant \beta$ *and* $\beta < \gamma$*, then* $\alpha < \gamma$ *(transitivity 3);*
(8) $\alpha < \mathrm{s}(\beta)$ *if and only if* $\alpha \leqslant \beta$ *(using Axiom 1 this gives* $\alpha < \mathrm{s}(\alpha)$*);*
(9) $\mathrm{s}(\beta) \leqslant \alpha$ *if and only if* $\beta < \alpha$*;*
(10) *if* $\alpha < \gamma$ *and* $\beta < \gamma$*, then* $\sup(\alpha, \beta) < \gamma$*;*
(11) *if* $\alpha < \sup(\alpha, \beta)$*, then* $\alpha < \beta$*;*
(12) *if* $\gamma < \alpha$ *and* $\alpha \leqslant \sup(\beta, \gamma)$*, then* $\alpha \leqslant \beta$*;*
(13) *for* $(\alpha_i)_{i\in I} \in \mathrm{Fam}(\mathfrak{F}, E^*)$ *and* $\beta \in E$*, we have*

$$\alpha_i \leqslant \beta \text{ for all } i \in I \text{ if and only if } \sup(\alpha_i)_{i\in I} \leqslant \beta,$$

(characteristic property of sup*);*
(14) *if* $\gamma < \beta$ *for all* $\gamma < \alpha$*, then* $\alpha \leqslant \beta$*;*
(15) *either* $\alpha \leqslant 0_E$ *or* $0_E < \alpha$*.*

The *category of $\mathfrak{F}$-orders* is defined by its morphisms

$$(E, <_E, \leqslant_E, 0_E, \sup_E, \mathrm{s}_E) \longrightarrow (F, <_F, \leqslant_F, 0_F, \sup_F, \mathrm{s}_F),$$

which are maps from E to F preserving the structure (in the usual meaning).

Comments.

(1) Let $\gamma \in E^*$ and $(\alpha_n)_{n\in\mathbb{N}}$ such that $\alpha_n = \gamma$ or $\alpha_n = \mathrm{s}(\gamma)$ for each n. The element $\sup(\alpha_n)_{n\in\mathbb{N}}$ hesitates between γ and $\mathrm{s}(\gamma)$. Thus, there is no hope that the disjunction "$\alpha \leqslant \beta$ or $\beta < \alpha$" can be constructive for arbitrary elements $\alpha, \beta \neq 0_E$. Consequently, we have introduced the sup map together with its axioms in order to best describe in what sense the order can be thought of as linear. Perhaps this is not optimal (reasonable axioms, satisfied for the set $\mathbf{Ord}_2$ of ordinals of the second class constructed in Section 3, might be missing).

(2) The irreflexivity is given a form that, instead of stating a negation, allows E to reduce to a singleton. This happens if and only if $0_E = \mathrm{s}(0_E)$, which implies $0_E < 0_E$ using Axiom 8.

(3) Axiom 15 expresses that $\{0_E\}$ is detachable. This contrasts with the fact that elements other than 0_E do not define detachable singletons. We have defined sup on E^* rather than on E in order to satisfy constructively the disjunction of Axiom 15.

(4) The characteristic property of sup shows that this law satisfies idempotence as well as generalised associativity and commutativity. ◇

2.3. *Some properties*

Proposition and Definition 2.2 (Generalising Definition 2.1). For $\alpha^1, \ldots, \alpha^r \in E$, we let

$$\sup(\alpha^1, \ldots, \alpha^r) \stackrel{\text{def}}{=} \begin{cases} 0_E & \text{if } \alpha^1 = \cdots = \alpha^r = 0_E \\ \text{the sup of the } \alpha^k \neq 0_E & \text{otherwise.} \end{cases}$$

The characteristic property of sup is satisfied:

$$\alpha^1 \leqslant \beta \text{ and } \ldots \text{ and } \alpha^r \leqslant \beta \text{ if and only if } \sup(\alpha^1, \ldots, \alpha^r) \leqslant \beta.$$

Fact 2.3. Let α, β be elements of E.

- $\mathrm{s}(\alpha) < \mathrm{s}(\beta)$ if and only if $\alpha < \beta$.
- $\mathrm{s}(\alpha) \leqslant \mathrm{s}(\beta)$ if and only if $\alpha \leqslant \beta$.

Proof. Use Axioms 8 and 9. □

Fact 2.4. Axioms 10 to 12 and 14 are in fact equivalences:

10. $\alpha < \gamma$ and $\beta < \gamma$ hold simultaneously if and only if $\sup(\alpha, \beta) < \gamma$;
11. $\alpha < \sup(\alpha, \beta)$ if and only if $\alpha < \beta$;
12. if $\gamma < \alpha$, then $\alpha \leqslant \sup(\beta, \gamma)$ holds if and only if $\alpha \leqslant \beta$;
14. $\alpha \leqslant \beta$ if and only if $\gamma < \beta$ for all $\gamma < \alpha$.

Proof. Use the transitivities and the characteristic property of sup. $\square$

Fact 2.5. (s commutes with finite sup's, notation as in Proposition and Definition 2.2). We have $\sup(\mathrm{s}(\alpha), \mathrm{s}(\beta)) = \mathrm{s}(\sup(\alpha, \beta))$ and more generally $\sup(\mathrm{s}(\alpha^1), \ldots, \mathrm{s}(\alpha^r)) = \mathrm{s}(\sup(\alpha^1, \ldots, \alpha^r))$.

In particular, if $\alpha^1 < \gamma$, …, $\alpha^r < \gamma$, then $\sup(\alpha^1, \ldots, \alpha^r) < \gamma$.

Proof. It suffices to prove $\mathrm{s}(\sup(\alpha, \beta)) = \sup(\mathrm{s}(\alpha), \mathrm{s}(\beta))$. We have the following chain of equivalences: $\mathrm{s}(\sup(\alpha, \beta)) \leqslant \gamma \iff \sup(\alpha, \beta) < \gamma \iff (\alpha < \gamma \text{ and } \beta < \gamma) \iff (\mathrm{s}(\alpha) \leqslant \gamma \text{ and } \mathrm{s}(\beta) \leqslant \gamma) \iff \sup(\mathrm{s}(\alpha), \mathrm{s}(\beta)) \leqslant \gamma$. $\square$

Proposition and Definition 2.6. (definition of infinitary s and its characteristic property). For any $(\alpha_i)_{i \in J} \in \mathrm{Fam}(\mathfrak{F}, E)$, we define $\mathrm{s}(\alpha_i)_{i \in J} = \sup(\mathrm{s}(\alpha_i))_{i \in J}$. Then we get the following equivalence:

$$\alpha_i < \beta \quad \text{for all } i \in J \ \text{ if and only if } \ \mathrm{s}(\alpha_i)_{i \in J} \leqslant \beta.$$

Proof. Use Axioms 13 and 9. $\square$

We write $\boxed{F \subseteq_f I}$ in order to express that F is a finitely enumerated subset of I.

Fact 2.7. Let $\alpha, \beta^1, \ldots, \beta^m \in E$.

(1) Assume that $\alpha = \mathrm{s}(\alpha_i)_{i \in J}$ with $(\alpha_i)_{i \in J} \in \mathrm{Fam}(\mathfrak{F}, E)$ and that $\alpha_i < \sup(\beta^1, \ldots, \beta^m)$ for all $i \in J$. Then $\alpha \leqslant \sup(\beta^1, \ldots, \beta^m)$.

(2) Assume that $\beta^k = \mathrm{s}((\beta^k)_i)_{i \in J_k}$ with $((\beta^k)_i)_{i \in J_k} \in \mathrm{Fam}(\mathfrak{F}, E)$ for $k \in [\![1..m]\!]$.
Let $F_1 \subseteq_f J_1, \ldots, F_m \subseteq_f J_m$ not all being empty. If

$$\alpha \leqslant \sup((\beta^k)_j)_{k \in [\![1..m]\!],\, j \in F_k},$$

then $\alpha < \sup(\beta^1, \ldots, \beta^m)$.

Proof. 1. This is Proposition and Definition 2.6.

2. Suppose, e.g., that F_1 is nonempty. Then $\alpha \leqslant \sup((\beta^1)_j)_{j \in F_1} < \beta^1 \leqslant \sup(\beta^1, \ldots, \beta^m)$. The strict inequality comes from Fact 2.5 because all $(\beta^1)_j$s are $< \beta^1$ by Proposition and Definition 2.6. □

3. Inductive Construction of Ordinals

In Sections 3 and 4, the set $\mathfrak{F}$ of index sets is fixed but often implied.

We shall define a set of ordinals **Ord** (more precisely $\mathbf{Ord}_{\mathfrak{F}}$) and we shall prove that it is an initial object in the category of $\mathfrak{F}$-orders.

First, we define a set **ord** of names for $\mathfrak{F}$-indexed ordinals by an inductive definition. The simplest inductive definition of an infinite set is that of $\mathbb{N}$: it admits an element 0 and a successor map $x \mapsto s(x)\colon \mathbb{N} \to \mathbb{N}$. The inductive definition of **ord** is very similar to that of $\mathbb{N}$. In $\mathbb{N}$, each element is either 0 or an $s(x)$ for an $x \in \mathbb{N}$. Similarly, in **ord**, each element is either $\underline{0}$ or the s of an $\mathfrak{F}$-indexed family in **ord**; we denote by **ord*** the set of elements of this second type.

Definition 3.1. The set **ord** (more precisely $\mathbf{ord}_{\mathfrak{F}}$) is defined in an inductive way: it is to admit a distinguished element $\underline{0}$ and a map

$$\mathrm{s}\colon \mathrm{Fam}(\mathfrak{F}, \mathbf{ord}) \to \mathbf{ord}.$$

N.b.: The only constraint in this inductive definition is that s be indeed a map from $\mathrm{Fam}(\mathfrak{F}, \mathbf{ord})$ to **ord**.

An element of **ord** will be called [*name of an*] *ordinal* in the sequel.

When $\mathfrak{F} = \mathfrak{F}_2$, we get the set of names of countable ordinals, denoted by $\mathbf{ord}_2$.

Remark 3.2. Each element $\alpha \in \mathbf{ord}^*$ is given with two data:

- the index set used in the definition of α: it will be denoted by In_α;
- the family $\chi_{\mathbf{ord}}(\alpha, i)_{i \in \mathrm{In}_\alpha}$ of its *definitional subordinals*, i.e. the element of $\mathrm{Fam}(\mathfrak{F}, \mathbf{ord})$ such that $\alpha = \mathrm{s}(\chi_{\mathbf{ord}}(\alpha, i))_{i \in \mathrm{In}_\alpha}$.

Thus, inductive definition of **ord** implies the existence of a map $\alpha \mapsto \mathrm{In}_\alpha\colon \mathbf{ord}^* \to \mathfrak{F}$ and the existence of a dependent family $(\alpha, i) \mapsto$

$\chi_{\mathbf{ord}}(\alpha, i)$ which is defined for $\alpha \in \mathbf{ord}^*$ and $i \in \mathrm{In}_\alpha$. In order to make the text more readable, we will perform a slight abuse of notation: we shall not mention the construction of the dependent family $\chi_{\mathbf{ord}}$, and the notation α_i will be an abbreviation for $\chi_{\mathbf{ord}}(\alpha, i)$. With these conventions we may write $\boxed{\alpha = \mathrm{s}(\alpha_i)_{i \in \mathrm{In}_\alpha}}$. ◇

For $\alpha^1, \ldots, \alpha^r \in \mathbf{ord}$, we define $\boxed{\mathrm{s}(\alpha^1, \ldots, \alpha^r) = \mathrm{s}(\alpha^i)_{i \in [\![1..r]\!]}}$.

In particular, if $\alpha \in \mathbf{ord}$, its *immediate successor* $\mathrm{s}(\alpha)$ is the element $\beta = \mathrm{s}(\beta_i)_{i \in \mathrm{In}_\beta}$ where $\mathrm{In}_\beta = \mathbb{N}_1 = \{0\}$ and $\beta_0 = \alpha$. The sequence $(\underline{m})_{m \in \mathbb{N}}$ in **ord** is defined inductively by $\underline{m+1} = \mathrm{s}(\underline{m})$. Then we can define $\boxed{\omega = \mathrm{s}(\underline{n})_{n \in \mathbb{N}}}$.

In order to prove a property of $\alpha = \mathrm{s}(\alpha_i)_{i \in \mathrm{In}_\alpha}$, it is sufficient to prove the property for each α_i. In a similar way, we can construct inductively a map whose domain is **ord**, or define inductively a predicate on **ord**. This is stated precisely in Fact 3.4.

3.1. *Subordinals*

Here is a correct inductive definition.

Definition 3.3. Let $\alpha = \mathrm{s}(\alpha_i)_{i \in \mathrm{In}_\alpha} \in \mathbf{ord}^*$. An element β of **ord** is a *definitional subordinal* of α if $\beta = \alpha_i$ for an $i \in \mathrm{In}_\alpha$: we write this $\beta \lessdot_1 \alpha$. An element γ is a *subordinal of* α if it is a definitional subordinal of α or a subordinal of a definitional subordinal of α. We write this $\gamma \lessdot \alpha$.

Thus, $\underline{0}$ is the only element of **ord** which has no subordinal.

Fact 3.4. The relations $\lessdot_1$ and $\lessdot$ on **ord** are well-founded.

Consequently, there is no infinite branch in the tree of subordinals of an element of **ord**, in the following sense.

Fact 3.5. A sequence $(\alpha^j)_{j=1,2,\ldots}$ in **ord**, where each α^{j+1} is a subordinal of α^j, reaches in a finite number of steps $\alpha^r = \underline{0}$.

Remark that in order to perform a construction (or a proof) by $\lessdot_1$-induction or by $\lessdot$-induction, the case $\underline{0}$ has to be dealt with separately since it has no subordinal. Nevertheless, we shall be able to avoid this case distinction until considering ordinal arithmetic on page 312.

3.2. *Definition of the* sup *law*

Definition 3.6.

(1) The law sup: $\mathrm{Fam}(\mathfrak{F}, \mathbf{ord}^*) \to \mathbf{ord}^*$ is defined in the following way. Let $(\alpha^j)_{j\in J}$ be a family in $\mathbf{ord}^*$ with $J \in \mathfrak{F}$. If $\alpha^j = \mathrm{s}((\alpha^j)_i)_{i\in I_j}$, then $\boxed{\sup(\alpha^j)_{j\in J}}$ is the element $\varepsilon = \mathrm{s}(\varepsilon_k)_{k\in K}$, where

- K is the disjoint union of the I_js;
- $(\varepsilon_k)_{k\in K}$ is the family defined by $\varepsilon_k = (\alpha^j)_i$ if $\iota_j(i) = k$,

(here $\iota_j : I_j \to K$ is the injective map from I_j to the disjoint union of the I_js). We shall write $\sup(\alpha^j)_{j\in[\![1..r]\!]} = \sup(\alpha^1, \ldots, \alpha^r)$.

(2) The sup of a finite family in **ord** is defined in the following way.

$$\sup(\alpha^1, \ldots, \alpha^r) \overset{\text{def}}{=} \begin{cases} \underline{0} & \text{if } \alpha^1 = \cdots = \alpha^r = \underline{0} \\ \text{the sup of the } \alpha^k \in \mathbf{ord}^* & \text{otherwise.} \end{cases}$$

We note that Item 2 is formally included in Item 1 if we adopt the convention $\mathrm{In}_{\underline{0}} = \mathbb{N}_0$. However, this convention would not allow us to define an arbitrary $\mathfrak{F}$-indexed sup in **ord**.

3.3. *Definition of* $\leqslant$ *and of* $<$

The main job remains to be done, i.e., to define two binary relations $\leqslant$ and $<$ on **ord** with the required properties, viz.

- the relation "$\alpha \leqslant \beta$ and $\beta \leqslant \alpha$" has to be an equivalence relation (we shall denote by **Ord** the quotient set);
- the relations $\leqslant$ and $<$ and the maps sup and s have to descend to the quotient (we shall not change their names), i.e., they have to be compatible with the equivalence relation;
- with these maps and relations, **Ord** has to be an $\mathfrak{F}$-order.

Moreover, since the map s: $\mathrm{Fam}(\mathfrak{F}, \mathbf{ord}) \to \mathbf{ord}^*$ is defined before the map sup: $\mathrm{Fam}(\mathfrak{F}, \mathbf{ord}^*) \to \mathbf{ord}^*$, we have to verify in our construction that Proposition and Definition 2.6 is satisfied in **Ord**. This will be a consequence of Fact 3.11 in the following.

For our job, we define inductively two asymmetric relations between, on the left side, an element of **ord** and, on the right side, *a nonempty finitely enumerated set* of elements of **ord**, written as a list:

$$\boxed{\alpha \leqslant \beta^1, \ldots, \beta^m \quad \text{and} \quad \alpha < \beta^1, \ldots, \beta^m \quad (m \geqslant 1).}$$

Conventions

- *The letters α, β, γ, ε, possibly with exponents, indices or primes, are used for elements of* **ord**.
- *If α is an element of* **ord** *and if F is a finite list, possibly empty, in In_α, we denote by α_F the list of the α_is with i in F.*

The two relations are defined by simultaneous induction in the following way.

Particular cases involving $\underline{0}$ are avoided by using the convention $\mathrm{In}_{\underline{0}} = \mathbb{N}_0$. Let m be an integer $\geqslant 1$.

$\alpha \leqslant \beta^1, \ldots, \beta^m$ is defined as $\alpha_i < \beta^1, \ldots, \beta^m$ for all $i \in \mathrm{In}_\alpha$.
$\alpha < \beta^1, \ldots, \beta^m$ is defined as there are $F_1 \subseteq_f \mathrm{In}_{\beta^1}, \ldots, F_m \subseteq_f \mathrm{In}_{\beta^m}$ not all empty with $\alpha \leqslant \beta^1_{F_1}, \ldots, \beta^m_{F_m}$.

This definition is correct since elements of **ord** are inductively defined and the pair of clauses is inductive.

Without the convention that $\mathrm{In}_{\underline{0}} = \mathbb{N}_0$, we would have had to include Fact 3.8 in what follows in the definition. This convention is a little miracle allowing us to avoid a case-by-case reasoning with respect to the disjunction "$\alpha = \underline{0}$ or $\alpha \in \mathbf{ord}^*$" in the proofs.

The meaning of the two relations is $\alpha \leqslant \sup(\beta^1, \ldots, \beta^m)$ and $\alpha < \sup(\beta^1, \ldots, \beta^m)$.

Lemma 3.7. *We have $\alpha < \beta^1, \ldots, \beta^m$ if and only if $\alpha <$ sup $(\beta^1, \ldots, \beta^m)$.*

Similarly, we have $\alpha \leqslant \beta^1, \ldots, \beta^m$ if and only if $\alpha \leqslant$ sup $(\beta^1, \ldots, \beta^m)$.

Proof. Let us write

$$\alpha \prec \beta^1, \ldots, \beta^m \quad \text{for } \alpha < \sup(\beta^1, \ldots, \beta^m),$$
$$\alpha \preccurlyeq \beta^1, \ldots, \beta^m \quad \text{for } \alpha \leqslant \sup(\beta^1, \ldots, \beta^m).$$

Let $\varepsilon = \sup(\beta^1, \ldots, \beta^m)$. Then $\alpha \prec \beta^1, \ldots, \beta^m$ if and only if $\alpha \leqslant \varepsilon_F$ with F a nonempty finitely enumerated subset of the disjoint union K of the In_{β^j}s and $\varepsilon_k = (\beta^j)_i$ if k is the image of i in K; letting $F_j = F \cap \mathrm{In}_{\beta^j}$, not all F_js are empty and this may be rewritten as $\alpha \leqslant \beta^1_{F_1}, \ldots, \beta^m_{F_m}$. This holds if and only if $\alpha < \beta^1, \ldots, \beta^m$.

We have $\alpha \preccurlyeq \beta^1, \ldots, \beta^m$ if and only if, for all $i \in \mathrm{In}_\alpha$, $\alpha_i < \varepsilon$, i.e., $\alpha_i \prec \beta^1, \ldots, \beta^m$, i.e., $\alpha_i < \beta^1, \ldots, \beta^m$; this holds if and only if $\alpha \leqslant \beta^1, \ldots, \beta^m$. □

The relation $\alpha =_{\mathbf{Ord}} \beta$ is defined as meaning "$\alpha \leqslant \beta$ and $\beta \leqslant \alpha$".

We shall show in Section 4 that the relation $\cdot =_{\mathbf{Ord}} \cdot$ is an equivalence relation and we shall define the set **Ord** as the quotient of **ord** by this relation.

Let us note that until Theorem 4.8, the symbol $=$ between two elements of **ord** is the equality in **ord** and has not the meaning of $=_{\mathbf{Ord}}$. Nevertheless, after having shown that the relations and the laws of **ord** descend to the quotient **Ord**, the statements with the symbol $=$ will also work for the symbol $=_{\mathbf{Ord}}$.

3.4. *Finite ordinals, bounded ordinals*

We start with a few properties of $\underline{0}$.

Fact 3.8. Let m be an integer $\geqslant 1$, $\alpha, \beta^1, \ldots, \beta^m \in \mathbf{ord}$, and $\gamma \in \mathbf{ord}^*$. We have

(1) $\underline{0} \leqslant \beta^1, \ldots, \beta^m$;
(2) $\underline{0} < \gamma, \beta^2, \ldots, \beta^m$;
(3) $\alpha < \underbrace{\underline{0}, \ldots, \underline{0}}_{m \text{ times}}$ is impossible.

Proof. This is straightforward from the definitions. □

Remark 3.9. Axiom 15 will be valid in **Ord** because every element of **ord** is given either as $\underline{0}$ or as an element $\gamma \in \mathbf{ord}^*$, so that always $\underline{0} < \gamma$ by Item 2 of Fact 3.8. ◇

Fact 3.10. Let $m, n \in \mathbb{N}$. Then

(1) $m \leqslant n$ if and only if $\underline{m} \leqslant \underline{n}$;
(2) $m < n$ if and only if $\underline{m} < \underline{n}$;
(3) $\underline{m} \leqslant \underline{n}$ and $\underline{n} < \underline{m}$ are incompatible.

Proof. Concerning the direct implications in (1) and (2), we write $n = m + r$ and we do an induction on r. For the reverse implications, cases $m = 0$ and $n = 0$ are already known. Next, we see that $\underline{m+1} \leqslant \underline{n+1}$ implies $\underline{m} \leqslant \underline{n}$, and that $\underline{m+1} < \underline{n+1}$ implies $\underline{m} < \underline{n}$. This allows us to conclude by induction on m.
Item 3 follows from Items 1 and 2. □

An element $\alpha \in \mathbf{ord}$ is said to be *finite* if $\alpha =_{\mathbf{Ord}} \underline{m}$ for an $m \in \mathbb{N}$, *bounded* if $\alpha \leqslant \underline{m}$ for an $m \in \mathbb{N}$. Bounded ordinals are much more complicated than finite ordinals (see Examples 3.17 and 3.18).

In Section 3.7, we shall discuss what the relations $\leqslant$ and $<$ on the set $\mathbf{ord}_{\mathfrak{F}}$ become in classical mathematics.

3.5. *First consequences*

The following fact shows that the s law will satisfy the characteristic property given in Proposition and Definition 2.6 when we shall know that it descends to the quotient **Ord**.

Fact 3.11 (`sdef`). We have $\alpha \leqslant \beta$ if and only if $\alpha_i < \beta$ for all $i \in \mathrm{In}_\alpha$.

Proof. This property is tautological: this is the definition of $\alpha \leqslant \beta$. □

Similarly, the following fact shows that the sup law will satisfy the characteristic property given in Axiom 13 when we shall know that it descends to the quotient **Ord**.

Fact 3.12 (`supdef`). Let $(\alpha^j)_{j \in J}$ be a family in $\mathbf{ord}^*$ with $J \in \mathfrak{F}$, $\gamma = \sup(\alpha^j)_{j \in J}$, and $\beta \in \mathbf{ord}$.

We have $\gamma \leqslant \beta$ if and only if $\alpha^j \leqslant \beta$ for all $j \in J$. In particular, $\sup(\alpha, \beta) \leqslant \beta$ if and only if $\alpha \leqslant \beta$.

N.b.: The result is equally true for the sup of a finite family in **ord**.

Proof. This is another linguistic tautology. We have $\alpha^j = \mathrm{s}((\alpha^j)_i)_{i \in I_j}$ for an $I_j \in \mathfrak{F}$. By the definition of γ and of $\leqslant$, the inequality $\gamma \leqslant \beta$ means that for each $j \in J$ and each $i \in I_j$ we have $(\alpha^j)_i < \beta$, i.e., that for each $j \in J$ we have $\alpha^j \leqslant \beta$. □

The following fact shows that Axioms 8 and 9 will be valid when we shall descend to the quotient **Ord**.

Fact 3.13.

(1) `ax8`. We have $\alpha < \mathrm{s}(\beta)$ if and only if $\alpha \leqslant \beta$.
(2) `ax9`. We have $\beta < \alpha$ if and only if $\mathrm{s}(\beta) \leqslant \alpha$.

Proof. Recall that the element $\gamma = \mathrm{s}(\beta)$ is defined by $\mathrm{In}_\gamma = \{0\}$ and $\gamma_0 = \beta$.

(1) By definition, $\alpha < \gamma$ means that $\alpha \leqslant \gamma_F$ for a nonempty list $F \subseteq_f \{0\}$. This forces $F = [0]$ and $\gamma_F = \beta$.
(2) By definition, $\gamma \leqslant \alpha$ means that $\gamma_0 < \alpha$, i.e., $\beta < \alpha$.

Thus, better than equivalences, these are tautologies. □

The following fact will allow us to shorten certain proofs by induction.

Fact 3.14.

(a) We have an inequality $\alpha \leqslant \beta$ if and only if for each $i \in \mathrm{In}_\alpha$ there exists a nonempty $F_i \subseteq_f \mathrm{In}_\beta$ such that $\alpha_i \leqslant \beta_{F_i}$.
(b) We have an inequality $\alpha < \beta$ if and only if there exists a nonempty $F \subseteq_f \mathrm{In}_\beta$ such that for each $i \in \mathrm{In}_\alpha$ we have $\alpha_i < \beta_F$.

Proof. Straightforward from the definitions. □

Now, we leave behind tautological proofs and turn to inductive proofs.

Fact 3.15.

- `weakening`. If $\alpha \leqslant \beta^1, \ldots, \beta^m$, then for each β we have $\alpha \leqslant \beta, \beta^1, \ldots, \beta^m$.
- `contraction`. If $\alpha \leqslant \beta^1, \beta^1, \beta^2, \ldots, \beta^m$, then $\alpha \leqslant \beta^1, \beta^2, \ldots, \beta^m$.
- The same properties hold with $<$ instead of $\leqslant$.

Proof. Use induction applying the definitions. □

The following lemma is a corollary of Fact 3.14. Item 1 (resp. 2) will imply that the s (resp. sup) map descends to the quotient in **Ord** (resp. **Ord***). Item 3 will imply that the relations $\leqslant$ and $=$ are reflexive in **Ord**; Items 5 and 7 will imply Axioms 3 and 14 for **Ord**.

Lemma 3.16.

(1) `s0`. *Let* $\alpha, \beta \in$ **ord** *with* $\mathrm{In}_\alpha = \mathrm{In}_\beta$ *and* $\alpha_i \leqslant \beta_i$ *for all* $i \in \mathrm{In}_\alpha$. *Then* $\alpha \leqslant \beta$.
(2) `sup0`. *Let* $\alpha, \beta \in$ **ord*** *with* $\mathrm{In}_\alpha = \mathrm{In}_\beta$ *and* $\alpha_i \leqslant \beta_i$ *for all* $i \in \mathrm{In}_\alpha$. *Then*

$$\sup(\alpha_i)_{i\in\mathrm{In}_\alpha} \leqslant \sup(\beta_i)_{i\in\mathrm{In}_\beta}.$$

The result works also for the sup *of a finite family in* **ord**.
(3) `rfl`. *For all* $\alpha \in$ **ord**, *we have* $\alpha \leqslant \alpha$. *A fortiori,* $\alpha \leqslant \alpha, \beta^1, \ldots, \beta^m$.
(4) `s1`. *For all* $\alpha \in$ **ord*** *and all* $i \in \mathrm{In}_\alpha$, *we have* $\alpha_i < \alpha$. *A fortiori,* $\alpha_i < \alpha, \beta^1, \ldots, \beta^m$.
(5) `irfl`. *For all* $\alpha \in$ **ord**, $\alpha < \alpha$ *is impossible.*
(6) $\alpha < \mathrm{s}(\alpha)$.
(7) `ax14`. *If* $\gamma < \beta$ *for all* $\gamma < \alpha$, *then* $\alpha \leqslant \beta$.

Proof.

(1) Straightforward from Fact 3.14a. We take $F = \{i\}$.
(2) Let $\gamma = \sup(\alpha_i)_{i\in\mathrm{In}_\alpha}$ and $\epsilon = \sup(\beta_i)_{i\in\mathrm{In}_\beta}$. By Fact 3.14a, for each $j \in \mathrm{In}_{\alpha_i}$ there exists a nonempty $F_{i,j} \subseteq_f \mathrm{In}_{\beta_i}$ with $(\alpha_i)_j \leqslant (\beta_i)_{F_{i,j}}$; $F_{i,j}$ is a fortiori in the disjoint union of the In_{β_i}s, so that $(\alpha_i)_j < \epsilon$ by definition of ϵ. By definition, $\alpha_i \leqslant \epsilon$, so that by Fact 3.12 $\gamma \leqslant \epsilon$.
(3) By induction: we use Fact 3.14a, we take $F = \{i\}$ and $\alpha \leqslant \alpha$ reduces to $\alpha_i \leqslant \alpha_i$.
(4) By induction: we use Fact 3.14b, we take $F = \{i\}$ and $\alpha_i < \alpha$ reduces to $(\alpha_i)_j < \alpha_i$.
(5) By induction: we use Fact 3.14b, we take $F = \{i\}$ and "$\alpha < \alpha$ is impossible" reduces to: "$\alpha_i < \alpha_i$ is impossible".
(6) Apply `s1` to $\beta = \mathrm{s}(\alpha)$.
(7) If $\alpha = \underline{0}$, the conclusion is clear. If $\alpha = \mathrm{s}(\alpha_i)_{i\in I}$, as $\alpha_i < \alpha$ for each $i \in \mathrm{In}_\alpha$ (Item 4), the hypothesis that $\gamma < \beta$ for all $\gamma < \alpha$ shows that $\alpha_i < \beta$ for all $i \in \mathrm{In}_\alpha$. We conclude by Fact 3.11 that $\alpha \leqslant \beta$.

□

3.6. *Ordinals and limited principles of omniscience*

Example 3.17. Let $(v_n)_{n\in\mathbb{N}}$ be a sequence in $\{0,1\}$ which takes at most once the value 1. The lesser limited principle of omniscience **LLPO** says that we have

$$\exists k \in \{0,1\}\ \forall n\ \ (v_n = 1 \Rightarrow\ n \equiv k \mod 2). \tag{$*$}$$

From such a sequence $(v_n)_{n\in\mathbb{N}}$, let us define ε, ε^1 and $\varepsilon^2 \in$ **ord** in the following way:

$$\varepsilon = \mathrm{s}(\underline{v_n})_{n\in\mathbb{N}},\quad \varepsilon^1 = \mathrm{s}(\underline{v_{2m}})_{m\in\mathbb{N}},\quad \varepsilon^2 = \mathrm{s}(\underline{v_{2m+1}})_{m\in\mathbb{N}}.$$

Then we have $\varepsilon \leqslant \sup(\varepsilon^1, \varepsilon^2)$. But $\varepsilon \leqslant \varepsilon^1$ gives $k = 0$ in $(*)$ and $\varepsilon \leqslant \varepsilon^2$ gives $k = 1$ in $(*)$. Thus, the disjunction $\varepsilon \leqslant \varepsilon^1$ or $\varepsilon \leqslant \varepsilon^2$ has no constructive proof: assuming the disjunction for an arbitrary (v_n) would imply **LLPO**.

Example 3.18. Let $(u_n)_{n\in\mathbb{N}}$ be a nondecreasing sequence in $\{0,1\}$. The limited principle of omniscience **LPO** says that such a sequence is eventually constant:

$$\exists n \in \mathbb{N}\ \forall m \in \mathbb{N},\quad u_m \leqslant u_n. \tag{$*$}$$

From such a sequence $(u_n)_{n\in\mathbb{N}}$, let us define α and $\beta \in$ **ord** in the following way:

$$\alpha = \mathrm{s}(\underline{u_n})_{n\in\mathbb{N}},\quad \beta = \mathrm{s}(\underline{u_n + 1})_{n\in\mathbb{N}}.$$

We note that the strict inequality $\alpha < \beta$ is equivalent (using Facts 3.10 and 3.14, and Lemma 3.7) to

$$\exists n \in \mathbb{N}\ \forall m \in \mathbb{N},\quad u_m < u_n + 1,$$

which amounts to $(*)$. In fact, α hesitates between 1 and 2, β hesitates between 2 and 3, and the inequality $\alpha < \beta$ is valid if we assume **LPO**. But asserting $\alpha < \beta$ for all sequences (u_n) implies **LPO** in constructive mathematics. Here, we see that hesitating between 1 and 2 for an infinite sequence has the same flavour as hesitating (in a classical setting) between bounded and unbounded for an infinite sequence of natural numbers: adding 1 to each term of the sequence increases strictly the sup only if the sequence is bounded.

3.7. *In classical mathematics*

Proposition 3.19 shows that the law of excluded middle (**LEM**) simplifies and/or obscures dramatically the structure of the set $\mathbf{ord}_{\mathfrak{F}}$ with respect to the relations $<$ and $\leqslant$.

Proposition 3.19. *Assume **LEM**. Then for* $\alpha, \beta \in \mathbf{ord}$, *we have* $\alpha \leqslant \beta$ *or* $\beta < \alpha$. *Moreover, if* $\beta < \alpha$, *there exists an* $i \in \mathrm{In}_\alpha$ *such that* $\beta \leqslant \alpha_i$.

Proof. We prove by simultaneous induction the two following properties.

$$\text{“}\alpha \leqslant \beta \text{ or } \beta < \alpha\text{”} \quad \text{and} \quad \text{“}\beta \leqslant \alpha \text{ or } \alpha < \beta\text{”}.$$

By induction hypothesis, we have for all $i \in \mathrm{In}_\alpha$ and all $j \in \mathrm{In}_\beta$, “$\alpha \leqslant \beta_j$ or $\beta_j < \alpha$”, and also “$\beta \leqslant \alpha_i$ or $\alpha_i < \beta$”.

The first disjunction implies by **LEM** that either $\beta_j < \alpha$ for all $j \in \mathrm{In}_\beta$ or there is $j \in \mathrm{In}_\beta$ such that $\alpha \leqslant \beta_j$. In the first case, we have $\beta \leqslant \alpha$ by definition of $\cdot \leqslant \cdots$. In the second case, we have $\alpha < \beta$ by definition of $\cdot < \cdots$, with for $F \subseteq_f \mathrm{In}_\beta$ the list $[j]$.

The symmetric reasoning yields the second disjunction. □

N.b.: For countable ordinals, the limited principle of omniscience (**LPO**) suffices to prove the proposition.

Corollary 3.20. *Assume **LEM**. Any ordinal* $\alpha \neq \underline{0}$ *is either an immediate successor or the* sup *of the ordinals* $\gamma < \alpha$.

Proof. Consider $\alpha = \mathrm{s}(\alpha_i)_{i \in \mathrm{In}_\alpha}$ and compare α with $\sup(\alpha_i)_{i \in \mathrm{In}_\alpha}$. The details are left to the reader. □

Corollary 3.21. *Assume **LEM**. Any bounded ordinal is finite.*

Proof. Left to the reader: use Fact 3.13. □

4. Fundamental Results

4.1. $\mathbf{Ord}_{\mathfrak{F}}$ *is an initial object in the category of* $\mathfrak{F}$*-orders*

Lemma 4.1. *For* $\alpha^1, \ldots, \alpha^r$ *in* $\mathbf{ord}$ $(r \geqslant 1)$, *we have*

$$\boxed{\sup(\alpha^j)_{j \in [\![1..r]\!]} < \mathrm{s}(\alpha^j)_{j \in [\![1..r]\!]}.}$$

Proof. Let us show, e.g., that $\epsilon = \sup(\alpha, \beta) < \gamma = \mathrm{s}(\alpha, \beta)$. We have $\mathrm{In}_\epsilon = \mathrm{In}_\alpha + \mathrm{In}_\beta$, with $\epsilon_k = \alpha_i$ if $\iota_1(i) = k$, and $\epsilon_k = \beta_j$ if $\iota_2(j) = k$. We have $\mathrm{In}_\gamma = \{1, 2\}$ with $\gamma_1 = \alpha$ and $\gamma_2 = \beta$. We apply Fact 3.14b with $F = \{1, 2\}$. For an arbitrary k in In_ϵ, we have $\epsilon_k < \alpha, \beta$ since ϵ_k is α_i or β_j and, by `s1`, we have $\alpha_i < \alpha$ (a fortiori $\alpha_i < \alpha, \beta$) and $\beta_j < \beta$ (a fortiori $\beta_j < \alpha, \beta$). □

Let us note that the preceding proof relies on the fact that the definitions of $\leqslant$ and $<$ have been given with lists on the right-hand side.

Lemma 4.2. (transitivities).

(1) `trans1`. *If* $\alpha \leqslant \beta^1, \ldots, \beta^m$ *and, for each* $j \in [\![1..m]\!]$, $\beta^j \leqslant \gamma^1, \ldots, \gamma^r$, *then* $\alpha \leqslant \gamma^1, \ldots, \gamma^r$.
(2) `trans2`. *If* $\alpha < \beta^1, \ldots, \beta^m$ *and, for each* $j \in [\![1..m]\!]$, $\beta^j \leqslant \gamma^1, \ldots, \gamma^r$, *then* $\alpha < \gamma^1, \ldots, \gamma^r$.
(3) `trans3`. *If* $\alpha \leqslant \beta^1, \ldots, \beta^m$ *and, for each* $j \in [\![1..m]\!]$, $\beta^j < \gamma^1, \ldots, \gamma^r$, *then* $\alpha < \gamma^1, \ldots, \gamma^r$.

As particular cases, Axioms 5 to 7 will be valid when we shall descend to the quotient **Ord**:

- if $\alpha \leqslant \beta$ and $\beta \leqslant \gamma$, then $\alpha \leqslant \gamma$;
- if $\alpha < \beta$ and $\beta \leqslant \gamma$, then $\alpha < \gamma$;
- if $\alpha \leqslant \beta$ and $\beta < \gamma$, then $\alpha < \gamma$.

Proof. The three transitivities are proved by simultaneous induction.

In order to prove `trans1`, we note that the hypothesis means that we have $\alpha_i < \beta^1, \ldots, \beta^m$ for all $i \in \mathrm{In}_\alpha$. Let us fix such an i. We use `trans2` with this α_i instead of α and we get $\alpha_i < \gamma^1, \ldots, \gamma^r$. Since this works for all $i \in \mathrm{In}_\alpha$, this gives the desired conclusion $\alpha \leqslant \gamma^1, \ldots, \gamma^r$.

In order to prove `trans2`, we note that the hypothesis implies that there are $G_j \subseteq_f \mathrm{In}_{\beta^j}$ not all empty such that $\alpha \leqslant \beta^1_{G_1}, \ldots, \beta^m_{G_m}$. We have also for $j \in [\![1..m]\!]$ and for all $h \in \mathrm{In}_{\beta^j}$, $\beta^j_h < \gamma^1, \ldots, \gamma^r$. A fortiori, this is true for the hs $\in G_j$. We use `trans3` with these β^j_hs instead of the β^js. This gives the desired conclusion $\alpha < \gamma^1, \ldots, \gamma^r$.

In order to prove `trans3`, we note that the hypothesis implies (by weakening) that there are $F_k \subseteq_f \mathrm{In}_{\gamma^k}$ not all empty such that

$\beta^j \leqslant \gamma^1_{F_1}, \ldots, \gamma^r_{F_r}$ for $j \in [\![1..m]\!]$. This time we use `trans1` with the γ^k_ℓs instead of the γ^ks and we deduce that $\alpha \leqslant \gamma^1_{F_1}, \ldots, \gamma^r_{F_r}$, which implies $\alpha < \gamma^1, \ldots, \gamma^r$. □

The following lemma shows that when descending to the quotient, Axiom 4 will be valid in **Ord**.

Lemma 4.3. (ax4). *Let* $\alpha, \beta^1, \ldots, \beta^m \in$ **ord**. *If* $\alpha < \beta^1, \ldots, \beta^m$, *then* $\alpha \leqslant \beta^1, \ldots, \beta^m$.

Proof. Proof by induction on α. We have $\alpha < \beta^1, \ldots, \beta^m$ if and only if we can find $F_k \subseteq_f \mathrm{In}_{\beta^k}$ not all empty such that, for each $i \in \mathrm{In}_\alpha$, we have $\alpha_i \leqslant \beta^1_{F_1}, \ldots, \beta^m_{F_m}$. Let us fix an $i \in \mathrm{In}_\alpha$. For $j \in F_k$, we have $\beta^k_j < \beta^k$, and by weakening $\beta^k_j < \beta^1, \ldots, \beta^m$. By `trans3`, we get $\alpha_i < \beta^1, \ldots, \beta^m$. Finally, since this is true for all $i \in \mathrm{In}_\alpha$, we have $\alpha \leqslant \beta^1, \ldots, \beta^m$. □

The following fact shows that Axiom 10 will be valid when we shall descend to the quotient **Ord**.

Lemma 4.4. (ax10). *If* $\alpha < \gamma$ *and* $\beta < \gamma$, *then* $\sup(\alpha, \beta) < \gamma$.

Proof. By definition, we have $\mathrm{s}(\alpha, \beta) \leqslant \gamma$. Lemma 4.1 gives $\sup(\alpha, \beta) < \mathrm{s}(\alpha, \beta)$. By transitivity, we get $\sup(\alpha, \beta) < \gamma$. □

Lemma 4.5. *Let n be a positive integer and* $\alpha^1, \ldots, \alpha^n \in$ **ord**. *It is impossible that, for each* $i \in [\![1..n]\!]$, *we have* $\alpha^i < \alpha^1, \ldots, \alpha^n$.

Proof. By induction. Using weakening, the hypothesis to be proven impossible gives finite lists

$$F_1 \subseteq_f \mathrm{In}_{\alpha_1}, \ \ldots, \ F_n \subseteq_f \mathrm{In}_{\alpha_n},$$

not all empty, such that

$$\alpha^i \leqslant \alpha^1_{F_1}, \ldots, \alpha^n_{F_n} \quad \text{for } i \in [\![1..m]\!].$$

In particular, for $j \in F_i$ (if F_i is nonempty), we have

$$\alpha^i_j < \alpha^1_{F_1}, \ldots, \alpha^n_{F_n}.$$

This reduces to the hypothesis with the nonempty list $\alpha^1_{F_1}, \ldots, \alpha^n_{F_n}$ instead of the list $\alpha^1, \ldots, \alpha^n$. □

Lemma 4.6. *Let* $\alpha^1, \dots, \alpha^n, \beta^1, \dots, \beta^m \in \mathbf{ord}$ $(n, m \geqslant 1)$.

(1) *If* $\alpha^i < \alpha^1, \dots, \alpha^n, \beta^1, \dots, \beta^m$ *for* $i \in [\![1..n]\!]$, *then* $\alpha^i < \beta^1, \dots, \beta^m$ *for each* i.
(2) *Let* $F_1 \subseteq_f \mathrm{In}_{\alpha_1}$, $\dots$, $F_n \subseteq_f \mathrm{In}_{\alpha_n}$. *If* $\alpha^i \leqslant \alpha^1_{F_1}, \dots, \alpha^n_{F_n}, \beta^1, \dots, \beta^m$ *for* $i \in [\![1..n]\!]$, *then* $\alpha^i \leqslant \beta^1, \dots, \beta^m$ *for each* i.

Proof.

(1) The hypothesis yields finite lists

$$F_1 \subseteq_f \mathrm{In}_{\alpha^1}, \dots, F_n \subseteq_f \mathrm{In}_{\alpha^n}, G_1 \subseteq_f \mathrm{In}_{\beta^1}, \dots, G_m \subseteq_f \mathrm{In}_{\beta^m},$$

not all empty, such that

$$\alpha^i \leqslant \alpha^1_{F_1}, \dots, \alpha^n_{F_n}, \beta^1_{G_1}, \dots, \beta^m_{G_m} \text{ for } i \in [\![1..n]\!]. \qquad (*)$$

Thus, we have for $i \in [\![1..n]\!]$ and $j \in \mathrm{In}_{\alpha^i}$

$$\alpha^i_j < \alpha^1_{F_1}, \dots, \alpha^i_{F_i}, \dots, \alpha^n_{F_n}, \beta^1_{G_1}, \dots, \beta^m_{G_m}.$$

Let us fix i and j: a fortiori, with $F_i' = F_i \cup \{j\}$

$$\alpha^i_j < \alpha^1_{F_1}, \dots, \alpha^i_{F_i'}, \dots, \alpha^n_{F_n}, \beta^1_{G_1}, \dots, \beta^m_{G_m}.$$

We have also by weakening, for $k \in [\![1..n]\!]$ and $\ell \in F_k$

$$\alpha^k_\ell < \alpha^1_{F_1}, \dots, \alpha^i_{F_i'}, \dots, \alpha^n_{F_n}, \beta^1_{G_1}, \dots, \beta^m_{G_m}.$$

Thus, by induction $\alpha^i_j < \beta^1_{G_1}, \dots, \beta^m_{G_m}$. Since j is arbitrary, we get $\alpha^i \leqslant \beta^1_{G_1}, \dots, \beta^m_{G_m}$. This gives the desired conclusion, $\alpha^i < \beta^1, \dots, \beta^m$, if at least one list G_k is nonempty, for an arbitrary $i \in [\![1..n]\!]$. If this is not the case, $(*)$ yields $\alpha^i \leqslant \alpha^1_{F_1}, \dots, \alpha^n_{F_n}$ for $i \in [\![1..n]\!]$, with lists F_i not all empty. By definition, this implies $\alpha^i < \alpha^1, \dots, \alpha^n$ for $i \in [\![1..n]\!]$, which is impossible by Lemma 4.5.
(2) We have for $i \in [\![1..n]\!]$ and $j \in \mathrm{In}_{\alpha^i}$

$$\alpha^i_j < \alpha^1_{F_1}, \dots, \alpha^i_{F_i}, \dots, \alpha^n_{F_n}, \beta^1, \dots, \beta^m.$$

Let us fix i and j: a fortiori, with $F_i' = F_i \cup \{j\}$,

$$\alpha^i_j < \alpha^1_{F_1}, \dots, \alpha^i_{F_i'}, \dots, \alpha^n_{F_n}, \beta^1, \dots, \beta^m.$$

We have also by weakening, for $k \in [\![1..n]\!]$ and $\ell \in F_k$

$$\alpha^k_\ell < \alpha^1_{F_1}, \dots, \alpha^i_{F_i'}, \dots, \alpha^n_{F_n}, \beta^1, \dots, \beta^m.$$

Item 1 then yields $\alpha^i_j < \beta^1, \ldots, \beta^m$. As j is arbitrary, we get what we want: $\alpha^i \leqslant \beta^1, \ldots, \beta^m$ for an arbitrary $i \in [\![1..n]\!]$. □

The following fact shows that Axioms 11 and 12 will be valid when we shall descend to the quotient **Ord**.

Lemma 4.7.

(1) `ax11`. *If* $\alpha < \sup(\alpha, \beta)$, *then* $\alpha < \beta$;
(2) `ax12`. *If* $\gamma < \alpha$ *and* $\alpha \leqslant \sup(\beta, \gamma)$, *then* $\alpha \leqslant \beta$.

Proof.

(1) Assume $\alpha < \sup(\alpha, \beta)$. Lemma 3.7 gives $\alpha < \alpha, \beta$. Item 1 of Lemma 4.6 gives $\alpha < \beta$.
(2) Assume $\gamma < \alpha$ and $\alpha \leqslant \sup(\beta, \gamma)$. The first hypothesis gives $\gamma \leqslant \alpha_F$ for a nonempty $F \subseteq_f \mathrm{In}_\alpha$. The second hypothesis gives $\alpha \leqslant \gamma, \beta$ (by Lemma 3.7). By transitivity, we have $\alpha \leqslant \alpha_F, \beta$. Item 2 of Lemma 4.6 gives $\alpha \leqslant \beta$. □

Theorem 4.8. *We have constructed* **Ord** *as an* $\mathfrak{F}$*-order.*

Proof. Using `rfl` and `trans1`, we first show that the equality is indeed an equivalence relation, and then that the relation $\leqslant$ descends to the quotient in **Ord**.

Similarly, `trans2` and `trans3` imply that the relation $<$ descends to the quotient in **Ord**.

The sup map descends to the quotient by Lemma 3.16, Item 2.

The unary s map descends to the quotient by Fact 3.13.

It remains to note that Axioms 1 to 15 of $\mathfrak{F}$-orders have been proved above. See, respectively: Lemma 3.16 (Item 3); Fact 3.8 (Item 1); Lemma 3.16 (Item 5); Lemma 4.3; Lemma 4.2; Fact 3.13; Lemma 4.4; Lemma 4.7; Fact 3.12; Lemma 3.16 (Item 7); Remark 3.9. □

The following theorem generalises Fact 3.10.

Theorem 4.9. *The set* **Ord** *is not reduced to a point. More precisely:*

- *for all* $\alpha, \beta \in \mathbf{Ord}$, $\beta \leqslant \alpha$ *and* $\alpha < \beta$ *are incompatible;*
- *the map* $n \mapsto \underline{n} \colon \mathbb{N} \to \mathbf{Ord}$ *is injective* ($m < n$ *if and only if* $\underline{m} < \underline{n}$);
- *for all* $\alpha \in \mathbf{Ord}$ *and* $n > m$ *in* $\mathbb{N}$, *it is impossible that* $\mathrm{s}^{(n)}(\alpha) =_{\mathbf{Ord}} \mathrm{s}^{(m)}(\alpha)$.

Proof. The first item is a consequence of `irfl` and of `trans2`. The rest follows. □

Theorem 4.10. **Ord** *is an initial object in the category of* $\mathfrak{F}$*-orders.*

Sketch of proof. The structure is purely algebraic and in order to construct **Ord**, we have only used the axioms of the structure.

In fact, let us consider an object $(E, <_E, \leqslant_E, 0_E, \sup_E, \mathrm{s}_E)$ in the category. Elements of **ord** do have their copies in E. Furthermore, the relations $\cdot < \cdots$ and $\cdot \leqslant \cdots$ defined in **ord** are valid in E by Fact 2.7 if interpreted in E with finite sups on the right-hand side (as we may by Lemma 3.7). This implies that there is a unique morphism from **Ord** to E in the category. □

4.2. *More properties*

Proposition 4.11. *The binary relation* $<$ *on* **Ord** *is well-founded.*

Proof. This is a direct consequence of Fact 3.4. □

Lemma 4.12. (weak forms of the disjunction "$\alpha \leqslant \beta$ or $\beta < \alpha$"). *Let* $r \geqslant 1$ *and* $\alpha, \beta^1, \ldots, \beta^r, \gamma \in$ **ord**.

(1) *If* $\alpha \leqslant \beta$ *and* $\beta < \alpha, \gamma^1, \ldots, \gamma^r$, *then* $\beta < \gamma^1, \ldots, \gamma^r$.
(2) *If* $\beta < \alpha$ *and* $\alpha \leqslant \beta, \gamma^1, \ldots, \gamma^r$, *then* $\alpha \leqslant \gamma^1, \ldots, \gamma^r$.

Proof. Introduce $\gamma = \sup(\gamma^1, \ldots, \gamma^r)$. Using Lemma 3.7, both items reduce to already established properties. □

Definition 4.13. An element $\beta = \mathrm{s}(\beta_i)_{i \in \mathrm{In}_\beta} \in$ **ord** is said to be *filtering* if for each $F \subseteq_f \mathrm{In}_\beta$ there exists $j \in \mathrm{In}_\beta$ such that $\sup(\beta_i)_{i \in F} \leqslant \beta_j$.

Lemma 4.14. *For each* $\alpha \in$ **ord**, *there exists* $\beta \in$ **ord** *such that* $\alpha =_{\mathbf{Ord}} \beta$ *and* β *is filtering.*

Proof. If $\alpha = \mathrm{s}(\alpha_i)_{i \in J}$, we let K be the set of finitely enumerated subsets of J, and for $F \subseteq_f J$, we let $\beta_F = \sup(\alpha_j)_{j \in F}$. Then $\beta =_{\mathbf{Ord}} \mathrm{s}(\beta_F)_{F \in K}$. □

4.3. *Elementary ordinal arithmetic*

4.3.1. *(Sequential) addition*

The sequential addition $\alpha + \beta$ (α followed by β: addition is not commutative) is defined by induction on β:

$$\alpha + \underline{0} = \alpha \quad \text{and} \quad \alpha + \beta = \mathrm{s}(\alpha + \beta_j)_{j \in \mathrm{In}_\beta} \ \text{ if } \ \beta = \mathrm{s}(\beta_j)_{j \in \mathrm{In}_\beta} \in \mathbf{ord}^*.$$

This formula works only for the case $\mathrm{In}_\beta \neq \mathbb{N}_0$: it would yield $\alpha + \underline{0} = \underline{0}$ for $\mathrm{In}_\beta = \mathbb{N}_0$. We also have $\alpha + \beta = \sup((\alpha + \beta_j) + \underline{1})_{j \in \mathrm{In}_\beta}$ if $\mathrm{In}_\beta \neq \mathbb{N}_0$.

The following properties can be proved by induction:

- if $\alpha \leqslant \alpha'$ and $\beta \leqslant \beta'$, then $\alpha + \beta \leqslant \alpha' + \beta'$;
- $(\alpha + \beta) + \gamma = \alpha + (\beta + \gamma)$;
- $\alpha + \underline{0} = \underline{0} + \alpha = \alpha$;
- $\alpha + \beta \leqslant \alpha + \gamma$ if and only if $\beta \leqslant \gamma$;
- $\alpha + \beta < \alpha + \gamma$ if and only if $\beta < \gamma$;
- $\alpha = \underline{1} + \alpha$ if and only if $\omega \leqslant \alpha$;
- if $\alpha \leqslant \gamma$, then there is β such that $\gamma = \alpha + \beta$;
- if $\alpha < \gamma$, then there is $\beta \neq \underline{0}$ such that $\gamma = \alpha + \beta$.

4.3.2. *Sequential sum*

Let $J \in \mathfrak{F}$ and consider a well-founded linear order relation $\prec$ on J with a detachable minimal element 0_J. Let $(\beta^j)_{j \in J}$ be an element of $\mathrm{Fam}(J, \mathbf{ord})$. The $\prec$-indexed sequential sum $\sum_{j \prec \ell} \beta^j$ is defined by induction on $\ell \in (J, \prec)$:

$$\sum_{j \prec 0_J} \beta^j = 0_J \quad \text{and}$$

$$\sum_{j \prec \ell} \beta^j = \sup \left(\Big(\sum_{j \prec k} \beta^j\Big) + \beta^k \right)_{k \prec \ell} \qquad \text{if } 0_J \prec \ell.$$

We show by induction on $\prec$ that, given two families $(\beta^j)_{j \in J}$ and $(\gamma^j)_{j \in J}$ such that $\beta^j \leqslant \gamma^j$ for all $j \in J$, we have $\sum_{j \prec \ell} \beta^j \leqslant \sum_{j \prec \ell} \gamma^j$ for all $\ell \in J$. This construction descends therefore to the quotient **Ord**.

Remark 4.15. This construction allows us to define a map $\mathbf{ord}_2^{\mathrm{Br}} \to \mathbf{ord}_2$, where $\mathbf{ord}_2^{\mathrm{Br}}$ is the set of names of Brouwer ordinals. See Troelstra [14] and Brouwer [7,15]. Troelstra only treats countable Brouwer ordinals.

4.3.3. *Multiplication*

We define $\alpha \cdot \beta$ by induction on $\beta \in \mathbf{ord}$:

$$\alpha \cdot \underline{0} = \underline{0} \quad \text{and} \quad \alpha \cdot \beta = \sup(\alpha \cdot \beta_j + \alpha)_{j \in \mathrm{In}_\beta} \text{ if } \beta = \mathrm{s}(\beta_j)_{j \in \mathrm{In}_\beta} \in \mathbf{ord}^*.$$

The following properties can be proved by induction:

- if $\alpha \leqslant \alpha'$ and $\beta \leqslant \beta'$, then $\alpha \cdot \beta \leqslant \alpha' \cdot \beta'$;
- $(\alpha \cdot \beta) \cdot \gamma = \alpha \cdot (\beta \cdot \gamma)$;
- $\alpha \cdot \underline{1} = \underline{1} \cdot \alpha = \alpha$;
- $\alpha \cdot (\beta + \gamma) = (\alpha \cdot \beta) + (\alpha \cdot \gamma)$;
- if $\underline{1} \leqslant \alpha$, then $\alpha \cdot \beta \leqslant \alpha \cdot \gamma$ if and only if $\beta \leqslant \gamma$;
- if $\underline{1} \leqslant \alpha$, then $\alpha \cdot \beta < \alpha \cdot \gamma$ if and only if $\beta < \gamma$.

4.3.4. *Exponentiation*

We define α^β by induction on $\beta \in \mathbf{ord}$, as follows:

$$\alpha^{\underline{0}} = \underline{1} \quad \text{and} \quad \alpha^\beta = \sup(\alpha^{\beta_j} \cdot \alpha)_{j \in \mathrm{In}_\beta} \text{ if } \beta = \mathrm{s}(\beta_j)_{j \in \mathrm{In}_\beta} \in \mathbf{ord}^*.$$

4.3.5. *Ackermann*

It is possible to continue this elementary arithmetic à la Ackermann. We define by induction an ordinal $\underline{\mathrm{Ack}}(\alpha, \beta, \gamma)$ that we get by iterating γ times the preceding map, initialised at α, i.e., more precisely

$$\begin{aligned}
\underline{\mathrm{Ack}}(\alpha, \beta, \underline{0}) &= \alpha + \beta \\
\underline{\mathrm{Ack}}(\alpha, \underline{0}, \gamma) &= \alpha \qquad \text{if } \gamma \in \mathbf{ord}^* \\
\underline{\mathrm{Ack}}(\alpha, \beta, \gamma) &= \sup\big(\sup(\underline{\mathrm{Ack}}(\underline{\mathrm{Ack}}(\alpha, \beta_j, \gamma), \alpha, \gamma_k))_{j \in \mathrm{In}_\beta}\big)_{k \in \mathrm{In}_\gamma} \\
&\qquad \text{if } \beta = \mathrm{s}(\beta_j)_{j \in \mathrm{In}_\beta} \text{ and } \gamma = \mathrm{s}(\gamma_k)_{k \in \mathrm{In}_\gamma}.
\end{aligned}$$

In particular, $\varepsilon_0 = \underline{\mathrm{Ack}}(\omega, \omega, \underline{4})$.

5. Countable Ordinals

5.1. *First steps*

As previously indicated, we get countable ordinals when we choose as set of index sets

$$\mathfrak{F}_2 = \{\, \mathbb{N}_k \mid k \in \mathbb{N}, k \geqslant 0 \,\} \cup \{\mathbb{N}\},$$

with convenient operations for the set of finite subsets of an $I \in \mathfrak{F}_2$ and for disjoint unions. We write $\mathbf{ord}_2$ and $\mathbf{Ord}_2$ for $\mathbf{ord}_{\mathfrak{F}_2}$ and $\mathbf{Ord}_{\mathfrak{F}_2}$. Thus, $\mathbf{Ord}_2$ is the set of ordinals of the second class and $\mathbf{ord}_2$ is a set of names for elements of $\mathbf{Ord}_2$.

Lemma 5.1. *Any countable ordinal is the* s *of a nondecreasing sequence of countable ordinals.*

Proof. This is Lemma 4.14. □

Proposition 5.2. *Assume* ***LPO****. Then, for $\alpha, \beta \in$* **Ord***, we have $\alpha \leqslant \beta$ or $\beta < \alpha$.*

Proof. Proceed as for Proposition 3.19, in the countable case. □

5.2. *Comparison with Martin-Löf ordinals*

We present a variation of the theory of ordinals in the book *Notes on Constructive Mathematics* [1, Chapter 3]. We write "variation" since Martin-Löf's theory is formulated in the framework of Markov's recursive mathematics, while we take as primitive intuitionistic logic with generalised inductive definitions, as does Heyting [11] (the fact that this setting can provide a more elegant treatment than the one in recursive mathematics is stressed in Kreisel's review [16] of this work).

5.2.1. *Martin-Löf's formal system*

In this system, ordinals are described inductively: if we have a finite or infinite sequence of ordinals $\sigma = \sigma_0, \ldots, \sigma_n, \ldots$ (maybe empty), then $\mathrm{s}(\sigma)$ is an ordinal.

The (classical) semantics of this operation is the following: to (σ_n) sequence of ordinals we associate the supremum of the sequence of the successors of the σ_ns.

In particular, $\underline{0}$ is defined as $\mathrm{s}(\sigma)$, where σ is the empty sequence.

We write simply $\mathrm{s}(\alpha)$ for $\mathrm{s}(\sigma)$, where σ is the sequence with one element $\sigma_0 = \alpha$.

In constructive mathematics, the set of all such ordinals is an example of a nondiscrete set.

As stated in the introduction, to any ordinal α we associate, by induction on α, a tree Tree(α): Tree(α) always contains the empty sequence, and Tree(s(σ)) contains $n \frown \ell$ if ℓ is in Tree(σ_n).

This set Tree(α) does not contain any infinite branch: if f is a numerical function, we can always find n such that $[f(0), \ldots, f(n-1)]$ is not in Tree(α). This is proved directly by induction on α. In other words, the tree Tree(α) is well-founded. The fact that we get in this way all well-founded trees is the content of Brouwer's bar theorem, which holds neither in Bishop's set theory nor in dependent type theory. This follows from the fact that both systems have an interpretation in recursive mathematics, while the bar theorem does not hold in recursive mathematics, as shown by an example due to Kleene (see [17]).

We define next what an atomic formula is: a formula of the form $\alpha < \beta$ or $\alpha \leqslant \beta$.

Finally, we can define when a sequent Γ is provable, where Γ is a finite *set* of atomic formulae. The formulation is quite elegant!

$$\frac{\Gamma, \alpha \leqslant \sigma_n}{\Gamma, \alpha < \mathrm{s}(\sigma)} \qquad \frac{\cdots\ \Gamma, \sigma_n < \beta\ \cdots}{\Gamma, \mathrm{s}(\sigma) \leqslant \beta}$$

Note that there is a direct proof of $\underline{0} \leqslant \beta$ by the second rule.

The intuitive meaning of a sequent is the classical disjunction of the atomic formulae it contains.

Martin-Löf shows that we can prove for instance $\alpha < \beta, \beta \leqslant \alpha$ by induction on β and α. He also shows that the following rule is admissible by induction on α:

$$\frac{\Gamma, \alpha < \alpha}{\Gamma},$$

which implies in particular that $\alpha < \alpha$ is not provable.

Let us give an example of such proofs by induction.

Lemma 5.3. *The sequents $\alpha \leqslant \alpha$ and $\alpha < \mathrm{s}(\alpha)$ are provable for all α.*

Proof. We prove $\alpha \leqslant \alpha$ by induction on α. If $\alpha = \mathrm{s}(\sigma)$, we have to show $\sigma_n < \mathrm{s}(\sigma)$ for all n, which follows from $\sigma_n \leqslant \sigma_n$, which holds by induction.

It follows that we have $\alpha < \mathrm{s}(\alpha)$ by the first derivation rule. □

5.2.2. *Comparison with our system*

Let us explain now why this definition does not coincide with ours by giving an example of the form $\alpha < \beta$, which is provable in this sequent calculus but implies **LPO** in our system.

Let us return to Example 3.18: define $\alpha = \mathrm{s}(\sigma)$, where $\sigma_n = \underline{u_n}$ with (u_n) a nondecreasing sequence of 0s and 1s, and $\beta = \mathrm{s}(\tau)$, where $\tau_n = s(\sigma_n) = \underline{u_n + 1}$.

Lemma 5.4. *The sequent $\alpha < \beta$ is provable.*

Proof. By the first rule, it is enough to show $\alpha < \mathrm{s}(\tau), \alpha \leqslant \tau_0$. And for this we have to show $\sigma_n < \tau_0, \alpha < \mathrm{s}(\tau)$ for all n. We fix n and we show $\sigma_n < \tau_0, \alpha < \mathrm{s}(\tau)$.

If we do have $\sigma_n < \tau_0 = \mathrm{s}(\sigma_0)$, this is fine. Note that we can *test* whether or not $\sigma_n < \tau_0$ holds since both σ_n and τ_0 are of the form $\underline{0}$ or $\underline{1}$ or $\underline{2}$.

Otherwise, we get explicitly n such that $\sigma_n \geqslant \mathrm{s}(\sigma_0)$ and we have then $\sigma_m \leqslant \sigma_n$ and so $\sigma_m < \tau_n$ for all m. We can prove $\sigma_n < \tau_0, \alpha < \mathrm{s}(\tau)$ by $\sigma_n < \tau_0, \alpha \leqslant \tau_n$ which holds since $\sigma_n < \tau_0, \sigma_m < \tau_n$ holds for all m. □

Note that we prove $\alpha < \mathrm{s}(\tau)$ by proving $\alpha < \mathrm{s}(\tau), \alpha \leqslant \tau_0$, and we have to "keep" $\alpha < \mathrm{s}(\tau)$: maybe $\alpha \leqslant \tau_0$ does not hold (it may happen that the sequence (σ_n) takes the value $\underline{1}$ and $\tau_0 = \underline{1}$).

In Example 3.18, we note that $\alpha < \beta$ implies **LPO** in our system. Therefore, in the set $\mathbf{Ord}_2^{\mathrm{ML}}$ of Martin-Löf ordinals, the equality is coarser than in the set $\mathbf{Ord}_2$, though both are based on the set $\mathbf{ord}_2$.

References

[1] P. Martin-Löf, *Notes on Constructive Mathematics*, Almqvist & Wiksell, Stockholm, 1970.

[2] N. Bourbaki, *Elements of Mathematics: Theory of Sets.* Hermann, Paris and Addison-Wesley, Reading, translated from the French, 1968.

[3] J.-L. Krivine, *Théorie des ensembles*, Cassini, Paris, 1998.

[4] P. Dehornoy, *La théorie des ensembles: introduction à une théorie de l'infini et des grands cardinaux*, Tableau Noir, vol. 106, Calvage et Mounet, Paris, 2017.

[5] R. Mines, F. Richman, and W. Ruitenburg, *A Course in Constructive Algebra*. Universitext, Springer-Verlag, New York, 1988. doi: 10.1007/978-1-4419-8640-5.

[6] Univalent Foundations Program, *Homotopy Type Theory: Univalent Foundations of Mathematics*, 2013. https://homotopytypetheory.org/book, Institute for Advanced Study, Princeton.

[7] L.E.J. Brouwer, Zur Begründung der intuitionistischen Mathematik. *Ann.* **96**, 451–488 (1926).

[8] G. Gentzen, Die Widerspruchsfreiheit der reinen Zahlentheorie, *Math. Ann.* **112**, 493–565, translation by M. Szabo: 'The consistency of elementary number theory', in [18], 132–201 (1936).

[9] A. Church, The constructive second number class, *Bull. Amer. Math. Soc.* **44**, 4, 224–232. (1938). doi: 10.1090/S0002-9904-1938-06720-1.

[10] S.C. Kleene, On notation for ordinal numbers, *J. Symb. Log.* **3**, 150–155 (1938).

[11] A. Heyting, Infinitistic methods from a finitist point of view. In *Infinitistic Methods: Proceedings of the Symposium on Foundations of Mathematics, Warsaw*, 2–9 September 1959. pp. 185–192. Pergamon, Oxford and Państwowe Wydawnictwo Naukowe, Warsaw, 1961.

[12] P. Aczel and M. Rathjen, *CST Book Draft*, https://www1.maths.leeds.ac.uk/~rathjen/book.pdf, 2010.

[13] N. Kraus, F. Nordvall Forsberg, and C. Xu, Connecting constructive notions of ordinals in homotopy type theory. In *46th International Symposium on Mathematical Foundations of Computer Science (MFCS 2021)*. LIPIcs. Leibniz Int. Proc. Inform., vol. 202, Schloss Dagstuhl. Leibniz-Zent. Inform., Wadern, 2021, pp. 70:1–70:16. doi: 10.4230/LIPIcs.MFCS.2021.70, arXiv:2104.02549 contains an appendix with proofs.

[14] A.S. Troelstra, *Principles of Intuitionism: Lectures Presented at the Summer Conference on Intuitionism and Proof Theory (1968) at SUNY at Buffalo, Lecture Notes in Mathematics, N.Y.*, vol. 95. Springer-Verlag, Berlin, 1969.

[15] L.E.J. Brouwer, Begründung der Mengenlehre unabhängig vom logischen Satz vom ausgeschlossenen Dritten. Erster Teil: Allgemeine Mengenlehre. *Verh. Nederl. Akad. Wetensch. Afd. Natuurk. Sect. 1* **12**(5) 3–43 (1918).

[16] G. Kreisel, Review of [11], *Math. Rev.* **26**, #2363 (MR0144822) (1963). https://mathscinet.ams.org/mathscinet-getitem?mr=144822.

[17] S.C. Kleene and R.E. Vesley, *The Foundations of Intuitionistic Mathematics, Especially in Relation to Recursive Functions*, North-Holland, Amsterdam, 1965.

[18] M.E. Szabo (ed.). *The Collected Papers of Gerhard Gentzen* (*Studies in Logic and the Foundations of Mathematics*), North-Holland, Amsterdam, 1969.

https://doi.org/10.1142/9789811245220_0013

Chapter 13

No Speedup for Geometric Theories*

Michael Rathjen
School of Mathematics, University of Leeds, England
michrathjen@snafu.de

Geometric theories based on classical logic are conservative over their intuitionistic counterparts for geometric implications. The latter result (sometimes referred to as Barr's theorem) is squarely a consequence of Gentzen's Hauptsatz. Prima facie though, cut elimination can result in superexponentially longer proofs. In this chapter, it is shown that the transformation of a classical proof of a geometric implication in a geometric theory into an intuitionistic proof can be achieved in a feasible way.

1. Introduction

For geometric theories, it is known that the existence of a classical proof of a geometric implication yields the existence of an intuitionistic proof. Existing effective proofs of this fact use cut elimination (see [1–5]) and thus are liable to effect a superexponential blow-up.

*Paper written in honour of Grisha Mints and Erik Palmgren.

When I visited Stanford in November 2013, Grisha Mints told me that he was aiming to show that the transformation can be achieved via a polynomial time algorithm. Sadly, this was the last time I saw him. Some of what he told me is reflected in his article [6], which presents a partial result in that a merely polynomially longer proof is achievable in an augmentation of the intutitionistic theory via relational Skolem axioms. The article, though, does not achieve what he had originally intended.[a] At the time, I had not started working on the problem, but the conversation with Grisha lingered in my mind over the years and for some reason I became convinced that there should be a simple argument based on an interpretation akin to the Friedman–Dragalin A-translation. The short chapter before the reader is an elaboration of that idea.[b]

As I was working on such a translation, it dawned on me that similar ideas must have been considered before. Indeed, Leivant in [7] used interpretations of classical logic into intuitionistic logic to obtain (partial) conservativity results for classical theories over their intuitionistic versions. He explained that the Friedman–Dragalin interpretation "is derived naturally from a trivial translation of I into M" ([7, p. 683]), where I and M signify intuitionistic and minimal logic, respectively, and "trivial translation" refers to Kolmogorov's 1925 translation [8]. The observation that the latter is actually an interpretation into Johanssons's 1937 minimal logic [9] emerges as the fulcrum for obtaining conservativity results, in that, in minimal logic falsum $\bot$ just acts as placeholder for an arbitrary formula. Crucial in the machinery of [7] are the definitions of three syntactical classes, namely, the *spreading, wiping and isolating* schemata and formulas with regard to the Kolmogorov interpretation (also see [10], Chapter 2, Section 3 for an exposition). The ideas underlying these classes have informed Definition 3.5 and Proposition 3.7. But alas, I couldn't see how to directly infer the conservativity of classical over intuitionistic geometric theories by reassembling results from [7]. However, Ishihara's article [11] actually furnishes what is needed.

[a]See also the comments on and an evaluation of [6] in [12].

[b]Barr's theorem is sometimes alleged to have miraculous powers that allow one to remove the axiom of choice from proofs of geometric implications. For a sceptical investigation of such claims, see [5].

A nice move in [11] is to introduce a new propositional constant into the language which can act as a placeholder for arbitrary formulas not only in minimal logic but also in intuitionistic logic.[c]

1.1. *Postscriptum*

A preprint version of this chapter was deposited on arXiv in May 2021. I gave a talk emanating from it at the Dagstuhl meeting "Geometric Logic, Constructivisation and Automated Theorem Proving" in November 2021. Only a couple of days before my talk, however, I chanced upon the 2002 article *An Intuitionistic Axiomatisation of Real Closed Fields* [18] by Erik Palmgren in which he also uses the Friedman–Dragalin to show the conservativity of classical over intuitionistic geometric theories. The procedure in [18] is somewhat different to the one in this chapter in that [18] uses certain normal forms for geometric implications. Furthermore, the current chapter shows conservativity for a larger syntactic class $\mathcal{Q}$ and the techniques could lend themselves to further generalisations.[d] It also elaborates on details as to why the transformation yields a polynomial time algorithm.

More importantly, though, I am glad that this chapter gives me a chance to dedicate it to the memory of two friends, Grisha Mints and Erik Palmgren, who are keenly missed.

2. Geometric Theories

Definition 2.1. *The* positive formulas *are constructed from atomic formulas and falsum* $\bot$ *by* $\wedge$ $\vee$ *and* $\exists$.

Geometric implications *are made up of the positive formulas, implications of positive formulas and the result of prefixing universal quantifiers to positive formulas and implications of positive formulas.*

---

[c]The translations of Kolmogorov, Gentzen, Gödel and Johansson belong to the oldest and simplest syntactic transformations in proof theory. They have been modified and mulled over many times and are still being discussed in relatively recent contemporary research, some of which is related to the theme of this chapter, e.g., [13–17].

[d]Besides, the current chapter is much indebted to Ishihara's article [11].

$\neg\varphi$ is defined as $\varphi \to \bot$. Thus, if φ is a positive formula, then $\neg\varphi$ is a geometric implication.

A theory is geometric *if all its axioms are geometric implications.*

In the following, we shall give several examples of geometric theories.

Examples 2.2.

(i) Robinson arithmetic formulated in the language with a constant 0, a unary successor function symbol s, binary function symbols $+$ and $\cdot$, and a binary predicate symbol $<$.

(ii) The theories of groups, rings, local rings and division rings have geometric axiomatisations. Local rings are commutative rings with $0 \neq 1$ having just one maximal ideal. On the face of it, the latter property appears to be of second order, but it can be rendered geometrically as follows:

$$\forall x\,(\exists y\, x \cdot y = 1 \ \vee\ \exists y\,(1-x)\cdot y = 1).$$

(iii) The theories of fields, ordered fields, algebraically closed fields and real closed fields have geometric axiomatisations. To express invertibility of nonzero elements, one uses $\forall x\,(x = 0 \vee \exists y\, x{\cdot}\mathrm{y}{=}1)$ rather than the nongeometric axiom $\forall x\,(x \neq 0 \to \exists y\, x \cdot y = 1)$. To express algebraic closure, replace axioms

$$s \neq 0 \to \exists x\, sx^n + t_1x^{n-1} + \cdots + t_{n-1}x + t_n = 0$$

by

$$s = 0 \vee \exists x\, sx^n + t_1x^{n-1} + \cdots + t_{n-1}x + t_n = 0,$$

where sx^k is short for $s \cdot x \cdot \cdots \cdot x$ with k many x.

Also the theory of *differential fields* has a geometric axiomatisation. This theory is written in the language of rings with an additional unary function symbol δ. The axioms are the field axioms plus $\forall x \forall y\, \delta(x+y) = \delta(x) + \delta(y)$ and $\forall x \forall y\, \delta(x \cdot y) = x \cdot \delta(y) + y \cdot \delta(x)$.

(iv) The theory of projective geometry has a geometric axiomatisation.

(v) The theories of equivalence relations, dense linear orders, infinite sets and graphs also have geometric axiomatisations.

It is also interesting that Kant's logic in his *Critique of Pure Reason* [19] and the *Jäsche Logik* [20] can be identified with geometric logic as shown by T. Achourioti and M. van Lambalgen in [21,22].

3. Conservativity

The best and most elegant proof system for proof-theoretic investigations is Gentzen's sequent calculus. With minor notational variations, this chapter will follow the presentation in Takeuti's book *Proof Theory* [23]. We will deviate, though, a bit from the setup in Chapter 1 of [23], in that we

- use $\bot$ as a propositional constant (or 0-ary predicate symbol) and define $\neg\varphi$ to be $\varphi \to \bot$;
- use the symbol $\to$ rather than $\supset$ for the implication symbol;
- use $\Rightarrow$ to separate the left and right part of a sequent, i.e., $\Gamma \Rightarrow \Delta$ rather than $\Gamma \to \Delta$;
- add sequents $\Gamma, \bot \Rightarrow \Delta$ as axioms[e] and omit the rules for $\neg$ (this axiom scheme for $\bot$ will be refereed to as $\mathsf{Ax}_\bot$).

Definition 3.1. Intuitionistic sequents $\Gamma \Rightarrow \Delta$ satisfy the extra requirement that the succedent Δ contains at most one formula. In the **intuitionistic** version of this sequent calculus, only intuitionistic sequents are allowed. In the **minimal logic** version, only intuitionistic sequents are permitted and the scheme $\mathsf{Ax}_\bot$ is omitted.

We convey derivability of a sequent $\Gamma \Rightarrow \Delta$ in classical, intuitionistic and minimal logic by writing $\vdash_c \Gamma \Rightarrow \Delta$, $\vdash_i \Gamma \Rightarrow \Delta$ and $\vdash_m \Gamma \Rightarrow \Delta$, respectively.

A theory T is a set of sentences. In derivability in T, one can use any sequent $\Gamma \Rightarrow \varphi$ with $\varphi \in T$ as an axiom (initial sequent). $T \vdash_c \Gamma \Rightarrow \Delta$, $T \vdash_i \Gamma \Rightarrow \Delta$ and $T \vdash_m \Gamma \Rightarrow \Delta$ are defined accordingly.

For a formula φ, we shall write $\vdash_c \varphi$, $\vdash_i \varphi$, and $\vdash_m \varphi$ to convey that $\vdash_c \emptyset \Rightarrow \varphi$, $\vdash_c \emptyset \Rightarrow \varphi$ and $\vdash_c \emptyset \Rightarrow \varphi$, respectively, where $\emptyset$ stands for the empty sequence of formulas.

Definition 3.2. *We shall use* E *as a symbol for a new propositional constant (or predicate symbol of arity 0). Its purpose will be to serve*

[e]In [23], axioms are called *initial sequents*.

as a placeholder for an arbitrary formula. Let $\neg_{\mathsf{E}}\varphi$ *be an abbreviation for* $\varphi \to \mathsf{E}$*. The* E*-negative translation* ${}^{\mathsf{E}}$ *is defined as follows:*

$$P^{\mathsf{E}} := \neg_{\mathsf{E}}\neg_{\mathsf{E}}P \text{ for } P \text{ prime and } P \not\equiv \bot;\ \bot^{\mathsf{E}} := \mathsf{E}$$

$$(\varphi \circ \psi)^{\mathsf{E}} := \varphi^{\mathsf{E}} \circ \psi^{\mathsf{E}} \text{for } \circ \in \{\wedge, \to\};\ (\varphi \vee \psi)^{\mathsf{E}} := \neg_{\mathsf{E}}\neg_{\mathsf{E}}(\varphi^{\mathsf{E}} \vee \psi^{\mathsf{E}})$$

$$(\forall x\, \varphi)^{\mathsf{E}} := \forall x\, \varphi^{\mathsf{E}};\ (\exists x\, \varphi)^{\mathsf{E}} := \neg_{\mathsf{E}}\neg_{\mathsf{E}}\exists x\, \varphi^{\mathsf{E}}.$$

The foregoing translation is basically the Gentzen–Gödel negative translation (see [10, 3.4, 3.5]), which engineers an interpretation of classical logic in minimal logic.

Corollary 3.3. *Given a theory* T*, the theory* T^{E} *has as axioms all formulas* ψ^{E} *with* ψ *an axiom of* T*.*

(i) $\vdash_m \neg_{\mathsf{E}}\neg_{\mathsf{E}}\varphi^{\mathsf{E}} \leftrightarrow \varphi^{\mathsf{E}}$.
(ii) $T \vdash_c \varphi \;\Rightarrow\; T^{\mathsf{E}} \vdash_m \varphi^{\mathsf{E}}$.

Proof. Since $\vdash_m \neg_{\mathsf{E}}\neg_{\mathsf{E}}(\theta_0 \vee \theta_1) \leftrightarrow \neg_{\mathsf{E}}(\neg_{\mathsf{E}}\theta_0 \wedge \neg_{\mathsf{E}}\theta_1)$ and $\vdash_m \neg_{\mathsf{E}}\neg_{\mathsf{E}}\exists x\theta(x) \leftrightarrow \neg_{\mathsf{E}}\forall x\neg_{\mathsf{E}}\theta(x)$ hold, it follows that the ${}^{\mathsf{E}}$-translation amounts to the same as the Gentzen–Gödel translation (the ^{g}-translation in [10, 3.4, 3.5]), taking into account that in minimal logic $\bot$ is an arbitrary propositional constant for which we can substitute E. Hence, (i) follows from [10, 2.3.3] and (ii) from [10, 2.3.5]. □

In the following, we shall frequently adopt the convention that a string of implications $\to$ is considered to be bracketed to the right, i.e., $\varphi_1 \to \varphi_2 \to \ldots \to \varphi_{n-1} \to \varphi_n$ is an abbreviation for $\varphi_1 \to (\varphi_2 \to (\ldots(\varphi_{n-1} \to \varphi_n)\ldots))$.

Lemma 3.4.

(1) $\vdash_m \varphi \to \neg_{\mathsf{E}}\neg_{\mathsf{E}}\varphi$;
(2) $\vdash_m (\varphi \to \psi) \to (\neg_{\mathsf{E}}\neg_{\mathsf{E}}\varphi \to \neg_{\mathsf{E}}\neg_{\mathsf{E}}\psi)$;
(3) $\vdash_m \neg_{\mathsf{E}}\neg(\varphi \wedge \psi) \to \neg_{\mathsf{E}}\neg\varphi \wedge \neg_{\mathsf{E}}\neg\psi$;
(4) $\vdash_m \neg_{\mathsf{E}}\neg_{\mathsf{E}}\varphi \wedge \neg_{\mathsf{E}}\neg_{\mathsf{E}}\psi \to \neg_{\mathsf{E}}\neg_{\mathsf{E}}(\varphi \wedge \psi)$;
(5) $\vdash_m \neg_{\mathsf{E}}\neg(\varphi \vee \psi) \to \neg_{\mathsf{E}}(\neg\varphi \wedge \neg\psi)$;
(6) $\vdash_m \neg_{\mathsf{E}}\neg_{\mathsf{E}}(\neg_{\mathsf{E}}\neg_{\mathsf{E}}\varphi \vee \neg_{\mathsf{E}}\neg_{\mathsf{E}}\psi) \to \neg_{\mathsf{E}}\neg_{\mathsf{E}}(\varphi \vee \psi)$;
(7) $\vdash_m \neg_{\mathsf{E}}\neg(\varphi \to \psi) \to (\neg_{\mathsf{E}}\neg_{\mathsf{E}}\varphi \to \neg_{\mathsf{E}}\neg\psi)$;
(8) $\vdash_i (\neg_{\mathsf{E}}\neg\varphi \to \neg_{\mathsf{E}}\neg_{\mathsf{E}}\psi) \to \neg_{\mathsf{E}}\neg_{\mathsf{E}}(\varphi \to \psi)$;
(9) $\vdash_m \neg_{\mathsf{E}}\neg\forall x\varphi(x) \to \forall x\neg_{\mathsf{E}}\neg\varphi(x)$;
(10) $\vdash_m \neg_{\mathsf{E}}\neg_{\mathsf{E}}\exists x\neg_{\mathsf{E}}\neg_{\mathsf{E}}\varphi(x) \to \neg_{\mathsf{E}}\neg_{\mathsf{E}}\exists x\varphi(x)$.

Proof. These claims are stated in [11, Lemma 2] without proofs (1), (2), (4), (6) and (10) are well known with E replaced by $\bot$ (see, e.g., [7, 1.2]), so it's clear that they hold in minimal logic. We now turn to the interesting cases that mix $\neg$ and $\neg_{\mathsf{E}}$.
For (3), note that $\vdash_m \neg\varphi \to \neg(\varphi \wedge \psi)$ and $\vdash_m \neg\psi \to \neg(\varphi \wedge \psi)$, and therefore,

$$\vdash_m (\neg(\varphi \wedge \psi) \to \mathsf{E}) \to (\neg\varphi \to \mathsf{E}) \wedge (\neg\psi \to \mathsf{E}).$$

(5) We have $\vdash_m [(\varphi \to\bot) \wedge (\psi \to\bot)] \to (\varphi \vee \psi) \to\bot$, yielding

$$\vdash_m [((\varphi \vee \psi) \to\bot) \to \mathsf{E}] \to ([(\varphi \to\bot) \wedge (\psi \to\bot)] \to \mathsf{E}).$$

(7) Successively, we see that:

$$\begin{aligned}
&\vdash_m \neg\psi \to (\varphi \to \neg(\varphi \to \psi))\\
&\vdash_m \neg\psi \to \varphi \to (\neg(\varphi \to \psi) \to \mathsf{E}) \to \mathsf{E}\\
&\vdash_m \neg\psi \to (\neg(\varphi \to \psi) \to \mathsf{E}) \to \varphi \to \mathsf{E}\\
&\vdash_m \neg\psi \to (\neg(\varphi \to \psi) \to \mathsf{E}) \to ((\varphi \to \mathsf{E}) \to \mathsf{E}) \to \mathsf{E}\\
&\vdash_m (\neg(\varphi \to \psi) \to \mathsf{E}) \to ((\varphi \to \mathsf{E}) \to \mathsf{E}) \to \neg\psi \to \mathsf{E}
\end{aligned}$$

(8) $\vdash_i \neg\varphi \to \varphi \to \psi$ and $\vdash_i ((\varphi \to \psi) \to \mathsf{E}) \to \neg\varphi \to \mathsf{E}$, so
(a) $\vdash_i ((\varphi \to \psi) \to \mathsf{E}) \to [(\neg\varphi \to \mathsf{E}) \to (\psi \to \mathsf{E}) \to \psi] \to (\psi \to \mathsf{E}) \to \mathsf{E}$
(b) $\vdash_i ((\varphi \to \psi) \to \mathsf{E}) \to \psi \to \mathsf{E}$
From (a) and (b), we obtain the desired

$$\vdash_i ((\varphi \to \psi) \to \mathsf{E}) \to [(\neg\varphi \to \mathsf{E}) \to (\psi \to \mathsf{E}) \to \psi] \to \mathsf{E}.$$

(9) We have $\vdash_m (\varphi(a) \to\bot) \to \forall x\varphi(x) \to\bot$, and hence

$$\begin{aligned}
&\vdash_m [(\forall x\varphi(x) \to\bot) \to \mathsf{E}] \to (\varphi(a) \to\bot) \to \mathsf{E}, \quad \text{whence}\\
&\vdash_m [(\forall x\varphi(x) \to\bot) \to \mathsf{E}] \to \forall x[(\varphi(x) \to\bot) \to \mathsf{E}].
\end{aligned}$$ □

The following syntactic classes bear some resemblance to the *spreading, wiping and isolating* schemata and formulas in [7] but are actually singled out in [11].

Definition 3.5. *We define syntactic classes of formulas $\mathcal{Q}$, $\mathcal{R}$ and $\mathcal{J}$ simultaneously by the following clauses:*

(1) $\bot$ *and every atomic formula* $Pt_1 \ldots t_n$ *belong to* $\mathcal{Q}$. *If* $Q, Q', \tilde{Q}(a) \in \mathcal{Q}$, *then so are* $Q \wedge Q'$, $Q \vee Q'$, $\exists x \tilde{Q}(x)$ *and* $\forall x \tilde{Q}(x)$. *If* $Q \in \mathcal{Q}$ *and* $J \in \mathcal{J}$, *then* $J \to Q \in \mathcal{Q}$.
(2) $\bot \in \mathcal{R}$. *If* $R, R', \tilde{R}(a) \in \mathcal{R}$, *then so are* $R \wedge R'$, *and* $\forall x \tilde{R}(x)$. *If* $R \in \mathcal{R}$ *and* $J \in \mathcal{J}$, *then* $J \to R \in \mathcal{R}$.
(3) $\bot$ *and every atomic formula* $Pt_1 \ldots t_n$ *belong to* $\mathcal{J}$. *If* J, J', $\tilde{J} \in \mathcal{J}$, *then so are* $J \wedge J'$, $J \vee J'$ *and* $\exists x \tilde{J}(x)$. *If* $J \in \mathcal{J}$ *and* $R \in \mathcal{R}$, *then* $R \to J \in \mathcal{J}$.

Corollary 3.6.

(i) *All positive formulas are in both,* $\mathcal{Q}$ *and* $\mathcal{J}$.
(ii) *All geometric implications are in* $\mathcal{Q}$.

Proof. Obvious. □

The following proposition is due to Ishihara [11].

Proposition 3.7.

(i) *For* $\varphi \in \mathcal{Q}$, $\vdash_i \varphi \to \varphi^{\mathsf{E}}$.
(ii) *For* $\psi \in \mathcal{R}$, $\vdash_i \neg_{\mathsf{E}}\neg\psi \to \psi^{\mathsf{E}}$.
(iii) *For* $\theta \in \mathcal{J}$, $\vdash_i \theta^{\mathsf{E}} \to \neg_{\mathsf{E}}\neg_{\mathsf{E}}\theta$.

Proof. We prove these derivabilities simultaneously by induction on the generation of the classes $\mathcal{Q}, \mathcal{R}, \mathcal{J}$. The proof given here is more detailed than the one for [11, Proposition 7].

(i) Obviously, we have $\vdash_i \bot \to \mathsf{E}$ and $\vdash_i \psi \to (\psi \to \mathsf{E}) \to \mathsf{E}$, which yields $\vdash_i A \to A^{\mathsf{E}}$ for atomic formulas A.

Now suppose $\vdash_i Q_i \to Q_i^{\mathsf{E}}$ for $i \in \{0,1\}$. Then $\vdash_i Q_0 \wedge Q_1 \to Q_0^{\mathsf{E}} \wedge Q_1^{\mathsf{E}}$, thus, $\vdash_i Q_0 \wedge Q_1 \to (Q_0 \wedge Q_1)^{\mathsf{E}}$. Likewise, one has $\vdash_i Q_0 \vee Q_1 \to Q_0^{\mathsf{E}} \vee Q_1^{\mathsf{E}}$ and, hence, $\vdash_i Q_0 \vee Q_1 \to \neg_{\mathsf{E}}\neg_{\mathsf{E}}(Q_0^{\mathsf{E}} \vee Q_1^{\mathsf{E}})$, i.e., $\vdash_i Q_0 \vee Q_1 \to (Q_0 \vee Q_1)^{\mathsf{E}}$.

Next assume $\vdash_i Q(a) \to Q(a)^{\mathsf{E}}$. Then $\vdash_i \forall x Q(x) \to \forall x Q(x)^{\mathsf{E}}$, so $\vdash_i \forall x Q(x) \to (\forall x Q(x))^{\mathsf{E}}$. Likewise, we have $\vdash_i \exists x Q(x) \to \exists x Q(x)^{\mathsf{E}}$, and so $\vdash_i \exists x Q(x) \to \neg_{\mathsf{E}}\neg_{\mathsf{E}}\exists x Q(x)^{\mathsf{E}}$, which is $\vdash_i \exists x Q(x) \to (\exists x Q(x))^{\mathsf{E}}$.

Finally, assume $\vdash_i J^{\mathsf{E}} \to \neg_{\mathsf{E}}\neg_{\mathsf{E}}J$ and $\vdash_i Q \to Q^{\mathsf{E}}$. Then, as $\vdash_i (J \to Q) \to (\neg_{\mathsf{E}}\neg_{\mathsf{E}}J \to \neg_{\mathsf{E}}\neg_{\mathsf{E}}Q)$ holds by Lemma 3.4(2),

$$(*) \quad (J \to Q) \to (J^{\mathsf{E}} \to \neg_{\mathsf{E}}\neg_{\mathsf{E}}Q).$$

We also obtain $\vdash_i \neg_{\mathsf{E}}\neg_{\mathsf{E}}Q^{\mathsf{E}} \to Q^{\mathsf{E}}$ from Corollary 3.3(i). As $\vdash_i Q \to Q^{\mathsf{E}}$ yields $\vdash_i \neg_{\mathsf{E}}\neg_{\mathsf{E}}Q \to \neg_{\mathsf{E}}\neg_{\mathsf{E}}Q^{\mathsf{E}}$, we have $\vdash_i \neg_{\mathsf{E}}\neg_{\mathsf{E}}Q \to Q^{\mathsf{E}}$, which yields $\vdash_i (J \to Q) \to (J^{\mathsf{E}} \to Q^{\mathsf{E}})$ by $(*)$, thus $\vdash_i (J \to Q) \to (J \to Q)^{\mathsf{E}}$.

(ii) Since $\vdash_i \neg\perp$ we have $\vdash_i \neg_{\mathsf{E}}\neg\perp \to \mathsf{E}$.

Now suppose that $\vdash_i \neg_{\mathsf{E}}\neg R_j \to R_j^{\mathsf{E}}$ holds for $j \in \{0,1\}$. According to Lemma 3.4(3), we have

$$\vdash_i \neg_{\mathsf{E}}\neg(R_0 \wedge R_1) \to \neg_{\mathsf{E}}\neg R_0 \ \wedge\ \neg_{\mathsf{E}}\neg R_1,$$

and thus $\vdash_i \neg_{\mathsf{E}}\neg(R_0 \wedge R_1) \to R_0^{\mathsf{E}} \wedge R_1^{\mathsf{E}}$, i.e., $\vdash_i \neg_{\mathsf{E}}\neg(R_0 \wedge R_1) \to (R_0 \wedge R_1)^{\mathsf{E}}$.

Next assume that $\vdash_i \neg_{\mathsf{E}}\neg R(a) \to R(a)^{\mathsf{E}}$. By Lemma 3.4(9), we have $\vdash_i \neg_{\mathsf{E}}\neg\forall x R(x) \to \forall x \neg_{\mathsf{E}}\neg R(x)$. Therefore, $\vdash_i \neg_{\mathsf{E}}\neg\forall x R(x) \to \forall x R(x)^{\mathsf{E}}$, i.e., $\vdash_i \neg_{\mathsf{E}}\neg\forall x R(x) \to (\forall x R(x))^{\mathsf{E}}$.

Finally, suppose that $\vdash_i J^{\mathsf{E}} \to \neg_{\mathsf{E}}\neg_{\mathsf{E}}J$ and $\vdash_i \neg_{\mathsf{E}}\neg R \to R^{\mathsf{E}}$. Then,

$$\vdash_i (\neg_{\mathsf{E}}\neg_{\mathsf{E}}J \to \neg_{\mathsf{E}}\neg R) \to (J^{\mathsf{E}} \to R^{\mathsf{E}}),$$

and hence, by Lemma 3.4(7), $\vdash_i \neg_{\mathsf{E}}\neg(J \to R) \to (J^{\mathsf{E}} \to R^{\mathsf{E}})$, i.e., $\vdash_i \neg_{\mathsf{E}}\neg(J \to R) \to (J \to R)^{\mathsf{E}}$.

(iii) We have $\vdash_i \perp^{\mathsf{E}} \to \neg_{\mathsf{E}}\neg_{\mathsf{E}}\perp$ and $\vdash_i A^{\mathsf{E}} \to \neg_{\mathsf{E}}\neg_{\mathsf{E}}A$ for atomic A since $A^{\mathsf{E}} \equiv \neg_{\mathsf{E}}\neg_{\mathsf{E}}A$.

Now assume that $\vdash_i J_i^{\mathsf{E}} \to \neg_{\mathsf{E}}\neg_{\mathsf{E}}J_i$ for $i \in \{0,1\}$. Then $\vdash_i (J_0 \wedge J_1)^{\mathsf{E}} \to (\neg_{\mathsf{E}}\neg_{\mathsf{E}}J_0 \wedge \neg_{\mathsf{E}}\neg_{\mathsf{E}}J_1)$, thus, $\vdash_i (J_0 \wedge J_1)^{\mathsf{E}} \to \neg_{\mathsf{E}}\neg_{\mathsf{E}}(J_0 \wedge J_1)$ follows by Lemma 3.4(4).

We also have $\vdash_i J_0^{\mathsf{E}} \vee J_1^{\mathsf{E}} \to \neg_{\mathsf{E}}\neg_{\mathsf{E}}J_0 \vee \neg_{\mathsf{E}}\neg_{\mathsf{E}}J_1$ and, hence,

$$\vdash_i \neg_{\mathsf{E}}\neg_{\mathsf{E}}(J_0^{\mathsf{E}} \vee J_1^{\mathsf{E}}) \to \neg_{\mathsf{E}}\neg_{\mathsf{E}}(\neg_{\mathsf{E}}\neg_{\mathsf{E}}J_0 \vee \neg_{\mathsf{E}}\neg_{\mathsf{E}}J_1),$$

from which $\vdash_i (J_0 \vee J_1)^{\mathsf{E}} \to \neg_{\mathsf{E}}\neg_{\mathsf{E}}(J_0 \vee J_1)$ follows by Lemma 3.4(6).

Assuming $\vdash_i J(a)^{\mathsf{E}} \to \neg_{\mathsf{E}}\neg_{\mathsf{E}}J(a)$, we have $\vdash_i \exists x J(x)^{\mathsf{E}} \to \exists x \neg_{\mathsf{E}}\neg_{\mathsf{E}}J(x)$, and therefore, $\vdash_i \neg_{\mathsf{E}}\neg_{\mathsf{E}}\exists x J(x)^{\mathsf{E}} \to \neg_{\mathsf{E}}\neg_{\mathsf{E}}\exists x J(x)$ by Lemma 3.4(10), i.e.,
$\vdash_i (\exists x J(x))^{\mathsf{E}} \to \neg_{\mathsf{E}}\neg_{\mathsf{E}}\exists x J(x)$.

Finally, assume $\vdash_i \neg_{\mathsf{E}}\neg R \to R^{\mathsf{E}}$ and $\vdash_i J^{\mathsf{E}} \to \neg_{\mathsf{E}}\neg_{\mathsf{E}}J$. Then,

$$\vdash_i (R^{\mathsf{E}} \to J^{\mathsf{E}}) \to \neg_{\mathsf{E}}\neg R \to \neg_{\mathsf{E}}\neg_{\mathsf{E}}J,$$

and hence, by Lemma 3.4(8), $\vdash_i (R^{\mathsf{E}} \to J^{\mathsf{E}}) \to \neg_{\mathsf{E}}\neg_{\mathsf{E}}(R \to J)$, i.e., $\vdash_i (R \to J)^{\mathsf{E}} \to \neg_{\mathsf{E}}\neg_{\mathsf{E}}(R \to J)$. □

We will make use of substitutions for variables and for the placeholder E. In the sequent calculi à la Gentzen and Takeuti [23] and in the Schütte calculi [24,25], one distinguishes syntactically between free $a, b, c, \ldots$ and bound $x, y, z, \ldots$ variables. As terms can contain only free variables, there will never be a problem of substitutability of terms for variables.

We use $\varphi(a/t)$ for the result of replacing every occurrence of the free variable a in φ by the term t. Similarly, for a sequent $\Gamma \Rightarrow \Delta$ and a derivation $\mathcal{D}$, we use $\Gamma(a/t) \Rightarrow \Delta(a/t)$ and $\mathcal{D}(a/t)$, respectively, for the result of replacing every occurrence of the free variable a by the term t. Note, however, that while $\varphi(t/a)$ will be a formula, too, $\mathcal{D}(a/t)$ may no longer be a derivation.

In a similar vein, for a propositional constant E we convey the result of replacing each of its occurrences in a formula, sequent and derivation via $\varphi(\mathsf{E}/\psi)$, $\Gamma(\mathsf{E}/\psi) \Rightarrow \Delta(\mathsf{E}/\psi)$ and $\mathcal{D}(\mathsf{E}/\psi)$, respectively. However, we have to add a caveat here. The formula formation rules in calculi that have different symbols for free and bound variables (such as [23]) allow to go from a formula $\varphi(a)$ to $\forall x \varphi(x)$ only if x does not already occur in $\varphi(a)$. Thus, if one substitutes a formula ψ for E in a formula, the resulting syntactic object may no longer be a formula. As a result, we tacitly require that before substitutions are made, bound variables in ψ have to be replaced by ones that avoid this clash.

Now, by obeying this additional requirement, $\varphi(\mathsf{E}/\psi)$ will be again a formula. However, $\mathcal{D}(\mathsf{E}/\psi)$ may no longer be a derivation as some eigenvariable conditions may have become violated in the process. So, we have to take care of that as well.

Two substitutability results will be useful.

Lemma 3.8. *Let T be a theory. If $\mathcal{D}(a)$ is a T-derivation of $\Gamma \Rightarrow \Delta$ and c is a variable that doesn't occur in $\mathcal{D}$, then $\mathcal{D}(a/c)$ is a T-derivation of $\Gamma(a/c) \Rightarrow \Delta(a/c)$.*

Proof. This is obvious as c is a completely new variable as far as $\mathcal{D}$ is concerned, so no eigenvariable conditions are affected by this substitution. Formally, one proves this by induction on the number of inferences of $\mathcal{D}$ (see [23, Ch. 1, Lemma 2.10]. □

Proposition 3.9. *Let T be a theory whose axioms do not contain* E. *If $\mathcal{D}$ is a T-derivation of $\Gamma \Rightarrow \Delta$ and ψ is an arbitrary formula, then there is a T-derivation $\mathcal{D}'$ of $\Gamma(\mathsf{E}/\psi) \Rightarrow \Delta(\mathsf{E}/\psi)$.*

Proof. Use induction on the number of inferences of $\mathcal{D}$. The only kinds of inferences we need to look at are $\forall$: right and $\exists$: left. So suppose the last inference of $\mathcal{D}$ was $\forall$: right with premise $\Gamma \Rightarrow \Delta', \varphi$ and conclusion $\Gamma \Rightarrow \Delta_0, \forall x\, \varphi(a/x)$, where Δ is $\Delta_0, \forall x\, \varphi(a/x)$ and a does not occur in $\Gamma \Rightarrow \Delta$. Let $\mathcal{D}_0$ be the immediate subderivation of $\mathcal{D}$ with end sequent $\Gamma \Rightarrow \Delta_0, \varphi$. Let c be a free variable that neither occurs in $\mathcal{D}$ nor in ψ. By Lemma 3.8, $\mathcal{D}_1 := \mathcal{D}_0(a/c)$ is a derivation, too. Note that $\mathcal{D}_1$ is a derivation of

$$\Gamma \Rightarrow \Delta_0, \quad \varphi(a/c),$$

owing to the eigenvariable condition satisfied by a.

Since $\mathcal{D}_1$ has fewer inferences than $\mathcal{D}$, we can apply the induction hypothesis to arrive at a derivation $\mathcal{D}_2$ of

$$\Gamma(\mathsf{E}/\psi) \Rightarrow \Delta_0(\mathsf{E}/\psi), \quad (\varphi(a/c))(\mathsf{E}/\psi).$$

As c does not occur in $\Gamma(\mathsf{E}/\psi) \Rightarrow \Delta_0(\mathsf{E}/\psi)$ and ψ, we can apply an inference $\forall$: right to obtain a derivation $\mathcal{D}'$ of

$$\Gamma(\mathsf{E}/\psi) \Rightarrow \Delta_0(\mathsf{E}/\psi), \quad \forall x\, (((\varphi(a/c))(\mathsf{E}/\psi))(c/x)),$$

which is the same as $\Gamma(\mathsf{E}/\psi) \Rightarrow \Delta_0(\mathsf{E}/\psi)$ since $\forall x\, (((\varphi(a/c))(\mathsf{E}/\psi))(c/x)) \equiv (\forall x\, \varphi(a/x))(\mathsf{E}/\psi)$. $\exists$: left is dealt with in a similar fashion. $\square$

We still haven't strictly shown that the proof length increases at most polynomially. This will be addressed in the next section.

Theorem 3.10. *If T is a geometric theory, i.e., the axioms of T are geometric implications, and φ is a geometric implication, then*

$$T \vdash_c \varphi \quad \text{yields } T \vdash_i \varphi.$$

Moreover, if $\mathcal{D}$ is a classical deduction of φ in T, then the size of the intuitionistic deduction of φ in T increases at most polynomially in the size of $\mathcal{D}$.

Proof. Suppose $T \vdash_c \varphi$. By Corollary 3.3(ii), we conclude that

$$T^{\mathsf{E}} \vdash_i \varphi^{\mathsf{E}}. \tag{1}$$

As $T \subseteq \mathcal{Q}$, it follows from (1) and Proposition 3.7(i) that

$$T \vdash_i \varphi^{\mathsf{E}}. \tag{2}$$

Now φ is of the form $\forall x_1 \ldots \forall x_r(\psi \rightarrow \theta)$ with $\psi, \theta \in \mathcal{Q} \cap \mathcal{J}$. Thus, by Proposition 3.7(i), we can conclude from (2) that

$$T \vdash_i \psi \rightarrow \theta^{\mathsf{E}}. \tag{3}$$

Fact (3) in conjunction with Proposition 3.7(iii) yields

$$T \vdash_i \psi \rightarrow \neg_{\mathsf{E}}\neg_{\mathsf{E}}\theta. \tag{4}$$

Now, since E is just a placeholder we may substitute θ for E everywhere in the derivation of (4) by Proposition 3.9, yielding a derivation showing that

$$T \vdash_i \psi \rightarrow ((\theta \rightarrow \theta) \rightarrow \theta). \tag{5}$$

As a result of (5), we have $T \vdash_i \psi \rightarrow \theta$ and hence $T \vdash_i \varphi$, as desired. □

From the proof of the foregoing theorem, it's clear that conservativity obtains for a wider collection of theories than just geometric ones.

Corollary 3.11. *If T is a theory whose axioms are in $\mathcal{Q}$ and φ is a geometric implication, then*

$$T \vdash_c \varphi \quad \text{yields } T \vdash_i \varphi.$$

Moreover, if $\mathcal{D}$ is a classical deduction of φ in T, then the size of the intuitionistic deduction of φ in T increases at most polynomially in the size of $\mathcal{D}$.

Proof. This follows from the proof of Theorem 3.10. □

4. Polynomial Time Bounds

In view of the foregoing results, it might be rather obvious that the transformation of a classical proof of a geometric implication in a geometric theory into an intuitionistic proof can be carried out in polynomial time. It might be in order, though, to be a bit more precise. The plan, however, is not to do this in detail but rather from a "higher" point of view.

It is a fact that the syntax of first-order logic can be recognised and manipulated by polynomial time algorithms in Buss's theory S^1_2. One place where the arithmetisation of metamathematics for the sequent calculus is carried out in detail is [26, Ch. 7]. Among other things, we require functions for the arithmetisation of substitutions in Lemma 3.8 and Proposition 3.9. They give rise to Σ^b_1-defined functions of S^1_2 (see [26, p. 130], where this is carried for substitution of a term into a formula). Moreover, all Σ^b_1-definable functions of S^1_2 are polynomial time computable functions (see [26, Corollary 8]). As all manipulation of proofs in this chapter can be carried out by Σ^b_1-definable functions of S^1_2, we have achieved our goal.

5. ∞-Geometric Theories

5.1. *The infinite geometric case*

Much more powerful notions of geometricity are available in infinitary logics. $\mathcal{L}_{\infty\omega}$-logic allows for the formation of infinite disjunctions $\bigvee \Phi$ and conjunctions $\bigwedge \Phi$, where Φ is an arbitrary set of (infinitary) formulae. In this richer syntax, a formula is said to be an ∞-*positive* formula if it can be generated from atoms and $\bot$ via $\vee, \wedge, \exists$ and $\bigvee$. More precisely, the latter means that the infinite disjunction $\bigvee \Phi$ is an ∞-positive formula whenever Φ is a set of ∞-positive formula.

The ∞-geometric implications are obtained in the same way from the ∞-positive formula as the geometric implications are obtained from the positive formulas. An ∞-geometric theory is one whose axioms are ∞-geometric implications.

Examples of such theories are the theories of *flat modules* over a ring (see [27]), *torsion groups*, fields of prime characteristic, Archimedean ordered fields and connected graphs. Even Peano arithmetic has an ∞-geometric axiomatisation (see [5, 2.4]).

∞-geometric classical theories are also conservative over their intuitionistic version for ∞-geometric formulas. This can be proved in Constructive Zermelo–Fraenkel set theory **CZF** (see [5, Theorem 7.9]) via cut elimination for $\mathcal{L}_{\infty\omega}$. The techniques of this chapter can also be extended to the $\mathcal{L}_{\infty\omega}$ context. As a result one can prove conservativity already in a much weaker fragment of **CZF**, namely, intuitionistic Kripke–Platek set theory with elementhood induction restricted to Σ-formulas, $\mathbf{IKP}^r$.

As the theories **CZF** and $\mathbf{IKP}^r$ allow for witness extraction from proofs of existential statements (see [28, 6.1], [29, 2.35], [30]), this offers the exciting prospect of extracting bounds from proofs in classical ∞-geometric theories. However, note that for this the classical proof must exist as an object in the constructive background theory. So it's not enough to know of its existence by appealing to principles such as the axiom of choice or Zorn's lemma.

Acknowledgement

The author acknowledges support via the John Templeton Foundation grant "A new dawn of intuitionism".

Thanks are owed to the anonymous referees of this chapter for valuable comments that helped to improve it.

References

[1] S. Negri, Contraction-free sequent calculi for geometric theories, with an application to Barr's theorem, *Arch. Math. Log.* **42**, 389–401 (2003).
[2] S. Negri, Glivenko sequent classes in the light of structural proof theory, *Arch. Math. Log.* **55**, 461–473 (2016).
[3] S. Negri and J. von Plato, *Proof Analysis.* Cambridge University Press, Cambridge, 2011.
[4] V.P. Orevkov, Glivenko's sequence classes. In V.P. Orevkov (ed.), *Logical-mathematical Calculus.* Part 1. Inst. Steklov, Leningrad, 1968, pp. 131–154.

[5] M. Rathjen, Remarks on Barr's theorem: Proofs in geometric theories. In D. Probst and P. Schuster (eds.), *Concepts of Proof in Mathematics, Philosophy, and Computer Science, vol. 6, Ontos Mathematical Logic.* De Gruyter, 2016, pp. 347–374.

[6] G. Mints, Classical and intuitionistic geometric logic, *IfCoLog J. Log. Appl.* **4**, 1227–1234 (2017).

[7] D. Leivant, Syntactic translations and provably recursive functions, *J. Symb. Log.* **50**, 682–688 (1985).

[8] A. Kolmogorov, O principe tertium non datur, *Matematiceskij Sbornik.* **32**, 646–667 (1925).

[9] I. Johansson, Der Minimalkalkül, ein reduzierter intuitionistischer Formalismus, *Compositio Mathematica.* **4**, 119–136 (1937).

[10] A. Troelstra and D. van Dalen, *Constructivism in Mathematics*, vols. I, II. North Holland, Amsterdam, 1988.

[11] H. Ishihara, A note on the Gödel-Gentzen translation, *Math. Log. Quart.* **46**, 135–137 (2000).

[12] R. Dyckhoff and S. Negri, Commentary on Grigori Mints' "classical and intuitionistic geometric logic", *IfCoLog J. Log. App.* **4**, 1235–1238 (2017).

[13] P. Aczel, The Russell–Prawitz modality, *Math. Structures Comput. Sci.* **11**(4), 541–554 (2001).

[14] M. Escardó and P. Oliva, The Peirce translation and the double negation shift. In F. Ferreira, B. Löwe, E. Mayordomo, and L. Mendes Gomes (eds.), *Programs, Proofs, Processes, vol. 6158, Lecture Notes in Computer Science*, Springer, 2010, pp. 151–161.

[15] G. Ferreira and P. Oliva, *On the Relation Between Various Negative Translations.* In U. Berger, H. Diener, P. Schuster, and M. Seisenberger (eds.), *Logic, Construction, Computation, vol. 3, Ontos Math. Log.* De Gruyter, 2012.

[16] R. Goldblatt, Grothendieck topology as geometric modality, *MLQ Math. Log. Q.* **27**(31–35), 495–529 (1981).

[17] F. Pfenning and R. Davies, A judgmental reconstruction of modal logic, *Math. Structures Comput. Sci.* **11**(4), 511–540 (2001).

[18] E. Palmgren. An intuitionistic axiomatisation of real closed fields, *Math. Log. Quart.* **48**, 297–299 (2002).

[19] I. Kant, *Critique of Pure Reason; Translated from the German by P. Guyer and A.W. Wood.* The Cambridge edition of the works of Immanuel Kant, Cambridge University Press, Cambridge, MA, 1998.

[20] I. Kant, *Lectures on Logic; Translated from the German by J.M. Young.* The Cambridge edition of the works of Immanuel Kant, Cambridge University Press, Cambridge, MA, 1992.

[21] T. Achourioti and M. van Lambalgen, A formalization of Kant's transcendental logic, *Rev. Symb. Log.* **4**, 254–289 (2011).
[22] T. Achourioti and M. van Lambalgen, Kant's logic revisited, *The IfCoLog J. Logics and their Applications.* **4**, 845–865 (2017).
[23] G. Takeuti, *Proof Theory.* North Holland, Amsterdam, 1987.
[24] K. Schütte, *Beweistheorie.* Springer, Berlin, 1960.
[25] K. Schütte, *Proof Theory.* Springer, Berlin, 1977.
[26] S. Buss, *Bounded Arithmetic.* Bibliopolis, Napoli, 1986.
[27] G.C. Wraith, Intuitionistic algebra: Some recent developments in topos theory. In *Proceedings of the International Congress of Mathematicians Helsinki.* Acad. Sci. Fennica, Helsinki, 1980, pp. 331–337.
[28] M. Rathjen, From the weak to the strong existence property, *Ann. Pure Appl. Log.* **163**, 1400–1418 (2012).
[29] J. Cook and M. Rathjen, Ordinal analysis of intuitionistic power and exponentiation Kripke Platek set theory. In R. Kahle, T. Strahm, and T. Studer (eds.), *Advances in Proof Theory, vol. 28, Progress in Computer Science and Applied Logic.* Birkhäuser 2016, pp. 79–179.
[30] M. Rathjen, Ordinal analysis and the existence property for intuitionistic set theories, *Philosophical Transactions A* (2023), to appear.

https://doi.org/10.1142/9789811245220_0014

Chapter 14

Domain Theory and Realisability over Scott's D_∞ in Constructive Set Theory

Eman Dihoum[*,‡], Michael Rathjen[*,§] and Avi Silterra[†,¶]

[*] *School of Mathematics, University of Leeds, Leeds LS2 9JT, England*

[†] *Parker, CO, USA*

[‡] *eman.dihoum@gmail.com*,
[§] *m.rathjen@leeds.ac.uk*
[¶] *avi.silterra@gmail.com*

The first nontrivial model of the untyped λ-calculus was Scott's D_∞ model, and it has exerted a great influence on the semantics of λ-calculi and the theory of programming languages. If one wants to develop domain theory in a constructive set theory such as **CZF**, though, one faces the problem that the notion of complete partial order implicitly uses the powerset axiom. However, most of domain theory can be attained with partial orders that are complete only with regard to denumerable ascending chains. In the chapter, it is shown that this development can be carried out on the basis of a fragment of **CZF**. The existence of various D_∞-models, however, is not provable in **CZF** alone but can be carried out in standard extensions of **CZF** with an appropriate form of the Regular Extension Axiom (which is still interpretable in Martin-Löf-type theory). Using the denizens of these models as realisers for realisability interpretations of set theory, it is shown that they provide universes of intuitionistic set theory in which there exists an infinite set A, comprising the naturals, that is, in one–one correspondence with the set of all functions from A to A, thereby furnishing a set-theoretic model of the λ-calculus, that is, one in which the λ-terms are interpreted as set-theoretic functions. We think that these realisability universes lend themselves to further interesting applications in the metamathematics of intuitionistic set theories.

1. Overview

Domain theory is a fascinating field with many facets and many challenges for the constructively minded researcher. The first three chapters (after the overview) of the present chapter develop crucial chunks of domain theory in Constructive Zermelo–Fraenkel set theory (**CZF**). The latter not merely uses intuitionistic logic, but is also based on set-theoretic axioms germane to a predicative foundation of mathematics in the sense of Martin–Löf-type theory. Since **CZF** does not have the Powerset Axiom, we avail ourselves of the insight that "most of domain theory can be formulated with partial orders that are complete only with respect to denumerable ascending chains" ([1, p. 3]). The approach here closely follows the presentation in [1, Chapter 3].

Beeson noted, quite some time ago, that "D_∞ models have not yet been applied to the metamathematics of constructive formal systems" ([2, p. 128]). In Chapter 5, we use the denizens of D_∞ as realisers for realisability interpretations of set theory. It is shown that they provide universes of intuitionistic set theory in which there exist infinite sets A (containing the set of naturals) that are in one–one correspondence with their own function space, that is, the set of all functions from A to A. In this way one obtains a set-theoretic model of the λ-calculus, that is, one in which the λ-terms are interpreted as set-theoretic functions.

In the last chapter, it is shown that realisability universes over D_∞ also validate certain Brouwerian principles, namely, various forms of the Fan Theorem and the Bar Induction Principle, provided that the same principles hold in the background universe.

Chapters 2 and 4 were originally written by Avi Silterra in the years 2004–2005 at the Ohio State University as part of a project supervised by Michael Rathjen. The material in Chapters 5 and 6 was originally adumbrated by Michael Rathjen and then worked out by Eman Dihoum in her PhD thesis [3] (supervised by the former).

2. Domain Theory Development in a Fragment of CZF

Large chunks of domain theory can be formalised in **CZF**. One does not even need the full strength of this theory. Rather than the full Set Induction scheme, one only needs induction on the naturals. What's more, Strong Collection can be replaced by Replacement and instead of Subset Collection, one only needs Exponentiation. We denote this fragment by $\mathbf{CZF}^w$. The latter theory is actually a predicative theory in the strict sense of Feferman and Schütte, having a proof-theoretic ordinal smaller than the Feferman–Schütte ordinal Γ_0 (this can be shown with the methods of [4]).

2.1. *Constructive Zermelo–Fraenkel set theory* CZF

We shall just recall the axioms of **CZF** (for details see [5–7]). The theory has the same first-order language as **ZF**, where the only non-logical symbol is $\in$ and it is based on the intuitionistic first-order logic with equality.

The axioms of **CZF** are *Extensionality*, *Pairing*, *Union*, *Set Induction Scheme* and *Infinity* in their usual forms and the following axiom schemas:

Bounded Separation Scheme

$$\forall x \exists y \forall a [a \in y \leftrightarrow a \in x \wedge \phi(a)],$$

for any bounded formula ϕ, where ϕ is bounded if all quantifiers occurring in ϕ are bounded.

Subset Collection Scheme

$$\begin{aligned}&\forall x \forall y \exists z \forall u [\forall a \in x \exists b \in y \psi(a,b,u)\\ &\quad \to \exists c \in z (\forall a \in x \exists b \in c \psi(a,b,u) \wedge \forall b \in c \exists a \in x \psi(a,b,u))],\end{aligned}$$

for any formula ψ.

Strong Collection Scheme

$$\forall x[\forall a \in x \exists b \phi(a, b) \rightarrow \exists y[\forall a \in x \exists b \in y \phi(a, b) \wedge \forall b \in y \exists a \in x \phi(a, b)]],$$

for any formula ϕ.

Strong Infinity Axiom

$$\exists a[\text{Ind}(a) \wedge \forall b[\text{Ind}(b) \rightarrow \forall x \in a(x \in b)]],$$

where we use the following abbreviations.

(1) $\text{Succ}(x, y)$ for $\forall z[z \in y \leftrightarrow z \in x \vee z = x]$,
(2) $\text{Ind}(a)$ for $(\exists y \in a)(\forall z \in y)\bot \wedge (\forall x \in a)(\exists y \in a)\text{Succ}(x, y)$. Here $\bot$ stands for falsum or "absurdity".[a]

2.2. *Preliminaries*

Let $(D, \leqslant)$ be a partially ordered set (poset). Usually we will write simply D with $\leqslant$ understood. That D is a set is needed in several instances so that we can use bounded instead of unbounded separation and iterate the steps of the construction. In this section, we shall work in $\mathbf{CZF}^w$. However, the next section will also feature ω-cpos and build models D_∞ that require additional axioms.

Definition 2.1 (Upper Bound). An upper bound for a class $S \subseteq D$ is an element $d \in D$ satisfying $UB_S(x) \equiv \forall s \in S\ s \leqslant x$.

Definition 2.2 (Least Upper Bound). A least upper bound (lub) for a class $S \subseteq D$ is an element d satisfying

$$LUB_S(x) \equiv UB_S(d) \wedge [\forall e \in D\ UB_S(e) \rightarrow d \leqslant e].$$

A lub, if it exists, is unique, and will be denoted by $\bigvee^D S$ or $\bigvee S$. Usually we will be concerned with S, a set.

Definition 2.3 (Least Element). A least element of D is an element $\bot \in D$ so that $\forall d \in D\ \bot \leqslant d$. Since a least element is necessarily unique, we will use $\bot_D$ or just $\bot$ for the least element of D if it exists.

[a]In intuitionistic logic, $\bot$ is used to express negation $\neg\phi$ as $\phi \rightarrow \bot$. Crucially, one has ex falso quodlibet, i.e., $\bot \rightarrow \psi$ for every formula ψ.

Definition 2.4 (Sequence). A sequence of elements in D is a function $f : \omega \to D$. We will usually write $f(n)$ as d_n and the image as $\langle x_n \rangle_{n<\omega}$, $\langle x_n \rangle$ or even just $\mathbf{x}$. The lub of a sequence, if it exists, will be denoted by $\bigvee_{n<\omega} x_n$ or $\bigvee x_n$.

Definition 2.5 (Ascending Chain). A sequence $\langle x_n \rangle_{n<\omega}$ of elements in D is ascending if $\forall n < \omega$, we have $x_n \leqslant x_{n+1}$.

Remark 2.6. Note that truncations of ascending chains with lubs have the same lub. In other words, for $p < \omega$, and an ascending sequence $\langle x_n \rangle$, we have

$$\bigvee_{n<\omega} x_n = \bigvee_{p<n<\omega} x_n,$$

when both exist, and when one does not exist, neither does the other.

Definition 2.7 (ω-dcpo). $(D, \leqslant)$ is an ω-dcpo if the image of every ascending sequence $\langle x_n \rangle$ has a least upper bound.

Definition 2.8 (ω-cpo). $(D, \leqslant)$ is an ω-cpo if $(D, \leqslant)$ is an ω-dcpo and D has a least element $\perp_D$.

Definition 2.9 (ω-continuity). A function $f : C \to D$ for ω-dcpos C, D is ω-continuous when for all ascending sequences $\langle c_n \rangle$, we have

$$\bigvee_{n<\omega}^{D} f(c_n) = f\left(\bigvee_{n<\omega}^{C} c_n\right),$$

where the lub on the left is in D and on the right is in C.

Example 2.10 (ω-cpos). Some classical examples of ω-cpos are any finite poset, and $\mathcal{P}(\omega)$, that is, the powerset of ω. In **CZF**, neither of these examples is acceptable without modification. In **CZF**, it is not possible to prove a nontrivial finite poset is an ω-cpo. This is because it would amount to solving the halting problem. The problem with unmodified $\mathcal{P}(\omega)$ is that $\mathcal{P}(\omega)$ is a proper class,[b] so that we cannot form the collection $[\mathcal{P}(\omega) \to \mathcal{P}(\omega)]$.

[b]To be more precise, **CZF** cannot prove that $\mathcal{P}(\omega)$ is a set, for if it did, one could interpret second-order Heyting arithmetic in it, but that is a theory that is astronomically stronger than **CZF** (see [8,9]).

One nontrivial example in an extension of **CZF** (that is still interpretable in Martin–Löf-type theory) of an ω-cpo is a $\bigcup$-closed, regular set which contains ω. These cannot be proved to exist in **CZF**. More will follow in the next section.

Remark 2.11. It requires a small proof to show that ω-continuity is indeed stronger than monotonicity.

Proposition 2.12 (ω-continuity implies monotonicity). *If $f : C \to D$ is ω-continuous, then f is monotone.*

Proof. Assume $f : C \to D$ as given, let $x \leqslant y$. Define an ascending sequence $\langle z_n \rangle$ by $z_0 = x$, $z_n = y$ for $n \geqslant 1$. Definition by cases is possible on naturals. So, $\bigvee z_n = y$, and hence, $f(y) \geqslant f(z_0) = f(x)$ as desired. □

Lemma 2.13 (Limit interchange). *Suppose $\langle a_{n,m} \rangle$ is an ascending sequence in each index (i.e., for each $n, m < \omega$, both $\langle a_{n,m} \rangle_{n<\omega}$ and $\langle a_{n,m} \rangle_{m<\omega}$ are ascending sequences). Then*

$$\bigvee_m \bigvee_n a_{n,m} = \bigvee_n \bigvee_m a_{n,m} = \bigvee_n a_{n,n}.$$

Proof. We'll show the first equality, the other is similar. Let $b_n = \bigvee_m a_{n,m}$. Note for all $n < \omega$, $b_n \geqslant a_{n,n}$ and so $\bigvee_n b_n \geqslant \bigvee_n a_{n,n}$.

For the reverse inequality, given $n < \omega$, we have

$$\begin{aligned} a_{n,m} &\leqslant a_{m,m} \quad \text{for } n \leqslant m \\ \bigvee_m a_{n,m} &\leqslant \bigvee_m a_{m,m} \\ b_n &\leqslant \bigvee_m a_{m,m}, \end{aligned}$$

and so $\bigvee_n b_n \leqslant \bigvee_m a_{m,m}$. □

Proposition 2.14. *Let $[C \to D]$ be the class of ω-continuous functions from C to D. We can order $[C \to D]$ pointwise, i.e., if $f, g \in [C \to D]$, then*

$$f \leqslant g \Longleftrightarrow \forall c \in C\ [f(c) \leqslant_D g(c)].$$

With this ordering, $[C \to D]$ is a ω-dcpo. Moreover, if D is a ω-cpo, so is $[C \to D]$.

Proof. Note that by Exponentiation and Bounded Separation, $[C \to D]$ is a set. Clearly, $[C \to D]$ is a poset. For an ascending sequence $\langle f_n \rangle$, the least upper bound is given by

$$\bigvee_n f_n = \lambda c. \bigvee_n^D f_n(c).$$

This is well defined since D is a ω-dcpo, and ω-continuous from the ω-continuity of f_n and Limit Interchange.

If D is a ω-cpo, so is $[C \to D]$ with $\lambda c \in C.\bot_D$ as a least element. □

Remark 2.15. The more common approach uses dcpos, where the lub exists for arbitrary directed sets and continuous functions, which preserve lubs of any directed set. Since collection of all directed subsets is a proper class from the perspective of **CZF**, the collection of continuous functions would in general require more than Bounded Separation to be a set.

Definition 2.16 (Triviality and satisfactory). An inhabited poset is trivial if it has exactly one element. An inhabited poset is satisfactory if it has at least two distinct elements. More rigorously, a poset $\mathbb{P}$ is satisfactory when there is an injection $f : 2 \to \mathbb{P}$.

Proposition 2.17 (Satisfactory). *If D is a satisfactory ω-dcpo and C is inhabited, then $[C \to D]$ is satisfactory as well.*

Proof. By stipulation, there is an injective $f : 2 \to D$. Note the constant functions are clearly ω-continuous. Define an injection $g : 2 \to [C \to D]$ by

$$g(0) = \lambda c \in C \cdot f(0),$$
$$g(1) = \lambda c \in C \cdot f(1).$$

Definition by cases is permitted from the natural numbers properties, and this is an injection since f is. □

Remark 2.18. There is a classical partial converse to Lemma 2.12. Classically, if D is a finite ω-cpo, then $[D \to D] = \{f \in D^D \mid f \text{ monotone}\}$. This is because any infinite ascending sequence of elements of D actually achieves its least upper bound. However, even this partial converse is not viable intuitionistically, as it would imply a solution to the Halting Problem.

Proposition 2.19 (ωDcpo). *The following forms a category called* ω**Dcpo**: *the objects are ω-dcpos, and the morphisms are ω-continuous functions.*

Proof. The identity functions are ω-continuous. The composition of ω-continuous functions is ω-continuous. □

Definition 2.20 (ωCpo). Let ω**Cpo** be the full subcategory of ω**Dcpo** whose objects are the ω-cpos.

Proposition 2.21 (Product Construction). *Both* ω**Cpo** *and* ω**Dcpo** *have categorical products.*

Proof. If C, D are ω-dcpos (ω-cpos), then defining $C \times D$, π_1, π_2 gives a categorical product in ω**Dcpo** (resp. ω-**Cpo**).

Define $C \times D$ by $\{\langle c, d\rangle \mid c \in C, d \in D\}$ as the underlying set, and ordering defined pointwise: $\langle c, d\rangle \leqslant \langle e, f\rangle$ when $c \leqslant_C e$ and $d \leqslant_D f$. The lubs are given pointwise as well.
Take π_1, π_2 to be the regular projections.

Bottom element is given by $\langle \bot_C, \bot_D\rangle$. □

Lemma 2.22 (Product Continuity). $f : C \times D \to E$ *is ω-continuous* $\iff$ $\lambda c.f(c, d_0), \lambda d.f(c_0, d)$ *are ω-continuous for all* $c_0 \in C, d_0 \in D$ *(in other words, f is continuous in each argument).*

Proof. $\Rightarrow$ Clear.

$\Leftarrow$ Let $\langle c_n, d_n\rangle$ be an ascending sequence in $C \times D$. Then we have

$$\begin{aligned} f\left(\bigvee \langle c_n, d_n\rangle\right) &= f\left(\bigvee c_n, \bigvee d_n\right) \\ &= \bigvee_m f(c_m, \bigvee d_n) \\ &= \bigvee_m \bigvee_n f(c_m, d_n) \\ &= \bigvee_n f(c_n, d_n), \end{aligned}$$

where the last equality holds by Lemma 2.13. □

Proposition 2.23 (Continuity of Abstraction). *Define abstraction (A) by*

$$\begin{aligned} A &: [(C \times D) \to E] \to [C \to [D \to E]] \\ f &\mapsto \lambda c \cdot \lambda d \cdot f(c, d) \end{aligned}$$

Abstraction is well defined, ω-continuous.

Proof. First, the function is well defined in that $A(f) \in [C \to [D \to E]]$. Let $\bigvee c_n = c$ with $\langle c_n \rangle$, an ascending sequence in C. Then

$$\begin{aligned} (A(f))(c) &= \lambda d \cdot f(c, d) \\ &= \lambda d \cdot f\left(\bigvee c_n, d\right) \\ &= \lambda d \cdot \bigvee f(c_n, d) \\ &= \bigvee \lambda d \cdot f(c_n, d) \\ &= \bigvee A(f)(c_n), \end{aligned}$$

as desired.

Next, ω-continuity. Let $\langle f_n \rangle$ be ascending in $[(C \times D) \to E]$. Then we have

$$\begin{aligned} A\left(\bigvee f_n\right) &= \lambda c \cdot \lambda d \cdot \bigvee_n f_n(c, d) \\ &= \bigvee_n \lambda c \cdot \lambda d \cdot f_n(c, d) \\ &= \bigvee_n A(f_n), \end{aligned}$$

which establishes ω-continuity. □

Proposition 2.24 (Exponential Construction). *Both* ω**Cpo** *and* ω**Dcpo** *have categorical exponential objects.*

Proof. If C, D are ω-dcpos (ω-cpos), then we are defining an object $[C \to D]$, and evaluation map ϵ.

We've already defined the exponential object, $[C \to D]$. Next, for the evaluation map ϵ, define

$$\epsilon : [C \to D] \times C \to D$$
$$\epsilon(f, x) = f(x).$$

By Lemma (2.22), we need to establish ω-continuity in each argument. ω-continuity for fixed $f \in [C \to D]$ is clear. Next, fix $c \in C$ and let $\langle f_n \rangle$ be ascending in $[C \to D]$. Then

$$\begin{aligned}\epsilon\left(\bigvee f_n, c\right) &= \left(\bigvee f_n\right)(c)\\ &= \bigvee f_n(c)\\ &= \bigvee \epsilon(f_n, c),\end{aligned}$$

showing ω-continuity with c fixed.

Abstraction is also ω-continuous by Proposition 2.23. □

2.3. D_∞ *construction*

In constructing D_∞, we mainly follow the exposition in [1,3.1]. The details are provided to ascertain that the proofs can be carried out on the basis of **CZF** and also to keep the chapter self-contained. A few more definitions first.

Definition 2.25 (Injection projection pair). An injection-projection (ip) pair between two ω-dcpos C and D is a pair $(i : C \to D, j : D \to C)$ such that

$$j \circ i = 1_C \quad \text{and} \quad i \circ j \leqslant 1_D,$$

under the pointwise ordering for $[C \to D], [D \to C]$, respectively. For such a pair, we will write $(i, j) : C \to_{ip} D$ if there is any danger of confusion with other maps. More usually, we will write simply $(i, j) : C \to D$. i is called the injection and j is called the projection.

Definition 2.26 (ip composition). Define composition $\circ$ of ip pairs componentwise. In other words, we define $\circ$ by

$$(i_1, j_1) \circ (i_0, j_0) = (i_1 \circ i_0, j_0 \circ j_1).$$

Definition 2.27 (ωDcpo$^{\rm ip}$, ωCpo$^{\rm ip}$). Define a category $\omega\mathbf{Dcpo}^{\rm ip}$ whose objects are the ω-dcpos and morphisms are ip pairs. Let $\omega\mathbf{Cpo}^{\rm ip}$ be the full subcategory of $\omega\mathbf{Dcpo}^{\rm ip}$ with objects the ω-cpos.

Proposition 2.28. *The objects and arrows of the above definition form a category.*

Proof. The identity map for a ω-dcpo D in either category is given by $(1_D, 1_D)$, the pair of set identity maps. Composition as given in Definition 2.26 results in another ip pair, and is associative. □

Proposition 2.29 (Basic properties of ip pairs). *If $(i,j){:}C{\to}D$ is an ip pair, then j is determined by i. Moreover,*

$$j(x) = \bigvee\{y \mid i(y) \leqslant x\}.$$

Proof. Letting (i, j') be another ip pair, we have

$$j' = 1_C \circ j' = j \circ i \circ j' \leqslant j \circ 1_D = j.$$

Thus, $j' \leqslant j$, and by symmetry $j \leqslant j'$, whence $j = j'$.

If $i(y) \leqslant x$, then $j(i(y)) \leqslant j(x)$, thus $y \leqslant j(x)$. Conversely, from $y \leqslant j(x)$ we get $i(y) \leqslant i(j(x))$, that is, $i(y) \leqslant x$. Hence, $i(y) \leqslant x \Leftrightarrow y \leqslant j(x)$, which yields $j(x) = \bigvee\{y \mid i(y) \leqslant x\}$. □

Definition 2.30 ($\to$ functor). Let $(i,j) : C \to D$ be an ip pair. Define

$$\begin{aligned}(i',j') \ &: \ [C \to C] \to_{ip} [D \to D]\\ i'(f) &= i \circ f \circ j\\ j'(g) &= j \circ g \circ i.\end{aligned}$$

We also use the notation

$$(i,j) \to (i,j),$$

in lieu of (i', j').

Proposition 2.31. *The construction of Definition 2.30 defines a functor H, with $H(D) = [D, D]$, $H((i,j)) = (i',j')$ in $\omega\mathbf{Dcpo}^{ip}$ and $\omega\mathbf{Cpo}^{ip}$.*

Proof. This is straightforward; it's also Exercise 3.1.9 in [1]. □

We now begin the construction itself. The definition is by recursion and then taking a limit.

Definition 2.32 (D_∞). Let D be a ω-cpo. Begin the base case of the recursion with

$$\begin{aligned}(i_0, j_0) &: D \to_{ip} [D \to D] \\ i_0(x) &= \lambda y \in D \cdot x \quad \text{and} \quad j_0(f) = f(\bot_D) \\ D_0 &= D.\end{aligned}$$

Define successor stages of the recursion by

$$\begin{aligned}D_{n+1} &= [D_n \to D_n] \\ (i_{n+1}, j_{n+1}) &= (i_n, j_n) \to (i_n, j_n).\end{aligned}$$

In other words, we form a sequence D_n, with $D_0 = D$ and $H(D_n) = D_{n+1}$. Now we take the limit to obtain D_∞.

$$D_\infty = \{\langle x_n \rangle \mid \forall m < \omega\ x_m \in D_m \wedge j_m(x_{m+1}) = x_m\}.$$

To simplify expressions, we'll often just write $\mathbf{x}$ for $\langle x_n \rangle$.

We will also notate the nth member of a sequence $\mathbf{x} \in D_\infty$ by $(\mathbf{x})_n$ or $[\mathbf{x}]_n$; thus if $\mathbf{x} = \langle x_n \rangle$, then $(\mathbf{x})_n = x_n$.

Definition 2.33 (Standard ip pair for D). Unlike the classical case, there may not be many initial choices for (i_0, j_0) since the classical multitude of step functions is not generally available. However, the above (i_0, j_0) are always available, and are called the standard ip pair for D.

Proposition 2.34 (Well defined). *D_∞ is a ω-cpo with the ordering defined pointwise. In other words, if $\mathbf{x}, \mathbf{y} \in D_\infty$, then*

$$\mathbf{x} \leqslant \mathbf{y} \iff \forall n < \omega\ x_n \leqslant_{D_n} y_n.$$

Proof. Replacement, Exponentiation and Bounded Separation give that D_∞ is a set. Let $\langle \mathbf{x}_k \rangle$ be an ascending sequence of elements of D_∞. We can then define $\mathbf{z} = \bigvee_k^{D_\infty} \langle \mathbf{x}_k \rangle$ componentwise, by $z_n = \bigvee_k^{D_n} \mathbf{x}_{k,n}$, where $\mathbf{x}_{k,n} := (\mathbf{x}_k)_n$. We need to show that this sequence is in D_∞. Let $n < \omega$, then by continuity of j_n we have

$$\begin{aligned} j_n(z_{n+1}) &= j_n\left(\bigvee_k^{D_{n+1}} x_{k,n+1}\right) \\ &= \bigvee_k^{D_n} j_n(x_{k,n+1}) \\ &= \bigvee_k^{D_n} x_{k,n} \\ &= z_n. \end{aligned}$$

The minimum element is $\langle \bot_{D_n} \rangle$. □

Definition 2.35 (ip from D_n to D_m). We'd like ip pairs from D_n to D_m for any $n, m < \omega$. We will use these functions as helpers to define a reasonable way of filling in a sequence into D_∞ given an element in D_n. Intuitively, we can define them as simply composing the needed functions. However, we would like a definition uniform in n, m so that we can collect them into a set (by Replacement). Define

$$k_{m,n} = \begin{cases} 1_{D_n} & \text{if } n = m \\ i_{n-1} \circ i_{n-2} \circ \ldots \circ i_{m+1} \circ i_m & \text{if } n > m \\ j_n \circ j_{n+1} \circ \ldots \circ j_{m-2} \circ j_{m-1} & \text{if } n < m, \end{cases}$$

which amounts to the same as the recursive definition

$$k_{m,n} = \begin{cases} 1_{D_n} & \text{if } n = m \\ k_{m+1,n} \circ i_m & \text{if } n > m \\ k_{m-1,n} \circ j_{m-1} & \text{if } n < m. \end{cases}$$

Then, we have an ip pair $(k_{m,n}, k_{n,m}) : D_m \to D_n$.

Definition by cases for natural numbers and recursion are valid in $\mathbf{CZF}^w$.

Definition 2.36 (ip from D_n to D_∞). Define

$$\begin{aligned} j_{n\infty} &: D_\infty \to D_n \\ j_{n\infty}(\mathbf{x}) &= x_n \quad (\text{thus } j_{n\infty}(\mathbf{x}) = (\mathbf{x})_n). \end{aligned}$$

Further, for each $n < \omega$, define

$$\begin{aligned} i_{n\infty} &: D_n \to D_\infty \\ i_{n\infty}(x) &= (k_{n,0}(x), \ldots, k_{n,n}(x), \ldots, k_{n,m}(x), \ldots) \\ i_{n\infty}(x) &= \langle k_{n,l}(x) \rangle_{l<\omega}. \end{aligned}$$

Intuitively, $i_{n\infty}$ fills in the rest of the sequence to match x. This definition is valid by Replacement. At times, for $x \in D_n$, we will overload notation and write x or $\mathbf{x}$ for $i_{n\infty}(x)$.

Proposition 2.37. *$i_{n\infty}$ and $j_{n\infty}$ are an injection projection pair.*

Proof. First, we need to check that both are ω-continuous. $j_{n\infty}$ is clearly ω-continuous. For $i_{n\infty}$, let $\langle x_p \rangle_{p<\omega}$ be ascending in D_n. So to start the base case of an induction, note

$$\begin{aligned} \left(i_{n\infty} \left(\bigvee_p^{D_n} x_p \right) \right)_n &= \bigvee_p^{D_n} x_p \\ &= \left(\bigvee_p^{D_\infty} i_{n\infty}(x_p) \right)_n . \end{aligned}$$

For components $m > n$ and $m < n$, we then have ω-continuity at component m by a Δ_0-induction and ω-continuity of $k_{n,m}$.

Last, we need to check that $j_{n\infty} \circ i_{n\infty} = 1_{D_n}$, and $i_{n\infty} \circ j_{n\infty} \leqslant 1_{D_\infty}$. The first condition is clear. For the second condition, we may proceed with a proof by cases on natural numbers. By induction, for $m \leqslant n$, $[(i_{n\infty} \circ j_{n\infty})(\mathbf{x})]_m = x_m$. However, for $m > n$, we no longer have equality of the components, but rather $[(i_{n\infty} \circ j_{n\infty})(\mathbf{x})]_m \leqslant x_m$.

This follows by a Δ_0-induction, with the base case being

$$\begin{aligned}[(i_{n\infty} \circ j_{n\infty})(\mathbf{x})]_{n+1} &= [i_{n\infty}(x_n)]_{n+1} \\ &= i_n(x_n) \\ &= i_n(j_n(x_{n+1})) \\ &\leqslant x_{n+1}.\end{aligned}$$

Together these establish $i_{n\infty} \circ j_{n\infty} \leqslant 1_{D_\infty}$ in the pointwise ordering. Hence, as desired, we have that $i_{n\infty}, j_{n\infty}$ is an ip pair. □

Lemma 2.38. *More properties that will be needed.*

(1) *If* $\mathbf{x} = \langle x_k \rangle$ *and* $\mathbf{y} = \langle y_k \rangle$ *are in* D_∞*, then* $\langle i_{n\infty}[x_{n+1}(y_n)] \rangle$ *is an ascending sequence in* D_∞.
(2) $\mathbf{x} = \bigvee_n i_{n\infty}(x_n)$. *Suppressing the* $i_{n\infty}$*, this reads more cleanly as* $\mathbf{x} = \bigvee_n x_n$.
(3) *If* $f : D_\infty \to D_\infty$ *is* ω*-continuous, then* $f(\mathbf{x}) = \bigvee_n f(i_{n\infty}[x_n])$.

Proof. Let $n < \omega$. We need that $i_{n\infty}[x_{n+1}(y_n)] \leqslant i_{n+1\infty}[x_{n+2}(y_{n+1})]$, which is equivalent to the statement that for all $p < \omega$,

$$(i_{n\infty}[x_{n+1}(y_n)])_p \leqslant (i_{n+1\infty}[x_{n+2}(y_{n+1})])_p\,.$$

One then proceeds by cases on the components, looking at $p \leqslant n$ and $p > n$ separately. Use the injection projection pair condition for $(k_{n,p}, k_{p,n})$.

Next, $\mathbf{x} = \bigvee_n i_{n\infty}(x_n)$.

Last, we need to show that if f is ω-continuous, then $f(\mathbf{x}) = \bigvee_n f(i_{n\infty}(x_n))$. We'll start using $i_{n\infty}$ implicitly to make the formulas easier to read.

$$\begin{aligned} f(\mathbf{x}) &= f(\bigvee_n x_n) \text{ by (2.38)} \\ &= \bigvee_n f(x_n) \text{ by } \omega\text{-continuity} \\ &= \bigvee_n \bigvee_m f(x_n)_m \text{ by (2.38)} \\ &= \bigvee_n f(x_n)_n \text{ Lemma 2.13.} \end{aligned}$$

□

We next develop isomorphisms between D_∞ and $[D_\infty \to D_\infty]$ (in $\omega\mathbf{Cpo}^{\mathrm{ip}}$).

Definition 2.39 (G, $\bullet$). We'll define $G : [D_\infty \to D_\infty] \to D_\infty$ and $\bullet : D_\infty \times D_\infty \to D_\infty$. Define, for $f \in [D_\infty \to D_\infty]$,

$$\begin{aligned} G_n &: [D_\infty \to D_\infty] \to D_{n+1} = [D_n \to D_n] \\ G_n(f) &= \lambda y \in D_n . f(y)_n \\ G(f) &= \bigvee_{n<\omega} i_{n\infty}\left(G_n(f)\right) \\ \mathbf{x} \bullet \mathbf{y} &= \bigvee_{n<\omega} i_{n\infty}\left(x_{n+1}(y_n)\right). \end{aligned}$$

These are functions by Union–Replacement. Verification of the claimed properties is the following lemma.

Lemma 2.40 (G, $\bullet$ properly defined). *We need to*

(1) *Check $\bullet$ is ω-continuous.*
(2) *Check that $G_n(f)$ with the above definition is in D_n: ω-continuous.*

Proof. First, check $\bullet$ is ω-continuous.

For ω-continuity, by Lemma 2.22 it is enough to establish ω-continuity in each argument.

We begin by fixing the first argument. Let $\mathbf{x}$ and an ascending sequence $\langle \mathbf{y_m} \rangle$ be given. Set $\mathbf{z} = \bigvee_m \mathbf{y_m}$. Unravelling the definitions, we need to check the condition that

$$\mathbf{x} \bullet (\bigvee_m \mathbf{y_m}) = \bigvee_m \mathbf{x} \bullet \mathbf{y_m},$$

which is equivalent to the condition

$$\bigvee_n i_{n\infty}[x_{n+1}(z_n)] = \bigvee_m \bigvee_n i_{n\infty}[x_{n+1}(y_{m,n})]$$

Now computing we have

$$\begin{aligned}
\bigvee_n i_{n\infty}[x_{n+1}(z_n)] &= \bigvee_n i_{n\infty}\left[x_{n+1}\left(\bigvee_m y_{m,n}\right)\right] \text{ definition of } z_n \\
&= \bigvee_n i_{n\infty}\left[\bigvee_m x_{n+1}(y_{m,n})\right] \ \omega\text{-continuity of } x_{n+1} \\
&= \bigvee_n \bigvee_m i_{n\infty}\left[x_{n+1}(y_{m,n})\right] \ \omega\text{-continuity of } i_{n\infty} \\
&= \bigvee_m \bigvee_n i_{n\infty}[x_{n+1}(y_{m,n})] \text{ Limit interchange} \\
&\quad \text{(Lemma 2.13).}
\end{aligned}$$

This establishes continuity in the first argument.

For continuity in the second argument, fix $\mathbf{y}$ and let $\mathbf{x_m}$ be an ascending sequence in D_∞, set $\mathbf{z} = \bigvee \mathbf{x_m}$. Then, we seek to show $\mathbf{z} \bullet \mathbf{y} = \bigvee \mathbf{z_n} \bullet \mathbf{y}$, which is equivalent to the condition

$$\bigvee_n i_{n\infty}[\bigvee_m x_{m,n+1}(y_n)] = \bigvee_m \bigvee_n i_{n\infty}[x_{m,n+1}(y_n)].$$

This condition holds, similar to above, by ω-continuity of $i_{n\infty}$ and the limit interchange lemma (Lemma 2.13). So, we have $\bullet$ is ω-continuous.

Case $n = 0$ for G_0.

Next, we need to check with $G_{n+1} : [D_\infty \to D_\infty] \to D_{n+1}$ as above, we have, for $n \geqslant 1$, that $G_{n+1}(f)$ is in $D_n = [D_n \to D_n]$ so ω-continuous with appropriate domain and codomain.

Third, we have that G is ω-continuous by Proposition 2.14. □

Definition 2.41 (F). Define $F : D_\infty \to [D_\infty \to D_\infty]$ by $F(\mathbf{x}) = \lambda \mathbf{y} \in D_\infty \cdot \mathbf{x} \bullet \mathbf{y}$. In other words, F is the curried version of $\bullet$.

Lemma 2.42. *F is well defined, ω-continuous.*

Proof. By abstraction, Lemma 2.23. □

Proposition 2.43 (Properties of $\bullet$). *Making repeated implicit use of $i_{n\infty}$, we have the following:*

(1) If $x \in D_{n+1}$, then $x \bullet \mathbf{y} = x(y_n)$ (with $i_{n\infty}$ explicit, the statement is $i_{n\infty}(x) \bullet \mathbf{y} = i_{n\infty}(x(y_n))$).
(2) If $x \in D_{n+1}$, then $x \bullet \mathbf{y} = (x \bullet \mathbf{y})_n$.
(3) If $y \in D_n$, then $(\mathbf{x} \bullet y)_n = x_{n+1}(y)$.
(4) Let $\mathbf{x} \in D_\infty$ and $\bot = \bot_{D_\infty}$ the bottom element of D_∞. Then, we have $\mathbf{x} \bullet \bot = x_0$.
(5) If $x \in D_{n+1}, y \in D_n$, then $x \bullet y = x(y)$.

Proof.

- If $x \in D_{n+1}$, then $x \bullet \mathbf{y} = x(y_n)$.
- If $x \in D_{n+1}$, then $x \bullet \mathbf{y} = (x \bullet \mathbf{y})_n$.
- If $y \in D_n$, then $(\mathbf{x} \bullet y)_n = x_{n+1}(y)$.
- $\mathbf{x} \bullet \bot = x_0$.
- If $x \in D_{n+1}, y \in D_n$, then $x \bullet y = x(y)$.

□

Remark 2.44. As noted in Exercise 18.4.5 in [10], it is not true that $(\mathbf{x} \bullet \mathbf{y})_n = x_{n+1}(y_n)$. This is because the definition involves taking the least upper bound as well. One counterexample is given by considering the interpretations of $\omega = \lambda x.xx$ and $I \equiv \lambda x.x$.

Theorem 2.45 (F, G are inverse isomorphisms). *F, G are inverse isomorphisms in ω***Cpo**.

Proof. First, $F \circ G = 1_{[D_\infty \to D_\infty]}$. We need to show that given $f \in [D_\infty \to D_\infty]$, for all $\mathbf{x} \in D_\infty$, we have

$$[F(G(f))](\mathbf{x}) = f(\mathbf{x}).$$

Computing, (making heavy implicit use of $i_{n\infty}$), we have

$$\begin{aligned}[F(G(f))](\mathbf{x}) &= \left(\bigvee_n G_n(f)\right) \bullet \mathbf{x} \\ &= \bigvee_n G_n(f) \bullet \mathbf{x}\end{aligned}$$

$$= \bigvee_n G_n(f)(x_n)$$

$$= \bigvee_n f(x_n)_n \text{ by definition of } G_n$$

$$= f(\mathbf{x}) \text{ Lemma 2.38.}$$

Second, $G \circ F = 1_{D_\infty}$. Let $\mathbf{x} \in D_\infty$. Then

$$\begin{aligned} F(\mathbf{x}) &= \lambda \mathbf{y}.\mathbf{x} \bullet \mathbf{y} \\ G_n(F(\mathbf{x})) &= i_{n\infty}[\lambda y \in D_n.(\mathbf{x} \bullet i_{n\infty}(y))_n] \\ &= x_{n+1} \\ G(F(\mathbf{x})) &= \bigvee x_{n+1} \\ &= \mathbf{x}, \end{aligned}$$

as desired. □

Theorem 2.46 (Extensionality of D_∞). *For any D an ω-cpo and $\mathbf{x}, \mathbf{y} \in D_\infty$, we have that*

(1) $\mathbf{x} \bullet \mathbf{z} \leqslant \mathbf{y} \bullet \mathbf{z}$ *for all* $\mathbf{z} \in D_\infty$ *if and only if* $\mathbf{x} \leqslant \mathbf{y}$.
(2) $\mathbf{x} \bullet \mathbf{z} = \mathbf{y} \bullet \mathbf{z}$ *for all* $\mathbf{z} \in D_\infty$ *if and only if* $\mathbf{x} = \mathbf{y}$.

Proof. For (1), $\Leftarrow$, recall that $\bullet$ being ω-continuous implies it is ω-continuous in each argument with the other argument fixed. In particular, it is monotonic with a fixed second argument, as desired.

For (1), $\Rightarrow$, we seek to show that $x_n \leqslant y_n$ for all $n < \omega$.

First consider the case $n = 0$. By Proposition 2.43(4), we have $x_0 \leqslant y_0$.

Next, consider the case $n = m + 1$. Let $z \in D_m$ be given. By the definition of pointwise ordering of D_n, we need to show that

$$x_{m+1}(z) \leqslant_{D_m} y_{m+1}(z).$$

By stipulation, we have $\mathbf{x} \bullet z \leqslant \mathbf{y} \bullet z$. By Proposition 2.43(3), we then have $x_{m+1}(z) \leqslant_{D_m} y_{m+1}(z)$ as desired.

(2) now follows from (1). □

3. How to Find an ω-cpo in CZF?

The previous section showed that D_∞ can be constructed from a given ω-cpo, using just means available in the fragment $\mathbf{CZF}^w$ of **CZF**. But is it possible to show the existence of a nontrivial ω-cpo in **CZF** so one can start building a D_∞ model?

It was already mentioned that in **CZF** it is not possible to prove that a nontrivial finite poset is a ω-cpo. This is because it would amount to solving the halting problem. This is even the case for full intutionistic Zermelo–Fraenkel set theory (**IZF**), as **IZF** is consistent with the assumption that all functions from naturals to naturals are recursive. There are some axioms, though, which have already been adopted in constructive set theory in addition to those of **CZF**, that yield the existence of ω-cpos.

Aczel gave an interpretation of **CZF** in Martin–Löf-type theory in a series of papers [5,11,12]. This embedding actually validated more than the axioms of **CZF** and one of the extra principles found in this way was the *Regular Extension Axiom*, (**REA**). The latter is very important if one wants to ensure the existence of inductively defined sets (see [6,7,12]).

As it turns out, a slight extension of **REA** enables one to generate a plethora of ω-cpos. So let's introduce these axioms.

Definition 3.1. A relation R from A to B is said to be *full* if $R \subseteq A \times B$ and $\forall x \in A\, \exists y \in B\, \langle x, y\rangle \in A \times B$. Let $\mathbf{mv}(^AB)$ be the class of all full relations from A to B.

A set C is said to be *regular* if it is transitive, inhabited (i.e., $\exists u\; u \in C$) and for any $u \in C$ and $R \in \mathbf{mv}(^uC)$ there exists a set $v \in C$ such that

$$\forall x{\in}u\, \exists y{\in}v\, \langle x, y\rangle{\in}R \;\wedge\; \forall y{\in}v\, \exists x{\in}u\, \langle x, y\rangle{\in}R.$$

We write $\mathbf{Reg}(C)$ to express that C is regular.

REA is the principle

$$\forall x\, \exists y\, (x \subseteq y \;\wedge\; \mathbf{Reg}(y)).$$

One is naturally led to consider strengthenings of the notion of a regular set, for instance, that the set should also be $\bigcup$-closed.

Definition 3.2. A class A is said to be $\bigcup$*-closed* if for all $x \in A$, $\bigcup x \in A$.

A transitive inhabited set C is said to be $\bigcup$*-regular* if C is regular and $\bigcup$-closed. The $\bigcup$*-Regular Extension Axiom* ($\mathbf{REA}_{\bigcup}$) is as follows:

Every set is a subset of a $\bigcup$*-regular set.*

With the help of $\mathbf{REA}_{\bigcup}$, the existence of ω-cpos can be guaranteed.

Theorem 3.3 ($\mathbf{CZF}^{w} + \mathbf{REA}_{\bigcup}$). *There are (class many) ω-cpos.*

Proof. Let D be a $\bigcup$-closed regular set containing ω. We can order D by $\subseteq$, this makes D a poset.

First, we claim that D is in fact an ω-cpo. Let $f : \omega \to D$ an ascending sequence be given. By regularity,[c] $f \in D$ and so $\mathrm{Im}(f) \in D$. Then by $\bigcup$-closure, $U := \bigcup \mathrm{Im}(f) \in D$. Let $Im(f) = \{f(u) \mid u \in \omega\}$. U is an upper bound for $\mathrm{Im}(f)$ since for all $n < \omega$, $f(n) \subseteq \bigcup \mathrm{Im}(f) = U$. Let V also be an upper bound for $\mathrm{Im}(f)$, then for all $n < \omega$ we have $f(n) \subseteq V$. Let $x \in U$ be given, so there exists $n < \omega$ so that $x \in f(n)$ by the definition of U. Then $f(n) \subseteq V$ implies $x \in V$, so as x was arbitrary, $U \subseteq V$. □

Example 3.4 ($\mathbf{CZF}^{w} + \mathbf{REA}_{\bigcup}$). Another similar example is the following. Let R be a $\bigcup$-closed regular set containing ω. Then we can define the *local power set* $\mathcal{P}_R(A)$ of a set A by

$$\mathcal{P}(A) \cap R \text{ or equivalently } \{x \in R \mid x \subseteq A\}.$$

Note that $\mathcal{P}_R(A)$ is then a ω-cpo, since for $f : \omega \to \mathcal{P}_R(A)$ we have that $U := \bigcup \mathrm{Im}(f) \subseteq A$ and regularity ensures that $U \in R$. In particular, taking $A = 2 = \{\emptyset, \{\emptyset\}\}$ gives a nontrivial ω-cpo $\mathcal{P}_R(A)$.

Finally, it is worth noting that the addition of **REA** and a fortiori $\mathbf{REA}_{\bigcup}$ to **CZF** or $\mathbf{CZF}^{w}$ results in a proof-theoretically much stronger theory. Measured in terms of fragments of second-order arithmetic, both **CZF** + **REA** and **CZF** + $\mathbf{REA}_{\bigcup}$ are of the same strength as Δ^1_2-comprehension plus bar induction. The exact results for **CZF** + **REA** can be found in [9], [8] Theorem 5.13 and for

[c] See [7, Lemma 11.1.5].

$\mathbf{CZF+REA}_{\bigcup}$ in [13] Theorem 4.7. Moreover, [8,9,13] show that both theories are closely related to systems of Martin–Löf-type theory.

There is actually a very simple method for constructing an ω-cpo from the natural numbers (also due to Scott).

Definition 3.5. Let $\bot$ be a set that is distinct from all natural numbers. Let

$$\mathbb{N}^+ = \mathbb{N} \cup \{\bot\},$$

and consider the following partial order $\preccurlyeq$ defined on $\mathbb{N}^+$ for all $x, y \in \mathbb{N}$

$$x \preccurlyeq y \iff (x = \bot \text{ and } y \in \mathbb{N}) \quad \text{or} \quad (x = y).$$

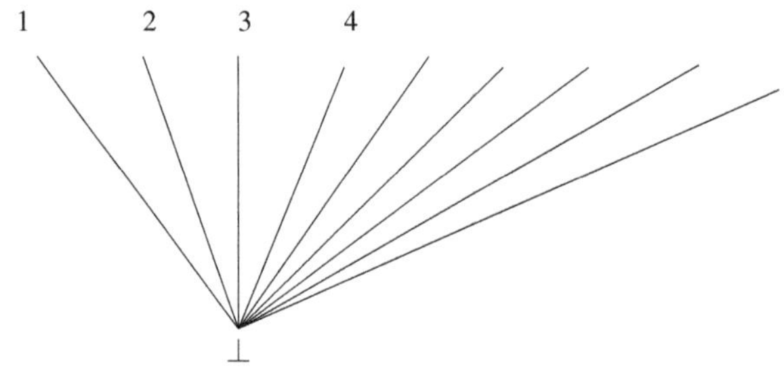

Notation. We will use $\mathbb{N}^+$ to mean $(\mathbb{N}^+, \preccurlyeq)$.

Lemma 3.6. $(\mathbb{N}^+, \preccurlyeq)$ *is an ω-cpo, actually a complete cpo.*

Proof. The proof is simple, but it uses classical logic. Given a $\preccurlyeq$-ascending sequence $(x_n)_n$, one wants to find an upper bound. If there exists an $i < \omega$ such that $x_i \in \mathbb{N}$, then $x_j = x_i$ for all $j \geqslant i$, whence $(x_n)_n$ has lub x_i. If on the other hand, $x_n = \bot$ for all n, then $(x_n)_n$ has lub $\bot$. So far our reasoning was constructive. However, to conclude that therefore $\mathbb{N}^+$ is an ω-cpo, it is required that the foregoing cases cover all possibilities, which follows with classical logic. □

It can easily be shown that the assumption that $\mathbb{N}^+$ is a ω-cpo solves the Halting Problem, thence this result isn't provable even in **IZF**.

The employment of classical logic in Lemma 3.6 is an example of what Bishop called the *limited principle of omniscience* (LPO). It is fairly benign when compared to the general law of excluded middle. This is borne out by the next result.

Theorem 3.7 (Rathjen). **CZF**+LPO *has the same proof-theoretic strength as* **CZF**.

Proof. This follows from Theorems 2.2 and 6.2 in [14]. □

To appreciate the difference between LPO and full excluded middle, note that **CZF** with classical logic is simply the same as classical Zermelo–Fraenkel set theory.

4. Soundness

The purpose of this section is to ascertain, working in the constructive background theory $\mathbf{CZF}^w$, that an ω-cpo gives rise to a model of the λ-calculus and, moreover, furnishes models of Feferman's theories $\mathbf{PCA}^+$ and **EON** (as described in [2, VI 2.3,2.4]).

Remark 4.1 (Syntactical Manipulations). Most of the following syntactical manipulations will be presented as usual, even if they appear classical. This is valid since the terms can be coded as numbers, and the syntactical operations are primitive recursive, so representable in $\mathbf{CZF}^w$. Hence, we can still argue by cases.

Alternatively, one could use the DeBruijn syntax, which avoids using variables and α equivalence entirely. However, it is harder to read.

The λ calculus syntax, i.e., its set of terms, $Terms$, is inductively generated:

(1) The variables x_i, $i < \omega$ are terms. We'll call the set of variables $Vars$. Note that $Vars$ is in bijection with ω by assumption. Often to enhance readability, we will use different letters in place of subscripts, but formally the subscripts are needed.
(2) If M is a term, then $\lambda x_i.M$ is a term for each $i < \omega$.
(3) If M, N are terms, then (MN) is a term.

Definition 4.2 (Free variables). Normal inductive definition for the class function $FV : Terms \to P^{fin}\omega$:

$$\begin{aligned} FV(x_i) &= \{i\}, \\ FV((MN)) &= FV(M) \cup FV(N), \\ FV(\lambda x_i.M) &= FV(M) \setminus \{i\}. \end{aligned}$$

Definition 4.3 (Closed terms). Let CT be the set of closed terms (terms M so that $FV(M) = \emptyset$).

Definition 4.4 (α congruence). In order to avoid variable conflicts, we define the notion of α congruence, written $=_\alpha$ or just $=$, which is equivalence up to a renaming of variables. We will reserve $\equiv$ for syntactical equality. More formally, two terms M, N are α congruent (or equivalent) if there is a permutation $f : \omega \to \omega$ so that $f[M] = N$, where $f[M]$ is defined in the natural, recursive way as renaming the bound variable indices according to f. Note, this definition requires Exponentiation if we wish $=_\alpha$ equivalence classes to be sets. At the cost of additional explanation, that can be weakened to only requiring FPA, the finite powers axiom,[d] in addition to **ECST**, since formulas will only involve finitely many variables.

Definition 4.5 (Substitution). Substitution is defined recursively by

(1) $x_i[x_i := N] = N$.
(2) $x_j[x_i := N] = x_j$, if $i \neq j$.
(3) $(\lambda x_j.M)[x_i := N] = \lambda x_j.(M[x_i := N])$.
(4) $(M\ P)[x := N] = ((M[x := N])\ (P[x := N]))$.

Where in clause 3, we choose $j \neq i$ and $x_j \notin FV(N)$ by α conversion. This definition is intuitionistically correct by the remark above.

Definition 4.6 (Conversion rules). We say $M \to_\beta N$ to be the relation generated by the rules (for all $M, N, Z \in Terms, x_i \in Vars$)

(1) $M =_\alpha N \Rightarrow M \to_\beta N$.
(2) $(\lambda x_i.M)N \to_\beta M[x_i := N]$.
(3) $M \to_\beta N \Rightarrow MZ \to_\beta NZ,\ ZM \to_\beta ZN, \lambda x_i.M \to_\beta \lambda x_i.N$.

We also take $\to_\beta^*$ to be the reflexive, transitive closure of $\to_\beta$.

Definition 4.7 (β nf). Let $M \in Terms$.

(1) M is in β normal form (nf) or just nf when M has no subterm $(\lambda x_i.R)\ S$.

[d]See [7, Definition 6.1.7].

(2) M has an nf if there exists $N \in Terms$ such that N is an nf and $N =_\beta M$.

Definition 4.8 ($\beta\eta$ nf). An apparently (but not really) stronger notion is $\beta\eta$ nf. Let $M \in Terms$.

(1) M is in $\beta\eta$ normal form ($\beta\eta$ nf) when M has no subterm $(\lambda x_i.R)\ S$ nor subterm $(\lambda x_i.RX)$ with $x_i \notin FV(R)$.
(2) M has an nf if there exists $N \in Terms$ such that N is a $\beta\eta$ nf and $N =_{\beta\eta} M$.

Remark 4.9. Having either a β or $\beta\eta$ normal form is not recursive.

Definition 4.10 (pre-reflexive). This definition requires some others first.

(1) For an applicative structure $(X, \bullet)$, we define the set of representable functions by

$$X \rightarrow_{rep} X = \{f \in {}^X X \mid \exists x \in X \forall y \in X f(y) = x \bullet y\}.$$

(2) (X, F, G) is said to be a **pre-reflexive domain** if $(X, \bullet)$ is an applicative structure and there are two functions $F : X \rightarrow (X \rightarrow X)$ and $G : (X \rightarrow X) \rightarrow X$ such that the following holds:

$$F(x)(y) = x \bullet y.$$

and $F \circ G = 1_{X \rightarrow X}$.

Note 4.11. Sometimes, we will present only (X, F, G), with $\bullet : X^2 \rightarrow X$ understood to be $x \bullet y = F(x)(y)$.

Definition 4.12 (reflexive). A ω-reflexive domain (D, F, G) is a pre-reflexive domain such that

(1) D is a ω-cpo.
(2) $D \rightarrow_{rep} D = [D \rightarrow D]$.
(3) F, G are ω-continuous.

Definition 4.13 (Assignments). For D any set, we can define a D-assignment (also known as an environment) to be a function $p^D : Vars \rightarrow D$. Usually, the D is omitted. We will be interested in the case when D is a ω-pre-reflexive domain, since then we can define the notion of an interpretation.

Definition 4.14 (Revised Assignments). Revisions to assignments are defined for ω-reflexive domains since we assume a syntax for the λ calculus where $Vars$ is in bijection with ω (as well as Exponentiation). A convenient notation is

$$p(x_j)_{x_i=z} \begin{cases} z & \text{if } j = i \\ [\![x_j]\!] & \text{if } j \neq i. \end{cases}$$

Definition 4.15 (λ model). A weaker notion than a ω-reflexive domain is a λ model. A λ model is a ω-pre-reflexive domain such that for all environments p and $i < \omega$, $\lambda d.[\![M]\!]_{p_{x_i:=d}} \in [D \to D]$. This ensures that the following notion of interpretation is well defined.

Definition 4.16 (Interpretations). With a λ model (D, F, G), we can define the notion of an interpretation. The interpretation induced by p is written $[\![-]\!]_p$ or $[\![-]\!]$ if the assignment is understood. We define an interpretation to be a function $Terms \to D$ given inductively by

$$\begin{aligned} [\![x_i]\!]_p &= p(x_i), \\ [\![(MN)]\!]_p &= F\left([\![M]\!]_p\right)\ [\![N]\!]_p, \\ [\![\lambda x_i.M]\!]_p &= G\left(\lambda z.[\![M]\!]_{p_{x_i:=z}}\right), \end{aligned}$$

where the last line uses the notation mentioned above for altering environments.

Proposition 4.17. *D_∞ is a ω-reflexive domain.*

Proof. We need to check that the set of representable functions equals to $[D_\infty \to D_\infty]$. Let $f \in [D_\infty \to D_\infty]$ be given, then $G(f)$ represents f by the proof of Theorem 2.45. □

Lemma 4.18. *Any ω-reflexive domain D is a λ model.*

Proof. Recall by stipulation that D is a ω-cpo. It is enough to show that for all terms M, and finite sequences of distinct variables $\vec{d}$, that $\lambda\vec{d}.[\![M]\!]_{p_{\vec{x}:=\vec{d}}}$ is ω-continuous.

We proceed by a bounded induction on the build of M. Note that the collection of finite sequences of variables forms a set by FPA.

Base Case: $M \equiv x_j$. We know $x_j \in \vec{x}$ or $x_j \notin \vec{x}$ by the results of Natural Numbers section.

Assume first $x_j \in \vec{x}$. Then

$$\lambda \vec{d} \cdot [\![M]\!]_{p_{\vec{x}:=\vec{d}}} = \lambda \vec{d} \cdot \vec{d},$$

which is ω-continuous (the identity function).

Next, assume $x_j \notin \vec{x}$. Then

$$\lambda \vec{d} \cdot [\![M]\!]_{p_{\vec{x}:=\vec{d}}} = \lambda \vec{d} \cdot p(x_j),$$

which is ω-continuous (a constant function).

Case: $M \equiv (N\ P)$. Computing,

$$\begin{aligned}\lambda \vec{d} \cdot [\![M]\!]_{p_{\vec{x}:=\vec{d}}} &= \lambda d \cdot [\![(N\ P)]\!]_{p_{\vec{x}:=\vec{d}}} \\ &= \lambda \vec{d} \cdot \left(F[\![N]\!]_{p_{\vec{x}:=\vec{d}}}\ [\![P]\!]_{p_{\vec{x}:=\vec{d}}}\right),\end{aligned}$$

which is ω-continuous by the IH and since evaluation is ω-continuous (Proposition 2.24).

Case: $M \equiv \lambda x_j.N$. By α conversion, we can ensure that $j \neq i$. Then

$$\begin{aligned}\lambda \vec{d}.[\![M]\!]_{p_{\vec{x}:=\vec{d}}} &= \lambda \vec{d}.[\![\lambda x_j.N]\!]_{p_{\vec{x}:=\vec{d}}} \\ &= \lambda \vec{d}.G\left(\lambda z.[\![N]\!]_{p_{\vec{x}:=\vec{d},x_j:=z}}\right).\end{aligned}$$

Note that $\vec{x}$ concatenated with x_j is a finite sequence of variables, so the IH applies. Hence, the right-hand side is ω-continuous by the IH and since evaluation is ω-continuous. □

Lemma 4.19 (Substitution Lemma). *For all assignments p, and variables x_i, we have*

$$[\![M[x_i := N]]\!]_p = [\![M]\!]_{p_{x_i:=[\![N]\!]_p}}.$$

Proof. Note the collection of all assignments is a set by Exponentiation, so we can proceed by a bounded induction on the build up of M.

Base Case: $M \equiv x_j$, then the statement is clearly true.

Case: $M \equiv (P\ Q)$. Computing,

$$\begin{aligned}
[\![M[x_i := N]]\!]_p &= [\![(PQ)[x_i := N]]\!]_p \\
&= [\![(P[x_i := N]\ Q[x_i := N])]\!]_p \\
&= F\left([\![P[x_i := N]]\!]_p\right)\ [\![Q[x_i := N]]\!]_p \\
&= F\left([\![P]\!]_{p_{x_i:=[\![N]\!]_p}}\right)\ [\![Q]\!]_{p_{x_i:=[\![N]\!]_p}} \\
&= [\![(P\ Q)]\!]_{p_{x_i:=[\![N]\!]_p}} \\
&= [\![M]\!]_{p_{x_i:=[\![N]\!]_p}}.
\end{aligned}$$

Case: $M \equiv \lambda x_j.P$. Note by α equivalence we ensure that $j \neq i$ and that $x_j \notin FV(N)$. Computing,

$$\begin{aligned}
[\![M[x_i := N]]\!]_p &= [\![(\lambda x_j.P)[x_i := N]]\!]_p \\
&= [\![\lambda x_j.P[x_i := N]]\!]_p \\
&= G\left(\lambda z.[\![P[x_i := N]]\!]_{p_{x_j:=z}}\right) \\
&= G\left(\lambda z.[\![P]\!]_{p_{x_j:=z, x_i:=[\![N]\!]_p}}\right) \quad \text{by IH} \\
&= [\![\lambda x_j.P]\!]_{x_i:=[\![N]\!]_p} \\
&= [\![M]\!]_{p_{x_i:=[\![N]\!]_p}}.
\end{aligned}$$

□

Theorem 4.20 (Soundness). *For any λ model D, we have soundness:*
If $M \to_\beta N$, then for all D environments p, $[\![M]\!]_p = [\![N]\!]_p$.

Proof. Proof by induction on the $\to_\beta$ derivation, one case for each derivation rule.

For the axiom scheme $(\lambda x.M)N \to M[x := N]$, we have

$$\begin{aligned}
[\![(\lambda x_i.M)\ N]\!]_p &= F(G(\lambda d.[\![M]\!]_{p_{x:=d}}))\ [\![N]\!]_p \ \text{definition of } [\![-]\!] \\
&= [\![M]\!]_{p_{x:=[\![N]\!]_p}} \ \text{as } F \circ G = 1. \\
&= [\![M[x := N]]\!]_p \ \text{Lemma 4.19.}
\end{aligned}$$

The case of α conversion is clear.

The other cases follow immediately by induction. □

4.1. *Structure of* D_∞

In this section, we assume existence of a D_∞, which is guaranteed in background theories such as $\mathbf{CZF}^w + \mathsf{LPO}$, $\mathbf{CZF}^w + \mathbf{REA}_\bigcup$. We will go through several of the exercises of [10] Chapter 18, in detail.

Continuing the discussion from Example 3.4, we now examine the interpretation in D_∞ of the following λ terms:

$$\begin{aligned} I &\equiv \lambda x_1 \cdot x_1 \\ K &\equiv \lambda x_1 x_2 \cdot x_1 \\ S &\equiv \lambda x_1 x_2 x_3 \cdot (x_1 x_2)\ (x_1 x_3) \\ \Omega &\equiv (\lambda x_1 \cdot x_1 x_1)\ (\lambda x_1 \cdot x_1 x_1). \end{aligned}$$

Surprisingly, the choice of D does not really effect the interpretations of these terms. However, using nonstandard i_0, j_0 would affect the interpretations strongly. We define some helper functions first.

Definition 4.21 (I^n, K^n, S^n, ω^n).

$$\begin{aligned} I^{n+1} &= \lambda x \in D_n.x \\ K^{n+2} &= \lambda x \in D_{n+1} \lambda y \in D_n.j_n(x) \\ S^{n+3} &= \lambda x \in D_{n+2} \lambda y \in D_{n+1} \lambda z \in D_n.x(i_n(z))\ (yz) \\ \omega^{n+1} &= \lambda x \in D_n.i_n(x)(x). \end{aligned}$$

Proposition 4.22 (Interpretation of I).

$$[\![I]\!]^{D_\infty} = \langle \bot, I, I^2, I^3, \ldots \rangle.$$

Proof. This will serve as an archetype for the other similar proofs, so will be done in detail.

First, we will determine $[\![I]\!]_{n+1}$. From that, $[\![I]\!]_0$ will be determined by the projection condition : $j_0\left([\![I]\!]_1\right) = [\![I]\!]_0$.

To begin, from the definition of interpretation we have

$$[\![I]\!]_{n+1} = \bigvee_m^{D_{n+1}} k_{m+1,n+1}\left(\lambda x \in D_m.x\right).$$

So, for the claim to hold, we will show by induction on m that

$$k_{m+1,n+1}\left(\lambda x \in D_m \cdot x\right) = \lambda x \in D_n \cdot x \quad \text{when } m \geqslant n,$$
$$k_{m+1,n+1}\left(\lambda x \in D_m \cdot x\right) \leqslant \lambda x \in D_n \cdot x \quad \text{when } m < n.$$

Consider the first statement. The base case follows from

$$\begin{aligned} k_{n+1,m+1}\left(\lambda x \in D_m \cdot x\right) &= j_{n+1,n+1}(\lambda x \in D_m \cdot x) \\ &= \lambda x \in D_n \cdot j_{n+1} i_{n+1}(x) \\ &= \lambda x \in D_n \cdot x \text{ by ip pair condition.} \end{aligned}$$

Next for the inductive step,

$$\begin{aligned} k_{m+1,n+1}\left(\lambda x \in D_m \cdot x\right) &= k_{m,n+1}\left[j_m\left(\lambda x \in D_m \cdot x\right)\right] \\ &= k_{m,n+1}\left[j_{m-1} \circ \left(\lambda x \in D_m \cdot x\right) \circ i_{m-1}\right] \\ &= k_{n,m+1}\left(\lambda x \in D_{m-1} \cdot x\right) \text{ by ip pair condition} \\ &= \lambda x \in D_n \cdot x \text{ by IH.} \end{aligned}$$

For the second statement, to establish the base case we have

$$\begin{aligned} k_{n+1,n+1}\left(\lambda x \in D_n \cdot x\right) &= i_{n-1,n}(\lambda x \in D_{n-1} \cdot x) \\ &= \lambda x \in D_n \cdot i_{n-1} j_{n-1}(x) \\ &\leqslant \lambda x \cdot D_n \cdot x. \end{aligned}$$

Next for the inductive step of the second statement,

$$\begin{aligned} k_{m,n+1}\left(\lambda x \in D_{m-1} \cdot x\right) &= k_{m,n+1}\left[i_{m,n+1}\left(\lambda x \in D_{m-1} \cdot x\right)\right] \\ &= k_{m,n+1}\left[i_{m-1} \circ \left(\lambda x \in D_{m-1} \cdot x\right) \circ j_{m-1}\right] \\ &= k_{m,n+1}\left[i_{m-1} \circ j_{m-1}\right] \\ &\leqslant k_{m+1,n+1}\left(\lambda x \in D_{m-1} \cdot x\right) \text{ by ip pair} \\ &\quad \text{condition} \\ &\leqslant \lambda x \in D_n \cdot x \text{ by IH.} \end{aligned}$$

Now, that $[\![I]\!]_0 = \bot$ is clear, $j_0(I^1) = \bot$. □

Proposition 4.23 (Interpretation of constant functions).

$$[\![\lambda x_2 \cdot x_1]\!]_p = \langle p(x_1)_0, \lambda y \in D_0 \cdot p(x_1)_0, \lambda y \in D_1 \cdot p(x_1)_1, \ldots\rangle.$$

Proof. Note as above, it suffices to show that

$$[\![\lambda x_2.x_1]\!]^p_{n+1} = \lambda y \in D_n.p(x_1)_n,$$

since then the first component is determined.

As before, we proceed by induction on n. □

Proposition 4.24 (Interpretation of *K*).

$$[\![K]\!]^{D_\infty} = \langle \bot, I, K^2, K^3, \ldots \rangle.$$

Proof. We have

$$[\![K]\!] = \bigvee_n G_n\left(\lambda d.[\![\lambda x_2.x_1]\!]_{x_1:=d}\right).$$

Proceeding as before, we need to show that

$$[\![K]\!]_{n+2} = \lambda x \in D_{n+1}.\lambda y \in D_n.j_n(x).$$

Then, the second and first components will be determined as I and $\bot$, respectively, by applying j_1, j_0.

For the second component, we have

$$\begin{aligned} j_1(K_2)(x) &= (j_0 \circ [\lambda x \in D_1 \lambda y \in D_0.j_0(x)] \circ i_0)(x) \\ &= j_0[\lambda y \in D_0.j_0(\lambda z \in D_0.x)] \\ &= j_0[\lambda y \in D_0.x] \\ &= x, \end{aligned}$$

and so $K_1 = I^1$, and hence, $K_0 = \bot$. □

Proposition 4.25 (Interpretation of *S*).

$$[\![S]\!]^{D_\infty} = \langle \bot, I, S^2, S^3, \ldots \rangle.$$

Proof. As earlier, we also proceed by induction.

We'll first establish

$$[\![S]\!]_{n+3} = \bigvee_p k_{p,n+3}.$$

Last, we'll apply the projections to determine the initial components. □

Proposition 4.26 (Interpretation of Ω).

$$[\![\Omega]\!]^{D_\infty} = \bot_{D_\infty}.$$

Proof. First, we show that

$$[\![\lambda x \cdot xx]\!]^{D_\infty} = \langle \bot, \omega^1, \omega^2, \ldots \rangle.$$

Next, we need to show that $\omega^{n+1}(\omega^n) = \bot_{D_n}$. We proceed by induction on n. □

Remark 4.27. Classically, the result of Proposition 4.26 follows from the more general fact that unsolvable terms are interpreted as $\bot$ in D_∞. The proof seems to be a lengthy application of Böhm trees and uses several proofs by contradiction. It is not clear if that more general proposition holds in a reasonable extension of **CZF**.

Next, we examine the development of D_∞ as a model of the theory $\mathbf{PCA}^+$ (a theory defined in [2, VI.2.3]) or an *applicative structure* as defined by Feferman [15]. One interesting step is deciding in which way D_∞ thinks of the natural numbers. An initial, brute force method is to ensure D_∞ contains a copy of ω, say by taking $D_0 = \mathcal{P}_R(\omega)$ where R is a $\bigcup$-closed regular set containing ω.

Beginning with this approach, we next need to show the following.

PCA Axioms — This follows since (D_∞, F, G) is a satisfactory λ model.

Pairing and Projection Axioms Here it does not seem that a similar approach to $\mathcal{P}_R(\omega)$ is sufficient, and instead we must use the usual artificial pairing and projection of combinatory algebra.

$$\begin{aligned} p &= [\![\lambda x_1 x_2 x_3 \cdot x_3 x_1 x_2]\!] \\ p_0 &= [\![\lambda x_1 \cdot x_1 K]\!] \\ p_1 &= [\![\lambda x_1 \cdot x_1 (KI)]\!]. \end{aligned}$$

Natural Number Predicate

$$N(\mathbf{x}) \iff \exists n \in \omega \cdot i_{0,\infty}(\{n\}) = \mathbf{x}.$$

Successor — Define $s_1 = \lambda X \in D_0 \cdot \{y+1 \mid y \in X\}$. Note that $s_1 \in [D_0 \to D_0]$. Now, let $S_\mathbb{N} = i_{1\infty}(s_1)$. By Proposition 2.43(5), we then have the desired behaviour on elements satisfying N.

If we start with the ω-cpo from Definition 3.5, then we can define $s_\perp(\perp) = \perp$ and $s_\perp(n) = n+1$, and also let $S_{\mathbb{N}} = i_{1\infty}(s_\perp)$.
Natural Number Axioms — Follows from Natural number section.
Definition by cases We also need a term d so that

$$(N(a) \wedge N(b) \wedge a = b) \rightarrow d(a,b,x,y) = x,$$

$$(N(a) \wedge N(b) \wedge a \neq b) \rightarrow d(a,b,x,y) = y.$$

Such a term can be engineered in any *pca* (see [2, VI.2.9]).

Definition 4.28. The theory **EON** (Elementary theory of Operations and Numbers, see [2, VI.2.4]) is $\mathbf{PCA}^+$ as well as the induction schema for numbers, i.e., the schema for all formulae ϕ of

$$[\phi(0) \wedge \forall x[N(x) \wedge \phi(x)] \rightarrow \phi(s_N(x))]\, \forall x(N(x) \rightarrow \phi(x)).$$

Another approach would be to consider the interpretations of the Church numerals

True $T \equiv \lambda xy \cdot x$
False $F \equiv \lambda xy \cdot y$
Pairing $[M, N] \equiv \lambda z \cdot zM\ N$
Zero $\overline{0} \equiv I$
Successor $\overline{n+1} \equiv [F, [n]]$
Church Zero Zero $\equiv \lambda x \cdot xT$
Church Numerals $c_n \equiv \lambda fx \cdot f^n(x)$
Church Successor $S_c^+ \equiv \lambda abc \cdot b(abc)$.

We can then interpret the natural number predicate as

$$N(\mathbf{x}) \iff \text{There exists } n \in \omega \text{ so that } [\![c_n]\!] = \mathbf{x}.$$

Of course, this predicate is not decidable, but that is not required.

Remark 4.29. Automorphisms of D_∞. It seems that classically, the number of automorphisms (elements of $[D_\infty \rightarrow D_\infty]$ with inverses) of D_∞ varies based on D_0 and i_0, j_0. For D_0 finite, the number of automorphisms is not at all clear, and may possibly be trivial. For D_0 infinite, it is believable there are many automorphisms. For example, classically, taking $D_0 = [0,1]$ or $D_0 = P\omega$, we'd likely have continuum many automorphisms. These would be given by defining the automorphisms first on the first component then lifting to other components by pre or post composition. It remains to be checked in detail that the result is an element of D_∞ and moreover that the mapping is w-continuous.

Remark 4.30 (Completeness). True completeness would be the statement that if $M, N \in CT$, and $[\![M]\!] = [\![N]\!]$, then $M =_\beta N$. This is not true, since the D_∞ models are extensional, which does not hold syntactically.

Another notion of completeness is that a model D is complete if for all $M, N \in Terms$ which have $\beta\eta$-normal forms, then $[\![M]\!] = [\![N]\!]$.

The proof in [1] uses several proofs by contradiction (Böhm's Theorem, and the Completeness theorem) to prove this.

5. Realisability over D_∞

For any notion of realisability, the point of departure is a domain of computation known as *a partial combinatorial algebra* (shorthand *pca*) and also known by the name *applicative structure.* This chapter studies the particular realisability models for intuitionistic set theories over D_∞ models.

In 1945, Kleene developed realisability semantics for intuitionistic arithmetic and later for other theories. Kreisel and Troelstra [16] gave a definition of realisability for higher order Heyting arithmetic which got extended to systems of set theory by Myhill [17] and subsequently by Friedman [18]. Later, further realisability models were developed by Beeson [2,19] for nonextensional set theories. The extensional version of this realisability, already indicated by Beeson, only required small additions for the atomic formulas and was implemented by McCarty [20]. McCarty [20] though, is mainly concerned with realisability for intuitionistic Zermelo–Fraenkel set theory (**IZF**). As this approach employs transfinite iterations of the powerset operation through all the ordinals in defining the realisability (class) structure $V(\mathcal{A})$ for any $\mathbf{PCA}^+$ structure $\mathcal{A}$, it was not clear whether this semantics could be developed internally in **CZF**. Moreover, in addition to the powerset axiom the approach in [20] also uses unrestricted separation axioms. As **CZF** lacks the powerset axiom and has only bounded separation, it was not clear whether **CZF** would suffice as background theory. The development of this kind of realisability on the basis of **CZF** was carried out in [21].

6. Applicative Structures

To define a realisability semantics, a notion of realising functions must be available. An especially elegant approach to realisability is built on domains of computation allowing for self-application and recursion that are known as *partial combinatory algebra* or *Schönfinkel algebras* or *applicative structures.* A partial combinatory algebras (*pca*) just has special constants for Schönfinkel's combinators **k** and **s** and obeys the pertaining laws for these combinators. It is usually useful to endow a *pca* with additional structure such as (a copy of) N of the natural numbers and special constants for pairing, definition by cases on N as well as successor and predecessor on N. These gadgets can in fact be constructed in any partial combinatory algebra; for details, see ([2], VI.2.9). These richer structures are known as *applicative structures.* The latter are best described as the models of a theory, **APP**. The following presents the main features of **APP**; for full details, cf. [2,15]. The language of **APP** is a first-order language with a ternary relation symbol App, a unary relation symbol N (for a copy of the natural numbers) and equality, $=$, as primitives. The language has an infinite collection of variables, denoted $x, y, z, \ldots$, and nine distinguished constants: $\mathbf{0}, \mathbf{s}_{\mathsf{N}}, \mathbf{p}_{\mathsf{N}}, \mathbf{k}, \mathbf{s}, \mathbf{d}, \mathbf{p}, \mathbf{p}_0, \mathbf{p}_1$ for, respectively, zero, successor on N, predecessor on N, the two basic combinators, definition by cases, pairing and the corresponding two projections. There is no arity associated with the various constants. The *terms* of **APP** are just the variables and constants. We write $t_1t_2 \simeq t_3$ for $\mathrm{App}(t_1, t_2, t_3)$.

Formulae are then generated from atomic formulae using the propositional connectives and the quantifiers.

In order to facilitate the formulation of the axioms, the language of **APP** is expanded definitionally with the symbol $\simeq$ and the auxiliary notion of an *application term* is introduced. The set of application terms is given by two clauses:

(1) all terms of **APP** are application terms; and
(2) if s and t are application terms, then (st) is an application term.

For s and t application terms, we have auxiliary, defined formulae of the following form:

$$s \simeq t \;:=\; \forall y(s \simeq y \leftrightarrow t \simeq y),$$

if t is not a variable. Here, $s \simeq a$ (for a a free variable) is inductively defined by

$$s \simeq a \quad \text{is} \quad \begin{cases} s = a, & \text{if } s \text{ is a term of } \mathbf{APP}, \\ \exists x, y[s_1 \simeq x \wedge s_2 \simeq y & \text{if } s \text{ is of the form} \\ \wedge \operatorname{App}(x, y, a)] & (s_1 s_2). \end{cases}$$

Some abbreviations are $t_1 t_2, \ldots, t_n$ for $((...(t_1 t_2)...)t_n)$; $t \downarrow$ for $\exists y(t \simeq y)$ and $\phi(t)$ for $\exists y(t \simeq y \wedge \phi(y))$.

Some further conventions are useful. Systematic notation for *n-tuples* is introduced as follows: (t) is t, (s,t) is $\mathbf{p}st$, and $(t_1, \ldots, t_n)$ is defined by $((t_1, \ldots, t_{n-1}), t_n)$. In this chapter, the **logic** of **APP** is assumed to be that of intuitionistic predicate logic with identity. **APP**'s **nonlogical axioms** are the following:

Applicative Axioms

(1) $\operatorname{App}(a, b, c_1) \;\wedge\; \operatorname{App}(a, b, c_2) \;\rightarrow\; c_1 = c_2$.
(2) $(\mathbf{k}ab) \downarrow \;\wedge\; \mathbf{k}ab \simeq a$.
(3) $(\mathbf{s}ab) \downarrow \;\wedge\; \mathbf{s}abc \simeq ac(bc)$.
(4) $(\mathbf{p}a_0a_1) \downarrow \;\wedge\; (\mathbf{p_0}a) \downarrow \;\wedge\; (\mathbf{p_1}a) \downarrow \;\wedge\; \mathbf{p_i}(\mathbf{p}a_0a_1) \simeq a_i$ for $i = 0, 1$.
(5) $\mathsf{N}(c_1) \;\wedge\; \mathsf{N}(c_2) \;\wedge\; c_1 = c_2 \rightarrow \mathbf{d}abc_1c_2 \downarrow \;\wedge\; \mathbf{d}abc_1c_2 \simeq a$.
(6) $\mathsf{N}(c_1) \;\wedge\; \mathsf{N}(c_2) \;\wedge\; c_1 \neq c_2 \rightarrow \mathbf{d}abc_1c_2 \downarrow \;\wedge\; \mathbf{d}abc_1c_2 \simeq b$.
(7) $\forall x\,\big(\mathsf{N}(x) \rightarrow \big[\mathsf{s_N}x \downarrow \;\wedge\; \mathsf{s_N}x \neq \mathbf{0} \;\wedge\; \mathsf{N}(\mathsf{s_N}x)\big]\big)$.
(8) $\mathsf{N}(\mathbf{0}) \;\wedge\; \forall x\,\big(\mathsf{N}(x) \;\wedge\; x \neq \mathbf{0} \rightarrow \big[\mathsf{p_N}x \downarrow \;\wedge\; \mathsf{s_N}(\mathsf{p_N}x) = x\big]\big)$.
(9) $\forall x\,\big[\mathsf{N}(x) \rightarrow \mathsf{p_N}(\mathsf{s_N}x) = x\big]$.
(10) $\varphi(\mathbf{0}) \;\wedge\; \forall x\big[\mathsf{N}(x) \;\wedge\; \varphi(x) \rightarrow \varphi(\mathsf{s_N}x)\big] \;\rightarrow\; \forall x\big[\mathsf{N}(x) \rightarrow \varphi(x)\big]$.

Let $\mathbf{1} := \mathsf{s_N}\,\mathbf{0}$. The applicative axioms entail that $\mathbf{1}$ is an application term that evaluates to an object falling under N but distinct from $\mathbf{0}$, i.e., $\mathbf{1} \downarrow$, $\mathsf{N}(\mathbf{1})$ and $\mathbf{0} \neq \mathbf{1}$.

Employing the axioms for the combinators $\mathbf{k}$ and $\mathbf{s}$, one can deduce an abstraction lemma yielding λ-terms of one argument. This can be generalised using n-tuples and projections.

Lemma 6.1 [(cf. [15]) (Abstraction Lemma)]. For each application term t, there is a new application term t^* such that the parameters of t^* are among the parameters of t minus $x_1, \ldots, x_n$ and such that

$$\mathbf{APP} \vdash t^* \downarrow \ \wedge \ t^* x_1 \ldots x_n \simeq t.$$

$\lambda(x_1, \ldots, x_n).t$ is written for t^*.

The most important consequence of the Abstraction Lemma is the Recursion Theorem. It can be derived in the same way as for the λ–calculus (cf. [15],[2, VI.2.7]).

Theorem 6.2 (The Recursion Theorem). *There is a term* $\mathbf{R}$ *such that* $\mathbf{APP}$ *proves:* $\mathbf{R}f \downarrow$ *and* $[g = \mathbf{R}f \to \forall x(gx \simeq fgx)]$.

Proof. Take $\mathbf{R}$ to be $\lambda f \cdot tt$ where $t = \lambda yx \cdot f(yy)x$. Then,

$$\begin{aligned} \mathbf{R}f &\simeq (\lambda f \cdot tt)f \\ &\simeq tt. \end{aligned}$$

But,

$$\begin{aligned} tt &\simeq (\lambda y \cdot (\lambda x \cdot f(yy)x))t \\ &\simeq (\lambda x \cdot f(tt))x \\ &\simeq f(tt). \end{aligned}$$

So, if $\mathbf{R}f \downarrow$ and $g = \mathbf{R}f$, then

$$g \simeq tt \simeq f(tt) \simeq fg.$$

Thus, $gx = fgx$. □

In order to facilitate the formulation of the axioms, the language of **APP** is expanded definitionally with the symbol $\simeq$ and the auxiliary notion of an *application term* is introduced. The set of application terms is given by two clauses:

(1) all terms of **APP** are application terms; and
(2) if s and t are application terms, then (st) is an application term.

For s and t application terms, we have auxiliary, defined formulae of the following form:

$$s \simeq t \;:=\; \forall y(s \simeq y \leftrightarrow t \simeq y),$$

if t is not a variable. Here, $s \simeq a$ (for a a free variable) is inductively defined by

$$s \simeq a \quad \text{is} \quad \begin{cases} s = a, & \text{if } s \text{ is a term of } \mathbf{APP}, \\ \exists x, y[s_1 \simeq x \wedge s_2 \simeq y & \text{if } s \text{ is of the form} \\ \wedge \operatorname{App}(x, y, a)] & (s_1 s_2). \end{cases}$$

Some abbreviations are $t_1 t_2, \ldots, t_n$ for $((\ldots(t_1 t_2)\ldots)t_n)$; $t \downarrow$ for $\exists y(t \simeq y)$ and $\phi(t)$ for $\exists y(t \simeq y \wedge \phi(y))$.

Some further conventions are useful. Systematic notation for n-*tuples* is introduced as follows: (t) is t, (s, t) is $\mathbf{p}st$, and $(t_1, \ldots, t_n)$ is defined by $((t_1, \ldots, t_{n-1}), t_n)$. In this chapter, the **logic** of **APP** is assumed to be that of intuitionistic predicate logic with identity.

6.1. *Axioms of choice*

Several choice principles propagate from the background universe to the realisability universe as was shown in [22,23].

Axioms of countable choice and dependent choices, in many texts on constructive mathematics, are considered as constructive principles.

The weakest choice axiom, denoted by $\mathbf{AC}^{\omega,\omega}$, asserts that there is a function $f : \omega \to \omega$ with $\forall i \in \omega\, \theta(i, f(i))$ whenever $\forall i \in \omega \exists j \in \omega\, \theta(i, j)$.

The *Axiom of Countable Choice* ($\mathbf{AC}_\omega$) asserts that if $\forall i \in \omega \exists x\, \theta(i, x)$, then there is a function f with domain ω such that $\forall i \in \omega\, \theta(i, f(i))$. Obviously, $\mathbf{AC}_\omega$ implies $\mathbf{AC}^{\omega,\omega}$.

The *Axiom of Dependent Choices* (**DC**) is a useful axiom to have in set theory which may be given by the following scheme:

For any formula ϕ, whenever $(\forall a \in x)(\exists b \in x)\phi(a, b)$ and $x_0 \in x$, then there exists a function $f : \omega \to x$ such that $f(0) = x_0$ and $(\forall n \in \omega)\, \phi(f(n), f(n+1))$.

A very useful axiom is the *Relativised Dependent Choices Axiom* (**RDC**), which states that for arbitrary formulas ϕ and ψ, if

$$\forall x[\phi(x) \to \exists y(\phi(y) \wedge \psi(x,y))],$$

and $\phi(a_0)$, then there exists a function f whose domain is ω with $f(0) = a_0$ and

$$(\forall n \in \omega)[\phi(f(n)) \wedge \psi(f(n), f(n+1))].$$

6.2. *The presentation axiom (*PAx*)*

A set P is said to be projective if for all P-indexed family $(X_i)_{i\in P}$ of inhabited sets X_i, there exists a function f with domain P such that for all $i \in P$, $f(i) \in X_i$.

The *presentation axiom* (**PAx**) asserts that each set is the surjective image of a projective set. Projective sets are often called *bases*.

6.3. *Realisability structures*

Realisability interpretations are a crucial semantics for intutionistic set theories, and to be able to define a realisability interpretation, a notion of realising functions (partial functions serving as realisers for the formulas of the theory) must be to hand. An elegant and general approach to realisability builds on *applicative structures* or *partial combinatory algebras.*

Assume that we can formalise (in **CZF**) the notion of an applicative structure and let $\mathcal{A}$ be an arbitrary fixed applicative structure which in fact is a set. Henceforth, $|\mathcal{A}|$ will denote the carrier set of $\mathcal{A}$.

Write $\langle x, y\rangle$ for the ordered pair of x and y.

Definition 6.3. Ordinals are transitive sets with transitive elements. Lower case Greek letters will be used to range over ordinals.

For $\mathcal{A} \models \mathbf{PCA}^+$, let

$$V(\mathcal{A})_\alpha = \bigcup_{\beta\in\alpha} \mathcal{P}(|\mathcal{A}| \times V(\mathcal{A})_\beta)$$

$$V(\mathcal{A}) = \bigcup_{\alpha} V(\mathcal{A})_\alpha.$$

In **CZF**, it is not clear whether $V(\mathcal{A})$ can be formalised since the power set axiom is not among its axioms. This is possible and follows from [21, Lemma 3.4]. Provably in **CZF**, we also have the following result.

Lemma 6.4.

(1) *For any* $\beta \in \alpha \quad V(\mathcal{A})_\beta \subseteq V(\mathcal{A})_\alpha$.
(2) *For a subset* $U \subseteq |\mathcal{A}| \times V(\mathcal{A})$, *we have* $U \in V(\mathcal{A})$.

Proof. Since $V(\mathcal{A})_\alpha = \bigcup_{\beta \in \alpha} \mathcal{P}(|\mathcal{A}| \times V(\mathcal{A})_\beta)$,

(i) Is immediate.
(ii) Let $U \subseteq |\mathcal{A}| \times V(\mathcal{A})$. Then,

$$\forall u \in U \exists \alpha \exists r \in| \mathcal{A} \mid \exists x \in V(\mathcal{A})_\alpha [u = \langle r, x \rangle].$$

Now, the scheme of strong collection implies

$$\exists O [\forall u \in U \exists \alpha \in O \exists r \in| \mathcal{A} \mid \exists x \in V(\mathcal{A})_\alpha u = \langle r, x \rangle],$$

where O is a set of ordinals.
Let $O' = \{\alpha + 1 : \alpha \in O\}$ where $\alpha + 1 = \alpha \cup \{\alpha\}$, and let $\gamma = \bigcup O'$. Note that γ is also an ordinal and that $\forall \alpha \in O[\alpha \in \gamma]$.
Therefore,

$$\forall u \in U \exists \alpha \in \gamma \exists r \in |\mathcal{A}| \exists x \in V(\mathcal{A})_\alpha u = \langle r, x \rangle,$$

and hence, $U \subset \bigcup_{\alpha \in \gamma} \mathcal{P}(|\mathcal{A}| \times V(\mathcal{A})_\alpha)$. Thus, $U \in \mathcal{P}(|\mathcal{A}| \times V(\mathcal{A})_\alpha) = V(\mathcal{A})_\gamma \subseteq V(\mathcal{A})$. $\square$

Now, we define the notion of realisability over $V(\mathcal{A})$. For $r \in |\mathcal{A}|$, we define what it means for r to realise a sentence ϕ whose parameters are in $V(\mathcal{A})$, which is written as $r \Vdash \phi$.

For $r \in |\mathcal{A}|$, write $(r)_0$ for $\mathbf{p}_0 r$ and $(r)_1$ for $\mathbf{p}_1 r$.

Definition 6.5. Bounded and unbounded quantifiers are syntactically considered as different types of quantifiers. If $r \in |\mathcal{A}|$ and $a, b \in V(\mathcal{A})$, then

(1) $r \Vdash a \in b \iff \exists c[\langle (r)_0, c \rangle \in b \wedge (r)_1 \Vdash a = c]$.
(2) $r \Vdash a = b \iff \forall g, f[(\langle g, f \rangle \in a \rightarrow (r)_0 g \Vdash f \in b) \wedge$
$(\langle g, f \rangle \in b \rightarrow (r)_1 g \Vdash f \in a)]$.
(3) $r \Vdash \phi \wedge \psi \iff (r)_0 \Vdash \phi$ and $(r)_1 \Vdash \psi$.

(4) $r \Vdash \phi \vee \psi \iff [(r)_0 = \mathbf{0} \wedge (r)_1 \Vdash \phi] \vee [(r)_0 = \mathbf{1} \wedge (r)_1 \Vdash \psi]$.
(5) $r \Vdash \neg\phi \iff \forall k \in |\mathcal{A}| \quad \neg k \Vdash \phi$.
(6) $r \Vdash \phi \to \psi \iff \forall k \in |\mathcal{A}|[k \Vdash \phi \to rk \Vdash \psi]$.
(7) $r \Vdash \forall x \in a\, \phi \iff \forall \langle k, h\rangle \in a \quad rk \Vdash \phi[x/h]$.
(8) $r \Vdash \exists x \in a\, \phi \iff \exists h(\langle (r)_0, h\rangle \in a \wedge (r)_1 \Vdash \phi[x/h])$.
(9) $r \Vdash \forall x\phi \iff \forall h \in V(\mathcal{A}) \quad r \Vdash \phi[x/h]$.
(10) $r \Vdash \exists x\phi \iff \exists h \in V(\mathcal{A}) \quad r \Vdash \phi[x/h]$.

Theorem 6.6 (The Soundness Theorem for CZF). For each axiom A of **CZF**, there is a closed application term t such that

$$\mathbf{CZF} \vdash (t \Vdash A).$$

Proof. This is shown in [21, Theorem 5.1]. □

Definition 6.7 (Representing ω in $V(\mathcal{A})$). Let $\mathcal{A}$ be any *pca*, and recall that the zero of $\mathcal{A}$ is denoted by $\mathbf{0}$ and $\mathbf{1} = S_N\mathbf{0}$. Then ω is represented in $V(\mathcal{A})$ by $\overline{\omega}$ defined as follows:

Set $\underline{0} = \mathbf{0}$, where 0 is the least element of ω (the empty set) and for every $n \in \omega$, let $\underline{n+1} = S_N\underline{n}$. Then define $\overline{n} := \{\langle \underline{m}, \overline{m}\rangle : m \in n\}$ and take $\overline{\omega} = \{\langle \underline{n}, \overline{n}\rangle : n \in \omega\}$.

6.4. *Absoluteness properties*

Definition 6.8. For $a, b \in V(\mathcal{A})$, define $\{a, b\}_{\mathcal{A}} := \{\langle \mathbf{0}, a\rangle, \langle \mathbf{1}, b\rangle\}$ and let

$$\langle a, b\rangle_{\mathcal{A}} := \{\langle \mathbf{0}, \{a, a\}_{\mathcal{A}}\rangle, \langle \mathbf{1}, \{a, b\}_{\mathcal{A}}\rangle\}.$$

Lemma 6.9 (Internal Pairing in $V(\mathcal{A})$). *If $a, b, x \in V(\mathcal{A})$, then*

(i) $V(\mathcal{A}) \Vdash x \in \{a, b\}_{\mathcal{A}} \leftrightarrow x = a \vee x = b$.
(ii) $V(\mathcal{A}) \Vdash x \in \langle a, b\rangle_{\mathcal{A}} \leftrightarrow x = \{a, a\}_{\mathcal{A}} \vee x = \{a, b\}_{\mathcal{A}}$.

Proof. When $\mathcal{A}$ is the first Kleene algebra, this was shown in Lemmata 3.2 and 3.4 of [20]. But these properties also hold for an arbitrary pca as shown in [22, Lemma 2.13]. □

Using transfinite recursion, for any set x, we define

$$x^{st} = \{\langle \mathbf{0}, y^{st}\rangle : y \in x\}.$$

$x^{st} \in V(\mathcal{A})$ is straightforwardly proved by $\in$-induction as follows.

Inductively assume that $y^{st} \in V(A)$ for all $y \in x$. As $\mathbf{0} \in \mathcal{A}$, this implies that $\{\langle \mathbf{0}, y^{st}\rangle \mid y \in x\} \subseteq |\mathcal{A}| \times V(\mathcal{A})$, and thus, $x^{st} \in V(\mathcal{A})$. To complete the proof, we need the following facts.

Proposition 6.10.

(i) $x = y$ iff $V(\mathcal{A}) \Vdash x^{st} = y^{st}$.
(ii) $x \in y$ iff $V(\mathcal{A}) \Vdash x^{st} \in y^{st}$.

Proof. This was shown to hold for the first Kleene algebra in [20, Ch. 3 10.2], but also works for an arbitrary pca by [22, Proposition 2.15]. □

6.5. $A \simeq A^A$ *in* $\mathtt{V}(D_\infty)$

For sets A, B, we use the notation $A \equiv B$ to convey that there exists a bijection between A and B. Moreover, in what follows we shall write $x \in D_\infty$ for $\mathbf{x} \in D_\infty$ for ease of notation since we are not mixing $\mathbf{x} \in D_\infty$ with its components x_n.

Theorem 6.11.

(i) *With*

$$A := \{\langle x, x^{st}\rangle : x \in D_\infty\},$$

we have

$$\mathtt{V}(D_\infty) \models A \equiv A \longrightarrow A \wedge \exists f\,[f : \mathbb{N} \longrightarrow A \wedge f \textit{ is injective}].$$

(ii) *If* $\mathtt{V} \models A$ *is a base, then* $\mathtt{V}(D_\infty) \models A$ *is a base.*

Proof. (i) First we'd like to find a realiser for the statement

$$\exists g\,[g : \mathbb{N} \longrightarrow A \wedge g \text{ is injective}].$$

Define $G : \mathbb{N} \longrightarrow A$ by

$$G := \{\langle \underline{n}, \langle \bar{n}, \underline{n}^{st}\rangle_{D_\infty}\rangle : n \in \mathbb{N}\},$$

(where perhaps $\underline{n}^{st}$ less ambiguously should be written as $(\underline{n})^{st}$).

Clearly, $G \in \mathtt{V}(D_\infty)$ as $\underline{n} \in D_\infty$ and $\langle \bar{n}, \underline{n}^{st}\rangle_{D_\infty} \in \mathtt{V}(D_\infty)$ by the internal pairing properties.

Now, we need to check that G is internally a function from $\mathbb{N}$ into A.

Firstly, using properties of internal pairing in $\mathtt{V}(D_\infty)$, we show that $\mathtt{V}(D_\infty)$ validates that G is a binary relation whose domain is $\bar{\omega}$ and range a subset of A. To show that G is realisably functional, assume that

$$h \Vdash \langle a, x\rangle_{D_\infty} \in G, \tag{1}$$

$$k \Vdash \langle a, y\rangle_{D_\infty} \in G. \tag{2}$$

By (1), there exists a b such that $\langle (h)_0, b\rangle \in G$ and $(h)_1 \Vdash \langle a, x\rangle_{D_\infty} = b$. As $\langle (h)_0, b\rangle \in G$, $(h)_0$ must be of the form $\underline{n}$ and $b = \langle \bar{n}, \underline{n}^{st}\rangle_{D_\infty}$ for some $n \in \mathbb{N}$. Thus,

$$(h)_1 \Vdash \langle a, x\rangle_{D_\infty} = \langle \bar{n}, \underline{n}^{st}\rangle_{D_\infty}. \tag{3}$$

Similarly, from (2) we infer the existence of an $l \in \mathbb{N}$ such that

$$(k)_1 \Vdash \langle a, y\rangle_{D_\infty} = \langle \bar{l}, \underline{l}^{st}\rangle_{D_\infty}. \tag{4}$$

Using the properties of internal pairing (Lemma 3.1.11), we derive from (3) and (4) that $\mathtt{V}(D_\infty) \models \bar{n} = a \wedge \bar{l} = a$ and thus $n = l$ by Lemma 3.1.4 and Proposition 3.1.8. Therefore, there is an application term t such that $t(h, k) \Vdash x = y$, showing that G is realisably functional.

Note also that from (1) alone, we deduced (3) for some $n \in \mathbb{N}$, and hence, there exists an application term s showing that $s \Vdash G \subseteq \bar{\omega} \times A$.

Finally, we need to verify that G is realisably injective. To this end, assume that

$$h \Vdash \langle a, x\rangle_{D_\infty} \in G, \tag{5}$$

$$k \Vdash \langle b, x\rangle_{D_\infty} \in G. \tag{6}$$

By (5), there exists a c such that $\langle (h)_0, c\rangle \in G$ and $(h)_1 \Vdash \langle a, x\rangle_{D_\infty} = c$. As $\langle (h)_0, c\rangle \in G$, $(h)_0$ must be of the form $\underline{n}$ and $c = \langle \bar{n}, \underline{n}^{st}\rangle_{D_\infty}$ for some $n \in \mathbb{N}$. Thus,

$$(h)_1 \Vdash \langle a, x\rangle_{D_\infty} = \langle \bar{n}, \underline{n}^{st}\rangle_{D_\infty}. \tag{7}$$

Similarly, from (6) we infer the existence of an $l \in \mathbb{N}$ such that

$$(k)_1 \Vdash \langle b, x\rangle_{D_\infty} = \langle \bar{l}, \underline{l}^{st}\rangle_{D_\infty}. \tag{8}$$

Using the properties of internal pairing (Lemma 3.1.11), we derive from (7) and (8) that $\mathtt{V}(D_\infty) \models \underline{n}^{st} = x \wedge \underline{l}^{st} = x$, and thus, $n = l$

by Lemma 3.1.4 and Proposition 3.1.13. Therefore, in view of (7) and (8), we can infer that $a = b$ is realised, i.e., we can define an application term t' such that $t'(h,k) \Vdash a = b$, showing that G is realisably injective.

Next, we need to show that there is a bijection between A and $A \longrightarrow A$ in $\mathtt{V}(D_\infty)$. First, for each $e \in D_\infty$, we define an object

$$f_e := \{\langle e, \langle x^{st}, (ex)^{st}\rangle_{D_\infty}\rangle \mid x \in D_\infty\}. \tag{9}$$

In the definition of f_e above we make use of the fact that D_∞ is a total applicative structure, namely, that ex is defined for all $x \in D_\infty$. One easily checks that $f_e \in \mathtt{V}(D_\infty)$ for all $e \in D_\infty$.

We first show that there is an application term t such that

$$t \Vdash f_e \text{ is a function from } A \text{ to } A \tag{10}$$

holds for all $e \in D_\infty$. To show that f_e is realisably functional, assume that

$$h \Vdash \langle a, u\rangle_{D_\infty} \in f_e, \tag{11}$$

$$k \Vdash \langle a, v\rangle_{D_\infty} \in f_e. \tag{12}$$

By (11), there exists a b such that $\langle (h)_0, b\rangle \in f_e$ and $(h)_1 \Vdash \langle a, u\rangle_{D_\infty} = b$. As $\langle (h)_0, b\rangle \in f_e$, we conclude that $(h)_0 = e$ and $b = \langle x^{st}, (ex)^{st}\rangle_{D_\infty}$ for some $x \in D_\infty$. Thus,

$$(h)_1 \Vdash \langle a, u\rangle_{D_\infty} = \langle x^{st}, (ex)^{st}\rangle_{D_\infty}. \tag{13}$$

Similarly, from (12) we infer the existence of a $y \in D_\infty$ such that

$$(k)_1 \Vdash \langle a, v\rangle_{D_\infty} = \langle y^{st}, (ey)^{st}\rangle_{D_\infty}. \tag{14}$$

Using the properties of internal pairing (Lemma 3.1.11), we derive from (13) and (14) that $\mathtt{V}(D_\infty) \models x^{st} = y^{st}$, and thus, $x = y$ by Proposition 3.1.13, and thus also $(ex)^{st} = (ey)^{st}$, which in view of (13) and (14) entails that $\mathtt{V}(D_\infty) \models u = v$. More precisely, we can distil an application term s such that $s(h,k) \Vdash u = v$.

Concentrating solely on (11), the arguments above also show that f_e realisably has domain and range A. Moreover, it is also clear that realisably every element of A is in the domain of f_e. As a result, there exists an application term t such that $t \Vdash$ f_e *is a function from* A *to* A holds for all $e \in D_\infty$, verifying (10).

Now let

$$F := \{\langle e, \langle e^{st}, f_e\rangle_{D_\infty}\rangle \mid e \in D_\infty\}. \tag{15}$$

Since $f_e \in \mathtt{V}(D_\infty)$ for all $e \in D_\infty$, we have $F \in \mathtt{V}(D_\infty)$. We want to show that F is a bijection between A and $A \longrightarrow A$ in $\mathtt{V}(D_\infty)$. Functionality of F in $\mathtt{V}(D_\infty)$ can be shown in the same vein as for f_e. It is also clear that F realisably has domain A and from (10) it follows that F realisably has a range consisting of functions from A to A. In sum, there is an application term r such that

$$r \Vdash F \quad \text{is a function from } A \textit{ to } A \to A.$$

It remains to verify that F realisably furnishes a bijection.

We first address injectivity. To this end, assume that

$$h \Vdash \langle a, u\rangle_{D_\infty} \in F, \tag{16}$$

$$k \Vdash \langle b, u\rangle_{D_\infty} \in F. \tag{17}$$

Then there exists an $e \in D_\infty$ such that $e = (h)_0$ and $(h)_1 \Vdash \langle a, u\rangle_{D_\infty} = \langle e^{st}, f_e\rangle_{D_\infty}$.

Similarly, from (17) we infer the existence of a $d \in D_\infty$ such that $d = (k)_0$ and $(k)_1 \Vdash \langle b, u\rangle_{D_\infty} = \langle d^{st}, f_d\rangle_{D_\infty}$. As a result of the above, we obtain $\mathtt{V}(D_\infty) \models f_e = f_d$. Unravelling the definitions of f_e and f_d, this then implies that $\mathtt{V}(D_\infty) \models (ex)^{st} = (dx)^{st}$, and hence, $ex = dx$ for all $x \in D_\infty$. Owing to the extensionality of D_∞ this can only hold if $e = d$, which by the above entails $s'(h, k) \Vdash a = b$ for some application term s'. Consequently, we can construct an application term s such that

$$s \Vdash F \quad \text{is injective.}$$

To show subjectivity, assume that we have

$$\tilde{d} \Vdash g \quad \text{is a function from } A \text{ to } A. \tag{18}$$

Then there is an application term t such that

$$t\tilde{d} \Vdash \forall u \in A\, \exists v \in A\, g(u) = v. \tag{19}$$

Let $d := t\tilde{d}$. Unravelling (19), we get that for all $x \in D_\infty$,

$$(dx)_1 \Vdash g(x^{st}) = ((dx)_0)^{st}. \tag{20}$$

Let $e := \lambda z.(dz)_0$ in D_∞. One then constructs an application term s such that for all $x \in D_\infty$,

$$s(\tilde{d}, x) \Vdash g(x^{st}) = f_e(x^{st}).$$

Since extensionality holds in $\mathtt{V}(D_\infty)$, one finds application terms s', s'' such that $s'\tilde{d} \Vdash g = f_e$ and furthermore $s''\tilde{d} \Vdash g = F(e^{st})$. As a result, we can concoct an application term r such that $r \Vdash F$ *maps onto* $A \to A$.

(ii) We need to find a realiser $\hat{e}$ such that $e \Vdash A$ is a base. Towards this goal, suppose that

$$e \Vdash \forall u \in A \exists v\, \phi(u, v) \text{ for some formula } \phi. \tag{21}$$

The aim is to construct a realiser $\hat{e}$ satisfying

$$\hat{e} \Vdash \exists H[\mathbf{Fun}(H) \wedge \mathbf{dom}(H) = A \wedge \forall u \in A\, \phi(u, H(u))]. \tag{22}$$

Since

$$\begin{aligned} e \Vdash \forall u \in A \exists v \phi(u, v) &\Leftrightarrow \forall \langle x, x^{st} \rangle \in A \quad ex \Vdash \exists v \phi(x^{st}, v) \\ &\Leftrightarrow \forall \langle x, x^{st} \rangle \in A \exists a \in \mathtt{V}(D_\infty) \quad ex \Vdash \phi(x^{st}, a). \end{aligned}$$

Hence, from (21) it follows that

$$\forall \langle x, x^{st} \rangle \in A \exists v \in \mathtt{V}(D_\infty) \quad ex \Vdash \phi(x^{st}, v).$$

Now, since A is a base in the background universe, $\mathtt{V}$, there is a function $F : A \longrightarrow \mathtt{V}(D_\infty)$ such that

$$\forall \langle x, x^{st} \rangle \in A \quad ex \Vdash \phi(x^{st}, F(\langle x, x^{st} \rangle)).$$

Next, we need to internalise F and show that this internalisation, $\tilde{F}$, defined in what follows, provides the function required to show that A is a base. Set

$$\tilde{F} := \{\langle \mathbf{p}(ex)(x), \langle x^{st}, F(\langle x, x^{st} \rangle)\rangle_{D_\infty} \rangle : \langle x, x^{st} \rangle \in A\}.$$

Since, $x, ex \in D_\infty$, we have $\mathbf{p}(ex)(x) \in D_\infty$. Moreover, x^{st}, $F(\langle x, x^{st} \rangle) \in \mathtt{V}(D_\infty)$ entails that $\langle x^{st}, F(\langle x, x^{st} \rangle)\rangle_{D_\infty}$, and hence, $\tilde{F} \in \mathtt{V}(D_\infty)$.

Next, we need to show that the following hold:

(1) $\mathbf{dom}(\tilde{F}) = A$.
To this end, suppose that $h \Vdash u \in A$. Then

$$h \Vdash u \in A \Leftrightarrow \exists c[\langle (h)_0, c\rangle \in A \wedge (h)_1 \Vdash u = c.$$

Now, let $r \Vdash \langle (h)_0^{st}, F(\langle (h)_0, (h)_0^{st}\rangle)\rangle_{D_\infty} \in \tilde{F}$. Then we can find a $b \in \mathtt{V}(D_\infty)$ such that $\langle (r)_0, b\rangle \in \tilde{F} \wedge (r)_1 \Vdash \langle (h)_0^{st}, F(\langle (h)_0, (h)_0^{st}\rangle)\rangle_{D_\infty} = b$. But $\langle (r)_0, b\rangle \in \tilde{F}$ implies that $(r)_0$ is of the form $\mathbf{p}(ex)(x)$ and b is of the form $\langle x^{st}, F(\langle x, x^{st}\rangle)\rangle_{D_\infty}$ for some $\langle x, x^{st}\rangle \in A$. By the internal pairing, we infer $\Vdash (h)_0^{st} = x^{st}$, which implies $\Vdash (h)_0 = x$ using Proposition 6.10. Therefore, we conclude that there is an application term, $\hat{h}$, constructible from h such that $\hat{h} \Vdash u \in \mathbf{dom}(\tilde{F})$.

Conversely, assume that $d \Vdash \langle u, v\rangle_{D_\infty} \in \tilde{F}$. Then this is equivalent to the existence of an $f \in D_\infty$ such that

$$\langle (d)_0, f\rangle \in \tilde{F} \wedge (d)_1 \Vdash \langle u, v\rangle_{D_\infty} = f.$$

$\langle (d)_0, f\rangle \in \tilde{F}$ entails that $(d)_0$ has the form $\mathbf{p}(ex)(x)$ and f is of the form $\langle x^{st}, F(\langle x, x^{st}\rangle)\rangle_{D_\infty}$ for some $\langle x, x^{st}\rangle \in A$. Hence, we have $d \Vdash \langle u, v\rangle_{D_\infty} \in \tilde{F}$ iff there exists $\langle x, x^{st}\rangle \in A$ such that $\langle (d)_0, \langle x^{st}, F(\langle x, x^{st}\rangle)\rangle_{D_\infty}\rangle \in \tilde{F}$ where $x = d_{01}$ and $(d)_1 \Vdash \langle u, v\rangle_{D_\infty} = \langle x^{st}, F(\langle x, x^{st}\rangle)\rangle_{D_\infty}$.

Let $s \Vdash x^{st} \in A$. Then this entails the existence of an element c of $\mathtt{V}(D_\infty)$ such that $\langle (s)_0, c\rangle \in A$ and $(s)_1 \Vdash x^{st} = c$. From $\langle (s)_0, c\rangle \in A$, it follows that $(s)_0 = x$ and $c = x^{st}$ so that $\mathbf{p}(x)(i_r) \Vdash x^{st} \in A$.

Thus, we infer the existence of an application term $\hat{d}$ with $\hat{d}d \Vdash u \in A$, and hence, we can cook up an application term, $\hat{e}$, with $\hat{e} \Vdash \mathbf{dom}(\tilde{F}) = A$.

(2) Functional realisability of $\tilde{F}$. To establish this, let

$$f \Vdash \langle u, v\rangle_{D_\infty} \in \tilde{F}, \tag{23}$$

$$k \Vdash \langle u, w\rangle_{D_\infty} \in \tilde{F}. \tag{24}$$

Then we have (23) iff $\exists c$ such that $\langle (f)_0, c\rangle \in \tilde{F} \wedge (f)_1 \Vdash \langle u, v\rangle_{D_\infty} = c$. Now, from $\langle (f)_0, c\rangle \in \tilde{F}$, we obtain $(f)_0 = \mathbf{p}(ex)(x)$ and $c = \langle x^{st}, F(\langle x, x^{st}\rangle)\rangle_{D_\infty}$ for some $\langle x, x^{st}\rangle \in A$, and hence,

(23) is equivalent to the existence of $\langle x, x^{st}\rangle \in A$ such that $f_{01} = x$ and $f_1 \Vdash \langle u, v\rangle_{D_\infty} = \langle x^{st}, F(\langle x, x^{st}\rangle)\rangle_{D_\infty}$.

Similarly, one can infer from (24) that there exists $\langle y, y^{st}\rangle \in A$ such that $k_{01} = y$ and $k_1 \Vdash \langle u, w\rangle_{D_\infty} = \langle y^{st}, F(\langle y, y^{st}\rangle)\rangle_{D_\infty}$. Thus, $\mathtt{V}(D_\infty) \models x^{st} = y^{st}$, which implies (by proposition 6.10) that $x = y$. Using this together with

(a) $r \Vdash \mathbf{dom}(\tilde{F}) = A$,

(b) $\forall \langle x, x^{st}\rangle \in A \quad ex \Vdash \phi(x^{st}, F(\langle x, x^{st}\rangle))$,

we can construct the realiser, $\hat{e}$, in (22) and this completes the proof of (ii). □

7. Brouwerian Principles

The purpose of this chapter is to study some key tenets of the *Brouwerian Mathematical World* known as *Brouwerian principles*, namely, the *Fan Theorem* (**FT**), and the principle of *Bar Induction* (**BI**).[e] The objective here is to determine whether these principles get realised in the realisability structures $\mathtt{V}(D_\infty)$.[f] As it turns out, the validity of these principles propagates from the background universe to $\mathtt{V}(D_\infty)$ if all functions from naturals to naturals are representable in D_∞. Except for the totality of the pca D_∞, no germane features of $\mathtt{V}(D_\infty)$ are exploited in the proof, that is, any other total pca satisfying $(*)$ could be used.

We first start by introducing these principles.

7.1. *The fan theorem* (FT)

FT is a crucial corollary of the *Bar Theorem* to be introduced shortly, and all of the known applications of the Bar Theorem already follow from **FT**.

[e]These principles were introduced by Brouwer to preserve parts of traditional analysis. In particular, he hoped to save the theorem that a continuous function from a closed interval of a finite length to the real line is uniformly continuous and he established this theorem as a consequence of his Bar Theorem [24].

[f]The study of these principles in realisability models of set theory started in [25] and was further pursued in [3] for the realisability structures $\mathtt{V}(D_\infty)$.

FT also holds classically and it is equivalent to *König's Lemma* [24].

Definition 7.1 ([25], Definition 4.2). Let $2^{\mathbb{N}}$ denote the set of all 01-sequences $\alpha : \mathbb{N} \longrightarrow \{0,1\}$ and set 2^* be the set of all finite such sequences. Write $s \subseteq t$ for $s, t \in 2^*$ to mean that s is an initial segment of t.

A bar $\mathbf{B}$ of 2^* is a subset of 2^* such that

$$\forall \alpha \in 2^{\mathbb{N}}\ \exists n \quad \bar{\alpha}n \in \mathbf{B},$$

where $\bar{\alpha}n + 1 := \langle \alpha(0), ..., \alpha(n) \rangle$ and $\bar{\alpha}0 := \langle\ \rangle$.

The bar $\mathbf{B}$ is *decidable* if it additionally satisfies

$$\forall t \in 2^* (t \in \mathbf{B} \vee t \notin \mathbf{B}).$$

The decidable Fan Theorem $\mathbf{FT_D}$ states that each decidable bar $\mathbf{B}$ of 2^* is uniform, in other words,

$$\exists m \in \mathbb{N}[\forall \alpha \in 2^{\mathbb{N}}\ \exists k \leqslant m \quad \bar{\alpha}k \in \mathbf{B}].$$

The statement of *The Full Fan Theorem* (**FT**) is the following:

Each bar $\mathbf{B}$ of 2^* is uniform.

7.2. *Bar induction principle* (BI)

Brouwer obtained $\mathbf{FT_D}$ by appealing to a principle called *decidable* Bar Induction, $\mathbf{BI_D}$.

Let $\mathbb{N}^*$ be the set of all finite sequences of $\mathbb{N}$. For $m \in \mathbb{N}$ and $s \in \mathbb{N}^*$ such that $s = \langle s_0, ..., s_k \rangle$, $s * \langle m \rangle$ denotes the sequence $\langle s_0, ..., s_k, m \rangle$.

A bar $\mathbf{B}$ of $\mathbb{N}^*$ is defined in the same way as that of 2^*, *i.e.*, $\mathbf{B} \subseteq \mathbb{N}^{\mathbb{N}}$ such that

$$\forall \alpha \in \mathbb{N}^{\mathbb{N}}\ \exists n \quad \bar{\alpha}n \in \mathbf{B}. \tag{25}$$

The bar $\mathbf{B}$ is *decidable* if

$$\forall t \in \mathbb{N}^* (t \in \mathbf{B} \vee t \notin \mathbf{B}). \tag{26}$$

7.3. *Decidable bar induction* ($\mathbf{BI_D}$)

$\mathbf{BI_D}$ states that for any decidable bar $\mathbf{B}$ of $\mathbb{N}^*$ and any arbitrary class G,

$$\text{if} \quad \forall t \in \mathbb{N}^*(t \in \mathbf{B} \longrightarrow t \in G), \tag{27}$$

$$\text{and} \quad \forall t \in \mathbb{N}^*[(\forall k \in \mathbb{N}, \ t * \langle k \rangle \in G) \longrightarrow t \in G], \tag{28}$$

$$\text{then} \quad \langle\ \rangle \in G. \tag{29}$$

$\mathbf{FT_D}$ can be viewed as a special case of $\mathbf{BI_D}$ (see [26], Chapter 1, Section 6.10 or [25]). $\mathbf{BI_D}$ follows classically from the dependent choice axiom.

7.4. *Monotone bar induction* (BI_M)

$\mathbf{BI_M}$ states that for any bar $\mathbf{B}$ of $\mathbb{N}^*$ and any arbitrary class G,

$$\text{if} \quad \forall s, t \in \mathbb{N}^*(s \in \mathbf{B} \longrightarrow s * t \in \mathbf{B}), \tag{30}$$

$$\text{and} \quad \forall t \in \mathbb{N}^*(t \in \mathbf{B} \longrightarrow t \in G), \tag{31}$$

$$\text{and} \quad \forall t \in \mathbb{N}^*[(\forall k \in \mathbb{N}, \ t * \langle k \rangle \in G) \longrightarrow t \in G], \tag{32}$$

$$\text{then} \quad \langle\ \rangle \in G. \tag{33}$$

$\mathbf{BI_M}$ entails $\mathbf{BI_D}$ (for details, see [27], Theorem 3.7). Furthermore, $\mathbf{BI_D}$ entails $\mathbf{FT_D}$ and $\mathbf{BI_M}$ entails $\mathbf{FT}$, as was shown in [26], Chapter 1, Section 6.10 (also see [25]).

To understand the idea behind this principle, consider the tree of all finite sequences of natural numbers $\mathbf{T}$. Then conditions (25–27) assert that any path through $\mathbf{T}$ meets the bar $\mathbf{B}$ at a node, n and hence G is hit at n as well. Condition 28 states that if G holds at all extensions of a node then it also holds at that node and this allows the property G to "sneak" towards the root of $\mathbf{T}$. Furthermore, we may think of this principle as an induction statement over a well-founded relation. Classically, one can argue as follows. Given condition (28), the failure of G to hold at the empty sequence, appealing to the axiom of dependent choice, would yield an infinite sequence of natural numbers

$$n_0, n_1, n_2, \ldots$$

such that at every initial segment $\langle n_0, n_1, n_2, \ldots \rangle$, G does not hold and therefore defining such a function α by

$$\alpha(k) := n_k$$

would contradict conditions (25) and (27).

7.5. *Fan theorem and realisability over D_∞*

The aim is to show that $\mathtt{V}(D_\infty)$ validates $\mathbf{FT_D}$ and $\mathbf{BI_D}$, respectively, providing they hold in the background universe $\mathtt{V}$. Because of the prominent role of Cantor and Baire space in these principles, the first step in the investigation will be to investigate *Baire space* from the point of view of $\mathtt{V}(D_\infty)$. As it turns out, it is instructive to look at the general case first, i.e., the internal $\mathtt{V}(\mathcal{A})$ presentation of Baire space for an arbitrary applicative structure, $\mathcal{A}$.

7.5.1. *Baire space in* $\mathtt{V}(\mathcal{A})$

Let $\underline{n}$ be the n-th numeral in the applicative structure $\mathcal{A}$, i.e., the interpretation of the term $S_{\mathbb{N}}, \ldots, S_{\mathbb{N}}0$ (with n-many symbols $S_{\mathbb{N}}$) in $\mathcal{A}$. In the following, we identify $\mathbb{N}$ with its interpretation in $\mathcal{A}$.

Recalling Definition (6.7), we have

$$\overline{n} := \{\langle \underline{m}, \overline{m}\rangle : m \in n\},$$

$$\overline{\omega} = \{\langle \underline{n}, \overline{n}\rangle : n \in \mathbb{N}\}.$$

And define

$$\begin{aligned}\langle a, b\rangle_{\mathcal{A}} &:= \{\langle \underline{0}, \{a, a\}_{\mathcal{A}}\rangle, \langle \underline{1}, \{a, b\}_{\mathcal{A}}\rangle\} \\ &= \{\langle \underline{0}, \{\langle \underline{0}, a\rangle, \langle \underline{1}, a\rangle\}, \langle \underline{1}, \{\langle \underline{0}, a\rangle, \langle \underline{1}, b\rangle\}\rangle\},\end{aligned}$$

where $a, b \in \mathtt{V}(\mathcal{A})$ and $\{a, b\}_{\mathcal{A}} := \{\langle \underline{0}, a\rangle, \langle \underline{1}, b\rangle\}$.

A realiser $e \in \mathcal{A}$ can be constructed such that

$$e \Vdash \langle a, b\rangle_{\mathcal{A}} \quad \text{is the ordered pair of } a, b,$$

and moreover e does not depend on a, b.

This can be seen from the proof of Lemma (6.9).

Definition 7.2. $f \in \mathcal{A}$ is said to be of Type 1 if for every $n \in \omega$, $f\underline{n}\downarrow$ and there is some $m \in \omega$ such that

$$f\underline{n} = \underline{m}.$$

Definition 7.3. Given a Type 1 $f \in \mathcal{A}$, we can construct $f_{\mathbb{B}}$ such that

$$f_{\mathbb{B}} := \{\langle \underline{n}, \langle \overline{n}, \overline{m}\rangle_{\mathcal{A}}\rangle : n \in \omega \text{ and } f\underline{n} = \underline{m}\}.$$

Proposition 7.4. *For every $f \in \mathcal{A}$ of Type 1, there is a realiser (in $\mathcal{A}$) for the statement that $f_{\mathbb{B}}$ is a function from ω to ω.*

Proof. Suppose that e is a realiser for the statement that $\langle a, b\rangle_{\mathcal{A}}$ is the ordered pair of a and b.

To show that $f_{\mathbb{B}}$ is (internally) a function from ω to ω, we need to show that $f_{\mathbb{B}} \subseteq \overline{\omega} \times \overline{\omega}$.

Towards this goal, let

$$\begin{aligned}
d \;\Vdash\; & \forall n \in \overline{\omega}\, \exists m \in \overline{\omega}\, \exists y \in f_{\mathbb{B}}\;\; y = \langle n, m\rangle \\
\iff & \forall \langle f_1, c\rangle \in \overline{\omega}\;\; df_1 \Vdash \exists m \in \overline{\omega}\, \exists y \in f_{\mathbb{B}}\;\; y = \langle c, m\rangle \\
\iff & \forall \langle f_1, c\rangle \in \overline{\omega}\, \exists c' \,(\langle (df_1)_0, c'\rangle \in \overline{\omega} \wedge (df_1)_1 \Vdash \exists y \in f_{\mathbb{B}}\;\; y = \langle c, c'\rangle) \\
\iff & \forall \langle f_1, c\rangle \in \overline{\omega}\, \exists c' \,\langle (df_1)_0, c'\rangle \in \overline{\omega}\, \exists c^* \,(\langle (df_1)_{10}, c^*\rangle \\
& \quad \times \in f_{\mathbb{B}} \wedge (df_1)_{11} \Vdash c^* = \langle c, c'\rangle).
\end{aligned}$$

Now, $\langle f_1, c\rangle \in \overline{\omega} \iff f_1 = \underline{n}$ and $c = \overline{n}$ for some $n \in \omega$, and hence,

$$d\underline{n} \Vdash \exists m \in \overline{\omega}\, \exists y \in f_{\mathbb{B}}\;\; y = \langle \overline{n}, m\rangle,$$

and, $\langle (d\underline{n})_0, c'\rangle \in \overline{\omega} \iff (d\underline{n})_0 = \underline{m}$ and $c' = \overline{m}$ for some $m \in \omega$ and

$$(d\underline{n})_{11} \Vdash c^* = \langle \overline{n}, \overline{m}\rangle,$$

$\langle (d\underline{n})_{10}, c^*\rangle \in f_{\mathbb{B}} \iff (d\underline{n})_{10} = \underline{n}$ and $c^* = \langle \overline{n}, \overline{m}\rangle_{\mathcal{A}}$ for some $n \in \omega$ and

$$(d\underline{n})_{11} \Vdash c^* = \langle c, c'\rangle.$$

Next,

$$\begin{aligned}(d\underline{n})_1 &= \mathbf{p}(d\underline{n})_{10}(d\underline{n})_{11}\\ &= \mathbf{p}(\underline{n})(e)\\ \text{and } d\underline{n} &= \mathbf{p}(d\underline{n})_0(d\underline{n})_1\\ &= \mathbf{p}(f\underline{n})(\mathbf{p}(\underline{n})(e)) \text{ as } \underline{m} = f\underline{n}.\end{aligned}$$

Thus, $d = \lambda x.\mathbf{p}(fx)\underline{(\mathbf{p}(x)(e))}$. Also, if $\langle \underline{n}, \langle \overline{n}, \overline{m}\rangle_{\mathcal{A}}\rangle, \langle \underline{n'}, \langle \overline{n'}, \overline{m'}\rangle_{\mathcal{A}}\rangle \in f_{\mathbb{B}}$ and $\mathtt{V}(\mathcal{A}) \models \overline{n} = \overline{n'}$, then $n = n'$ by Proposition 6.10.

This yields that $m = m'$ and, therefore, $f_{\mathbb{B}}$ is realisably functional and one easily constructs a realiser for the statement

$$\forall x, y \in f_{\mathbb{B}}[(x)_0 = (y)_0 \longrightarrow (x)_1 = (y)_1].$$

We can combine these realisers to obtain a realiser for the statement that $f_{\mathbb{B}}$ is a function. □

Proposition 7.5. *Suppose* $\mathtt{V}(\mathcal{A}) \models f$ *is a function from* $\overline{\omega}$ *to* $\overline{\omega}$. *Then, there is a Type 1* $g \in \mathcal{A}$ *such that*

$$\mathtt{V}(\mathcal{A}) \models f = g_{\mathbb{B}}.$$

Proof. As there is a realiser stating that f is a function, we can find realisers a, b and $c \in \mathcal{A}$ such that

$$a \Vdash \forall n \in \overline{\omega}\ \exists m \in \overline{\omega}\ \langle n, m\rangle \in f, \tag{34}$$

$$b \Vdash \forall x \in f\ \exists n, m \in \overline{\omega}\ x = \langle n, m\rangle, \tag{35}$$

$$c \Vdash \forall n, m, m'\,[\langle n, m\rangle \in f \wedge \langle n, m'\rangle \in f \longrightarrow m = m']. \tag{36}$$

So, we have

$$\begin{aligned}(34) &\iff \forall \langle f_1, c\rangle \in \overline{\omega}\quad af_1 \Vdash \exists m \in \overline{\omega}\ \langle c, m\rangle \in f\\ &\iff \forall \langle f_1, c\rangle \in \overline{\omega}\ \exists c'\,(\langle (af_1)_0, c'\rangle \in \overline{\omega} \wedge ((af_1)_1)_1 \Vdash \langle c, c'\rangle \in f).\end{aligned}$$

Now, $\langle f_1, c\rangle \in \overline{\omega} \iff f_1 = \underline{n}$ for some $n \in \omega$ and $c = \bar{f_1} = \overline{n}$.

$$\langle (af_1)_0, c'\rangle \in \overline{\omega} \iff (af_1)_0 \quad \text{and} \quad c' = \overline{(af_1)_0}.$$

Let $g := \lambda x.(ax)_0$. Then by the above g must be of Type 1. Next, each element of $g_{\mathbb{B}}$ is of the form

$$\langle \underline{n}, \langle \overline{n}, \overline{g\underline{n}}\rangle_{\mathcal{A}}\rangle = \langle \underline{n}, \langle \overline{n}, \overline{(a\underline{n})_0}\rangle_{\mathcal{A}}\rangle \quad \text{and} \quad (a\underline{n})_1 \Vdash \langle \overline{n}, \overline{(a\underline{n})_0}\rangle_{\mathcal{A}} \in f.$$

Thus, we obtain a realiser for $g \subseteq f$.
Next, we also get

$$\begin{aligned}(35) \iff & \forall \langle l, k\rangle \in f \quad bl \Vdash \exists n, m \in \overline{\omega}\; k = \langle n, m\rangle \\ \iff & \forall \langle l, k\rangle \in f\; \exists q\,(\langle (bl)_0, q\rangle \in \overline{\omega} \wedge (bl)_1 \Vdash \exists m \in \overline{\omega}\; k = \langle q, m\rangle) \\ \iff & \forall \langle l, k\rangle \in f\; \exists q\,(\langle (bl)_0, q\rangle \in \overline{\omega} \wedge \\ & \qquad \exists t\,(\langle (bl)_{10}, t\rangle \in \overline{\omega} \wedge (bl)_{11} \Vdash k = \langle q, t\rangle)).\end{aligned}$$

Then for all $\langle l, x\rangle$ in f, we have

$$bl \Vdash \exists u, v \in \overline{\omega}\; x = \langle u, v\rangle.$$

Let $n = (bl)_0$, $m = (bl)_{10}$. Then,
$(a\underline{n})_1 \Vdash \langle \overline{n}, \overline{(a\underline{n})_0}\rangle_{\mathcal{A}} \in f$ and also, $(bl)_{11} \Vdash \langle \overline{(bl)_0}, \overline{(bl)_{10}}\rangle_{\mathcal{A}} = \langle \overline{n}, \overline{m}\rangle_{\mathcal{A}} \in f$.

Next, we need to find a realiser for $\overline{m} = \overline{(a\underline{n})_0}$, which gives a realiser for $x = \langle \overline{n}, \overline{(a\underline{n})_0}\rangle_{\mathcal{A}}$, but by (36) a realiser for $\overline{m} = \overline{(a\underline{n})_0}$ can be constructed and thus one can compute a realiser for $f \subseteq g$.

Therefore, $\mathtt{V}(\mathcal{A}) \models f = g_{\mathbb{B}}$ as required. □

Proposition 7.6. *$\mathtt{V}(\mathcal{A}) \models \{\langle f, f_{\mathbb{B}}\rangle : f \in \mathcal{A}$ and $\forall x \in \mathbf{N}\; fx \in \mathbf{N}\}$ is the set of all functions from ω to ω.*

Proof. The proof is immediate using the previous two propositions. □

As a result, based on what the internal *Baire Space* of $\mathtt{V}(\mathcal{A})$ for any applicative structure $\mathcal{A}$ looks like, the investigation of **FT** in $\mathtt{V}(D_\infty)$ reduces to investigating the set

$$\{\langle f, f_{\mathbb{B}}\rangle : f \in \mathcal{A} \text{ and } \forall x \in \mathbf{N}\; f \bullet \mathbf{x} \in \mathbf{N}\},$$

where $\mathbf{N}$ are the natural numbers of $\mathcal{A}$. A bit later we shall investigate for which D_∞ models the condition

$$\forall g : \mathbf{N} \longrightarrow \mathbf{N}\; \exists f \in \mathcal{A} \;\; \forall x[x \in \mathbf{N} \quad f \bullet x = g(x)] \tag{37}$$

obtains.

7.6. $\mathbf{FT_D}$ *and* $\mathbf{BI_D}$ *in* $\mathtt{V}(D_\infty)$

A proof of the next theorem appeared first in [3], adapting [25, Theorem 6.5].

Theorem 7.7. *Assume that*

$$(*) \qquad \forall g : \mathbf{N} \longrightarrow \mathbf{N}\ \exists f \in \mathcal{A}\ \ \forall x[x \in \mathbf{N} \ \rightarrow\ f \bullet x = g(x)]$$

holds for D_∞*, where* $\mathbf{N}$ *stands for the choice of naturals of* D_∞ *as an applicative structure.*

(1) *If* $\mathtt{V} \models \mathbf{FT_D}$*, then* $\mathtt{V}(D_\infty) \models \mathbf{FT_D}$*.*
(2) *If* $\mathtt{V} \models \mathbf{BI_D}$*, then* $\mathtt{V}(D_\infty) \models \mathbf{BI_D}$*.*

Proof. We shall concentrate on finding a realiser for $\mathbf{BI_D}$.

First, since finite sequences of natural numbers (elements of $\mathbb{N}^*$) can be coded by the elements of $\mathbb{N}$ and for all natural number n in ω, n codes a finite sequence s in $\mathbb{N}^*$, we may therefore identify $\mathbb{N}^*$ with $\mathbb{N}$.

Now, suppose there exist realisers e_1, e_2, e_3 and $e_4 \in D_\infty$ such that

$$e_1 \Vdash \forall \alpha \in \mathbb{N}^{\mathbb{N}}\ \exists n \in \mathbb{N}\ \mathbf{B}(\bar{\alpha}n), \tag{38}$$

$$e_2 \Vdash \forall s \in \mathbb{N}\ (\mathbf{B}(s) \vee \neg\mathbf{B}(s)), \tag{39}$$

$$e_3 \Vdash \forall s \in \mathbb{N}\ \ (\mathbf{B}(s) \longrightarrow G(s)), \tag{40}$$

$$e_4 \Vdash \forall s \in \mathbb{N}\ \ [(\forall k \in \mathbb{N}\ G(s * \langle k \rangle)) \longrightarrow G(s)]. \tag{41}$$

By the realisability definition, (38) is equivalent to

$$\forall \langle f, c\rangle \in (\mathbb{N}^{\mathbb{N}})^{\mathtt{V}(D_\infty)}\ \ e_1 f \Vdash \exists n \in \mathbb{N}\ \ \mathbf{B}(\bar{c}n)$$
$$\Longleftrightarrow\ \forall \langle f, f_{\mathbb{B}}\rangle \in (\mathbb{N}^{\mathbb{N}})^{\mathtt{V}(D_\infty)}\ \ e_1 f \Vdash \exists n \in \mathbb{N}\ \ \mathbf{B}(\bar{f_{\mathbb{B}}}n),$$

since $\langle f, c\rangle$ is in the internal Baire Space of $\mathtt{V}(D_\infty)$, and thus has the form

$$\langle f, f_{\mathbb{B}}\rangle = \langle \underline{n}, \langle \overline{n}, \overline{m}\rangle_{D_\infty}\rangle, \quad \text{for some Type 1 } f \in D_\infty$$

and some $n, m \in \omega$.

$$\Longleftrightarrow\ \forall \langle f, f_{\mathbb{B}}\rangle \in (\mathbb{N}^{\mathbb{N}})^{\mathtt{V}(D_\infty)} \exists d \in \mathtt{V}(D_\infty)\ \ (\langle (e_1 f)_0, d\rangle \in (\mathbb{N})^{\mathtt{V}(D_\infty)} \wedge (e_1 f)_1 \Vdash \mathbf{B}(\bar{f_{\mathbb{B}}}d).$$

However, $\langle (e_1 f)_0, d\rangle \in (\mathbb{N})^{\mathbf{V}(D_\infty)}$ entails that $d = \overline{(e_1 f)_0}$ for some $m \in \mathbb{N}$ and $d = \overline{m}$. Thus, we arrive at

$$(e_1 f)_1 \Vdash \mathbf{B}(\bar{f}_{\mathbb{B}}\overline{(e_1 f)_0}).$$

And, (39) is equivalent to
$\forall \langle g, d\rangle \in (\mathbb{N})^{\mathbf{V}(D_\infty)} \quad e_2 g \Vdash (\mathbf{B}(d) \vee \neg\mathbf{B}(d))$.
$\langle g, d\rangle \in (\mathbb{N})^{\mathbf{V}(D_\infty)} \iff \langle g, d\rangle$ is of the form $\langle \underline{n}, \overline{n}\rangle$ for some $n \in \omega$.
Thus, for all $\langle \underline{n}, \overline{n}\rangle \in (\mathbb{N})^{D_\infty}$, we have

$$\begin{aligned} &e_2\underline{n} \Vdash \mathbf{B}(\overline{n}) \ \vee \ \neg\mathbf{B}(\overline{n}), \\ &[(e_2\underline{n})_0 = \mathbf{0} \wedge (e_2\underline{n})_1 \Vdash \mathbf{B}(\overline{n})] \ \vee \ [(e_2\underline{n})_0 \\ &\qquad = \mathbf{1} \wedge (e_2\underline{n})_1 \Vdash \neg\mathbf{B}(\overline{n})]. \end{aligned} \tag{42}$$

Define $\tilde{\mathbf{B}}(s) :\iff (e_2\underline{s})_0 = \mathbf{0}$, and define a function $\psi : \mathbb{N} \longrightarrow D_\infty$ as follows:

$$\psi(\underline{s}) = \begin{cases} (e_3\underline{s})(e_2\underline{s})_1 & \text{if } (e_2\underline{s})_0 = \mathbf{0} \\ (e_4\underline{s})(\lambda u.\psi(\underline{s * \langle u\rangle}) & \text{if } (e_2\underline{s})_0 \neq \mathbf{0}. \end{cases}$$

It is actually not clear that the above definition of ψ, makes sense. The definition is obviously self-referential, so we need to appeal to some kind of recursion theorem. First, we know how the primitive recursive functions $*$ and $k \mapsto < k >$ operate on the naturals. Given any applicative structure $\mathcal{A}$, they can be defined by the recursion theorem for applicative structures in such a way that they are total on the copy of $\mathbb{N}$ of $\mathcal{A}$, however, cannot be guaranteed to be total on the whole applicative structure. In our case, though, we are dealing with the total applicative structure D_∞, so these functions have total extensions. Now, instead of the above definition of ψ, we commence by defining a function $\tilde{\psi} : D_\infty \longrightarrow D_\infty$, using the recursion theorem for applicative structures, by letting

$$\tilde{\psi}(x) = \begin{cases} (e_3 x)(e_2 x)_1 & \text{if } (e_2 x)_0 = \mathbf{0} \\ (e_4 x)(\lambda u.\psi(x * \langle u\rangle) & \text{if } (e_2 x)_0 \neq \mathbf{0}, \end{cases}$$

where $*$ and $u \mapsto \langle u \rangle$ stand for some total extensions of the familiar functions to D_∞. The case distinction is also alright when we use the case distinction constant d. Note that $\tilde{\psi}$ will then be a total function on D_∞.

Finally, define $\psi : \mathbb{N} \longrightarrow D_\infty$ by $\tilde{\psi}(\underline{s}$. Let $\tilde{G}(s) : \Longleftrightarrow \quad \psi(\underline{s}) \Vdash G(\overline{s})$.

Claim 7.8. *We claim that the above defined* $\tilde{\mathbf{B}}$ *and* $\tilde{G}$ *are the bar and the class which ensure the validity of* $\mathbf{BI_D}$ *in* $\mathtt{V}(D_\infty)$*. In other words, the following hold:*

(1) $\forall \alpha \in \mathbb{N}^{\mathbb{N}}\, \exists n \in \mathbb{N} \;\; \tilde{\mathbf{B}}(\bar{\alpha} n)$.
(2) $\forall s \in \mathbb{N} \;\; (\tilde{\mathbf{B}}(s) \vee \neg \tilde{\mathbf{B}}(s))$.
(3) $\forall s \in \mathbb{N} \;\; (\tilde{\mathbf{B}}(s) \longrightarrow \tilde{G}(s))$.
(4) $\forall s \in \mathbb{N} \;\; [(\forall k \in \mathbb{N}\, \tilde{G}(s * \langle k \rangle)) \longrightarrow \tilde{G}(s)]$.

Proof of the claim.

(1) By the assumption $(*)$, we have that for all $\alpha \in \mathbb{N}^{\mathbb{N}}$ there exists $f \in D_\infty$ such that $\underline{\bar{\alpha} n} = f \bullet \underline{n}$ for all n in $\mathbb{N}$. Now, $\mathbf{B}(\overline{\bar{\alpha} n})$ holds in $\mathtt{V}(D_\infty)$. Thus, by (42), $(e_2 \underline{\bar{\alpha} n})_0 = \mathbf{0}$ so that $\tilde{\mathbf{B}}(\bar{\alpha} n)$ holds.
(2) Since $\tilde{\mathbf{B}}(s) \; : \Longleftrightarrow \quad (e_2 \underline{s})_0 \;=\; \mathbf{0}$, it is immediate that $\forall s \in (\mathbb{N})^{\mathtt{V}(D_\infty)} (\tilde{\mathbf{B}}(\underline{s}) \;\vee\; \neg \tilde{\mathbf{B}}(\underline{s}))$ using (42).
(3) Suppose that $\tilde{\mathbf{B}}(s)$, then $(e_2 \underline{s})_0 = \mathbf{0}$ holds, and hence, $(e_2 \underline{s})_1 \Vdash \mathbf{B}(\overline{s})$, so it follows from (40) that $(e_3 \underline{s})(e_2 \underline{s})_1 \Vdash G(\overline{s})$. Consequently, by definition of ψ it follows $\psi(\underline{s}) \Vdash G(\overline{s})$, and therefore, by definition of $\tilde{G}(s)$ we have $\tilde{G}(s)$.
(4) Suppose that $\forall k \in \mathbb{N}\; \tilde{G}(s * \langle k \rangle)$, then the definition of $\tilde{G}$ entails

$$\underline{s * \langle u \rangle} \downarrow \;\wedge\; \psi(\underline{s * \langle u \rangle}) \Vdash G(\overline{s * \langle u \rangle}).$$

Moreover, unravelling ((41)) yields

$$\begin{aligned}
(41) \;\Longleftrightarrow\;& \forall \langle f, c \rangle \in (\mathbb{N})^{\mathtt{V}(D_\infty)}\, e_4 f \Vdash [(\forall k \in \mathbb{N}\, G(\overline{c} * \langle k \rangle)) \\
& \longrightarrow G(c)] \\
\Longleftrightarrow\;& e_4 \underline{s} \Vdash [(\forall k \in \mathbb{N}\, G(\overline{s} * \langle k \rangle)) \longrightarrow G(\overline{s})], \\
& \text{since } \langle f, c \rangle \in (\mathbb{N})^{\mathtt{V}(D_\infty)} \text{ has to be of the form} \\
& \langle \underline{s}, \overline{s} \rangle \text{ for some } s \text{ in } \omega \\
\Longleftrightarrow\;& \forall r \in D_\infty [(r \Vdash \forall k \in \mathbb{N}\, G(\overline{s} * \langle k \rangle)) \longrightarrow (e_4 \underline{s}) r \Vdash G(\overline{s})].
\end{aligned}$$

However, $r = \lambda \underline{u}.\psi(\underline{s * \langle u \rangle})$. Therefore,

$$(e_4 \underline{s})(\lambda \underline{u}.\psi(\underline{s * \langle u \rangle}) \Vdash G(\overline{s}),$$

and thus $\psi(\underline{s}) \Vdash G(\overline{s})$ for all s in ω, which by definition of $\tilde{G}$ entails $\tilde{G}(s)$.

Invoking $\mathbf{BI_D}$ in $\mathtt{V}$ and using the claim above, we can infer $\tilde{G}(\langle\ \rangle)$, which by the definition of $\tilde{G}$ entails that $\psi(\underline{\langle\ \rangle}) \Vdash G(\overline{\langle\ \rangle})$. Thus, $\mathbf{BI_D}$ holds in the special realisability model $\mathtt{V}(D_\infty)$. □

The foregoing proof can be adapted to monotone Bar Induction as well.

Theorem 7.9. *Assume that*

$$(*) \qquad \forall g : \mathbf{N} \longrightarrow \mathbf{N}\ \exists f \in \mathcal{A}\ \ \forall x[x \in \mathbf{N} \to f \bullet x = g(x)]$$

holds for D_∞, where $\mathbf{N}$ stands for the choice of naturals of D_∞ as an applicative structure. If $\mathbf{BI_M}$ holds in the background universe, then $\mathtt{V}(D_\infty) \models \mathbf{BI_M}$. Likewise, if $\mathbf{FT}$ holds in the background universe, then $\mathtt{V}(D_\infty) \models \mathbf{FT}$.

Proof. We concentrate on finding a realiser for $\mathbf{BI_M}$. Again, since finite sequences of natural numbers (elements of $\mathbb{N}^*$) can be coded by the elements of $\mathbb{N}$ such that for all natural numbers n in ω, n codes a finite sequence s in $\mathbb{N}^*$, we may therefore identify $\mathbb{N}^*$ with $\mathbb{N}$.

Now, suppose there exists realisers e_1, e_2, e_3 and $e_4 \in D_\infty$ such that

$$e_1 \Vdash \forall \alpha \in \mathbb{N}^{\mathbb{N}}\ \exists n \in \mathbb{N}\ \ \mathbf{B}(\bar{\alpha}n), \tag{43}$$

$$e_2 \Vdash \forall s, t \in \mathbb{N}\ \ (\mathbf{B}(s) \to \mathbf{B}(s * t)), \tag{44}$$

$$e_3 \Vdash \forall s \in \mathbb{N}\ \ (\mathbf{B}(s) \longrightarrow G(s)), \tag{45}$$

$$e_4 \Vdash \forall s \in \mathbb{N}\ \ [(\forall k \in \mathbb{N}\ G(s * \langle k \rangle)) \longrightarrow G(s)]. \tag{46}$$

Then we can also find a realiser e_5 such that

$$e_5 \Vdash \forall \alpha \in \mathbb{N}^{\mathbb{N}}\ \exists n \in \mathbb{N}\, \forall t \in \mathbb{N}\, G((\bar{\alpha}n) * t)), \tag{47}$$

since the latter formula is a logical-arithmetical consequence of the formulas in (43)–(45). Using the Recursion Theorem for D_∞, one can define a D_∞ computable function $\tilde{\varphi} : D_\infty \longrightarrow D_\infty$ such that

$$\tilde{\varphi}(x) = \eta(e_1, e_2, e_3, e_4, e_5, \lambda u.\tilde{\varphi}(x * u)),$$

where η is an appropriate auxiliary D_∞ computable function such that by employing $\mathbf{BI_M}$ in the background universe,

$$\forall s \in \mathbb{N}, \tilde{\varphi}(\underline{s}) \Vdash G(\overline{s}). \qquad \square$$

Acknowledgement

The second author acknowledges support via the John Templeton Foundation grant, "A new dawn of intuitionism".

References

[1] R. Amadio and P.-L. Curien, *Domains and Lambda-Calculi.* Cambridge University Press, 1998.
[2] M.J. Beeson, *Foundations of Constructive Mathematics, vol. 6, Ergebnisse der Mathematik und ihrer Grenzgebiete (3) [Results in Mathematics and Related Areas (3)].* Springer-Verlag, Berlin, 1985. ISBN 3-540-12173-0. doi: 10.1007/978-3-642-68952-9.
[3] E. Dihoum, *Realizability Interpretations for Intuitionistic Set Theories.* Thesis (Ph.D.) — The University of Leeds, 2016.
[4] M. Rathjen, The natural numbers in constructive set theory, *Math. Log. Quart.* **54**, 83–97 (2008).
[5] P. Aczel, The type theoretic interpretation of constructive set theory. In A. MacIntyre, L. Pacholski, and J. Paris (eds.), *Logic Colloquium '77.* North Holland, 1978, pp. 55–66.
[6] P. Aczel and M. Rathjen, Notes on constructive set theory. Technical Report 40, Institut Mittag-Leffler, The Royal Swedish Academy of Sciences, Stockholm, 2001.
[7] P. Aczel and M. Rathjen, Notes on constructive set theory. Available at http://www1.maths.leeds.ac.uk/$\sim$rathjen/book.pdf, 2010.
[8] E. Griffor and M. Rathjen, The strength of some Martin-Löf type theories, *Arch. Math. Log.* **33**(5), 347–385 (1994).
[9] M. Rathjen, The strength of some Martin-Löf type theories. Preprint, Department of Mathematics, Ohio State University, available at http://www1.maths.leeds.ac.uk/~rathjen/typeOHIO.pdf, 1993.
[10] H. Barendregt, *The Lambda Calculus, it's Syntax and Semantics.* College Publications, 2012.

[11] P. Aczel, The type theoretic interpretation of constructive set theory: Choice principles. In A. Troelstra and D. van Dalen (eds.), *The L.E.J. Brouwer Centenary*. North Holland, 1982, pp. 1–40.

[12] P. Aczel, The type theoretic interpretation of constructive set theory: Inductive definitions. In R.M. *et al.* (ed.), *Logic, Methodology and Philosophy of Science VII*. North Holland, 1986, pp. 17–49.

[13] M. Rathjen, The anti-foundation axiom in constructive set theories. In G. Mints and R. Muskens (eds.), *Games, Logic, and Constructive Sets*. CSLI Publications Stanford, 2003.

[14] M. Rathjen, Constructive Zermelo-Fraenkel set theory and the limited principle of omniscience, *Ann. Pure Appl. Log.* **165**, 563–572 (2014).

[15] S. Feferman, A language and axioms for explicit mathematics. In J. Crossley (ed.), *Algebra and Logic, vol. 450. Lecture Notes in Mathematics*, Springer, Berlin, 1975, pp. 87–139.

[16] G. Kreisel and A. Troelstra, Formal systems for some branches of intuitionistic analysis, *Ann. Math. Log.* **1**, 229–387 (1970).

[17] J. Myhill, Some properties of intuitionistic Zermelo-Frankel set theory. In *Cambridge Summer School in Mathematical Logic*. Springer, 1973, pp. 206–231.

[18] H. Friedman, Some applications of Kleene's methods for intuitionistic systems. In *Cambridge summer school in mathematical logic*. Springer, 1973, pp. 113–170.

[19] M.J. Beeson, Continuity in intuitionistic set theories. In *Studies in Logic and the Foundations of Mathematics*, vol. 97. Elsevier, 1979, pp. 1–52.

[20] D.C. McCarty, *Realizability and Recursive Mathematics*. Thesis (Ph.D.) — The University of Edinburgh, 1985.

[21] M. Rathjen, Realizability for Constructive Zermelo-Fraenkel set theory. In J. Väänänen and V. Stoltenberg-Hansen (eds.), *Logic Colloquium '03, vol. 24, Lecture Notes in Logic*, A.K. Peters, Wellesley, 2006, pp. 228–314.

[22] E. Dihoum and M. Rathjen, Preservation of choice principles under realizability, *Logic J. IGPL.* **27**(5), 746–765 (02, 2019). ISSN 1367-0751. doi: 10.1093/jigpal/jzz002.

[23] M. Rathjen, Metamathematical properties of intuitionistic set theories with choice principles. In S. Cooper, B. Löwe, and A. Sorbi (eds.), *New Computational Paradigms*. Springer, 2008, pp. 287–312.

[24] A. Troelstra and D. van Dalen, *Constructivism in Mathematics*, vols. I, II. North Holland, Amsterdam, 1988.

[25] M. Rathjen, Constructive set theory and Brouwerian principles, *J. Univ. Comput. Sci.* **11**, 2008–2033 (2005).

[26] S. Kleene and R. Vesley, *The Foundations of Intuitionistic Mathematics. Especially in Relation to Recursive Functions.* North-Holland, Amsterdam, 1965.

[27] M. Dummett, *Elements of Intuitionism.* Oxford University Press, Oxford, 2000.

https://doi.org/10.1142/9789811245220_0015

Chapter 15

Proof Complexity of Quantified Boolean Logic — A Survey

Olaf Beyersdorff

Friedrich Schiller University Jena,
Institute of Computer Science, Jena, Germany

olaf.beyersdorff@uni-jena.de

We provide an overview of proof complexity for quantified Boolean formulas (QBFs). While traditionally the complexity of propositional proofs has been at the centre of research, the past two decades have witnessed a surge in proof complexity of QBF.

We cover major proof systems (including QBF resolution, Frege and further systems), important sample formulas, and techniques for proving lower bounds for the size of QBF proofs. Focusing on the latter, we explain tight connections to circuit complexity and strategy extraction.

1. Introduction

We start our survey with some general remarks on proof complexity.

The main objective in proof complexity is to understand the size of proofs of specific formulas in various formal systems. Historically, the possibly first mention of this question was in a rediscovered letter of Gödel to von Neumann [1]. The first paper dedicated to proof complexity was written by Tseitin [2].

1.1. *Some remarks on propositional proof complexity*

The foundations of proof complexity were laid in the 1970s in the work of Cook and Reckhow [3,4], who formalised the notion of a proof system and set out a programme to show lower bounds for the proof size in strong calculi, in particular for Frege systems [4].[a] The first exponential lower bound for propositional resolution was shown by Haken [5] followed by lower bounds for bounded-depth Frege systems [6–8].

Since then, proof complexity has studied a number of different proof systems originating from logic, geometry and algebra. Arguably even more valuable than the proof systems themselves, we have a number of *lower-bound techniques* that can be used in different systems and for various formulas. In particular for propositional resolution — arguably the best-studied propositional proof system — a host of techniques and classes of hard formulas exist [9–11].

While the study of proof size lower bounds is a technically quite challenging endeavour that is interesting and justified in its own right, proof complexity enjoys close relations to a number of other disciplines. The relation to computational complexity was already laid out in the work [4], establishing that proof size lower bounds are related to separating complexity classes. A second connection goes to first-order logic and in particular to theories of bounded arithmetic [12–14], which are tightly linked to upper and lower bounds in central calculi [9,15].

Last but not least, proof complexity connects to practical SAT solving [125]. In particular, most modern SAT solvers use conflict-driven clause learning (CDCL) [127], which corresponds to resolution [17–19]. Thus, proof complexity acts as a theoretical underpinning of solving, and lower bounds for resolution proofs yield lower bounds for the running time of SAT solvers.

Despite a lot of progress in proof complexity and no known intrinsic barriers to even stronger proof size bounds,[b] many challenging

[a]These systems are also often referred to as Hilbert systems in a first-order context.

[b]This is different in computational complexity and circuit complexity, where a number of barriers have been identified [30–32].

problems remain, with the problem to show lower bounds for Frege — posed already in [4] — still open.

1.2. *Why QBF proof complexity?*

While the complexity of propositional proofs has received central attention in the community in the past five decades, there has been a growing body of work on proof complexity of various other logics (cf. [20] for an overview). These logics include modal logic for which exciting results are known for modal Frege [21,22] and initial results on modal resolution [23], intuitionistic logic [24], first-order logic and quantified Boolean logic.

The first proof systems for quantified Boolean formulas (QBF) were already developed in the 1980s and 1990s in the form of quantified sequent systems [25] and quantified resolution [26] and relations of these calculi to theories of bounded arithmetic were explored [13,14,27]. However, little was known even by 2010 on the complexity of quantified proofs and almost no proof-size lower bounds had been shown.

This has drastically changed in the past decade, which witnessed a surge of activity in QBF proof complexity. The process has mainly been driven by exciting developments in QBF solving, starting with lifting CDCL to QCDCL for the computationally more challenging setting of QBF [28].[c] While QBF solving is still less advanced than SAT, it has constantly been covering ground (see, e.g., [126]) and reaches to a number of applications [29].

While CDCL is the dominant approach in SAT solving, there are a number of conceptually quite different, competing paradigms in QBF solving [33]. This has led to the development of many different QBF resolution systems, aiming to model aspects of solving. Further proof systems for QBF were defined, including strong QBF Frege systems. In parallel, a number of works have aimed to lay rigorous foundations for the field, including developing dedicated lower-bound techniques for quantified proofs.

[c]While SAT is NP complete, deciding QBF is complete for the presumably much larger class PSPACE, encompassing all of the polynomial hierarchy.

Given the growing literature on different QBF calculi (of which possibly already confusingly many exist), hard formulas and lower-bound techniques, we feel that the time is ripe for a first survey on QBF proof complexity.[d]

In this survey, we will emphasise conceptual approaches and lower-bound techniques. We will also highlight differences to propositional proof complexity. While some notions are straightforwardly generalised from the propositional level (i.e., the existential fragment of QBF) to higher quantification order, some QBF lower bound methods are conceptually quite different. This manifests in particular in a close connection between QBF lower bounds and circuit complexity.

1.3. *Organisation*

We start in Section 2 with some background on the syntax and semantics of QBFs. In Section 3, we introduce three QBF families that will serve as running examples throughout the text. Central QBF calculi are explained in Section 4. The main part of this survey is Section 5, where we describe current lower-bound techniques for QBF proof systems, which we illustrate on our running examples. In Section 6, we review propositional lower-bound techniques and their applicability for QBFs.

2. Quantified Boolean Formulas

We assume basic familiarity with logic, including the notions of clauses, conjunctive normal form (CNF), (satisfying) assignments, substitutions, etc. (cf. [9]). We also use basic concepts from computational complexity [34] and circuit complexity [35].

Quantified Boolean formulas (QBF) use Boolean quantifiers $\exists$ and $\forall$ that range over propositional variables taking 0/1 values. *Prenex QBFs* — which we will mostly consider — have the form

$$\exists X_1 \forall U_1 \exists X_2 \cdots \forall U_{d-1} \exists X_d \; \phi(X_1, U_1, X_2, \ldots, U_{d-1}, X_d), \tag{1}$$

[d]While the recent second edition of the *Handbook of Satisfiability* [36] contains an exposition of QBF solving and its relations to QBF proof systems [33], there is no survey yet focussing on proof complexity of QBF.

with a *quantifier prefix* consisting of blocks of existential quantifiers X_i (the blocks X_1 and X_d are possibly empty) and universal quantifiers U_i and a propositional formula ϕ, called the *matrix* of the QBF. If the matrix is in CNF — which is what we will often consider — the formula is called a (prenex) QCNF.

A *hierarchy* Σ_i^b, Π_i^b of quantified formulas is defined by counting the number of quantifier alternations, where Σ_i^b (respectively Π_i^b) consists of all prenex QBFs that have $i-1$ quantifier alternations and start with an existential (respectively, universal) block.

We mostly consider QBFs that are *fully quantified*, i.e., there are no free variables. Such formulas must be either true or false. For a fully quantified prenex QBF, the truth value can be determined by a *2-player game* between an existential and a universal player. In this game, the players assign values to variables according to their type and in the order of the prefix from left to right, i.e., in the QBF (1) the existential player would start to assign all variables from X_1, after which the universal player assigns U_1, and so on. The existential (respectively, universal) player wins if the matrix evaluates to true (respectively, false) at the end of the game.

A *winning strategy* of the universal player is guaranteed to falsify the matrix against all moves of the existential player. Consider a QBF Φ as in (1). Let $x_1, \dots, x_m$ be all the existential variables of Φ (i.e., $X_1 \cup \dots \cup X_d = \{x_1, \dots, x_m\}$) and $u_1, \dots, u_n$ be all its universal variables. A universal winning strategy for Φ is a collection of functions $t_i(x_1, \dots, x_m) \mapsto u_i$ for $i \in [n]$ with the conditions: (i) the value of function t_i only depends on the values of the existential variables left of u_i in the prefix of Φ, and (ii) for each assignment α to $x_1, \dots, x_m$, the assignment $\alpha \cup t_1(\alpha) \cup \dots \cup t_n(\alpha)$ falsifies the matrix ϕ.

A winning strategy for the existential player is defined analogously. A true (respectively, false) QBF admits an existential (respectively, universal) winning strategy.

The collection of functions t_i can also be combined into a single function $t : (x_1, \dots, x_m) \mapsto (u_1, \dots, u_n)$ where $t(x_1, \dots, x_m) = (t_1(x_1, \dots, x_m), \dots, t_n(x_1, \dots, x_m))$, and in this case t is also referred to as a *countermodel* for the QBF Φ.

Throughout, we will only consider false QBFs (following the convention in propositional proof complexity that concentrates on unsatisfiable formulas) and refutational calculi that prove the falseness of QBFs. Hence, all QBFs we consider have a universal winning strategy, which we will often simply refer to as a winning strategy. Focussing on false QBFs is not really a limitation as the negation of a false QBF is a true QBF and most calculi can be easily reformulated as proof systems for true QBFs.[e]

3. Some Important Formulas

We now present three classes of QBFs, which we will use as examples throughout the remainder. These formulas appear frequently in the QBF literature and form the basis of many hardness and separation results between QBF proof systems.

3.1. *The equality formulas*

Our first example is the *equality formulas* from [37]. It is one of the simplest QBFs that is interesting from a proof complexity point of view (one can argue that at least a Σ_3^b prefix for fully quantified QBFs is needed for anything interesting to happen).

The equality formulas are defined as

$$\exists x_1 \cdots x_n \forall u_1 \cdots u_n \exists t_1 \cdots t_n \left(\bigwedge_{i=1}^{n} ((x_i \leftrightarrow u_i) \rightarrow \bar{t}_i) \right) \wedge \left(\bigvee_{i=1}^{n} t_i \right).$$

As can be easily seen, the only universal winning strategy for these formulas is to play $u_i = x_i$ for $i = 1, \ldots, n$. The formulas can be understood as being composed from an easy unsatisfiable propositional formula, namely,

$$\bar{t}_1 \wedge \cdots \wedge \bar{t}_n \wedge (t_1 \vee \cdots \vee t_n),$$

which is obfuscated by the easy "QBF gadgets" $\exists x_i \forall u_i (x_i \leftrightarrow u_i)$, enforcing the universal player to play $x_i = u_i$.[f]

[e] This is in contrast to propositional proof complexity where considering only unsatisfiable formulas (or alternatively propositional tautologies) is somewhat of a limitation and propositional proof complexity does not apply to satisfiable formulas.

[f] A rigorous treatment on how to compose hard QBFs from easy propositional formulas and simple QBF gadgets is contained in [38].

Often it is convenient to state the equality formulas as a prenex QCNF, for which they can be reformulated as follows:

$$\exists x_1 \cdots x_n \forall u_1 \cdots u_n \exists t_1 \cdots t_n \left(\bigwedge_{i=1}^{n} (x_i \vee u_i \vee \bar{t}_i) \wedge (\bar{x}_i \vee \bar{u}_i \vee \bar{t}_i) \right) \wedge \left(\bigvee_{i=1}^{n} t_i \right).$$

We denote this formula by EQ_n.

3.2. *The parity formulas*

As a second example, we consider the parity formulas $QParity_n$ from [39]

$$QParity_n = \exists x_1 \cdots x_n \, \forall u \, \exists t_1 \cdots t_n \quad (x_1 \leftrightarrow t_1) \wedge \bigwedge_{i=2}^{n} ((t_{i-1} \oplus x_i) \leftrightarrow t_i) \wedge (u \not\leftrightarrow t_n).$$

Informally, $QParity_n$ reasons about $x_1 \oplus \cdots \oplus x_n$, where the variables t_i encode the prefix sums $x_1 \oplus \cdots \oplus x_i$. Hence, t_n encodes $x_1 \oplus \cdots \oplus x_n$. Through the only universal variable u, the formula expresses the obvious contradiction that $x_1 \oplus \cdots \oplus x_n$ is neither 0 nor 1. Crucially, the only strategy of the universal player to win on this false QBF is to play $u = x_1 \oplus \cdots \oplus x_n$.

The matrix in the formulation above is not a CNF. Should this be desired (as is the case, e.g., for QBF resolution systems), the formula can be easily turned into a prenex QCNF by using simple clause sets that encode the parity computations (see, e.g., [40]).

3.3. *The KBKF formulas*

Our third running example is the KBKF family, named after its creators Kleine Büning, Karpinski, and Flögel [26]. The QBF $KBKF_n$ has the prefix

$$\exists x_1 y_1 \forall u_1 \cdots \exists x_n y_n \forall u_n \exists z_1 \cdots z_n,$$

and matrix clauses

$$\begin{array}{ll}
(\bar{x}_1 \vee \bar{y}_1)\,, & \\
(x_i \vee u_i \vee \bar{x}_{i+1} \vee \bar{y}_{i+1})\,, & \text{for } i \in [n-1]\,, \\
(y_i \vee \bar{u}_i \vee \bar{x}_{i+1} \vee \bar{y}_{i+1})\,, & \text{for } i \in [n-1]\,, \\
(x_n \vee u_n \vee \bar{z}_1 \vee \cdots \vee \bar{z}_n)\,, & \\
(y_n \vee \bar{u}_n \vee \bar{z}_1 \vee \cdots \vee \bar{z}_n)\,, & \\
(u_i \vee z_i)\,, & \text{for } i \in [n]\,, \\
(\bar{u}_i \vee z_i)\,, & \text{for } i \in [n]\,.
\end{array}$$

In comparison to our first two examples, the meaning of the KBKF formulas is perhaps somewhat more difficult to comprehend. We will argue that they are false QBFs by exhibiting a winning strategy for the universal player. We claim that the universal player can win the evaluation game by playing $u_i = x_i$.

Let us give some intuition on this strategy. The existential player starts by setting one of x_1 and y_1 to 0, otherwise he falsifies $(\bar{x}_1 \vee \bar{y}_1)$ and loses immediately. Assume that the existential chooses $x_1 = 0$ and $y_1 = 1$. If the universal player tries to win, he will counter with $u_1 = 0$, thus forcing the existential player again to set one of x_2 and y_2 to 0. This continues for n rounds, leaving in each round a choice of $x_i = 0$ or $y_i = 0$ to the existential player, to which the universal responds by setting u_i accordingly. Finally, the existential player is forced to set one of $z_1, \ldots, z_n$ to 0. This will falsify either of the clauses $(u_i \vee z_i)$ or $(\bar{u}_i \vee z_i)$ for some $i \in [n]$, and the universal player wins.

It is clear from this explanation that the existential player has exponentially many choices, and that the universal player needs to counter all of these choices in any winning strategy.

4. Central QBF Proof Systems

In this section, we review some of the most important QBF proof systems. In propositional logic, resolution is one of the simplest and best-understood proof systems and a large part of the proof complexity research effort has been spent on that system. The situation is not much different in QBF, so we start with describing QBF resolution systems.

4.1. *QU-Resolution*

Propositional resolution [41,42] is a refutational system operating with clauses, i.e., it demonstrates the unsatisfiability of a given CNF. It has only one rule, the *resolution rule*

$$\frac{C \vee x \qquad D \vee \bar{x}}{C \vee D}, \tag{2}$$

for clauses C, D and a propositional pivot variable x.

QBF resolution systems work with fully quantified prenex QCNFs $\mathcal{Q}\,\phi$. As in propositional resolution, these QBF systems are refutational calculi, i.e., they refute false QBFs.

The simplest QBF resolution system QU-Resolution (QU-Res) can be obtained by augmenting the propositional resolution system using rule (2) by just one new rule, the *universal reduction rule*,

$$\frac{C}{\sigma(C)}, \tag{3}$$

where σ is a partial substitution that allows to substitute a universal variable from a clause C by either 0 or 1, provided u appears right of all variables in C in the prefix $\mathcal{Q}$. Intuitively, this means that a universal variable u can be deleted from a clause C if u is rightmost in C with respect to $\mathcal{Q}$, i.e., no variable in C depends on u.[g]

The soundness of the rule can be seen by considering the two-player game played on $\mathcal{Q}\,\phi$. If the existential player — who aims to satisfy ϕ — wants to win, he must satisfy the clause C (regardless of whether C is contained in ϕ or derived). But since the $\forall$ player has the choice on $u \in C$ after the $\exists$ player has made all his moves, the clause C must already be satisfied by the assignment in the game after the $\exists$ player finishes his assignments on C and regardless of how the $\forall$ player sets u. This means that the $\exists$ player will have to satisfy $\sigma(C)$.

[g]This is how the system is originally defined in [43]. However, in this formulation one needs the extra condition that u appears in only one polarity in C, i.e., that C is not tautological. In our formulation above, tautological clauses such as $u \vee \bar{u}$ can be derived, but reducing them via the universal reduction rule yields $\top$. Our definition of universal reduction originates from [44] and has the advantage of generalising to stronger systems.

As a simple example, consider the first instance EQ_1 of the equality formulas

$$\exists x \forall u \exists t \, (x \vee u \vee \bar{t}) \wedge (\bar{x} \vee \bar{u} \vee \bar{t}) \wedge t.$$

The formula can be refuted in QU-Res as follows:

$$\dfrac{\dfrac{\dfrac{x \vee u \vee \bar{t} \qquad t}{x \vee u}}{x}\,(\text{red}) \qquad \dfrac{\dfrac{\bar{x} \vee \bar{u} \vee \bar{t} \qquad t}{\bar{x} \vee \bar{u}}}{\bar{x}}\,(\text{red})}{\bot},$$

where the universal reduction steps are labelled (red). The reduction steps are allowed because the only other variable x in the clause does not depend on u.

We have already mentioned that the equality and parity formulas are exponentially hard for QU-Res [37,40] (i.e., they require proofs of exponential size), but the KBKF formulas are easy for QU-Res (i.e., they have polynomial-size proofs). We will discuss the hardness results and the techniques to obtain them in some depth in Section 5.

4.2. *Turning propositional proof systems into QBF systems*

Propositional proof complexity considers many more and stronger calculi than resolution. As it turns out, almost all of these systems can be straightforwardly lifted to quantified Boolean logic by adding a slightly generalised version of the universal reduction rule (3) above. This means that starting with a line-based propositional proof system P, we obtain a QBF proof system $P+$, which operates on fully quantified QBFs and uses the rules of the system P together with a similar universal reduction rule as (3). The only difference is that the systems P and $P+$ might not operate with clauses, but use other lines, e.g., formulas in Frege systems [4], polynomial equations in polynomial calculus (PC) [45] or inequalities in cutting planes (CP) [46]. The universal reduction rule is then applied to a line F as

$$\frac{F}{\sigma(F)},$$

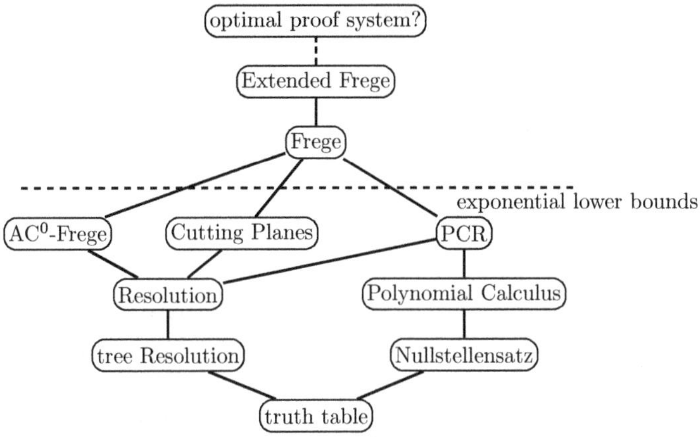

Fig. 1. The simulation order of central propositional proof systems. Lines denote simulations, where the upper system simulates the lower one. Where lines are missing, the systems are incomparable, witnessed by mutual separations, except for those lines that follow by transitivity of simulations. For all systems below the dotted line, exponential lower bounds for proof size are known and simulations are also separations. No super-polynomial lower bounds are known for Frege and stronger proof systems.

with the same side condition as in QU-Res, i.e., σ is a partial substitution that can only substitute those universal variables with 0/1 that appear right of all other variables in F with respect to the quantifier prefix. In particular, the quantifier prefix remains unchanged throughout the proof and only the lines are manipulated through rules from P and the rule.

With this general definition of $P+$ (introduced in [47]), Resolution+ corresponds exactly to QU-Res as defined above. However, we can also use the construction with all the other proof systems in propositional logic as depicted in the simulation order in Figure 1 (except for the static truth table and Nullenstellensatz proof systems that do not operate with lines).[h]

[h]A proof system P p-simulates a proof system Q if every Q proof can be efficiently transformed into a P proof of the same formula. An optimal proof system simulates all other proof systems [48], but their existence is open [49,50].

We briefly explain some of these systems. *Frege systems* are the usual logic textbook systems [4], working with propositional formulas and using a set of axioms and rules (see, e.g., [9]). Axioms are typically simple formulas such as $x \rightarrow x$. An example of a rule is modus ponens

$$\frac{x \quad x \rightarrow y}{y}.$$

In both axioms and rules, variables can be substituted by arbitrary formulas. The actual choice of the axioms and rules does not matter, as all sound and implicationally complete Frege systems are equivalent with respect to simulations [4]. In a Frege proof, the last formula is the proven formula. Therefore, Frege is not a refutational calculus, but can be equivalently cast as one. Not much is known on the complexity of proofs in Frege and in particular no superpolynomial lower bounds are known.

QBF Frege systems are discussed in detail in [44] and we will comment in Section 5 on their proof complexity. We already remark here that all three of our running examples of QBF families, EQ_n, $QParity_n$ and $KBKF_n$, are easy for QBF Frege.

Extended Frege (EF) is a possibly even stronger system than Frege [4,9], allowing the introduction of new extension variables that can abbreviate complex propositional formulas. Equivalently, EF can be seen as a Frege system working with Boolean circuits instead of Boolean formulas (cf.[51]). EF can be also lifted to quantified Boolean formulas in the same way as Frege [44].

Cutting Planes (CP) [46] is a geometric system working with linear inequalities of the form $\sum x_i a_i \geqslant A$ for integers a_i and A. To enforce 0/1 values, Boolean axioms $x \geq 0$ and $-x \geq -1$ for all variables x are added. Rules of CP allow addition of inequalities as well as multiplication and division by positive integers as rules. Translating propositional clauses into inequalities, CP derives the contradiction $0 \geqslant 1$, thereby demonstrating that the original set of inequalities has no solution. CP is a proof system of intermediate strength: it simulates resolution, but allows short proofs for the pigeonhole formulas hard for resolution [5], while it is simulated by and strictly weaker than Frege [52,53].

The *quantified version of cutting planes* was investigated in [54]. As in the propositional case, QBF CP simulates QU-Res, but is exponentially stronger than that system. For the separation one can use similar formulas as $QParity_n$, but based on the majority function [54]. On the other hand, the equality formulas remain hard for QBF cutting planes [37].

Polynomial Calculus (PC) is an algebraic proof system that works with polynomials [45]. The rules of PC comprise addition of polynomials and multiplication by variables, together with Boolean axioms. A variant PCR of PC is defined such that it naturally simulates resolution.

Polynomial calculus in its QBF variant was introduced in [37,47] and a number of lower bounds for the system, including for the equality formulas, were obtained in [37]. We will comment on this in Section 5.4. We also mention that there are quite a few more algebraic proof systems (cf. [55]), some of which could also be lifted to QBF.

4.3. *Variants of QBF resolution*

QU-Resolution is by far not the only QBF resolution system in the literature. The historically first resolution calculus for QBF is *Q-Resolution* (Q-Res), introduced in 1995 by Kleine Büning, Karpinski and Flögel [26]. It is a fragment of QU-Resolution as defined above, with the restriction that in the resolution rule (2), the pivot x must be an existential variable.

QU-Resolution is known to be exponentially stronger than Q-Resolution, where the separation is provided by the KBKF formulas which are easy in QU-Resolution [43], but exponentially hard in Q-Resolution [26,56]. We remark that the KBKF formulas are of unbounded quantifier complexity and, interestingly, the separation has to use unbounded quantification, as Q-Resolution and QU-Resolution are polynomially equivalent on formulas of bounded quantifier complexity [57,58].

Another extension of Q-Resolution is *long-distance Q-Resolution* (LD-Q-Res). This system originates from QBF CDCL solving [28] and was only later turned into a formal calculus [59]. As Q-Resolution, the system LD-Q-Res only allows existential pivots in resolution steps. However, LD-Q-Res permits to derive and reduce certain universal tautologies that are useless in Q-Resolution (or forbidden in the original formulation of [26]).[i] As for Q-Resolution vs. QU-Resolution, also LD-Q-Res is exponentially stronger than Q-Res, with the KBKF formulas serving again as the separating formulas [59].

The question might arise, which extension of Q-Res is better with respect to proof size. Comparing the systems QU-Resolution and LD-Q-Res, it turns out that these are exponentially incomparable. For such an incomparability result, one needs to exhibit QBFs easy in one system, but hard in the other, and vice versa. Specifically, this can be done using the parity or equality formulas (both easy in LD-Q-Res [60,61], but hard in QU-Res [37,40]) and versions of the KBKF formulas (easy for QU-Res, but hard for LD-Q-Res [62]). In fact, both separations can be shown with different versions of KBKF [62], which are quite versatile in QBF proof complexity.

Suitably combining the extensions used in QU-Res and LD-Q-Res yields the system LQU$^+$-Res [62], which is exponentially stronger than both QU-Res and LD-Q-Res, the separations witnessed again by modifications of KBKF [62].

While all of the above QBF proof systems utilise universal reduction, a completely different paradigm for QBF proof systems comes through the concept of *expanding universal variables.* Basically, this uses the fact that a QBF $\exists x \forall y \exists z\, \phi(x, y, z)$ is semantically equivalent to $\exists x \exists z^{y/0} \exists z^{y/1}\, \phi(x, 0, z^{y/0}) \wedge \phi(x, 1, z^{y/1})$ with two fresh variables $z^{y/0}$ and $z^{y/1}$. This universal expansion of y eliminates the universal variable y by substituting it with $0/1$ (and creates new copies of existential variables right of y).

The idea of the proof system ∀Exp+Res [63] is to use this universal expansion to successively eliminate all universal variables, leading to a purely existentially quantified formula. This prenex QCNF can then be refuted using propositional resolution. Of course, in the presence of n universal variables, the expansion phase will in general already create 2^n new existential copies for each innermost existential variable. While this will inevitably happen on some QBFs,

[i]We refrain from giving the full definition of LD-Q-Res, for which we would need to redefine the universal reduction rule as in [26], as there is a subtle interplay between the resolution and reduction rules in LD-Q-Res. The formal details can be found, e.g., in [33].

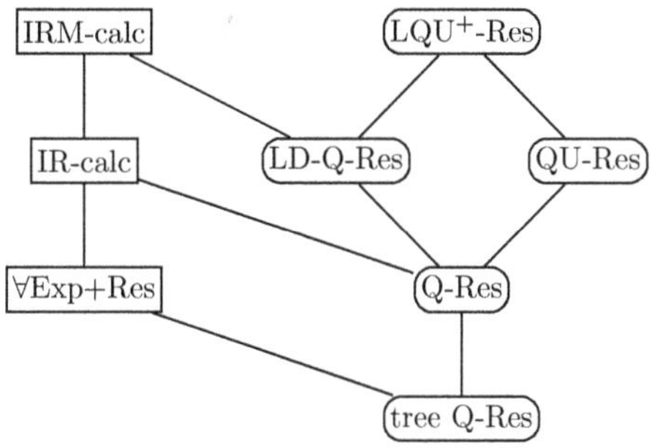

Fig. 2. The simulation order of QBF resolution systems. Lines denote simulations and exponential separations, where the upper system simulates the lower one. Where lines are missing, the systems are incomparable, witnessed by mutual exponential separations, except for those lines that follow by transitivity of simulations. Rectangular boxes indicate expansion calculi, whereas boxes with rounded-off corners denote variants of Q-Res, modelling aspects of QCDCL.

it might be possible for other QBFs to expand variables in just one polarity and still arrive at an unsatisfiable CNF. Consider the QBF $\forall u \exists x\, (u \vee x) \wedge (u \vee \bar{x})$, where just expanding with $u = 0$ results in the false existential formula $\exists x^{u/0}\, x^{u/0} \wedge \bar{x}^{u/0}$.

In ∀Exp+Res, proofs are separated into two distinct phases: the expansion phase, eliminating all universal variables, and the resolution phase. A more flexible approach was designed in [40,64] in the form of the system IR-calc,[j] where partial expansion steps are allowed and can be interleaved with resolution steps. This results in an exponentially stronger system. An even stronger calculus IRM-calc[k] incorporates ideas of long-distance resolution into the expansion paradigm [40,64].

From our running examples, the equality formulas and KBKF are hard for ∀Exp+Res and IR-calc [40,65], while the parity formulas become easy in ∀Exp+Res and stronger expansion systems [40]. On the other hand, it is easy to design formulas that are hard for ∀Exp+Res (they require an exponential expansion phase), but are easy for Q-Res [63]. Thus again, the expansion paradigm and Q-Res are incomparable.

The full simulation order of the QBF resolution systems described so far is depicted in Figure 2.

[j]IR stands for **i**nstantiation and **r**esolution.

[k]IRM indicates **i**nstantiation, **r**esolution, and **m**erging.

For completeness, we mention two further QBF resolution paradigms: Merge Resolution (M-Res) [60,66] and QBF resolution systems with a symmetry rule [67,68]. *Merge Resolution* incorporates explicit information on strategies in proofs via a special format of merge maps.

Symmetry systems augment resolution calculi with a new rule that allows to apply a symmetry of the original formula to a derived clause. This significantly strengthens the system and allows short proofs of, e.g., the equality, parity and KBKF formulas [67,68].

4.4. *QBF proof systems and solving*

After going through the multitude of QBF resolution systems above, the reader might wonder why we care to study so many different versions of QBF resolution. The answer is that many of these systems were designed to capture ideas in QBF solving. In SAT solving, conflict driven clause learning (CDCL) is the prominent solving approach, which was pioneered in the late 1990s [16,69,70]. A few years later, this approach was lifted to quantified formulas in the form of QCDCL [28]. From runs of CDCL on unsatisfiable formulas, resolution proofs can be efficiently extracted [18]. Likewise, QCDCL runs on false QBFs can be translated into long-distance Q-Resolution proofs [28,59], and this was the reason for the proof system LD-Q-Res to be developed in [59], after it had been employed algorithmically in QCDCL solving [28]. In some solving approaches, proofs can even be extracted in Q-Resolution, avoiding the need for long-distance steps [71].

Different from SAT solving, where CDCL is the ruling paradigm, QBF solving employs a host of different techniques. In addition to QCDCL, a second major paradigm is expansion solving using the counterexample guided refinement (CEGAR) approach [72,73], as implemented, e.g., in the solvers RAReQS [73] and Ijtihad [74]. Expansion solving is modelled by the proof system $\forall$Exp+Res [63,74].

Given these relations between proof systems and solving, proof complexity results as shown in Figure 2 can shed light on the relative performance of solving approaches [75].

We mention though, that the correspondence between solving and proof systems is not always perfect. While for SAT solving,

CDCL — at least when viewed in a nondeterministic model — is equivalent to propositional resolution [17–19], this correspondence fails for QCDCL [76,77].

Firstly, QCDCL is exponentially incomparable to Q-Res [76,77]. One separation is provided by the parity formulas which are hard for Q-Res, but turn out to be easy in (nondeterministic) QCDCL. For the other separation, we can use formulas CR_n based on a completion principle from [63], which is easy for Q-Res, but hard for QCDCL [77,78].

Secondly, QCDCL is exponentially weaker than LD-Q-Res as exemplified by the equality formulas [76]. Therefore, QCDCL, is not characterised by any of the QBF resolution systems discussed above. The paper [76] casts QCDCL as a precisely defined (nondeterministic) proof system, for which some lower bounds are shown in [76,78,79].

Dependency schemes are yet another solving approach [80,81], implemented, e.g., in the solver DepQBF [82] and modelled by QBF resolution systems using dependency schemes [16,83,84]. For an overview, we refer to [85]. The proof complexity of dependency calculi is investigated in [86,87].

We also mention that the calculus QRAT [88] — building on the RAT system for SAT [89] — is a general proof format that allows to model most current QBF techniques, including preprocessing.

For the remainder of this overview we will not get into more detail on the relations between solving and proof complexity. A recent overview of QBF solving approaches and their corresponding proof systems can be found in [33].

4.5. *QBF sequent calculi*

All the QBF systems above share the property that they work with the original quantifier prefix, which is fixed throughout the proof. Only the matrix of the QBF is manipulated and new lines (clauses, formulas, inequalities, etc.) are derived. In contrast, QBF sequent systems allow to also manipulate the prefix and derive new quantified formulas. This makes these systems far more powerful.

In fact, the very first proof system for quantified Boolean logic was a sequent system G, introduced by Krajíček and Pudlák [25].

A slightly modified version of G was later defined by Cook and Morioka [27], which we will sketch here. It builds on Gentzen's sequent calculus LK [90] and adds the following four rules for quantifiers to LK: ∀-introduction:

$$\frac{\phi(x/\psi), \Gamma \longrightarrow \Delta}{\forall x\, \phi, \Gamma \longrightarrow \Delta}\ (\forall\text{-l}) \qquad \frac{\Gamma \longrightarrow \Delta, \phi(x/p)}{\Gamma \longrightarrow \Delta, \forall x\, \phi}\ (\forall\text{-r}),$$

and ∃-introduction:

$$\frac{\phi(x/p), \Gamma \longrightarrow \Delta}{\exists x\, \phi, \Gamma \longrightarrow \Delta}\ (\exists\text{-l}) \qquad \frac{\Gamma \longrightarrow \Delta, \phi(x/\psi)}{\Gamma \longrightarrow \Delta, \exists x\, \phi}\ (\exists\text{-r}).$$

For the rules ∀-l and ∃-r, $\phi(x/\psi)$ is the result of substituting ψ for all free occurrences of x in ϕ. The formula ψ may be any quantifier-free formula that is free for substitution for x in ϕ (i.e., no free occurrence of x in ϕ is within the scope of a quantifier Qy such that y occurs in ψ). The variable p in the rules ∀-r and ∃-l must not occur free in the bottom sequent.

In addition to the sequent system G for all QBFs, there are also subsystems G_i (defined in two different versions in [25] and [27]) that operate with $\Sigma_i^b \cup \Pi_i^b$ formulas. Further, weak cut-free sequent calculi for QBF were introduced and investigated in [91].

Different from the propositional case, where Frege and LK are equivalent with respect to simulations [4], G is a much stronger system than Frege+ [44].

5. Lower-bound Techniques for QBF Proof Systems

In what follows, we will review the main techniques for proving lower bounds for QBF proof systems. While some of the lower bounds, in particular for the QBF resolution systems sketched above, are somewhat ad hoc, we also have some generic techniques that work for a number of different calculi. It is also quite interesting that some of these techniques differ considerably from approaches in the propositional domain. Before we get to the techniques themselves, we need to explain strategy extraction.

5.1. *Strategy extraction*

Strategy extraction is a distinctive feature of QBF proof systems for which no propositional analogue exists. Interestingly, most QBF lower-bound techniques employ strategy extraction.

From a proof π of a false QBF Φ in existential variables $x_1, \ldots, x_m$ and universal variables $u_1, \ldots, u_n$, strategy extraction allows to efficiently compute a winning strategy for the universal player on Φ. As explained in Section 2, a strategy is a collection of functions $t_i : (x_1, \ldots, x_m) \mapsto u_i$ for $i \in [n]$ from existential variables to universal variables, specifying the action of the universal player in response to the existential player's moves.

If we want to extract such a strategy from a proof, we need to fix a computational model in which to represent the functions t_i. As it turns out, the simple model of *decision lists* is most useful for this purpose. Decision lists were first suggested by Rivest [92] in the context of machine learning. A decision list for function $t(x_1, \ldots, x_m) \mapsto u$ is a sequence of instructions of the form

$$
\begin{array}{l}
\textsc{If } t_1 \textsc{ Then } u = b_1 \\
\textsc{Else If } t_2 \textsc{ Then } u = b_2 \\
\qquad\qquad\vdots \\
\textsc{Else If } t_k \textsc{ Then } u = b_k \\
\textsc{Else } u = b_{k+1}
\end{array}
$$

where t_i are terms (i.e., conjunctions of literals) in variables $x_1, \ldots, x_m$ and b_i are constants 0/1.

It is easy to see that decision lists can be transformed into bounded-depth circuits (in fact, depth 3 is sufficient).

As an example, let us consider the equality formulas. We already argued that the only winning strategy for the universal player is to play $u_i = x_i$ for $i \in [n]$. This strategy can be very easily represented as a collection of decision lists

$$
\begin{array}{l}
\textsc{If } x_i \textsc{ Then } u_i = 1 \\
\textsc{Else } u_i = 0.
\end{array}
$$

In contrast, the only winning strategy for the parity formulas is to compute $u = x_1 \oplus \cdots \oplus x_n$. Again, this can be represented as a decision list. However, unlike in the equality example, there are no small decision lists that compute the parity function, because parity is hard to compute for bounded-depth circuits [93].

As indicated above, strategy extraction means that decision lists computing winning strategies for universal players can be extracted from QBF resolution proofs. This was first realised by Balabanov and Jiang [94] for Q-Res (though stated in slightly different terms in that paper). We will sketch strategy extraction for QU-Res here.

Theorem 5.1 (Balabanov and Jiang [94]). Given a QU-Res refutation π of a QBF Φ, we can compute in linear time a sequence of decision lists (one for each universal variable of Φ), which computes a winning strategy for the universal player on Φ.

Proof. We sketch the proof of this result. Let Φ contain universal variables $u_1, \ldots, u_n$. Let $\pi = (C_1, \ldots, C_k = \bot)$ be a QU-Res derivation and let $\Phi_i = \Phi \wedge \bigwedge_{j=1}^{i} C_j$. We construct decision lists for Φ_i by reverse induction on i, starting with $i = k$. For $i = 0$, this gives the theorem.

For $i = k$, the formula Φ_k contains the empty clause $\bot$ on which the universal player trivially wins. So we simply take empty lists which compute a trivial winning strategy for Φ_k.

For the inductive step, assume that lists L_i^j have already been constructed for the universal variables u_j with $j \in [n]$ in the QBF Φ_i. We construct lists L_{i-1}^j as follows.

If C_i was derived by resolution, we set $L_{i-1}^j = L_i^j$. We need to argue that the output computed by these lists still falsifies Φ_{i-1}. If the clause falsified by the lists L_i is still part of Φ_{i-1}, then this trivially holds. If, on the other hand, the falsified clause was C_i, then by the correctness of the resolution rule, one of the parents of C_i is falsified, both of which are contained in Φ_{i-1}.

If C_i was derived by a universal reduction step, say

$$\frac{C \vee u_j}{C},$$

i.e., C was derived by universal reduction using substitution $u_j = 0$, then we construct the list L_{i-1}^j as

IF $\neg C$ THEN $u_j = 0$
ELSE L_i^j.

Note that $\neg C$ is a term and the old list L_i^j is appended to L_{i-1}^j. In case the reduction starts from a clause $C \vee \bar{u}_j$, then u_j is set to 1.

For all other universal variables, the list remains unchanged, i.e., for $l \neq j$, we set $L^l_{i-1} = L^l_i$.

To argue again that the output of the lists falsifies Φ_{i-1}, consider the case that the clause falsified by the assignment computed by the lists L^j_i is C. In this case, the lists L_{i-1} will falsify $C \vee u_j$ by construction. Otherwise, a clause $D \in \Phi_i \setminus \{C\}$ was falsified, and D is still falsified by the lists L_{i-1} by inductive hypothesis. □

Such a strategy extraction result is also interesting for QBF solving, as proofs and consequently winning strategies can be extracted from runs of solvers. These winning strategies can then be used to certify the correctness of the solver's answer [95,96].

5.2. *Proof size lower bounds through circuit lower bounds*

Of prime importance for the following is the observation that strategy extraction allows to prove lower bounds. As a first example, we can immediately see that the parity formulas are exponentially hard for QU-Res.

Theorem 5.2 (Beyersdorff, Chew, Janota [40]). The parity formulas require exponential-size proofs in QU-Res.

Proof. From QU-Res refutations of the parity formulas, we can extract decision lists computing winning moves for the universal player on the parity formulas. The only winning strategy for the universal variable u is to compute the parity $x_1 \oplus \cdots \oplus x_n$. The decision list extracted for u can be turned into a bounded-depth circuit for the parity function. Such circuits must be of exponential size due to the circuit lower bound from [93]. Consequently, the QU-Res proofs must be of exponential size as well. □

Strategy extraction is not only effective for QU-Res, but also works for many more QBF proof systems. What changes is the computational model in which strategies can be extracted. As a rule of thumb, when the proof system gets stronger (and proofs potentially shorter), the computational model also needs to be strengthened. More precisely, the computational model needed for strategy extraction appears to hinge on the complexity of lines in the proof system.

This can be best understood in the hierarchy of Frege systems. Let $\mathcal{C}$-Frege denote the Frege system working with lines from the circuit class $\mathcal{C}$ (cf. [51]). With this notation, Frege corresponds to NC^1-Frege (using NC^1 circuits, i.e., formulas, as lines) and extended Frege to P/poly-Frege (using P/poly, i.e., polynomial-size circuits). Strategy extraction for these systems then precisely works with the circuit class $\mathcal{C}$.

Theorem 5.3 (Beyersdorff, Bonacina, Chew, Pich [44]). Given a $\mathcal{C}$-Frege+ proof of a false QBF Φ, winning strategies on Φ for the universal player, specified as $\mathcal{C}$ circuits, can be computed in linear time.

The idea for the computation of strategies is similar to the method sketched above in the proof of Theorem 5.1.

With this strategy extraction result at hand, we can readily lift existing circuit lower bounds to proof size lower bounds in QBF Frege systems. This needs a general method of how to construct QBFs from a hard function. We do this similarly as for the parity formulas. Let $f(x_1, \ldots, x_n)$ be a total Boolean function computable by polynomial-size circuits. We construct a formula Q-f as

$$\begin{aligned} Q\text{-}f = \exists x_1 \ldots \exists x_n \forall u \exists t_1 \ldots \exists t_{p(n)} \, (u \not\leftrightarrow \mathrm{comp}_f \\ \times (x_1, \ldots, x_n, t_1, \ldots, t_{p(n)})), \end{aligned}$$

where $\mathrm{comp}_f(x_1, \ldots, x_n, t_1, \ldots, t_{p(n)})$ describes the circuit computation of the function $f(x_1, \ldots, x_n)$. The auxiliary variables $t_1, \ldots, t_{p(n)}$ abbreviate the gates of the circuits computing f. Informally, Q-f states that there is an input for which $f(x_1, \ldots, x_n)$ is undefined, a clear contradiction as f is total. As in the parity formulas, the only winning strategy for the universal player is to play $u = f(x_1, \ldots, x_n)$.

Using circuit lower bounds for f against a circuit class $\mathcal{C}$, we can infer proof size lower bounds for Q-f in QBF proof systems that allow strategy extraction by $\mathcal{C}$ circuits. This paradigm allows to prove strong lower bounds for QBF proof systems.

Theorem 5.4 (Beyersdorff, Bonacina, Chew, Pich [44]). Let f be a function computable in PSPACE/poly, i.e., the function can be described by a nonuniform sequence of QBFs.

(1) If f is hard for Boolean circuits of depth 2, then Q-f is hard for QU-Res.
(2) If $f \notin AC^0$ (i.e., not computable by bounded-depth circuits), then Q-f is hard for AC^0-Frege+.
(3) If $f \notin AC^0[p]$ (i.e., not computable by bounded-depth circuits with mod p gates), then Q-f is hard for $AC^0[p]$-Frege+.
(4) If $f \notin$ P/poly (i.e., not computable by polynomial-size Boolean circuits), then Q-f is hard for extended Frege+.

We note that explicit functions meeting requirements of Items (5.4) to (5.4) are known in circuit complexity [93,97–99], while the assumption of the existence of functions as in Item (5.4) is very plausible.

This provides a very direct, sort of black-box transfer from circuit lower bounds to QBF proof size lower bounds. Such a direct transfer has been much speculated upon in proposition proof complexity (cf., e.g., [100]) but has not manifested so far, despite intense research. In particular, lower bounds for $AC^0[p]$-Frege are not known and present a problem at the frontier of current proof complexity research, despite known circuit lower bounds for $AC^0[p]$ and intense efforts to lift these to propositional proof-size bounds.

5.3. *Genuine QBF lower bounds under an NP-oracle*

Concerning the relation between propositional proofs and quantified proofs, it is apparent that *any* lower bound for a propositional proof system P immediately lifts to a lower bound in the QBF proof system P+. This holds because proof size of a propositional formula $\phi(x_1, \ldots, x_n)$ in P is identical to the proof size of the existentially quantified version $\exists x_1 \ldots x_n\, \phi(x_1, \ldots, x_n)$ in P+. For example, the existentially quantified pigeonhole principle (PHP) is hard for QU-Res as the propositional lower bound for resolution from [5] straightforwardly carries over to QU-Res.

Though perhaps a matter of opinion, this might not be the phenomenon one wants to study in QBF proof complexity. Instead, we would argue, that it is more interesting to identify reasons for hardness that clearly stem from quantifier alternation. This view was first proposed in [101] and further substantiated in [102].

One way to eliminate propositional hardness is to allow proof checking access to an NP oracle. Though this takes us outside the Cook–Reckhow framework of polynomially-verifiable proofs [4], several interesting frameworks for proof system utilising stronger computational resources for proof checking have been previously considered in the literature [103–105].

In the model of [102], given any QBF proof system P, the QBF proof systems P^{NP} is granted access to an NP oracle to perform inference steps of the kind

$$\frac{F_1 \quad \dots \quad F_k}{G}, \tag{4}$$

where $F_1, \dots, F_k$ and G are lines of the system P and $F_1 \wedge \dots \wedge F_k \models G$, i.e., $F_1, \dots, F_k$ propositionally entail G.

In each system P^{NP}, false QBFs with only existential quantifiers are trivial to refute with just one step of the rule (4). For example, the existentially quantified PHP is easy for QU-Res^{NP}.

Showing lower bounds for proof size in the presence of the NP-oracle rule (4) therefore becomes more challenging, and we will refer to such lower bounds as "genuine" QBF lower bounds (in contrast to lower bounds such as PHP for QU-Res, lifted from propositional proof systems). Interestingly, the lower bounds we have seen so far in Theorems 5.2 and 5.4 as well as almost all lower bounds mentioned in the following are genuine in this sense.

For $P+$ systems, the presence of NP oracles essentially means that any P-subderivations come for free (they can be derived with one application of (4) each) and we are only counting the universal reduction steps. For example, the lower bound in Theorem 5.2 is genuine as the strategy extraction algorithm in Theorem 5.1 computes the decision lists only from the universal reduction steps in the refutations, whereas resolution steps do not impact the computed decision lists.

5.4. *The size-cost-capacity technique*

We will now describe another general lower-bound technique for QBF proof systems originating from [37], which is grounded in the semantics of QBF. Again this technique will yield genuine lower bounds. As in the circuit lower-bound technique described in Section 5.2, the method also utilises strategy extraction, albeit in a different way. Remember that a winning strategy for the universal player for a false QBF is a function from assignments to existential variables to assignments to universal variables (cf. Section 2). For the method in Section 5.2, we used QBFs requiring winning strategies that are hard to compute in some circuit class.

Now, we will use QBFs that do not have small strategies, where size is simply measured by the cardinality of the range of the winning strategy. These conditions are incomparable. For example, the parity formulas are hard to compute by bounded-depth circuits (and hence, decision lists), but have a countermodel with range 2 (as there is just one universal variable in the formula). In contrast, the unique winning strategy for the equality formulas is trivial to compute with decision lists (each universal variable u_i is simply set equal to x_i), but has range 2^n.

As it turns out, the *minimal size of the range of a countermodel* is a lower bound for proof size in *each* system of the form $P+$, albeit only in their *tree-like* versions.[1] This was established in the following result:

Theorem 5.5 (Beyersdorff and Hinde [106]). For any QBF Φ, if π is a tree-like $P+$ proof of Φ, then $|\pi|$ is at least the size of the range of the smallest countermodel of Φ.

This means that the EQ_n as well as the $KBKF_n$ formulas will be hard for all tree-like proof systems of the form $P+$.

The proof of Theorem 5.5 relies again on efficient strategy extraction by decision lists as outlined in the proof of Theorem 5.1 (and stated for further systems $P+$ in Theorem 5.3). In that procedure, strategies are extracted from universal reduction steps in $P+$ refutations. Intuitively, if many responses by the universal player are needed (i.e., countermodels are large), then there must be many universal reduction steps in the proofs, from which these responses are extracted (i.e., (tree-like) $P+$ proofs must be large). From this intuition, it also becomes clear that we are just counting universal

[1]Depicted as graphs, tree-like proofs have the form of a tree, i.e., each derived line can be used at most once for an inference step.

reduction steps, hence the lower bounds will be genuine (as discussed in Section 5.3).

To obtain lower bounds in the general *dag-like* proof model, we need to refine (or more precisely, weaken) the measure of size of a countermodel. This is done with the notion of *cost* in [37]. Cost is similar to the minimal cardinality of the range of countermodels, but only measures the cardinality of the range of countermodels on *one* block of universal quantifiers. The choice of the universal block is arbitrary, but all strategies and all blocks are considered.

For Σ_3^b formulas, such as the equality formulas, this does not make any difference: the cost of EQ_n is 2^n, as is the cardinality of the range of the unique countermodel. However, for the $KBKF_n$ formulas, the minimal range is exponential, while the cost is 2, as there is only one universal variable per quantifier block.

For QU-Res, cost yields an absolute lower bound:

Theorem 5.6 (Beyersdorff, Blinkhorn, Hinde [37]). For any QBF Φ, if π is a QU-Res proof of Φ, then $|\pi|$ is at least the cost of Φ.

Hence, EQ_n is exponentially hard for QU-Res. Interestingly, $KBKF_n$, which have small cost, admit short proofs in QU-Res [43].

To obtain a lower bound for general systems $P+$ in the spirit of Theorem 5.5, we need the second notion of the *capacity* of a line. We will not give the formal definition (which can be found in [37]). Intuitively, the capacity of a line again considers the 2-player evaluation game, but this time only for one particular line L, i.e., for a clause in the case of QU-Res. The capacity of L then counts how many different universal moves are needed to falsify the line L, if this is possible at all given the existential player's moves.

For example, consider a clause $C = x \vee \bar{y} \vee u \vee \bar{w}$ in existential variables x, y and universal variables u, w. Regardless of how the existential plays on x, y, the only sensible choice for the universal player is to set $u = 0$ and $w = 1$ as otherwise C is satisfied. Therefore, the capacity of C (and of any other clause) is 1.

This is different in Frege+ systems. Consider, e.g., the line $L = x \leftrightarrow u$ with existential variable x and universal u. The capacity of this line is 2 as the universal player needs to set $u = \bar{x}$ to falsify L. It is easy to construct Frege lines that even have exponential capacity.

The capacity of a proof π is defined as the maximum of the capacity of the lines in π. The size–cost–capacity theorem then gives a lower bound for $P+$ systems.

Theorem 5.7 (Beyersdorff, Blinkhorn, Hinde [37]). Let π be a $P+$ refutation of Φ. Then

$$|\pi| \geqslant \frac{\text{cost}(\Phi)}{\text{capacity}(\pi)}.$$

Intuitively, this holds as cost(Φ) counts how many universal responses are needed on Φ, while capacity counts how many responses can be read off from one line in π. In strategy extraction, each response is extracted from a universal reduction step. Hence, there must be at least $\frac{\text{cost}(\Phi)}{\text{capacity}(\pi)}$ many reduction steps in π. Again, it is apparent that lower bounds shown by Theorem 5.7 are always genuine.

As the capacity of clauses is always 1, Theorem 5.6 is a special case of Theorem 5.7.

Similarly, it can be seen that the capacity of linear inequalities as used in the QBF version CP+ of cutting planes is always 1. Therefore, we obtain from Theorem 5.7:

Corollary 5.8 (Beyersdorff, Blinkhorn, Hinde [37]). If π is a CP+ refutation of Φ, then $|\pi| \geqslant \text{cost}(\Phi)$.

For polynomial calculus (i.e., for PC+ in QBF), the capacity of a line (i.e., a polynomial) can be arbitrarily large. However, it can be shown that a PC+ proof, where each line contains at most m monomials, has capacity at most m. As a proof π can contain at most $|\pi|$ monomials, we infer from Theorem 5.7 that $|\pi| \geqslant \frac{\text{cost}(\pi)}{|\pi|}$, and hence:

Corollary 5.9 (Beyersdorff, Blinkhorn, Hinde [37]). If π is a PC+ refutation of Φ, then $|\pi| \geqslant \sqrt{\text{cost}(\Phi)}$.

Thus, the equality formulas (and all other QBFs with high cost) are hard in both QBF CP and PC.

The situation is different for QBF Frege systems. While Theorem 5.7 still applies, we cannot in general bound the capacity of proofs anymore, and in particular the equality formulas have refutations of linear size and exponential capacity in Frege+ (cf. [37]).

5.5. *Circuit characterisations for QBF Frege systems*

The two lower-bound techniques described in Sections 5.2 and 5.4 have one common theme: they both use strategy extraction and argue that certain QBFs only have "complex" strategies. In Section 5.2, "complex" means hard to compute in the circuit class underlying the proof system, while in Section 5.4, "complex" is interpreted as having no small countermodels. The question arises whether these two interpretations of "complex" can be combined such that a *characterisation* of (genuine) hardness can be achieved for the QBF calculi in question. In a sense, this would mean to "ultimately understand" lower bounds for these systems.

Indeed, such characterisations can be obtained for the hierarchy of QBF Frege systems. For QBF Frege, it turns out that proof size is characterised completely by circuit size (in the NP-oracle model).

Theorem 5.10 (Beyersdorff, Bonacina, Chew, Pich [44]). For a family of false polynomial-size QBFs Φ_n, the following are equivalent:

(1) Φ_n have polynomial-size proofs in $(\mathcal{C}\text{-Frege+})^{\mathrm{NP}}$.
(2) The universal player has winning strategies on Φ_n that are computable in $\mathcal{C}$.

The direction (5.10)⇒(5.10) follows by strategy extraction for $\mathcal{C}$-Frege (Theorem 5.4). For the direction (5.10)⇒(5.10), we start with the winning strategy for Φ_n computable in $\mathcal{C}$ and formally prove the correctness of the strategy in $\mathcal{C}$-Frege+, yielding short proofs of Φ_n.

Formulated differently, this means that showing genuine lower bounds for $\mathcal{C}$-Frege+ is equivalent to showing lower bounds for the circuit class $\mathcal{C}$. Unfortunately, we do not have such lower bounds for strong enough circuit classes $\mathcal{C}$, e.g., for choosing $\mathcal{C}$ as nonuniform NC^1 or P/poly.[m]

Nevertheless, we can conclude from Theorem 5.10 the equivalence of the following problems:

[m]Note that while many problems outside P/poly exist by a simple counting argument, we need such problems to be also in PSPACE to express them as QBFs. The existence of such problems is open.

Theorem 5.11 (Beyersdorff, Bonacina, Chew, Pich [44]). The following are equivalent:

(1) $(\text{Frege+})^{\text{NP}}$ is not polynomially bounded.[n]
(2) PSPACE $\not\subseteq$ NC^1.

For simplicity, we just stated the theorem for Frege and NC^1, but an analogous result holds for other natural circuit classes as well, including P/poly corresponding to extended Frege. Note that by NC^1 we mean the nonuniform version of that class.

Removing NP oracle access, we can also state the result alternatively as follows:

Theorem 5.12 (Beyersdorff, Bonacina, Chew, Pich [44]). Frege+ is not polynomially bounded if and only if

(1) Propositional Frege is not polynomially bounded, or
(2) PSPACE $\not\subseteq$ NC^1.

The equivalent result also holds for extended Frege and P/poly. This characterisation of QBF Frege lower bounds therefore combines the major problems from proof complexity (Frege lower bounds) and circuit complexity (lower bounds for unrestricted circuits). Both statements (5.12) and (5.12) from Theorem 5.12 are wide open. A similar connection between circuit lower bounds (for e.g., P/poly) and proof-size lower bounds for propositional extended Frege has been much speculated upon [100], but has not been made formal yet.

5.6. *Circuit characterisations for QU-Res*

Given the tight circuit characterisation for QBF Frege from the last section, it is natural to ask whether such characterisations can be obtained for other QBF calculi as well, and in particular for the QBF resolution systems discussed above. Indeed, such a characterisation is known for QU-Res. For this we have to slightly change the decision list model from Section 5.1.

---

[n] A proof system P for a language L is polynomially bounded if there exists a polynomial p such that every $x \in L$ has a P proof of size $\leqslant p(|x|)$.

Classical decision lists [92] compute Boolean functions with one output bit. We used a collection of these decision lists in the strategy extraction procedure in Theorem 5.1, computing one decision list per universal variable.

Instead of using one list per universal variable, we can tweak the model slightly to allow multiple outputs, such that one multi-output decision list is enough to compute the whole universal strategy. These multi-output decision lists were introduced in [57] and termed *unified decision lists* (UDL).

For a QBF Φ in existential variables $x_1, \ldots, x_m$ and universal variables $u_1, \ldots, u_n$, a *UDL for* Φ is required to compute a countermodel for Φ and to be of the syntactic form

$$\begin{array}{l}
\text{IF } t_1 \text{ THEN } u_1 = b_{1,1},\ u_2 = b_{1,2},\ \ldots,\ u_n = b_{1,n} \\
\text{ELSE IF } t_2 \text{ THEN } u_1 = b_{2,1},\ u_2 = b_{2,2},\ \ldots,\ u_n = b_{2,n} \\
\qquad\qquad \vdots \\
\text{ELSE IF } t_k \text{ THEN } u_1 = b_{k,1},\ u_2 = b_{k,2},\ \ldots,\ u_n = b_{k,n} \\
\text{ELSE } u_1 = b_{k+1,1},\ u_2 = b_{k+2,2},\ \ldots,\ u_n = b_{k+1,n}.
\end{array}$$

Here, t_i are terms in $x_1, \ldots, x_m$ (they might not contain all x variables) and each line in the UDL specifies a *total* assignment $b_{i,1}, \ldots, b_{i,n}$ to the universal variables $u_1, \ldots, u_n$.

As an example, note that each UDL for the equality formulas EQ_n must be of size 2^n as there has to be a line in the UDL for each total assignment to the universal variables (all of these assignments appear in the unique universal winning strategy). In contrast, EQ_n has a collection of constant-size decision lists (with single outputs).

UDLs can be viewed as a circuit model and they provide a characterisation of proof size in $(\text{QU-Res})^{\text{NP}}$.

Theorem 5.13 (Beyersdorff, Blinkhorn, Mahajan [57]). For QBFs of bounded quantifier complexity, minimal proof size in $(\text{QU-Res})^{\text{NP}}$ and minimal UDL size are polynomially equivalent.

Thus, for bounded-quantifier complexity (which is needed for the efficient transformation of QU-Res proofs into UDLs), UDLs tightly characterise genuine QU-Res proof size, similarly to Theorems 5.10 and 5.11 for Frege above.

Omitting the NP oracle, this can be phrased as follows. For QBFs Φ_n of bounded quantifier complexity, the formulas Φ_n are hard for QU-Res if and only if (i) Φ_n require large UDLs, or (ii) Φ_n contain propositional resolution hardness. Additionally, in case (ii) the propositional hardness can be precisely identified.

As Q-Res is equivalent to QU-Res on bounded alternation [57,58], the above characterisation holds for Q-Res as well.

As mentioned, it is easy to prove the hardness of the equality formulas via Theorem 5.13. Also, the parity formulas can be shown to require exponential-size UDLs [57], thus providing an alternative proof of their genuine hardness for QU-Res. The method does not apply to KBKF though, as these formulas have unbounded quantifier complexity.

It is also interesting to mention that in the QBF Frege characterisation in Section 5.5, the characterising circuit class exactly corresponds to the lines in the proof system. This is different in QU-Res. Here the lines are clauses and the system operates with CNFs (which could be seen as depth-2 circuits). However, by [107], the complexity of decision lists (and UDLs) is strictly intermediate between depth-2 and depth-3 circuits, and hence, does not match the complexity of the lines.

6. Propositional Lower-bound Techniques: Size-width, Interpolation, Games

Propositional proof complexity has a host of lower-bound techniques that have been applied to many systems, in particular propositional resolution. We will give an overview of their applicability to QBF proof systems.

6.1. *The size-width technique*

Arguably the most prominent technique is the size-width technique [108]. It uses the notion of *width* which for a clause just counts the number of literals in it. Ben-Sasson and Wigderson [108] then show how to transform short resolution proofs into narrow resolution proofs (i.e., containing only clauses of small width). This allows to infer size lower bounds from width lower bounds, which are often easier to show. The size-width result of [108] can be stated for propositional formulas ϕ as

$$S(\phi \vdash \bot) = \exp\left(\Omega\left(\frac{(w(\phi \vdash \bot) - w(\phi))^2}{n}\right)\right), \tag{5}$$

where

- $S(\phi \vdash \bot)$ denotes the minimal size of a resolution refutation of ϕ;
- $w(\phi \vdash \bot)$ is the minimal width of a resolution refutation of ϕ;
- $w(\phi)$ is the initial width of the formula ϕ; and
- n is the number of variables in ϕ.

Interestingly, an analogous size-width relation was ruled out for QU-Res and other QBF resolution systems in [109,110] as specific counterexamples to (5) can be constructed. Also the proof method of [108] can be shown to fail for QBF resolution in one precisely identified step [109].

However, it turns out that the idea to show size lower bounds via width still works in QBF. This uses the UDL characterisation for QU-Res from Theorem 5.13 and combines this with a size-width result for UDLs. The latter can be obtained in a similar spirit as [108] (Bshouty [111] proves a size-width result for decision lists which is slightly improved in [57]). This yields the following result for QCNFs Φ, shown in [57]:

$$S(\Phi \vdash \bot) = \exp\left(\Omega\left(\frac{(w_{\exists}(\Phi \vdash \bot))^2}{d^3 n}\right)\right), \tag{6}$$

where

- $S(\Phi \vdash \bot)$ denotes the minimal size of a $(\text{QU-Res})^{\text{NP}}$ refutation of Φ;
- $w_{\exists}(\Phi \vdash \bot)$ is the minimal existential width of a $(\text{QU-Res})^{\text{NP}}$ resolution refutation of Φ, where only existential literals are counted for the width;
- d is the number of quantifier alternations of Φ; and
- n is the number of existential variables in Φ.

Note that the relation (6) is somewhat similar to (5), but also has important differences: the initial width is absent in (6) and the quantifier complexity enters.

Using (6), it is quite easy to show lower bounds for our running examples. For example, linear UDL width lower bounds for parity are very easy to show as are width lower bounds for equality [57], implying exponential hardness of both formulas for $(\text{QU-Res})^{\text{NP}}$.

6.2. *Feasible interpolation*

Feasible interpolation is another widely used propositional technique. It was first shown for propositional resolution [112], but also works for cutting planes [53]. The technique also uses a link to circuit complexity, albeit less direct than the relations that manifest in QBF.

The starting point of feasible interpolation is Craig's interpolation theorem [113], stating that for each tautological implication

$$\phi(\vec{x}, \vec{y}) \to \psi(\vec{x}, \vec{z}), \tag{7}$$

where $\vec{x}$ are the common variables of ϕ and ψ and $\vec{y}$, $\vec{z}$ are further disjoint tuples of variables, also disjoint from $\vec{x}$, there exists an *interpolant* $\theta(\vec{x})$ in only the variables $\vec{x}$ such that

$$\phi(\vec{x}, \vec{y}) \to \theta(\vec{x}) \quad \text{and} \quad \theta(\vec{x}) \to \psi(\vec{x}, \vec{z})$$

are both tautological.

A proof system P is said to have the *feasible interpolation property* if we can efficiently compute an interpolant from a P-proof π of the implication (7) [112]. While small interpolants might in general not exist [114], the point is here that the interpolant is computed from the proof π, which itself might not be small.

This can be used for a lower bound argument: if there are sequences $\phi_n \to \psi_n$ of the form (7) without small interpolants, then $\phi_n \to \psi_n$ cannot have small proofs in every proof system P with the feasible interpolation property.

Such formulas are easy to construct using suitable assumptions from computational complexity. To show proof-size lower bounds without such assumptions, one can prove a stricter version of interpolation, where the interpolant is computed by *monotone* Boolean circuits, provided the formulas (7) meet certain monotonicity conditions. This is called *monotone feasible interpolation* and was shown to hold for propositional resolution [112] and cutting planes [53].

In [115], this technique was lifted to QBF and shown to be effective for almost all the QBF resolution systems mentioned in Section 4, including Q-Res, QU-Res, LD-Q-Res and the expansion resolution systems ∀Exp+Res, IR-calc and IRM-calc. It also holds for the QBF variant of cutting planes [54].

In addition to the technique, suitable formulas need to be constructed on which the technique can be applied. In [53,112,115], these formulas originate from graph theory and express — in different ways, also utilising quantifiers [115] — that graphs with large cliques cannot have small chromatic number.

6.3. *Prover-delayer games*

A third propositional technique that we mention is via prover–delayer games. Such games come in different flavours. Here, we will sketch the model introduced in [116] and refined in [117,118] for propositional *tree-like* resolution. In this setting, a delayer claims to know a satisfying assignment for an unsatisfiable CNF. In each round the prover questions the delayer about the value of one variable in his assignment. The delayer can either answer 0/1 or leave the choice to the prover and score 1 point.

The game ends when a clause is falsified by the assignment thus constructed. While it is clear that the prover always wins, the question is how many points the delayer can score in the game. The relation to proof size is given in the following result:

Theorem 6.1 (Pudlák and Impagliazzo [116]). Let ϕ be an unsatisfiable CNF. If the delayer has a strategy to score at least n points during each game, then tree-like resolution refutations of ϕ must be of size at least 2^n.

Thus, by exhibiting good delayer strategies, we can show lower bounds on proof size.

The game was refined in [117] to an *asymmetric* prover–delayer game, where the delayer scores points according to some probability distribution (in the game of [106] described above, the delayer always scores 1 point). This version of the game *characterises* tree-like resolution size [118], i.e., the proof method allows to always obtain the optimal lower bounds (cf. [119] for an example of the optimal lower bound for the pigeonhole formulas).

This technique can be lifted to Q-Res by slightly adapting the above rules [120]. Indeed, an asymmetric version of this adapted game tightly characterises tree-like proof size in Q-Res. A small further modification of the game yields a characterisation of QU-Res [120]. As mentioned, this allows to always construct optimal proofs. Examples of delayer strategies for the parity and KBKF formulas are given in [120].

7. Conclusion

This survey provides a snapshot of QBF proof complexity at the time of writing (early 2022). We concentrated on central QBF proof systems and lower-bound techniques available for them.

Given the wealth of research during the past 20 years, a selection was inevitable. Apart from the only brief mentioning of the relation to QBF solving (covered in depth in the survey [33]), the most serious omissions probably are the relations between theories of bounded arithmetic to QBF sequent systems [9,13,25,27] and QBF Frege [44] along with translations to further fragments of first-order logic [121,122], the strong proof system QRAT [88] and its relation to other QBF calculi [123], the incorporation of dependency schemes into QBF proof systems [16,83,84] and their proof complexity [86,87], and the generalisation of QBF proof systems to dependency QBFs (DQBF) [16,124].

Acknowledgements

I thank Uwe Egly, Agnes Schleitzer and Helen Wötzel for helpful comments on a draft of this chapter.

This work was supported by DFG grant BE 4209/3-1. Part of the research results explained in this survey were supported by grants from the John Templeton Foundation and the Carl–Zeiss Foundation.

References

[1] K. Gödel, A letter to John von Neumann, March 20, 1956. In P. Clote and J. Krajíček (eds.), *Arithmetic, Proof Theory, and Computational Complexity.* Oxford University Press, 1993, pp. 7–9.

[2] G.C. Tseitin, On the complexity of derivations in propositional calculus. In A.O. Slisenko (ed.), *Studies in Mathematics and Mathematical Logic*, Part II, 1968, pp. 115–125.

[3] S.A. Cook, Feasibly constructive proofs and the propositional calculus. In *Proc. 7th Annual ACM Symposium on Theory of Computing*, 1975, pp. 83–97.

[4] S.A. Cook and R.A. Reckhow, The relative efficiency of propositional proof systems, *J. Symb. Log.* **44**(1), 36–50 (1979).

[5] A. Haken, The intractability of resolution, *Theor. Comput. Sci.* **39**, 297–308 (1985).
[6] M. Ajtai, The complexity of the pigeonhole-principle, *Combinatorica.* **14**(4), 417–433 (1994).
[7] J. Krajíček, P. Pudlák, and A. Woods, Exponential lower bounds to the size of bounded depth Frege proofs of the pigeonhole principle, *Random Struct. Algorithms.* **7**(1), 15–39 (1995).
[8] T. Pitassi, P. Beame, and R. Impagliazzo, Exponential lower bounds for the pigeonhole principle, *Comput. Compl.* **3**, 97–140 (1993).
[9] J. Krajíček, *Proof complexity. vol. 170, Encyclopedia of Mathematics and Its Applications.* Cambridge University Press, 2019.
[10] N. Segerlind, The complexity of propositional proofs, *Bull. Symb. Log.* **13**(4), 417–481 (2007).
[11] A. Urquhart, The complexity of propositional proofs, *Bull. Symb. Log.* **1**, 425–467 (1995).
[12] S.R. Buss. First order proof theory of arithmetic. In S.R. Buss (ed.), *Handbook of Proof Theory.* Elsevier, Amsterdam, 1998, pp. 79–147.
[13] S.A. Cook and P. Nguyen, *Logical Foundations of Proof Complexity.* Cambridge University Press, 2010.
[14] J. Krajíček, *Bounded Arithmetic, Propositional Logic, and Complexity Theory, vol. 60, Encyclopedia of Mathematics and Its Applications.* Cambridge University Press, Cambridge, 1995.
[15] O. Beyersdorff, On the correspondence between arithmetic theories and propositional proof systems — a survey, *Math. Log. Quart.* **55**(2), 116–137 (2009).
[16] O. Beyersdorff, J. Blinkhorn, L. Chew, R.A. Schmidt, and M. Suda, Reinterpreting dependency schemes: Soundness meets incompleteness in DQBF, *J. Autom. Reason.* **63**(3), 597–623 (2019).
[17] A. Atserias, J.K. Fichte, and M. Thurley, Clause-learning algorithms with many restarts and bounded-width resolution, *J. Artif. Intell. Res.* **40**, 353–373 (2011).
[18] P. Beame, H.A. Kautz, and A. Sabharwal, Towards understanding and harnessing the potential of clause learning, *J. Artif. Intell. Res. (JAIR).* **22**, 319–351 (2004).
[19] K. Pipatsrisawat and A. Darwiche, On the power of clause-learning SAT solvers as resolution engines, *Artif. Intell.* **175**(2), 512–525 (2011).
[20] O. Beyersdorff and O. Kutz. Proof complexity of non-classical logics. In N. Bezhanishvili and V. Goranko (eds.), *Lectures on Logic and Computation - ESSLLI 2010 /ESSLLI 2011, Selected Lecture Notes,* Springer-Verlag, Berlin Heidelberg, 2012, pp. 1–54.

[21] P. Hrubeš, On lengths of proofs in non-classical logics, *Ann. Pure Appl. Logic.* **157**(2–3), 194–205 (2009).
[22] E. Jeřábek, Substitution Frege and extended Frege proof systems in non-classical logics, *Ann. Pure Appl. Logic.* **159**(1–2), 1–48 (2009).
[23] S. Sigley and O. Beyersdorff, Proof complexity of modal resolution, *J. Autom. Reason.* **66**(1), 1–41 (2022).
[24] P. Hrubeš, A lower bound for intuitionistic logic, *Ann. Pure Appl. Logic.* **146**(1), 72–90 (2007).
[25] J. Krajíček and P. Pudlák, Quantified propositional calculi and fragments of bounded arithmetic, *Zeitschrift für Mathematische Logik und Grundlagen der Mathematik.* **36**, 29–46 (1990).
[26] H. Kleine Büning, M. Karpinski, and A. Flögel, Resolution for quantified Boolean formulas, *Inf. Comput.* **117**(1), 12–18 (1995).
[27] S.A. Cook and T. Morioka, Quantified propositional calculus and a second-order theory for NC^1, *Arch. Math. Log.* **44**(6), 711–749 (2005).
[28] L. Zhang and S. Malik, Conflict driven learning in a quantified Boolean satisfiability solver. In *Proc. IEEE/ACM International Conference on Computer-aided Design (ICCAD)*, 2002, pp. 442–449.
[29] A. Shukla, A. Biere, L. Pulina, and M. Seidl, A survey on applications of quantified Boolean formulas. In *Proc. IEEE International Conference on Tools with Artificial Intelligence (ICTAI)*, 2019, pp. 78–84.
[30] S. Aaronson and A. Wigderson, Algebrization: A new barrier in complexity theory, *ACM Trans. Comput. Theory.* **1**(1) (2009).
[31] T. Baker, J. Gill, and R. Solovay, Relativizations of the P=?NP question, *SIAM J. Comput.* **4**, 431–442 (1975).
[32] A.A. Razborov and S. Rudich, Natural proofs, *J. Comput. Syst. Sci.* **55**(1), 24–35 (1997).
[33] O. Beyersdorff, M. Janota, F. Lonsing, and M. Seidl. Quantified Boolean formulas. In A. Biere, M. Heule, H. van Maaren, and T. Walsh (eds.), *Handbook of Satisfiability*, Frontiers in Artificial Intelligence and Applications. IOS Press, 2021, pp. 1177–1221.
[34] S. Arora and B. Barak, *Computational Complexity — A Modern Approach.* Cambridge University Press, 2009. ISBN 978-0-521-42426-4.
[35] H. Vollmer, *Introduction to Circuit Complexity — A Uniform Approach.* Texts in Theoretical Computer Science, Springer Verlag, Berlin Heidelberg, 1999.
[36] A. Biere, M. Heule, H. van Maaren, and T. Walsh (eds.), *Handbook of Satisfiability*, Frontiers in Artificial Intelligence and Applications. IOS Press, 2021.

[37] O. Beyersdorff, J. Blinkhorn, and L. Hinde, Size, cost, and capacity: A semantic technique for hard random QBFs, *Log. Meth. Comput. Sci.* **15**(1) (2019).
[38] A. Schleitzer and O. Beyersdorff, Classes of hard formulas for QBF resolution, *Electron. Colloquium Comput. Complex.* (TR22-036) (2022). to appear in Proc. of SAT'22.
[39] O. Beyersdorff, L. Chew, and M. Janota, Proof complexity of resolution-based QBF calculi. In *Proc. Symposium on Theoretical Aspects of Computer Science (STACS'15)*. LIPIcs, 2015, pp. 76–89.
[40] O. Beyersdorff, L. Chew, and M. Janota, New resolution-based QBF calculi and their proof complexity, *ACM Trans. Computa. Theory.* **11**(4), 26:1–26:42 (2019).
[41] A. Blake, *Canonical Expressions in Boolean Algebra.* PhD thesis, University of Chicago, 1937.
[42] J.A. Robinson, A machine-oriented logic based on the resolution principle, *J. ACM.* **12**, 23–41 (1965).
[43] A. Van Gelder, Contributions to the theory of practical quantified Boolean formula solving. In *Proc. Principles and Practice of Constraint Programming (CP)*, 2012, pp. 647–663.
[44] O. Beyersdorff, I. Bonacina, L. Chew, and J. Pich, Frege systems for quantified Boolean logic, *J. ACM.* **67**(2), 9:1–9:36 (2020).
[45] M. Clegg, J. Edmonds, and R. Impagliazzo. Using the Groebner basis algorithm to find proofs of unsatisfiability. In *Proc. 28th ACM Symposium on Theory of Computing*, 1996, pp. 174–183.
[46] W. Cook, C.R. Coullard, and G. Turán, On the complexity of cutting-plane proofs, *Discrete Appl. Math.* **18**(1), 25–38 (1987).
[47] O. Beyersdorff, I. Bonacina, and L. Chew, Lower bounds: From circuits to QBF proof systems. In *Proc. ACM Conference on Innovations in Theoretical Computer Science (ITCS'16)*. ACM, 2016, pp. 249–260.
[48] J. Krajíček and P. Pudlák, Propositional proof systems, the consistency of first order theories and the complexity of computations, *J. Symb. Log.* **54**(3), 1063–1079 (1989).
[49] O. Beyersdorff and Z. Sadowski, Do there exist complete sets for promise classes?, *Math. Log. Quart.* **57**(6), 535–550 (2011).
[50] J. Köbler, J. Messner, and J. Torán, Optimal proof systems imply complete sets for promise classes, *Infor. Comput.* **184**(1), 71–92 (2003).
[51] E. Jeřábek, *Weak Pigeonhole Principle, and Randomized Computation.* PhD thesis, Faculty of Mathematics and Physics, Charles University, Prague, 2005.

[52] A. Goerdt. Cutting plane versus Frege proof systems. In *Proc. 4th Workshop on Computer Science Logic*, 1991, pp. 174–194.
[53] P. Pudlák, Lower bounds for resolution and cutting planes proofs and monotone computations, *J. Symb. Logic.* **62**(3), 981–998 (1997).
[54] O. Beyersdorff, L. Chew, M. Mahajan, and A. Shukla, Understanding cutting planes for QBFs, *Inf. Comput.* **262**, 141–161 (2018).
[55] T. Pitassi and I. Tzameret, Algebraic proof complexity: Progress, frontiers and challenges, *SIGLOG News.* **3**(3), 21–43 (2016).
[56] O. Beyersdorff and J. Blinkhorn, A simple proof of QBF hardness, *Infor. Proc. Lett.* **168** (2021).
[57] O. Beyersdorff, J. Blinkhorn, and M. Mahajan. Hardness characterisations and size-width lower bounds for QBF resolution. In *Proc. ACM/IEEE Symposium on Logic in Computer Science (LICS).* ACM, 2020, pp. 209–223.
[58] J. Clymo. *Proof Complexity for Quantified Boolean Formulas.* PhD thesis, School of Computing, University of Leeds, 2021.
[59] U. Egly, F. Lonsing, and M. Widl, Long-distance resolution: Proof generation and strategy extraction in search-based QBF solving. In *Proc. Logic for Programming, Artificial Intelligence, and Reasoning (LPAR)*, 2013, pp. 291–308.
[60] O. Beyersdorff, J. Blinkhorn, and M. Mahajan, Building strategies into QBF proofs, *J. Autom. Reasoning* (2020).
[61] L. Chew, *QBF Proof Complexity.* PhD thesis, University of Leeds, Leeds, 2017.
[62] V. Balabanov, M. Widl, and J.-H.R. Jiang, QBF resolution systems and their proof complexities. In *Proc. Theory and Applications of Satisfiability Testing (SAT)*, 2014, pp. 154–169.
[63] M. Janota and J. Marques-Silva, Expansion-based QBF solving versus Q-resolution, *Theor. Comput. Sci.* **577**, 25–42 (2015).
[64] O. Beyersdorff, L. Chew, and M. Janota, On unification of QBF resolution-based calculi. In *Proc. 39th Symposium on Mathematical Foundations of Computer Science, vol. 8635, Lecture Notes in Computer Science*, 2014, pp. 81–93.
[65] O. Beyersdorff and J. Blinkhorn, Lower bound techniques for QBF expansion, *Theory Comput. Syst.* **64**(3), 400–421 (2020).
[66] O. Beyersdorff, J. Blinkhorn, M. Mahajan, T. Peitl, and G. Sood, Hard QBFs for merge resolution. In *Annual Conference on Foundations of Software Technology and Theoretical Computer Science (FSTTCS), vol. 182, LIPIcs*, 2020, pp. 12:1–12:15.
[67] M. Kauers and M. Seidl, Short proofs for some symmetric quantified Boolean formulas, *Inf. Process. Lett.* **140**, 4–7 (2018).

[68] A. Shukla, F. Slivovsky, and S. Szeider, Short Q-resolution proofs with homomorphisms. In L. Pulina and M. Seidl (eds.), *Theory and Applications of Satisfiability Testing (SAT), vol. 12178, Lecture Notes in Computer Science*, Springer, 2020, pp. 412–428.

[69] J.P. Marques Silva and K.A. Sakallah, GRASP — a new search algorithm for satisfiability. In *Proc. IEEE/ACM International Conference on Computer-aided Design (ICCAD)*, 1996, pp. 220–227.

[70] L. Zhang, C.F. Madigan, M.W. Moskewicz, and S. Malik, Efficient conflict driven learning in Boolean satisfiability solver. In *Proc. IEEE/ACM International Conference on Computer-Aided Design (ICCAD)*, 2001, pp. 279–285.

[71] F. Lonsing, U. Egly, and A.V. Gelder, Efficient clause learning for quantified Boolean formulas via QBF pseudo unit propagation. In *Proc. International Conference on Theory and Applications of Satisfiability Testing (SAT)*, 2013, pp. 100–115.

[72] E.M. Clarke, O. Grumberg, S. Jha, Y. Lu, and H. Veith, Counterexample-guided abstraction refinement for symbolic model checking, *J. ACM.* **50**(5), 752–794 (2003).

[73] M. Janota, W. Klieber, J. Marques-Silva, and E.M. Clarke, Solving QBF with counterexample guided refinement, *Artif. Intell.* **234**, 1–25 (2016).

[74] R. Bloem, N. Braud-Santoni, V. Hadzic, U. Egly, F. Lonsing, and M. Seidl, Two SAT solvers for solving quantified boolean formulas with an arbitrary number of quantifier alternations, *Formal Meth. Syst. Des.* **57**(2), 157–177 (2021).

[75] O. Beyersdorff, L. Pulina, M. Seidl, and A. Shukla. QBFFam: A tool for generating QBF families from proof complexity. In C. Li and F. Manyà (eds.), *Theory and Applications of Satisfiability Testing (SAT), vol. 12831, Lecture Notes in Computer Science*, Springer, 2021, pp. 21–29.

[76] O. Beyersdorff and B. Böhm, Understanding the Relative Strength of QBF CDCL Solvers and QBF Resolution. In *12th Innovations in Theoretical Computer Science Conference (ITCS 2021), vol. 185, Leibniz International Proceedings in Informatics (LIPIcs)*, 2021, pp. 12:1–12:20.

[77] M. Janota, On Q-Resolution and CDCL QBF solving. In *Proc. International Conference on Theory and Applications of Satisfiability Testing (SAT)*, 2016, pp. 402–418.

[78] B. Böhm and O. Beyersdorff, Lower bounds for QCDCL via formula gauge. In C.-M. Li and F. Manyà (eds.), *Theory and Applications of Satisfiability Testing — SAT 2021*, Springer International Publishing, Cham, 2021, pp. 47–63.

[79] O. Beyersdorff and B. Böhm, QCDCL with cube learning or pure literal elimination — What is best? *Electron. Colloquium Comput. Complex.* 131 (2021).

[80] F. Lonsing, *Dependency Schemes and Search-Based QBF Solving: Theory and Practice.* PhD thesis, Johannes Kepler University Linz, 2012.

[81] M. Samer and S. Szeider, Backdoor sets of quantified Boolean formulas, *J. Automat. Reason.* **42**(1), 77–97 (2009).

[82] F. Lonsing and U. Egly, DepQBF 6.0: A search-based QBF solver beyond traditional QCDCL. In *Proc. International Conference on Automated Deduction (CADE)*, 2017, pp. 371–384.

[83] T. Peitl, F. Slivovsky, and S. Szeider, Long-distance Q-resolution with dependency schemes, *J. Autom. Reasoning.* **63**(1), 127–155 (2019).

[84] F. Slivovsky and S. Szeider, Soundness of Q-resolution with dependency schemes, *Theor. Comput. Sci.* **612**, 83–101 (2016).

[85] J. Blinkhorn. *Quantified Boolean Formulas: Proof Complexity and Models of Solving.* PhD thesis, University of Leeds, Leeds, 2019.

[86] O. Beyersdorff and J. Blinkhorn, Dynamic QBF dependencies in reduction and expansion, *ACM Trans. Comput. Log.* **21**(2), 8:1–8:27 (2020).

[87] O. Beyersdorff, J. Blinkhorn, and T. Peitl, Strong (D)QBF dependency schemes via tautology-free resolution paths. In *Theory and Applications of Satisfiability Testing (SAT).* Springer, 2020, pp. 394–411.

[88] M.J.H. Heule, M. Seidl, and A. Biere, Solution validation and extraction for QBF preprocessing, *J. Autom. Reason.* **58**(1), 97–125 (2017).

[89] M. Heule, Proofs of unsatisfiability. In A. Biere, M. Heule, H. van Maaren, and T. Walsh (eds.), *Handbook of Satisfiability, 2nd edition,* Frontiers in Artificial Intelligence and Applications. IOS press, 2021, pp. 635–668.

[90] G. Gentzen, Untersuchungen über das logische Schließen, *Mathematische Zeitschrift.* **39**, 68–131 (1935).

[91] U. Egly, On sequent systems and resolution for QBFs. In *Theory and Applications of Satisfiability Testing (SAT)*, 2012, pp. 100–113.

[92] R.L. Rivest, Learning decision lists, *Machine Learning.* **2**(3), 229–246 (1987).

[93] J. Håstad, *Computational Limitations of Small Depth Circuits.* MIT Press, Cambridge, 1987.

[94] V. Balabanov and J.-H.R. Jiang, Unified QBF certification and its applications, *Form. Methods Syst. Des.* **41**(1), 45–65 (2012).

[95] T. Peitl, F. Slivovsky, and S. Szeider. Polynomial-time validation of QCDCL certificates. In O. Beyersdorff and C.M. Wintersteiger (eds.),

Theory and Applications of Satisfiability Testing (SAT), vol. 10929, Lecture Notes in Computer Science, Springer, 2018, pp. 253–269.

[96] M.N. Rabe and L. Tentrup, CAQE: A certifying QBF solver. In R. Kaivola and T. Wahl (eds.), *Formal Methods in Computer-Aided Design (FMCAD)*, IEEE, 2015, pp. 136–143.

[97] J. Håstad, Almost optimal lower bounds for small depth circuits. In S. Micali (ed.), *Randomness and Computation, Advances in Computing Reasearch*, vol. 5. JAI Press, 1989, pp. 143–170.

[98] A.A. Razborov, Lower bounds for the size of circuits of bounded depth with basis $\{\wedge, \oplus\}$, *Math. Notes Acad. Sci. USSR.* **41**(4), 333–338 (1987).

[99] R. Smolensky, Algebraic methods in the theory of lower bounds for boolean circuit complexity. In *Proc. ACM Symposium on Theory of Computing (STOC)*. ACM Press, 1987, pp. 77–82.

[100] P. Beame and T. Pitassi, Propositional proof complexity: Past, present, and future. In G. Paun, G. Rozenberg, and A. Salomaa (eds.), *Current Trends in Theoretical Computer Science: Entering the 21st Century*. World Scientific Publishing, 2001, pp. 42–70.

[101] H. Chen, Proof complexity modulo the polynomial hierarchy: Understanding alternation as a source of hardness, *ACM Trans. Comput. Theory.* **9**(3), 15:1–15:20 (2017).

[102] O. Beyersdorff, L. Hinde, and J. Pich, Reasons for hardness in QBF proof systems, *ACM Trans. Comput. Theory.* **12**(2), 10:1–10:27 (2020).

[103] O. Beyersdorff, J. Köbler, and S. Müller, Proof systems that take advice, *Infor. Computa.* **209**(3), 320–332 (2011).

[104] O. Beyersdorff and S. Müller. A tight Karp-Lipton collapse result in bounded arithmetic, *ACM Tran. Comput. Log.* **11**(4), (2010).

[105] S.A. Cook and J. Krajíček, Consequences of the provability of NP $\subseteq$ P/poly, *J. Symb. Log.* **72**(4), 1353–1371 (2007).

[106] O. Beyersdorff and L. Hinde, Characterising tree-like Frege proofs for QBF, *Inf. Comput.* **268** (2019).

[107] M. Krause, On the computational power of Boolean decision lists, *Comput. Complex.* **14**(4), 362–375 (2006).

[108] E. Ben-Sasson and A. Wigderson, Short proofs are narrow — resolution made simple, *J. ACM.* **48**(2), 149–169 (2001).

[109] O. Beyersdorff, L. Chew, M. Mahajan, and A. Shukla, Are short proofs narrow? QBF resolution is *not* so simple, *ACM Trans. Comput. Log.* **19**(1), 1:1–1:26 (2018).

[110] J. Clymo and O. Beyersdorff, Relating size and width in variants of Q-resolution, *Inf. Process. Lett.* **138**, 1–6 (2018).

[111] N.H. Bshouty, A subexponential exact learning algorithm for DNF using equivalence queries, *Inf. Process. Lett.* **59**(1), 37–39 (1996).

[112] J. Krajíček, Interpolation theorems, lower bounds for proof systems and independence results for bounded arithmetic, *J. Symb. Log.* **62**(2), 457–486 (1997).

[113] W. Craig, Three uses of the Herbrand-Gentzen theorem in relating model theory and proof theory, *J. Symb. Log.* **22**(3), 269–285 (1957).

[114] D. Mundici, Tautologies with a unique Craig interpolant, uniform vs. nonuniform complexity, *Ann. Pure Appl. Logic.* **27**, 265–273 (1984).

[115] O. Beyersdorff, L. Chew, M. Mahajan, and A. Shukla, Feasible interpolation for QBF resolution calculi, *Logical Meth. Comput. Sci.* **13**(2) (2017).

[116] P. Pudlák and R. Impagliazzo, A lower bound for DLL algorithms for SAT. In *Proc. 11th Symposium on Discrete Algorithms (SODA)*, 2000, pp. 128–136.

[117] O. Beyersdorff, N. Galesi, and M. Lauria, Parameterized complexity of DPLL search procedures, *ACM Trans. Computati. Log.* **14**(3), 20:1–20:21 (2013).

[118] O. Beyersdorff, N. Galesi, and M. Lauria, A characterization of tree-like resolution size, *Infor. Proc. Lett.* **113**(18), 666–671 (2013).

[119] O. Beyersdorff, N. Galesi, and M. Lauria, A lower bound for the pigeonhole principle in tree-like resolution by asymmetric prover-delayer games, *Infor. Proc. Lett.* **110**(23), 1074–1077 (2010).

[120] O. Beyersdorff, L. Chew, and K. Sreenivasaiah, A game characterisation of tree-like Q-Resolution size, *J. Comput. Syst. Sci.* **104**, 82–101 (2019).

[121] U. Egly, On stronger calculi for QBFs. In N. Creignou and D.L. Berre (eds.), *Theory and Applications of Satisfiability Testing (SAT), vol. 9710, Lecture Notes in Computer Science*, Springer, 2016, pp. 419–434.

[122] M. Seidl, F. Lonsing, and A. Biere, qbf2epr: A tool for generating EPR formulas from QBF. In P. Fontaine, R.A. Schmidt, and S. Schulz (eds.), *Third Workshop on Practical Aspects of Automated Reasoning (PAAR), vol. 21, EPiC Series in Computing.* EasyChair, 2012, pp. 139–148.

[123] L. Chew and F. Slivovsky, Towards uniform certification in QBF. In P. Berenbrink and B. Monmege (eds.), *International Symposium on Theoretical Aspects of Computer Science (STACS), vol. 219, LIPIcs.* Schloss Dagstuhl — Leibniz-Zentrum für Informatik, 2022, pp. 22:1–22:23.

[124] O. Beyersdorff, L. Chew, R.A. Schmidt, and M. Suda, Lifting QBF resolution calculi to DQBF. In N. Creignou and D.L. Berre (eds.),

Theory and Applications of Satisfiability Testing (SAT), vol. 9710, Lecture Notes in Computer Science, Springer, 2016, pp. 490–499.

[125] S. Buss and J. Nordström. Proof complexity and SAT solving. In A. Biere, M. Heule, H. van Maaren, and T. Walsh (eds.), *Handbook of Satisfiability*, Frontiers in Artificial Intelligence and Applications. IOS Press, 2021, pp. 233–350.

[126] P. Faymonville, B. Finkbeiner, M.N. Rabe, and L. Tentrup, Encodings of bounded synthesis. In *Tools and Algorithms for the Construction and Analysis of Systems (TACAS)*, pp. 354–370 (2017).

[127] J.P. Marques Silva, I. Lynce, and S. Malik, Conflict-driven clause learning SAT solvers. In A. Biere, M. Heule, H. van Maaren, and T. Walsh (eds.), *Handbook of Satisfiability*, Frontiers in Artificial Intelligence and Applications. IOS Press, 2021.

https://doi.org/10.1142/9789811245220_0016

Chapter 16

Subject Reduction in Multi-Universe Type Theories

Marco Benini
Department of Science and High Technology,
University of Insubria,
Italy
marco.benini@uninsubria.it

The Subject Reduction property states that if t is a term of type T in the context Γ, so is any term s to which t reduces. In this contribution the property is shown to be false in dependent-type theories with multiple universes and η-reduction by providing a judgement $\Gamma \vdash t : T$ and a one-step η-reduction $t \rhd s$ such that $\Gamma \vdash s:T$ is impossible to derive. While the counterexample is immediate to check, the impossibility result requires setting up some machinery having interest of its own. The chapter also proposes an enhancement to the type system in which the claim that the Subject Reduction property holds has a strong justification.

1. Introduction

The Subject Reduction Theorem says that: if $\Gamma \vdash t:T$ and $t \rhd s$, then $\Gamma \vdash s:T$. Thus, it relates deduction with computation in a direct way. Its importance comes from the fact that *proving* and *calculating* are two of the most fundamental activities in Mathematics.

Proving, that is, deducing valid consequences from supposedly valid premises, is the prominent subject of Logic, and in a type system it means to derive a statement, usually expressed as a type, in the formal calculus exhibiting a term which inhabits the type.

However, calculating is equally essential: every time a complex expression is encountered, it is usually simplified before starting to reason about it. In a type system, simplification is directly rendered by the reduction relation.

The interplay between proving and calculating is so natural that it is usually taken for granted, not even thinking it should be justified. But in a foundational approach it must, and providing such a justification is precisely the purpose of the Subject Reduction Theorem. In other terms, the Subject Reduction Theorem tells that calculation preserves provability. Hence, the Subject Reduction property fails to hold exactly when the reduction system is incompatible with the proving/typing system.

The counterexample in Section 2 requires four interrelated features to contradict the statement of the Subject Reduction property: multiple universes; function spaces; judgemental equivalence; η-reduction. In particular, the η-reduction requires the presence of an abstraction (λ) operator to be expressed, which happens when function spaces are also in the formal system. Judgemental equality is needed to describe inside the formal system how distinct types of the same term are linked via convertibility. But the key feature, from which is derived the title of this contribution, is having multiple universes, which allows for a term to have potentially multiple nonequivalent types.

Natural and significant examples of type systems exhibiting those features are many variants of Martin–Löf-type Theory [1], Homotopy-type Theory [2], and in general type theories with universes and dependent function spaces.

In [1], the notion of reduction, from which convertibility arises, is a bit peculiar: specifically, η-reduction is absent and β-reduction does not operate inside λ-abstractions. These limitations allow to deduce a Strong Normalisation Theorem and, in turn to prove the Subject Reduction Theorem. On the contrary, in Homotopy Type Theory it is suggested[a] that β-reductions may freely operate whenever an abstraction is applied to some term, and η-reduction is present in the form of the $\Pi-\mathsf{uniq}$ rule. And, of course, universes are essential to express Univalence. Although the normalisation problem has been essentially fixed in the case of a single universe, see [3], a normalisation result is still lacking despite it is sometimes regarded as folklore. As a consequence, many corollaries [1] become unjustified, like the Church–Rosser property and the Subject Reduction property.

In this contribution, a counterexample to the Subject Reduction property is presented in a formal system with the above-mentioned features. The syntax of this system is roughly based on Appendix A.2 of [2]: with inessential modifications[b] this syntax can be adapted to most type theories having the mentioned features, thus allowing to replicate the counterexample in them, unless some additional flexibility is introduced to manage universes.

The usual way to manage universes is to allow for an infinite, cumulative and open-ended hierarchy of them. This approach is simple and intuitive, but it becomes questionable in the light of the counterexample of Section 2: in some way, the hierarchy of universes should be reflected in the way in which function spaces are allowed to be formed and related to each other. This is the purpose of Section 5.

In between, the typed formal system of Section 2 is coupled with a pure system, see Section 3, which is proved to be confluent in Section 4. From the confluence result, the impossibility part of the counterexample follows.

[a]This fact can be inferred noting the rather liberal use of β-reductions in proofs in [2].

[b]Indeed, the type of natural numbers is not necessary, but it is still useful to clarify how the recursive inductive types can be formally rendered in the pure system of Section 3.

2. A Counterexample

The formal calculus is shown in Figures 1–3: it contains the structural rules and the types of dependent function spaces and natural numbers. As usual, $A \to B$ abbreviates $\Pi x : A.\, B$ when x does not occur in B, and the hierarchy of universes is indexed by the natural numbers. Also, the variable x introduced by the ctx$-$EXT rule has to be new.

In this calculus, it is routine to derive the judgement $f : \mathbb{N} \to \mathcal{U}_0, x : \mathbb{N}\,\mathsf{ctx}$, where f is assumed to be a function mapping a natural to a type in the lowest universe and x is assumed to be a natural, from which one can prove

$$\frac{}{\bullet\ \mathsf{ctx}}\ \mathsf{ctx{-}EMP} \qquad \frac{\Gamma \vdash A : \mathcal{U}_i}{\Gamma, x : A\ \mathsf{ctx}}\ \mathsf{ctx{-}EXT} \qquad \frac{x_1 : A_1, \ldots, x_n : A_n\ \mathsf{ctx}}{x_1 : A_1, \ldots, x_n : A_n \vdash x_i : A_i}\ \mathsf{Vble}$$

$$\frac{\Gamma \vdash a : A}{\Gamma \vdash a \equiv a : A}\ \equiv\!\mathsf{-refl} \qquad \frac{\Gamma \vdash a \equiv b : A}{\Gamma \vdash b \equiv a : A}\ \equiv\!\mathsf{-sym} \qquad \frac{\Gamma \vdash a \equiv b : A \quad \Gamma \vdash b \equiv c : A}{\Gamma \vdash a \equiv c : A}\ \equiv\!\mathsf{-trans}$$

$$\frac{\Gamma \vdash a : A \quad \Gamma \vdash A \equiv B : \mathcal{U}_i}{\Gamma \vdash a : B}\ \equiv\!\mathsf{-subst} \qquad \frac{\Gamma \vdash a \equiv b : A \quad \Gamma \vdash A \equiv B : \mathcal{U}_i}{\Gamma \vdash a \equiv b : B}\ \equiv\!\mathsf{-subst{-}eq}$$

$$\frac{\Gamma\ \mathsf{ctx}}{\Gamma \vdash \mathcal{U}_i : \mathcal{U}_{i+1}}\ \mathcal{U}\mathsf{-intro} \qquad \frac{\Gamma \vdash A : \mathcal{U}_i}{\Gamma \vdash A : \mathcal{U}_{i+1}}\ \mathcal{U}\mathsf{-cumul} \qquad \frac{\Gamma \vdash A \equiv B : \mathcal{U}_i}{\Gamma \vdash A \equiv B : \mathcal{U}_{i+1}}\ \mathcal{U}\mathsf{-cumul{-}eq}$$

Fig. 1. The formal calculus I.

$$\frac{\Gamma \vdash A : \mathcal{U}_i \quad \Gamma, x : A \vdash B : \mathcal{U}_i}{\Gamma \vdash \Pi x : A.\, B : \mathcal{U}_i}\ \Pi\mathsf{-form} \qquad \frac{\Gamma, x : A \vdash b : B}{\Gamma \vdash \lambda x : A.\, b : \Pi x : A.\, B}\ \Pi\mathsf{-intro}$$

$$\frac{\Gamma \vdash f : \Pi x : A.\, B \quad \Gamma \vdash a : A}{\Gamma \vdash f\, a : B[a/x]}\ \Pi\mathsf{-elim}$$

$$\frac{\Gamma, x : A \vdash b : B \quad \Gamma \vdash a : A}{\Gamma \vdash (\lambda x : A.\, b)\, a \equiv b[a/x] : B[a/x]}\ \Pi\mathsf{-comp} \qquad \frac{\Gamma \vdash f : \Pi x : A.\, B}{\Gamma \vdash \lambda x : A.\, f\, x \equiv f : \Pi x : A.\, B}\ \Pi\mathsf{-uniq}$$

$$\frac{\Gamma \vdash A \equiv A' : \mathcal{U}_i \quad \Gamma, x : A \vdash B \equiv B' : \mathcal{U}_i}{\Gamma \vdash \Pi x : A.\, B \equiv \Pi x : A'.\, B' : \mathcal{U}_i}\ \Pi\mathsf{-form{-}eq}$$

$$\frac{\Gamma, x : A \vdash b \equiv b' : B \quad \Gamma \vdash A \equiv A' : \mathcal{U}_i}{\Gamma \vdash \lambda x : A.\, b \equiv \lambda x : A'.\, b' : \Pi x : A.\, B}\ \Pi\mathsf{-intro{-}eq}$$

$$\frac{\Gamma \vdash f \equiv g : \Pi x : A.\, B \quad \Gamma \vdash a \equiv b : A}{\Gamma \vdash f\, a \equiv g\, b : B[a/x]}\ \Pi\mathsf{-elim{-}eq}$$

Fig. 2. The formal calculus II.

$$
\dfrac{\dfrac{\dfrac{\dfrac{\vdots}{f:\mathsf{N}\to\mathcal{U}_0,x:\mathsf{N}\ \mathsf{ctx}}}{f:\mathsf{N}\to\mathcal{U}_0,x:\mathsf{N}\vdash f:\mathsf{N}\to\mathcal{U}_0}\,\mathsf{Vble}\quad\dfrac{\dfrac{\vdots}{f:\mathsf{N}\to\mathcal{U}_0,x:\mathsf{N}\ \mathsf{ctx}}}{f:\mathsf{N}\to\mathcal{U}_0,x:\mathsf{N}\vdash x:\mathsf{N}}\,\mathsf{Vble}}{\dfrac{f:\mathsf{N}\to\mathcal{U}_0,x:\mathsf{N}\vdash f\,x:\mathcal{U}_0}{f:\mathsf{N}\to\mathcal{U}_0,x:\mathsf{N}\vdash f\,x:\mathcal{U}_1}\,\mathcal{U}\!-\!\mathsf{cumul}}\,\Pi\!-\!\mathsf{elim}}{f:\mathsf{N}\to\mathcal{U}_0\vdash\lambda x:\mathsf{N}.\,f\,x:\mathsf{N}\to\mathcal{U}_1}\,\Pi\!-\!\mathsf{intro}
$$

$$
\dfrac{\Gamma\ \mathsf{ctx}}{\Gamma\vdash\mathsf{N}:\mathcal{U}_i}\,\mathsf{N\!-\!form}\qquad\dfrac{\Gamma\vdash\mathsf{N}:\mathcal{U}_i}{\Gamma\vdash 0:\mathsf{N}}\,\mathsf{N\!-\!intro}_1\qquad\dfrac{\Gamma\vdash\mathsf{N}\to\mathsf{N}:\mathcal{U}_i}{\Gamma\vdash\mathsf{succ}:\mathsf{N}\to\mathsf{N}}\,\mathsf{N\!-\!intro}_2
$$

$$
\dfrac{\begin{array}{l}\Gamma\vdash\Pi P:\mathsf{N}\to\mathcal{U}_i.\,\Pi c_1:P\,0.\\ \quad\Pi c_2:(\Pi x:\mathsf{N}.\,\Pi r:P\,x.\,P\,(\mathsf{succ}\,x)).\,\Pi e:\mathsf{N}.\,P\,e:\mathcal{U}_{i+1}\end{array}}{\begin{array}{l}\Gamma\vdash\mathsf{ind}_{\mathsf{N}}:\Pi P:\mathsf{N}\to\mathcal{U}_i.\,\Pi c_1:P\,0.\\ \quad\Pi c_2:(\Pi x:\mathsf{N}.\,\Pi r:P\,x.\,P\,(\mathsf{succ}\,x)).\,\Pi e:\mathsf{N}.\,P\,e\end{array}}\,\mathsf{N\!-\!elim}
$$

$$
\dfrac{\begin{array}{c}\Gamma\vdash c_1:P\,0\\ \Gamma\vdash P:\mathsf{N}\to\mathcal{U}_i\quad\Gamma\vdash c_2:\Pi x:\mathsf{N}.\,\Pi r:P\,x.\,P\,(\mathsf{succ}\,x)\end{array}}{\Gamma\vdash\mathsf{ind}_{\mathsf{N}}\,P\,c_1\,c_2\,0\equiv c_1:P\,0}\,\mathsf{N\!-\!comp}_1
$$

$$
\dfrac{\begin{array}{c}\Gamma\vdash P:\mathsf{N}\to\mathcal{U}_i\qquad\Gamma\vdash c_1:P\,0\\ \Gamma\vdash n:N\qquad\Gamma\vdash c_2:\Pi x:\mathsf{N}.\,\Pi r:P\,x.\,P\,(\mathsf{succ}\,x)\end{array}}{\Gamma\vdash\mathsf{ind}_{\mathsf{N}}\,P\,c_1\,c_2\,(\mathsf{succ}\,n)\equiv c_2\,n\,(\mathsf{ind}_{\mathsf{N}}\,P\,c_1\,c_2\,n):P\,(\mathsf{succ}\,n)}\,\mathsf{N\!-\!comp}_2
$$

Fig. 3. The formal calculus III.

Then, $\lambda x:\mathbb{N}.\,f\,x \rhd f$ by the usual η-reduction [4], so the Subject Reduction property would allow to derive $f:\mathbb{N}\to\mathcal{U}_0\vdash f:\mathbb{N}\to\mathcal{U}_1$.

However, this is impossible: indeed, the conclusion is necessarily derived by Vble after a possibly empty sequence ζ of instances of $\mathcal{U}$−cumul and $\equiv$−subst as it follows by the Inversion Lemma 2.1. Since $\mathbb{N}\to\mathcal{U}_0$ is syntactically different from $\mathbb{N}\to\mathcal{U}_1$, ζ has to be nonempty. If at least one application of $\mathcal{U}$−cumul appears in ζ, then appropriately composing the elements in ζ, it holds $f:\mathbb{N}\to\mathcal{U}_0\vdash\mathbb{N}\to\mathcal{U}_1\equiv\mathcal{U}_j:\mathcal{U}_i$ and $f:\mathbb{N}\to\mathcal{U}_0\vdash\mathcal{U}_k\equiv\mathbb{N}\to\mathcal{U}_0:\mathcal{U}_i$ for some $k<j<i$. Since $\Gamma\vdash A\equiv B:R$ implies $A\sim B$ in the pure system, see Proposition 3.5, it should follow that $\mathbb{N}\to\mathcal{U}_0\sim\mathcal{U}_k$, which is impossible by Corollary 4.16. Thus, necessarily ζ is a nonempty sequence of instances of $\equiv$−subst only, leading to $f:\mathbb{N}\to\mathcal{U}_0\vdash\mathbb{N}\to\mathcal{U}_1\equiv\mathbb{N}\to\mathcal{U}_0:\mathcal{U}_i$ for some i. Hence, $\mathbb{N}\to\mathcal{U}_1\sim\mathbb{N}\to\mathcal{U}_0$ by Proposition 3.5, that is impossible by Corollary 4.14, both terms being irreducible.

The rest of the chapter is devoted to prove the mentioned properties. In fact, with the exception of the Inversion Lemma 2.1, they are all consequences of the confluence of the associated *Pure System*, see Sections 3 and 4.

The Inversion Lemma reduces just to an observation:

Lemma 2.1. *A derivation of* $\Gamma \vdash t : T$ *is necessarily such that it contains a deduction of* $\Gamma \vdash t : N$ *as a subproof, and either* $\Gamma \vdash T \equiv N : \mathcal{U}_i$, *or* $\Gamma \vdash T \equiv \mathcal{U}_j : \mathcal{U}_i$ *and* $\Gamma \vdash N \equiv \mathcal{U}_k : \mathcal{U}_i$ *with* $k < j$, *for a* natural *type* N *obtained as follows by cases on* t:

(1) *If* $\Gamma \vdash x{:}T$ *with* x *a variable, then* $\Gamma \vdash x{:}N$ *by* Vble *with* $x{:}N \in \Gamma$.
(2) *If* $\Gamma \vdash \mathbb{N} : T$, *then* $\Gamma \vdash \mathbb{N} : \mathcal{U}_h$ *by* $\mathbb{N}$−form *for some* h.
(3) *If* $\Gamma \vdash 0 : T$, *then* $\Gamma \vdash 0 : \mathbb{N}$ *by* $\mathbb{N}-\mathsf{intro}_1$.
(4) *If* $\Gamma \vdash \mathsf{succ} : T$, *then* $\Gamma \vdash \mathsf{succ} : \mathbb{N} \to \mathbb{N}$ *by* $\mathbb{N}-\mathsf{intro}_2$.
(5) *If* $\Gamma \vdash \mathsf{ind}_{\mathbb{N}} : T$, *then* $\Gamma \vdash \mathsf{ind}_{\mathbb{N}} : N$ *by* $\mathbb{N}$−elim.
(6) *If* $\Gamma \vdash \mathcal{U}_h : T$, *then* $\Gamma \vdash \mathcal{U}_h : \mathcal{U}_{h+1}$ *by* $\mathcal{U}$−intro.
(7) *If* $\Gamma \vdash \Pi x{:}A.\, B{:}T$, *then* $\Gamma \vdash \Pi x{:}A.\, B{:}\mathcal{U}_h$ *by* Π−form *for some* h.
(8) *If* $\Gamma \vdash \lambda x : A.\ b : T$, *then* $\Gamma \vdash \lambda x.\ b : \Pi x : A. B$ *by* Π−intro *for some* B.
(9) *If* $\Gamma \vdash f\, c : T$, *then* $\Gamma \vdash f\, c : N$ *by* Π−elim.

Proof. It suffices to observe that in each case the only ways to derive $\Gamma \vdash t : T$ are by $\mathcal{U}$−cumul, $\equiv$−subst, or the indicated rule, and the first two possibilities can be iterated to generate a sequence whose composition satisfies the statement. □

3. Pure System

The pure formal system is defined by a collection of *terms* and their *reductions*. Here, the adjective *pure* means with no types. The pure system is, in principle, nothing but the formal system of Section 1, where the constraints in the formation of judgements given by the types have been removed.

Definition 3.1 (Term). Let V be a denumerable set of symbols, called *variables*. Then a *term*, with its *free variables*, is inductively defined as

(1) If $x \in V$, then x is a term and $\mathsf{FV}(x) = \{x\}$.
(2) The constant $\mathcal{U}_i$ for every i, and the constants $\mathbb{N}$, 0, succ, $\mathsf{ind}_{\mathbb{N}}$ are terms and $\mathsf{FV}(\mathcal{U}_i) = \mathsf{FV}(\mathbb{N}) = \mathsf{FV}(0) = \mathsf{FV}(\mathsf{succ}) = \mathsf{FV}(\mathsf{ind}_{\mathbb{N}}) = \emptyset$.

(3) If A and B are terms and $x \in V$ such that $x \notin \mathsf{FV}(A)$, then $\Pi x : A.\, B$ and $\lambda x : A.\, B$ are terms and $\mathsf{FV}(\Pi x : A.\, B) = \mathsf{FV}(\lambda x : A.\, B) = \mathsf{FV}(A) \cup (\mathsf{FV}(B) \setminus \{x\})$.
(4) If A and B are terms, then $A\, B$ is a term and $\mathsf{FV}(A\, B) = \mathsf{FV}(A) \cup \mathsf{FV}(B)$.

Two terms are said to be *α-equivalent* or *α-convertible* when they differ just in the name of *bound*, i.e., nonfree variables. From now on, terms are considered equal when they are α-equivalent.

A fundamental operation on terms is *substitution*: it replaces a variable with a term inside another term. It is important to remark that substitution is a *purely syntactical* operation: it does not generate new terms, and $b[a/x]$ is just a convenient writing to denote the result of the operation. Substitution is defined, as usual, by replacing the occurrence of the variable x with the term a in b, unless x is bound. When replacing x with a in a term b, care has to be taken to avoid variable capturing.

Pure terms, as defined above, have the only purpose to support *reductions*. In particular, a reduction is the minimal reflexive and transitive congruence which contains the so-called *contractions*.

Definition 3.2 (Contraction). The *contraction* $\blacktriangleright$ is a binary relation on terms defined as follows:

- $(\lambda x : A.\, b)\, a \blacktriangleright b[a/x]$, called *$\beta$-reduction* or *$\beta$-contraction*.
- $\lambda x{:}A.\, f\, x \blacktriangleright f$ when $x \notin \mathsf{FV}(f)$, called *η-reduction* or *η-contraction*.
- $\mathsf{ind}_{\mathbb{N}}\, P\, c_1\, c_2\, 0 \blacktriangleright c_1$, and $\mathsf{ind}_{\mathbb{N}}\, P\, c_1\, c_2\, (\mathsf{succ}\, n) \blacktriangleright c_2\, n\, (\mathsf{ind}_{\mathbb{N}}\, P\, c_1\, c_2\, n)$, called ind-*reductions* or ind-*contractions*.

The left-hand side is called *redex* and the right-hand side is called *contractum*.

The $\blacktriangleright^X$ contraction with X being β, η or ind is defined admitting X-contractions only, while $\blacktriangleright^{X,Y}$ is defined admitting X- and Y-contractions only. Particularly, $\blacktriangleright^\eta$ and $\blacktriangleright^{\beta,\mathsf{ind}}$ will be important in the following.

Contraction models the action on terms performed by the computation and uniqueness rules. These contractions naturally generate a congruence.

Definition 3.3 (One-step reduction). The *one-step reduction* $\rhd_1$ is the minimal congruence extending $\blacktriangleright$, inductively defined as

- If $a \blacktriangleright b$, then $a \rhd_1 b$.
- If $a \rhd_1 b$, then $f\, a \rhd_1 f\, b$, $a\, e \rhd_1 b\, e$, $\lambda x{:}e.\, a \rhd_1 \lambda x{:}e.\, b$, $\lambda x{:}a.\, e \rhd_1 \lambda x{:}b.\, e$, $\Pi x : a.\, e \rhd_1 \Pi x : b.\, e$, $\Pi x : e.\, a \rhd_1 \Pi x : e.\, b$ whenever these pair of terms fall in the domain of $\rhd_1$.

The $\rhd_1^X$ and $\rhd_1^{X,Y}$ one-step reductions are defined from $\blacktriangleright^X$ and $\blacktriangleright^{X,Y}$.

The obvious next step is to perform a sequence of one-step reductions. This notion generates the concept of reduction. Also, a notion of equivalence naturally arises by saying that two terms are equivalent exactly when they can be joined by a sequence of reductions.

Definition 3.4 (Reduction). The *reduction* $\rhd$ relation is the reflexive and transitive closure of $\rhd_1$, while the *equivalence* relation $\sim$ is the reflexive, symmetric and transitive closure of $\rhd$.

It is evident that $\rhd$ is the minimal reflexive and transitive relation containing $\rhd_1$ and $\blacktriangleright$, thus it is a congruence. Also it is obvious that $\sim$ is the minimal equivalence relation containing $\blacktriangleright$ and $\rhd_1$, and it is a congruence. The notation $\rhd^=$ is used to denote the reflexive closure of $\rhd_1$, while $\rhd^X$, $\rhd^{X,Y}$, $\sim^X$ and $\sim^{X,Y}$ are the reductions and equivalences associated to the corresponding one-step reductions. Also, $\rhd^{=X}$, $\rhd^{=X,Y}$ are the reflexive closures of $\rhd_1^X$ and $\rhd_1^{X,Y}$, respectively.

Reduction $\rhd$ models the process of simplification in the pure system, and *conversion* $\sim$ as it is traditionally called, intends to model the *judgemental equality* of the typed system. In fact, a straightforward induction on the derivation of $\Gamma \vdash a \equiv b : T$ allows to prove

Proposition 3.5. *If* $\Gamma \vdash a \equiv b : T$, *then* $a \sim b$.

In particular, the $\Pi-$uniq and the computation rules map to contractions, $\equiv-$refl, $\equiv-$sym and $\equiv-$trans model the fact that $\sim$ is an equivalence relation, while the other rules model the fact that $\sim$ is a congruence, with the exception of $\equiv-$subst$-$eq, which does not affect the pure system. The converse, convertibility of terms having the same type in the same context, implies they are judgementally equal in the type and context, would be a consequence of the Church–Rosser Theorem in the typed calculus, which fails because of the counterexample in Section 2.

However, the Church–Rosser Theorem holds in the pure system. The following properties pave the way towards its proof. Following the guideline of Lemma 2.1.16 in [4], it is easy to prove

Lemma 3.6. *Let $x \neq y$ and assume $x \notin \mathsf{FV}(B)$. Then $M[A/x][B/y] = M[B/y][A[B/y]/x]$.*

The next properties show how substitution and reduction behave together. They are proved by induction on the one-step reduction and the substitution.

Proposition 3.7. *If $A \rhd_1 B$, then $A[Q/x] \rhd_1 B[Q/x]$.*

Proposition 3.8. *If $A \rhd_1 B$, then $P[A/x] \rhd P[B/x]$.*

It is worth noting, by closely inspecting the proofs, how Propositions 3.7 and 3.8 specialise to $\rhd_1^\eta$ in particular, and to any other previously defined one-step reduction in general. Also, if x occurs just once in P, then Proposition 3.8 can be strengthened.

Corollary 3.9. *If $A \rhd_1 B$ and x occurs just once in P, then $P[A/x] \rhd_1 P[B/x]$.*

Moreover, by a trivial induction, Propositions 3.7 and 3.8 can be extended assuming $\rhd$ instead of $\rhd_1$ in both the hypothesis and the conclusion.

Reminding that two binary relations ρ_1 and ρ_2 on the same domain *commute* when $\rho_1 \circ \rho_2 = \rho_2 \circ \rho_1$, it holds that

Proposition 3.10. *The $\rhd^\eta$ and $\rhd^{\beta,\mathsf{ind}}$ reductions commute.*

Proof. The result comes by Lemma 3.3.6 in [4], so it suffices to show that

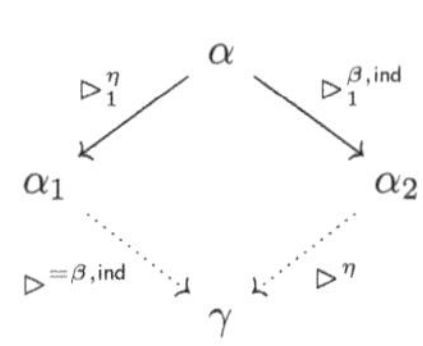

The possible cases are as follows:

(1) The reductions do not intersect.
(2) The β-contraction happens inside the redux of the η-contraction.
(3) The η-contraction happens inside the redux of the β-contraction.
(4) The ind-contraction happens inside the redux of the η-contraction.
(5) The η-contraction happens inside the redux of the ind-contraction.

In all of them, ϕ_1 and ϕ_2 exist by Propositions 3.7, 3.8 and Corollary 3.9. □

4. Confluence in the Pure System

The proof of confluence for the $\rhd^{\beta,\mathsf{ind}}$ reduction uses the well-known technique by Tait and Martin–Löf, described in Section 3.2 of [4]. The idea is to define a notion of reduction $\gg$ which performs contractions in parallel, so as to satisfy the diamond property. Since the transitive closure of $\gg$ is $\rhd^{\beta,\mathsf{ind}}$, the result follows.

Definition 4.1. The *TML-reduction*[c] $\gg$ is a binary relation on pure terms inductively defined as

(1) $A \gg A$.
(2) If $A \gg A'$, then $\lambda x.\, A \gg \lambda x.\, A'$.
(3) If $A \gg A'$ and $T \gg T'$, then $\lambda x : T.\, A \gg \lambda x : T'.\, A'$.
(4) If $A \gg A'$ and $B \gg B'$, then $\Pi x : A\, B \gg \Pi x : A'.\, B'$.
(5) If $A \gg A'$ and $B \gg B'$, then $A\, B \gg A'\, B'$.
(6) If $A \gg A'$ and $B \gg B'$, then $(\lambda x.\, A)\, B \gg A'[B'/x]$.
(7) If $A \gg A'$ and $B \gg B'$, then $(\lambda x : T.\, A)\, B \gg A'[B'/x]$.
(8) If $c_1 \gg c_1'$, then $\mathsf{ind}_{\mathbb{N}}\, P\, c_1\, c_2\, 0 \gg c_1'$.
(9) If $P \gg P'$, $c_1 \gg c_1'$. $c_2 \gg c_2'$, $n \gg n'$, then

$$\mathsf{ind}_{\mathbb{N}}\, P\, c_1\, c_2\, (\mathsf{succ}\, n) \gg c_2'\, n'\, (\mathsf{ind}_{\mathbb{N}}\, P'\, c_1'\, c_2'\, n')\ .$$

[c]So-called after the inventors, Tait and Martin–Löf.

Proposition 4.2. $\rhd_1^{\beta,\mathsf{ind}} \subseteq \gg \subseteq \rhd^{\beta,\mathsf{ind}}$.

Proof. Clearly, $\blacktriangleright^{\beta,\mathsf{ind}} \subseteq \gg$ by Definition 4.1, thus $\rhd_1^{\beta,\mathsf{ind}} \subseteq \gg$ by checking the definitions. Then $\gg \subseteq \rhd^{\beta,\mathsf{ind}}$ follows by induction on $A \gg A'$. □

With respect to substitution, $\gg$ is well behaved:

Proposition 4.3. *If $A \gg A'$ and $B \gg B'$, then $A[B/x] \gg A'[B'/x]$.*

Proof. The proof follows the guideline of Lemma 3.2.4 in [4]. □

If the left-hand side of $\gg$ is an abstraction, or a Π-term, or an application, the shapes of the right-hand side can be predicted. Note how these are the only kinds of terms which are subject to contractions.

Proposition 4.4. *If $\lambda x:T.\,A \gg B$, then $B = \lambda x:T'.\,A'$ and $A \gg A'$, $T \gg T'$.*

Proof. The only applicable cases of Definition 4.1 are (1), (2) or (3). □

Proposition 4.5. *If $\Pi x:A.\,B \gg C$, then $C = \Pi x:A'.\,B'$ and $A \gg A'$, $B \gg B'$.*

Proof. The only applicable cases of Definition 4.1 are (1) or (4). □

Proposition 4.6. *If $A\,B \gg C$, then exactly one of the following cases holds:*

(1) $C = A'\,B'$ *and* $A \gg A'$, $B \gg B'$.
(2) $A = \lambda x:T.\,D$, $C = D'[B'/x]$ *and* $B \gg B'$, $D \gg D'$.
(3) $A = \mathsf{ind}_{\mathbb{N}}\,P\,c_1\,c_2$, $B = 0$, *and* $c_1 \gg C$.
(4) $A = \mathsf{ind}_{\mathbb{N}}\,P\,c_1\,c_2$, $B = \mathsf{succ}\,n$, *and* $C = c_2'\,n'\,(\mathsf{ind}_{\mathbb{N}}\,P'\,c_1'\,c_2'\,n')$, *with* $P \gg P'$, $c_1 \gg c_1'$, $c_2 \gg c_2'$, *and* $n \gg n'$.

Proof. The only applicable cases of Definition 4.1 are: (1), which corresponds to (4.6) in the statement; (5), which is (4.6) in the statement; (6), which is (4.6) in the statement; (7), which is (4.6) in the statement; (8), which is (4.6) in the statement; and (9), which is (4.6) in the statement. □

Proposition 4.7. *If $A_i \gg A'_i$ for every $1 \leq i \leq n$, then $A_1 \cdots A_n \gg A'_1 \cdots A'_n$.*

The previous properties are instrumental to prove that $\gg$ is confluent. The proof roughly follows the pattern of Lemma 3.2.6 in [4].

Theorem 4.8. *The $\gg$ relation satisfies the diamond property: if $M \gg M_1$ and $M \gg M_2$, then there is M_3 such that $M_1 \gg M_3$ and $M_2 \gg M_3$.*

Proof. By induction on $M \gg M_1$ to show that for all $M \gg M_2$ there is M_3 such that $M_1 \gg M_3$ and $M_2 \gg M_3$. By using the previous propositions and Definition 4.1, the proof is long but straightforward. □

Proposition 4.9. *The transitive closure of $\gg$ is $\rhd^{\beta,\mathsf{ind}}$.*

Proof. By Proposition 4.2, $\rhd_1^{\beta,\mathsf{ind}} \subseteq \gg \subseteq \rhd^{\beta,\mathsf{ind}}$, thus the transitive closure of $\rhd_1^{\beta,\mathsf{ind}}$, which is $\rhd^{\beta,\mathsf{ind}}$, is contained in the transitive closure of $\gg$, and also the transitive closure of $\gg$ is contained in the transitive closure of $\rhd^{\beta,\mathsf{ind}}$, which is itself. □

Theorem 4.10. *$\rhd^{\beta,\mathsf{ind}}$ satisfies the diamond property.*

Proof. Remind Lemma 3.2.2 in [4]:

> If ρ is a binary relation satisfying the diamond property and ρ^* is its transitive closure, then ρ^* satisfies the diamond property.

Applying it to $\gg$, which satisfies the diamond property by Theorem 4.8, and whose transitive closure is $\rhd^{\beta,\mathsf{ind}}$ by Proposition 4.9, the result follows. □

Proving the Church–Rosser Theorem for $\rhd$ is almost immediate now, by the previous results. A number of consequences could be derived.

Theorem 4.11 (Church–Rosser). *The $\rhd$ reduction satisfies the diamond property.*

Proof. By Proposition 3.10, $\rhd^{\eta}$ and $\rhd^{\beta,\mathsf{ind}}$ commute. By Theorem 4.10, $\rhd^{\beta,\mathsf{ind}}$ satisfies the diamond property. By Lemma 3.3.7 in [4], $\rhd^{\eta}$ has the diamond property. Therefore, by the Hindley–Rosen Lemma, see Proposition 3.3.5 in [4], stating

> Let ρ_1 and ρ_2 be two binary relations on the same domain both satisfying the diamond property. If they commute, then the reflexive and transitive closure of $\rho_1 \cup \rho_2$ satisfies the diamond property.

the reflexive and transitive closure of $\rhd^{\eta} \cup \rhd^{\beta,\mathsf{ind}}$ has the diamond property. Since $\rhd^{\eta} \cup \rhd^{\beta,\mathsf{ind}} \subseteq \rhd$ and $\rhd$ is contained in the reflexive and transitive closure of $\rhd^{\eta} \cup \rhd^{\beta,\mathsf{ind}}$, this relation is $\rhd$. □

Proposition 4.12. *If $A \sim B$, then there is C such that $A \rhd C$ and $B \rhd C$.*

Proof. The relation $A \sim B$ can be written as

$$A = D_0 \lhd D_1 \rhd D_2 \cdots D_{n-2} \lhd D_{n-1} \rhd D_n = B,$$

with $n \geq 1$. By iterating Theorem 4.11, the result follows. □

Two immediate corollaries are obtained when one or both the elements in $A \sim B$ are irreducible.

Corollary 4.13. *If $A \sim B$ and B is irreducible, then $A \rhd B$.*

Corollary 4.14. *If $A \sim B$ and both A and B are irreducible, then $A = B$.*

Noting that variables, constants and universes are irreducible, and that function spaces reduce to expressions of the same shape, it holds that

Corollary 4.15. *If x is a variable*

- *If y is a variable, $x \sim y$ if and only if $x = y$.*
- *For every i, $x \not\sim \mathcal{U}_i$.*
- $x \not\sim \mathbb{N}$.
- $x \not\sim 0$ *and* $x \not\sim \mathsf{succ}\, n$ *for every term* n.
- *For every t_1, t_2 terms, $x \not\sim \Pi x : t_1 . t_2$.*

Corollary 4.16. *Using the same naming conventions as above,*

- $\mathcal{U}_i \sim \mathcal{U}_j$ *if and only if* $i = j$.
- *For every i, $\mathcal{U}_i \not\sim \mathbb{N}$.*
- $\mathcal{U}_i \not\sim 0$ *and* $\mathcal{U}_i \not\sim \mathsf{succ}\, n$ *for every term* n.
- *For every i, for every t_1, t_2 terms, $\mathcal{U}_i \not\sim \Pi x : t_1 . t_2$.*

Corollary 4.17. *Using the same naming conventions as above,*

- $\mathbb{N} \not\sim 0$ *and* $\mathbb{N} \not\sim \mathsf{succ}\, n$ *for every term* n.
- *For every t_1, t_2 terms, $\mathbb{N} \not\sim \Pi x : t_1 . t_2$.*

Corollary 4.18. *Using the same naming conventions as above,*

- *If* $\mathsf{succ}\, m \sim \mathsf{succ}\, n$*, then* $n \sim m$.
- $0 \not\sim \mathsf{succ}\, n$ *for every term* n.
- *For every* t_1, t_2, n *terms,* $0 \not\sim \Pi x : t_s.\, s_2$ *and* $\mathsf{succ}\, n \not\sim \Pi x : t_2.\, t_2$.

The case of Π-expressions has been left as the last one because it is particularly relevant: it states that Π is *injective* on pure terms

Corollary 4.19. *If* $\Pi x : A_1.\, B_1 \sim \Pi x : A_2.\, B_2$*, then* $A_1 \sim A_1$ *and* $B_1 \sim B_2$.

As a final remark, a λ-expression may reduce only in its arguments or as a whole by η-reduction; an application may reduce only in its arguments or as a whole when its first argument is a λ-expression by β-reduction, or when its first argument is $\mathsf{ind}_{\mathbb{N}}$ and it has the right number of arguments, with the last one being 0 or $\mathsf{succ}\, n$ for some term n. In this case, the ind-reduction can be applied.

5. How to Fix it?

The counterexample in Section 2 naturally suggests a way to overcome the problem: Subject Reduction fails because an object f has two distinct types that are neither equivalent nor universes in contrast with the Inversion Lemma 2.1. Hence, the natural enhancement would be to relate these two types within the calculus, so that a lift via $\mathcal{U}-\mathsf{cumul}$ in the codomain would result in a lift of the whole type: if $\Gamma \vdash f : \mathbb{N} \to \mathcal{U}_i$, then $\Gamma \vdash f : \mathbb{N} \to \mathcal{U}_{i+1}$. More in general, one would like to say that when $\Gamma \vdash t : A$ implies $\Gamma \vdash t : B$ for every term t satisfying the antecedent, then $\Gamma \vdash A \uparrow B$, *A extends to B*.

This intuition is captured in the calculus in Figures 2–4. It drops the $\equiv-\mathsf{subst}$, $\equiv-\mathsf{subst-eq}$, $\mathcal{U}-\mathsf{cumul}$, and $\mathcal{U}-\mathsf{cumul-eq}$ rules, and it introduces the $\equiv\uparrow$, $\mathcal{U}\uparrow$, $\Pi\uparrow$, $\uparrow-\mathsf{trans}$, $\uparrow-\mathsf{subst}$, and $\uparrow-\mathsf{subst-eq}$ rules. By combining $\equiv\uparrow$ and $\mathcal{U}\uparrow$ with $\uparrow-\mathsf{subst}$ and $\uparrow-\mathsf{subst-eq}$, the dropped rule can be shown to be derivable, so the enhanced calculus is a proper extension of the usual one. The new rules model a few very natural facts:

- two judgementally equal types are one the extension of the other;
- every universe extends to the next one in the hierarchy of universes;
- extension is transitive;

$$\frac{}{\bullet\ \mathsf{ctx}}\ \text{ctx}-\text{EMP} \qquad \frac{\Gamma \vdash A : \mathcal{U}_i}{\Gamma, x : A\ \mathsf{ctx}}\ \text{ctx}-\text{EXT} \qquad \frac{x_1 : A_1, \ldots, x_n : A_n\ \mathsf{ctx}}{x_1 : A_1, \ldots, x_n : A_n \vdash x_i : A_i}\ \text{Vble}$$

$$\frac{\Gamma \vdash a : A}{\Gamma \vdash a \equiv a : A}\ \equiv-\text{refl} \qquad \frac{\Gamma \vdash a \equiv b : A}{\Gamma \vdash b \equiv a : A}\ \equiv-\text{sym}$$

$$\frac{\Gamma \vdash a \equiv b : A \quad \Gamma \vdash b \equiv c : A}{\Gamma \vdash a \equiv c : A}\ \equiv-\text{trans} \qquad \frac{\Gamma\ \mathsf{ctx}}{\Gamma \vdash \mathcal{U}_i : \mathcal{U}_{i+1}}\ \mathcal{U}-\text{intro}$$

$$\frac{\Gamma \vdash A \equiv B : \mathcal{U}_i}{\Gamma \vdash A \uparrow B}\ \equiv\uparrow \qquad \frac{\Gamma\ \mathsf{ctx}}{\Gamma \vdash \mathcal{U}_i \uparrow \mathcal{U}_{i+1}}\ \mathcal{U}\uparrow$$

$$\frac{\Gamma \vdash A \uparrow A' \quad \Gamma, x : A' \vdash B \uparrow B'}{\Gamma \vdash \Pi x : A'.\, B \uparrow \Pi x : A.\, B'}\ \Pi\uparrow \qquad \frac{\Gamma \vdash A \uparrow B \quad \Gamma \vdash B \uparrow C}{\Gamma \vdash A \uparrow C}\ \uparrow-\text{trans}$$

$$\frac{\Gamma \vdash a : A \quad \Gamma \vdash A \uparrow B}{\Gamma \vdash a : B}\ \uparrow-\text{subst} \qquad \frac{\Gamma \vdash a \equiv b : A \quad \Gamma \vdash A \uparrow B}{\Gamma \vdash a \equiv b : B}\ \uparrow-\text{subst}-\text{eq}$$

Fig. 4. The enhanced calculus

- if a term belongs to a type, it belongs to any extension of the type;
- if two terms are judgementally equal in a type, they are so in any extension of the type.

The only rule which requires some explanation is $\Pi\uparrow$. It tells that a function space extends to another function space exactly when the codomain of the first space extends the codomain of the second one, and the domain of the first space restricts the domain of the second one. Thinking of term f of type $\Pi x : A.\, B$ as a programme, the rule becomes natural: given an input of type A, f computes a value of type B, so f also yields a value of type B' with $B \uparrow B'$, and f operates on the subset of inputs A' with $A' \uparrow A$, too, with no change in the algorithmic behaviour of f.

The claim is that the enhanced calculus enjoys the Subject Reduction property. However, proving this fact is complex, and the details are still to be fixed and cleared. In the first place, it is worth showing that the enhanced calculus prevents the counterexample to happen. In fact,

$$\dfrac{\dfrac{\dfrac{\dfrac{\dfrac{\vdots}{f : \mathbb{N} \to \mathcal{U}_0\ \mathsf{ctx}}}{f : \mathbb{N} \to \mathcal{U}_0 \vdash \mathbb{N} : \mathcal{U}_0}\ \mathbb{N}-\text{form}}{f : \mathbb{N} \to \mathcal{U}_0 \vdash \mathbb{N} \equiv \mathbb{N} : \mathcal{U}_0}\ \equiv-\text{refl}}{f : \mathbb{N} \to \mathcal{U}_0 \vdash \mathbb{N} \uparrow \mathbb{N}}\ \equiv\uparrow \qquad \dfrac{\dfrac{\vdots}{f : \mathbb{N} \to \mathcal{U}_0, x : \mathbb{N}\ \mathsf{ctx}}}{f : \mathbb{N} \to \mathcal{U}_0, x : \mathbb{N} \vdash \mathcal{U}_0 \uparrow \mathcal{U}_1}\ \mathcal{U}\uparrow}{f : \mathbb{N} \to \mathcal{U}_0 \vdash \mathbb{N} \to \mathcal{U}_0 \uparrow \mathbb{N} \to \mathcal{U}_1}\ \Pi\uparrow$$

thus $f{:}\mathbb{N} \to \mathcal{U}_0 \vdash f{:}\mathbb{N} \to \mathcal{U}_1$ by $\uparrow$−subst, overcoming the impossibility part of the counterexample.

The Subject Reduction Theorem states

Theorem 5.1 (Subject Reduction). *If $t \rhd s$ and $\Gamma \vdash t : T$ in the enhanced calculus, then $\Gamma \vdash s : T$ and $\Gamma \vdash t \equiv s : T$.*

The proof is a straightforward induction on the reduction $t \rhd s$, in which the hard bit is the one-step reduction. This is proved by induction on the derivation of $\Gamma \vdash t : T$. Again, the proof proceeds smoothly, with the aid of a few auxiliary properties which pose no difficulties, until one hits the cases involving contractions. Here, two problems arise: to treat η-conversion, the Thinning Lemma is needed; to deal with β- and ind-conversions, one needs to know that Π is injective.

The Thinning Lemma says that, if $\Gamma, x{:}S \vdash t{:}T$ and x is not free in t and T, then $\Gamma \vdash t{:}T$. The difficulty lies in the notion of free variable: in fact, expressing the condition as $x \notin \mathsf{FV}(t) \cup \mathsf{FV}(T)$ is not enough. For example, consider $z : \mathcal{U}_0, f : z \to \mathbb{N}, x : z \vdash f\,x : \mathbb{N}$, which is easily derivable. Clearly, $z \notin \mathsf{FV}(f\,x) \cup \mathsf{FV}(\mathbb{N})$, but $f : z \to \mathbb{N}, x : z \vdash f\,x : \mathbb{N}$ is not even a judgement, since its context is invalid. The right notion to use is that of *active variables*: $\mathsf{AV}(\Gamma \vdash t : T) = \bigcup_{0 \leq i \leq |\Gamma|} \mathsf{AV}_i(\Gamma \vdash t : T)$ where

$$\begin{aligned}
\mathsf{AV}_0(\Gamma \vdash t : T) &= \mathsf{FV}(t) \cup \mathsf{FV}(T), \\
\mathsf{AV}_{i+1}(\Gamma \vdash t : T) &= \mathsf{AV}_i(\Gamma \vdash t : T) \\
&\quad \cup \bigcup \{\mathsf{FV}(A) \colon x : A \in \Gamma \text{ and } x \in \mathsf{AV}_i(\Gamma \vdash t : T)\}\,.
\end{aligned}$$

Hence, the Thinning Lemma becomes:

Lemma 5.2. *If $\Gamma, x : S \vdash t : T$ and $x \notin \mathsf{AV}(\Gamma, x : S \vdash t : T)$, then $\Gamma \vdash t : T$.*

The big difficulty in proving the lemma lies in the fact that t does not contain enough information to effectively reconstruct a proof of $\Gamma, x : S \vdash t : T$. Specifically, when $t = f\,a$, the type A of the domain of f is missing in t and T, and thus a proof may contain an A such that $x \in \mathsf{AV}(\Gamma \vdash A : \mathcal{U}_i)$. Hence, one has to show that there is an A' allowing to prove $\Gamma, x : S \vdash f\,a : B[a/x]$ in which $x \notin \mathsf{AV}(\Gamma \vdash f : \Pi y : A'.\, B) \cup \mathsf{AV}(\Gamma \vdash a : A')$.

Although the proof of the Thinning Lemma appears hard, it is essential to remark that its statement is completely natural and intuitive. In fact, it would be very surprising if the Thinning Lemma turns out to be false: if $t : T$ depends directly or indirectly on a set of assumptions, it should be possible to prove $t : T$ from that set with no additional hypothesis because of the propositions-as-types interpretation.

The need to show that Π is injective comes from the Inversion Lemma. In the enhanced calculus, it takes the form: if $\Gamma \vdash t : T$, then $\Gamma \vdash t : N$ for a natural type N as in Lemma 2.1, and $\Gamma \vdash N \uparrow T$. In the proof of Theorem 5.1, when dealing with the β- and the ind-conversions, and after applying the Inversion Lemma, one needs

Proposition 5.3. *If* $\Gamma \vdash \Pi x : A'.\, B' \uparrow \Pi x : A.\, B$*, then* $\Gamma \vdash A \uparrow A'$ *and* $\Gamma, x : A \vdash B' \uparrow B$*.*

The proof of Proposition 5.3 is a simple induction on the derivation of the premise, but the hard part occurs in the case $\Gamma \vdash \Pi x : A'.\, B' \equiv \Pi x : A.\, B : \mathcal{U}_i$. Precisely, in this point injectivity of Π comes in:

Proposition 5.4. *If* $\Gamma \vdash \Pi x : A'.\, B' \equiv \Pi x : A.\, B : \mathcal{U}_i$*, then* $\Gamma \vdash A' \equiv A : \mathcal{U}_i$ *and* $\Gamma, x : A' \vdash B' \equiv B : \mathcal{U}_i$*.*

With the aid of Proposition 5.4, the proof of Proposition 5.3 easily follows. However, Proposition 5.4 is challenging: the transitivity of $\equiv$ is the source of problems. The path to the proof uses:

Theorem 5.5 (Church–Rosser). *If* $\Gamma \vdash A \equiv B : T$*, then there is* C *such that* $A \rhd C \lhd B$ *and* $\Gamma \vdash A \equiv C : T$, $\Gamma \vdash B \equiv C : T$*.*

In turn, the Church–Rosser Theorem is a corollary of the Subject Reduction Theorem 5.1, creating a vicious circle. Avoiding such a vicious circle is the problem, and the still incomplete solution at the moment is to observe how the injectivity of Π can be coded as a pair of inference rules, which could be removed after proving the Subject Reduction Theorem 5.1.

Nevertheless, despite the sketchy description of a complex proof still under development, Proposition 5.4 is very natural, and its pure counterpart, Corollary 4.19, strongly supports its truth. If Π fails to be injective in the enhanced calculus, this would be surprising: ultimately, it means that judgemental equality is not the formal counterpart of convertibility.

At this point, the reader may wonder why to describe the difficulties in showing that the enhanced calculus satisfies the Subject Reduction Theorem instead of completing and illustrating its proof. The answer is three-fold.

In the first place, the proof of Theorem 5.1 will be long, involved and full of subtleties: presenting it in this chapter would obscure the crucial fact that the type system supporting Homotopy Type Theory and many variants of Martin-Löf's-type Theory actually **has** a structural problem. And the enhanced calculus is a promising attempt to solve it.

In the second place, a type theory like the ones considered in Section 1 should stand on its own legs, not relying on an intended semantics unless when strictly necessary. Indeed, coupling the type system with a pure one enables to prove the impossibility part of the counterexample without passing through semantics. This technique makes the search and the proof of negative results more solid and robust, since it does not involve semantics.

In the last place, Section 5 shows how the structural proof theory of dependent types is largely underdeveloped. While it is clear from the described difficulties that it is not a subject for faint heart researchers, there is an immense mathematical space which is still unexplored, with a large potential for insights and applications.

References

[1] P. Martin-Löf, An intuitionistic theory of types: Predicative part. In H. Rose and J. Shepherdson (eds.), *Logic Colloquium '73, vol. 80, Studies in Logic and the Foundations of Mathematics*. Elsevier Science Publisher B.V., 1975, pp. 73–118.

[2] The Univalent Foundations Program, *Homotopy Type Theory: Univalent Foundations of Mathematics*. `https://homotopytypetheory.org/book`, Institute for Advanced Study, 2013.

[3] A. Abel, T. Coquand, and P. Dybjer. Normalization by evaluation for Martin–Löf type theory with typed equality judgements. In *Proceedings of the 22nd Annual IEEE Symposium on Logic in Computer Science*, LICS '07. IEEE Computer Society, 2007, pp. 3–12.

[4] H.P. Barendregt, *The Lambda Calculus: Its Syntax and Semantics, 2nd edn. vol. 103, Studies in Logic and the Foundations of Mathematics*. North Holland, 1984.

https://doi.org/10.1142/9789811245220_bmatter

Author Index

B

Benini, M., 441
Berger, J., 133, 143
Berger, U., 223
Beyersdorff, O., 397
Bridges, D.S., 1

C

Capucci, M., 195
Coquand, Th., 287

D

Dihoum, E., 335

E

Enes, J., 251

F

Fernández-Duque, D., 165
Ferreira, F., 251

G

Ghani, N., 195
Gjetaj, O., 165

I

Ishihara, H., 143

K

Kupke, C., 195

L

Ledent, J., 195
Lombardi, H., 287

N

Nemoto, T., 143
Neuwirth, St., 287
Nordvall Forsberg, F., 195

R

Rathjen, M., 319, 335
Richman, F., 25

S

Schwichtenberg, H., 39
Seisenberger, M., 223
Silterra, A., 335
Spreen, D., 71, 223
Svindland, G., 133

T

Towsner, H., 273
Tsuiki, H., 223

W

Weiermann, A., 165

www.ingramcontent.com/pod-product-compliance
Lightning Source LLC
Chambersburg PA
CBHW061516060426
42492CB00005B/184
9789811245213